Enhanced Oil Recovery

Oil recovery efficiency can be increased by applying the enhanced oil recovery (EOR) processes, which are based on the improvement of mobility ratio, reduction of interfacial tension between oil and water, wettability alteration, reduction of oil viscosity, formation of oil banks, and so forth. This book describes the different EOR methods and their mechanisms, which are traditionally used after conventional primary and secondary processes. The present scenario of different EOR processes, at both the field application stage and research stage, is also covered. Further, it discusses some of the recent advances in EOR processes such as low-salinity water flooding, the application of nanotechnology in EOR, microbial EOR, carbonated water injection, etc.

Features:

- Comprehensive coverage of all enhanced oil recovery (EOR) methods
- Discussion of reservoir rock and fluid characteristics
- Illustration of steps in design and field implementation as well as the screening criteria for process selection
- Coverage of novel topics of nanotechnology in EOR and hybrid EOR method and low-salinity waterfloods
- Emphasis on recent technologies, feasibility, and implementation of hybrid technologies

This book is aimed at graduate students, professionals, researchers, chemists, and personnel involved in petroleum engineering, chemical engineering, surfactant manufacturing, polymer manufacturing, oil/gas service companies, and carbon capture and utilization.

Enhanced Oil Recovery

Mechanisms, Technologies and Feasibility Analyses

Ajay Mandal and Keka Ojha

CRC Press
Taylor & Francis Group
Boca Raton London New York

CRC Press is an imprint of the
Taylor & Francis Group, an **informa** business

Designed cover image: © Shutterstock

First edition published 2024
by CRC Press
6000 Broken Sound Parkway NW, Suite 300, Boca Raton, FL 33487-2742

and by CRC Press
4 Park Square, Milton Park, Abingdon, Oxon, OX14 4RN

CRC Press is an imprint of Taylor & Francis Group, LLC

© 2024 Ajay Mandal and Keka Ojha

ISBN: 978-0-367-56667-8 (hbk)
ISBN: 978-0-367-56669-2 (pbk)
ISBN: 978-1-003-09885-0 (ebk)

DOI: 10.1201/9781003098850

Typeset in Times
by Apex CoVantage, LLC

Contents

About the Authors

Dr. Ajay Mandal is presently working as a professor in the Department of Petroleum Engineering, Indian Institute of Technology (Indian School of Mines), Dhanbad. Earlier, he served as the head of the department and associate dean (research and development) of the institute. He has more than 18 years of experience as a faculty of petroleum engineering in this institute. Dr. Mandal received his BSc in chemistry (hons.) with a bronze medal from Calcutta University and his BTech in chemical engineering from the same university. He was awarded the gold medal for being the best student in the master's degree program in chemical engineering from Jadavpur University, India. He obtained his PhD degree from Indian Institute of Technology, Kharagpur, in chemical engineering.

Currently, Dr. Mandal is involved in intense research in the field of reservoir engineering, with a specialization in enhanced oil recovery and gas hydrates. The outcome of his research is reflected in 250 research papers, books, and book chapters, with more than 10,500 citations. Dr. Mandal has worked on 20 sponsored R&D projects, mostly on enhanced oil recovery, and 11 consultancy projects from different government funding agencies as well as E&P companies operating in India. He is also a reviewer of more than 20 journals in his specialized area.

Dr. Mandal has been conferred various national and international awards like the 2015 SPE South Asia Regional Distinguished Achievement Award for Petroleum Engineering Faculty; IIChE (Indian Institute of Chemical Engineers) Award for the Year 2017 (ONGC Award for Excellence in Design and Development of Oil/Gas Related Process Plant and/or Chemicals); 2019 SPE Reservoir Dynamics and Description Regional Award for South Asia and Pacific region; and Inder Mohan Thapar Foundation (IMTF) Research Award from IIT (ISM) for the years 2016–2017, 2017–2018, and 2018–2019 (jointly with PhD students). He was also a recipient of the prestigious DAAD fellowship of Germany in 2008.

Dr. Mandal has guided 24 doctoral students, and 6 students are currently pursuing their PhD program. He was directly involved in the field during deputation to the Institute of Reservoir Studies, ONGC, under the Industry-Academic interaction program for one year. Dr. Mandal carried out collaborative research work with the Department of Petroleum Engineering, Curtin University of Technology, under the Australia-India Joint Research Program, in 2016. He has rich and diverse experience through collaborative research with Universiti Teknologi PETRONAS, Malaysia, the National University of Singapore, and King Abdulla University of Science and Technology, Saudi Arabia. At present, he is working on an international collaborative project on chemically enhanced oil recovery under the Scheme for Promotion of Academic and Research Collaboration (SPARC), Government of India with the National University of Singapore (NUS), and Technion-Israel Institute of Technology as collaborating Institutes. Dr. Mandal is a member of the Enhanced Recovery Screening Committee of the Ministry of Petroleum and Natural Gas, Government of India. He was ranked within the top 1% in the field of energy by an independent study done by Stanford University scientists in 2020. Dr. Mandal was also admitted as a Fellow of the Royal Society of Chemistry for his outstanding performance in petroleum chemistry in the year 2020. He is also an active member of the Society of Petroleum Engineers (SPE), United States, and Indian Institute of Chemical Engineers (IIChE), India.

Dr. Keka Ojha, Professor & Head, Petroleum Engineering Department, IIT (ISM), Dhanbad, has been involved in teaching and research in various areas, including the areas of reservoir engineering, EOR coalbed methane, and shale gas, since 2004.

Prof. Ojha holds PhD and MTech degrees in chemical engineering from Indian Institute of Technology, Kharagpur. She obtained her BChE degree (Honors) from Jadavpur University, India in chemical engineering. She worked as a postdoctoral fellow from the University of Notre Dame, United States, before joining IIT (ISM), Dhanbad.

Dr. Ojha has published peer-reviewed papers in journals of repute besides conference publications of similar volume and a book—the outcome of more than 25 R&D and consultancy projects on EOR, fracturing fluids, coal-bed methane, water shutoff jobs, and shale gas, in addition to a number of EDP courses offered to the executives of various E&P companies operating in India. She is an active member of various national and international professional bodies, including SPE-USA, MGMI, and IIChE. Dr. Ojha is an active reviewer of various petroleum engineering–related journals of her research area under various other publishers. Her academic knowledge is rich with field experience through academic and industrial collaborations with Curtin University of Technology, Perth, Australia; IRS, ONGC Ltd; Universiti Teknologi PETRONAS, Malaysia; King Abdulla University of Science and Technology, Saudi Arabia; and the National University of Singapore.

The contribution of Dr. Ojha in petroleum engineering academics was recognized by SPE through the "2019 SPE South Asia Regional Distinguished Achievement Award for Petroleum Engineering Faculty." She and her research team bagged the first prize for presentation in national and international conferences on a number of occasions.

Preface

Petroleum-bearing reservoirs encompass a vast hydrocarbon pool in porous and permeable formations, which may be extracted by the application of various recovery methods. Extraction of oil requires very complex technology because of its properties and interaction with the reservoir's rock. Applications of primary (under natural drives) and secondary recovery (mostly, pressure maintenance) techniques generally produce a limited fraction of original oil in place (OOIP), while a larger part of it remains untapped due to the complexity of the reservoir as well as the inefficacy of the applied technologies, in addition to economic constraints. Here, enhanced oil recovery (EOR) methods become the savior, with significant potential to enhance oil recovery, through a remarkable reduction in the residual oil saturation; thus, the technology contributes enormously to achieving energy security for the human civilization.

However, the implementation of EOR technology to a field demands a detailed understanding of the reservoir and screening of the best-suited economic methods. Moreover, depletion of easy oil and requirement of petroleum production from complex, mature fields containing heavy oil throws up challenges to the implementation of EOR technologies for producing significant incremental oil with notable profit.

This book is written primarily for the undergraduate and postgraduate students of petroleum engineering, who need to gain overall knowledge about EOR processes before joining the industry. However, the book should be also equally helpful to practicing professionals working in the field of petroleum production or enhanced oil recovery projects. We assume that the readers of this book have fundamental knowledge of reservoir engineering. The book contains ten chapters covering the existing and emerging EOR technologies and their principles and applications.

The book starts with the history and fundamentals of EOR techniques and their classifications and screening criteria. The fundamentals of immiscible fluid displacement processes are discussed in Chapter 2. The basic principles, mechanisms, and oil recovery calculations by immiscible water-flooding are described in this chapter, in addition to the coverage of the pore-level microscopic displacement efficiency, linear displacement theory, and volumetric macroscopic displacement efficiency.

The different EOR processes—starting from conventional chemical flooding, miscible-immiscible fluid injection, and thermal EOR to the latest technology of low-salinity or smart waterflood, including the application of nanotechnology—are described in detail in the chapters. The fundamental concepts, along with the scientific principles involved in each step of the individual processes, are elaborated in detail to give a clear understanding of the process and its application. While describing the various stages, experimental data from laboratory studies, in addition to the published data from field applications, are used for analyzing the processes. Examples of case studies for various EOR technologies are discussed in detail to understand the actual field application and their merits and limitations.

EOR processes involve the thermal and/or nonthermal means of changing the properties of crude oil in reservoirs—such as density, IFT, and viscosity—which ensures improved oil displacement in the reservoir and, consequently, better recovery. The chemical flooding method, which involves the injection of surfactants, polymers, and alkalis in different combinations to improve the capillary number (N_C), is described in Chapter 3. In EOR, a high N_C value is desirable to achieve favorable oil displacement by decreasing the strength of capillary forces and increasing the magnitude of viscous forces. The fundamentals of other EOR methods—miscible flooding, including CO_2 injection; thermal recovery methods (hot waterflooding, cyclic steam injection, steam flooding, in situ combustion, etc.); microbial oil recovery—are discussed in Chapters 4, 5, and 7, respectively.

Nanotechnology has a predominant role in various industries, and especially in the petroleum industry, it is an emerging aspect. In petroleum engineering, this modification can be attributed to the nanomaterials' unique properties, including high surface to volume ratio, wettability control, and interfacial tension reduction. In Chapter 6, the application of nanotechnology in the oil and gas industry is discussed, with special emphasis on nanoparticle- and nanoemulsion-based enhanced oil recovery. Low-salinity waterflooding (LSW) is a promising new technique for enhancing oil recovery (EOR) in both sandstone and carbonate reservoirs. The potential of LSW has drawn the attention of the oil industry in the past decade. The fundamental mechanisms of LSW and smart waterflooding for enhanced oil recovery are discussed in Chapter 8.

The planning of an EOR project demands meticulous attention to many problems, thus requiring considerable lead time for studies, evaluations, project design, and most of all, the economics of these high-cost EOR projects. The techno-economic feasibility analyses of different EOR processes are reported in Chapter 9.

The authors have tried their best to include the latest technologies developed so far. The technological world is versatile; the scientific and engineering community involved in the petroleum industry, like in any other industry, is involved in developing and innovating new technologies for producing hydrocarbons at a lower cost from conventional and unconventional reservoirs. Thus, the students, scientists, and engineers involved in EOR processes must update themselves with the latest literature published by different sources. We also hope that enhanced oil recovery technologies will contribute significantly to making petroleum widely available as a cheap, primary energy source.

Acknowledgments

The authors wish to express their sincere thanks to the IIT (ISM) authority, especially Prof. Rajiv Shekhar, director, Indian Institute of Technology (Indian School of Mines), Dhanbad, for granting permission and continuous encouragement to publish this work.

The excellent library facility, with the accessibility of a large number of books and journal volumes in the field of petroleum engineering, helped us enormously in collecting the required information for various topics of the book. We have also been privileged to teach different courses on petroleum engineering, especially "Enhanced Oil Recovery" and "Reservoir Engineering," to our UG/PG students for the last 15 years, which helped us to grow in confidence to take up this project. We express our sincere acknowledgments to the Institute of Reservoir Studies (IRS), ONGC, Ahmedabad; Oil India Ltd.; Oil Industry Development Board, Government of India, New Delhi; Council of Scientific & Industrial Research (CSIR), New Delhi; and Science & Engineering Research Board (SERB), New Delhi, for providing technical and financial supports to various research activities.

In the process of preparing this book, we also acknowledge the contributions of our PhD scholars, Dr. Sunil Kumar, Dr. Neha Saxena, Dr. Abhijit Samanta, Dr. Achinta Bera, Dr. Nilanjan Pal, Dr. Narendra Kumar Rawat, Dr. Amit Kumar, Dr. Prathibha Pillai, Dr. Mohammad Yunus Khan, Dr. Rohit Kumar Saw, Dr. Rakesh Kumar Pandey, Dr. Moumita Maiti, Mr. Shubham Prakash, Mr. Dinesh Joshi, Kumar Abhijeet Raj, Milan Mandal, Abhinav Kumar, and Mr. Chirag Garg, for the typing and drawing of figures, review of some sections, and elaborate discussions on specific topics incorporated in this book. During the writing of this book, we had very close interaction with our colleagues and friends from the oil and gas industry, who provided us with innovative information on various topics in this book.

The authors also gratefully acknowledge receipt of permission to use the materials and figures from different published journal papers. These are duly cited in the book.

Special appreciation and gratitude are deserved by Aneek and Ankit, our sons, for their patience and cooperation and for sparing part of their valuable time from their childhood activities for the required professional preparation and writing of this book. Finally, thanks are due to our parents and other family members for their continuous support, encouragement, and cooperation.

We sincerely hope that this book will be very useful to the students of petroleum engineering and engineers who are working in the oil and gas industry.

1 Introduction to Enhanced Oil Recovery

1.1 INTRODUCTION

Crude oil is a nonrenewable fossil fuel, generated as a result of continuous chemical and thermal degradation processes of organic matters (plankton) over millions of years under subsurface conditions. The produced hydrocarbons are migrated and trapped into porous "reservoir rock" formations. Depending on the source and maturity of the kerogen (converted organic matter) and deposition condition, oil and gas may be formed. Recovery of oil from the reservoir rock is comparatively difficult because of its viscous nature and complex interaction within the formation rock. With the discovery of many new fields and the application of new technologies, the world petroleum, especially oil, reserves attained a very high peak in the mid-20th century to the end of the 20th century. Initially, Hubbert (1956) developed a "peak oil" theory, characterized by an initial increase, followed by the attainment of a peak, and eventually, a steady decline of oil production. Since then, this model has been undergoing alterations as a result of increasing consumption, improved production capacity, discovery of new oil fields, and technological developments (Campbell and Laherrère, 1998; Maggio and Cacciola, 2012; Chapman, 2014). The overall contribution of crude oil to the global energy markets is increasing steadily and can comply with future requirements, provided the current policies remain in place. Renewable energy sources are thought to contribute considerably to the world's energy requirements; however, they are yet to take a step forward, leaving the world economy to rely on oil as a long-term source in the coming decades. Figure 1.1 depicts an analysis and projection of world energy consumption. A 28% increase in the use of energy is projected by 2040, with the greatest contribution being from petroleum, natural gas, and other fuels.

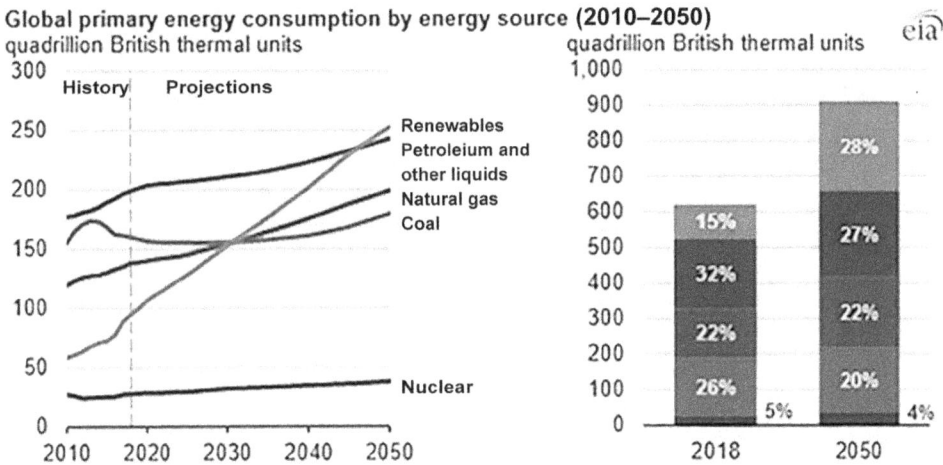

FIGURE 1.1 Global energy consumption during 2010–2050.

Source: U.S. Energy Information Administration, International Energy Outlook, 2019.

DOI: 10.1201/9781003098850-1

1.2   OIL FORMATION, MIGRATION, AND ACCUMULATION

The formation of crude oil has been theorized and disputed to have either abiogenic origins or biogenic origins (Speight, 2015). As per the abiogenic origin theory, also known as abiotic theory, the hydrocarbons were formed without any biotic activity and the generation of crude oil had occurred in the mantle or deep crust of the earth (Kutcherov and Krayushkin, 2010). However, the biogenic origin theory, also known as biotic theory, states that the hydrocarbons were formed by the remains of buried biological matter (Walters, 2006). The evidence and scientific literature supporting both abiogenic and biogenic theories can be found; however, the biogenic genesis of petroleum has been widely accepted (Speight, 2015). Regardless of a common consensus on the origin of petroleum, it had been accepted that the formation of hydrocarbons occurred in a source rock and migrated through capillaries and pores of source and carrier rocks (Tissot and Welte, 1978). The migration of crude oil and gases after expulsion from the source rock gets accumulated in the porous and permeable sedimentary rocks, which are enclosed by impermeable rocks. These impermeable caprocks are mostly shale or evaporites that form a seal around the permeable reservoir rocks, where the pool of gas and crude oil are found at structural highs due to their lower density in comparison to water, which was prefilled in the pores of the reservoir rocks. Thus, a reservoir is defined as "hydrocarbon accumulation in a single trap with a unified pressure system and oil (gas)-water interface" (Zhao et al., 2017). Figure 1.2 shows the hydrocarbon formation, migration, and accumulation in a reservoir. It can be seen from the figure that the hydrocarbon migrates from the source rocks to the reservoir rocks, where the reservoir of petroleum is formed due to the structural formations such as anticlines, faults, salt domes, and other stratigraphic traps (Aminzadeh and Dasgupta, 2013).

1.3   CLASSIFICATION OF CRUDE OIL

Crude oil can be classified as light, medium, or heavy, based on its density. The American Petroleum Institute gravity, commonly shortened to API gravity, compares the density of crude to water. An API gravity higher than 10 means the oil is lighter than water and will float on it. An API gravity lower than 10 means the oil is denser than water and will sink in it.

When referring to oil, an API gravity greater than 31.1° is considered light. An API gravity between 22.3° and 31.1° is considered medium. An API gravity between 10.0° and 22.3° is considered heavy.

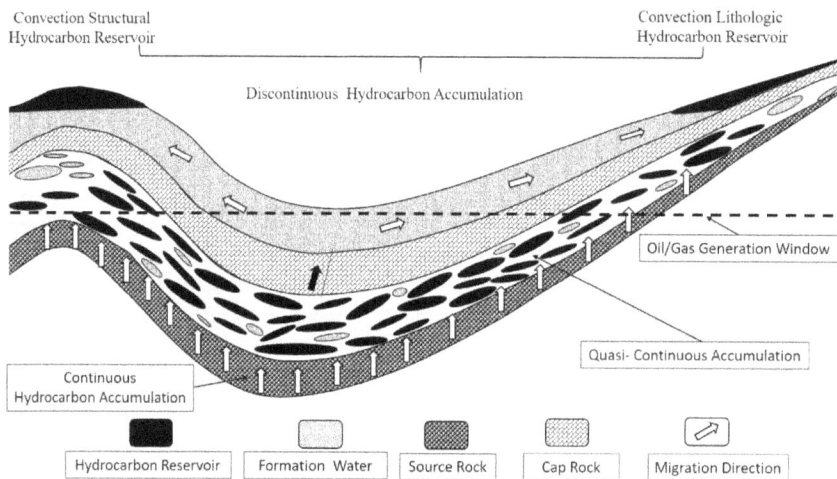

FIGURE 1.2   Formation and distribution models of the continuous, quasi-continuous, and discontinuous hydrocarbon accumulations.

Source: Zhao et al., 2017.

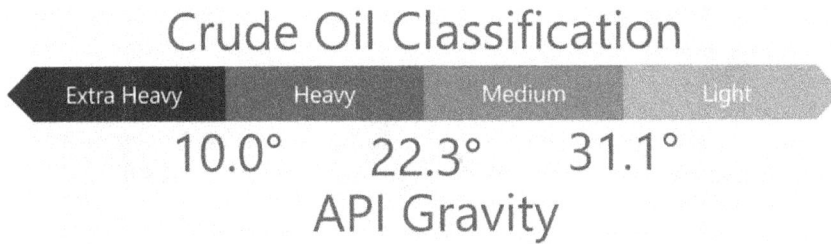

Crude Oil Classification

| Extra Heavy | Heavy | Medium | Light |

10.0°   22.3°   31.1°

API Gravity

FIGURE 1.3 Classifications of crude oil.

Finally, an API gravity of less than 10.0° would be considered extra heavy. In addition to API, viscosity and gas-oil ratio (GOR) are also considered to greatly contribute to the classification of crude. Figure 1.3 shows the classifications for crude oil density:

1.4 METHODS OF OIL RECOVERY

The recovery of crude oil and gas from the reservoir occurs through wells, drilled through the cap-rock, which acts as a faucet for petroleum products. The drilled wells allow the flow of petroleum products from the underground reservoir to surface facilities. The initial production of petroleum from the reservoir to the wellbore is caused by the natural energy of the reservoir and is regarded as the primary oil recovery. These drive energies are provided by the innate reservoir properties such as compressibility of reservoir rock and fluids, composition of reservoir fluids, presence of a gas cap or active aquifer, and gravitational force resulting from the density difference of reservoir fluids (Speight, 2016). The expansion drive mechanism is the dominant one when the reservoir pressure is above the bubble-point pressure, and the production of reservoir fluids through the wells causes a reduction in reservoir pressure and subsequent expansion of reservoir fluids. The average recovery by the expansion drive mechanism is very low, with a maximum up to 3% (Iglauer et al., 2010; Satter and Iqbal, 2016). The reduction of reservoir pressure below the bubble-point pressure leads to the release of the dissolved gases from the crude oil. As the released gases have greater compressibility in comparison to the other reservoir fluids (oil and water) and rock, the expansion of gases becomes the dominant drive mechanism for the production of crude oil. This drive mechanism is termed as solution gas drive or depletion drive and has an average oil recovery of 20% (Satter and Iqbal, 2016). The crude oil production from the reservoir caused by the expansion of reservoir fluids has lower recovery factors and leads to significant reduction in reservoir pressure. When the reservoir is in communication with a gas cap, present at the top of the crude oil–bearing zone, or an aquifer, present at the bottom or peripheral of the crude oil–bearing zone, the pressure differential between the borehole and reservoir is maintained, leading to significantly higher oil recovery and production life of the reservoir.

The natural capacity of a reservoir to produce oil gets depleted continuously with production and requires additional substitutes through external fluid injection. The external energy is usually provided in the form of pressure maintenance or waterflooding. The pressure maintenance methodology mainly includes gas injection above gas-oil contact or water injection below oil-water contact, which maintains or increases the pressure of reservoir (Vishnyakov et al., 2020). The injection of water in the oil zone of the reservoir is termed as waterflooding, where the injected water sweeps the hydrocarbons from the reservoir toward the production well. The oil recovery from the reservoir by the use of external aid is termed as secondary or improved oil recovery. The primary and secondary oil recovery stages do not incorporate any alteration of reservoir rock and fluid properties, leading to the trapping of crude oil in the reservoir. The distribution of residual oil after waterflooding is shown in Figure 1.4.

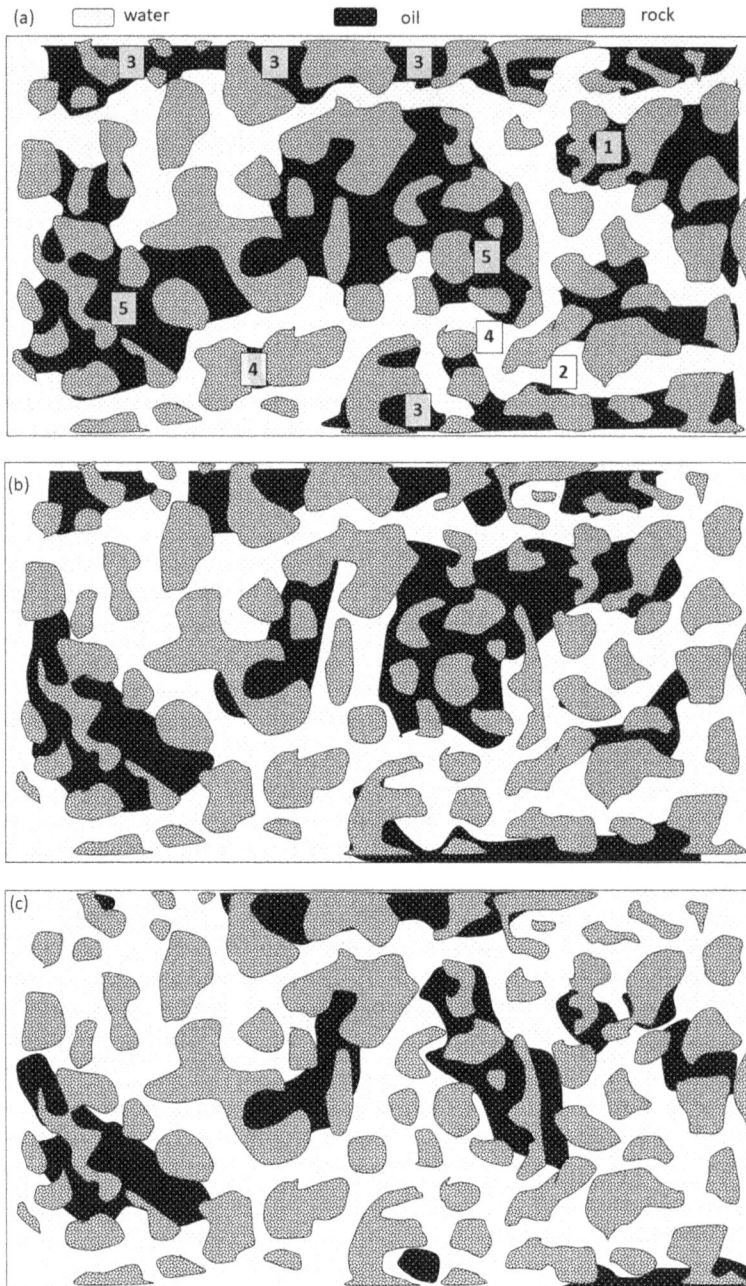

FIGURE 1.4 Residual oil distribution in the (a) oil-wet, (b) neutral-wet, and (c) water-wet porous medium when the water cut is 98%. (1) Isolated oil droplet, (2) Oil film, (3) Residual oil in dead ends, (4) Residual oil in pore throats, (5) Cluster residual oil.

Source: Guo et al., 2019.

The unrecovered crude oil is trapped in the pores of the reservoir due to various factors such as high capillary forces acting on the oil droplets caused due to the high interfacial tension (IFT) of the oil-aqueous interface. The trapping of crude oil due to high IFT is more prominent in

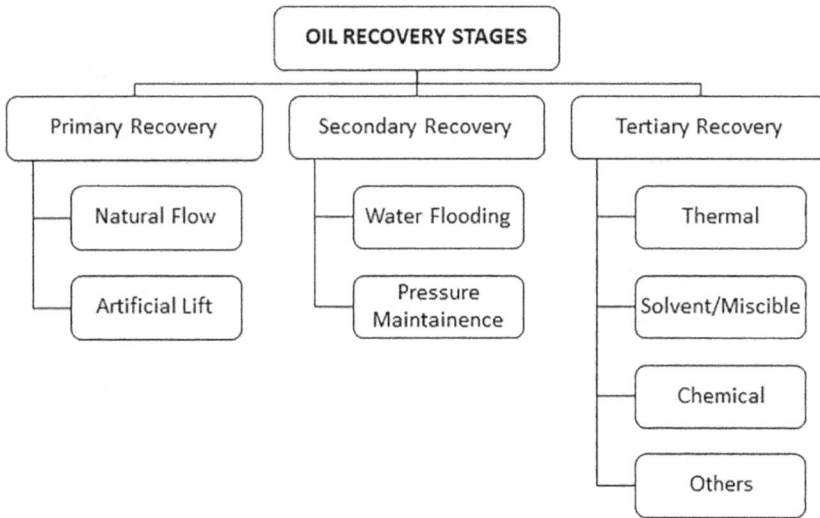

FIGURE 1.5 Classification of oil recovery.

water-wet reservoirs, whereas in oil-wet reservoirs, the residual oil is present as oil films coated on rock surfaces. Thus, in an oil-wet reservoir, the oil is stuck in the smaller pores, leading to greater residual oil saturation. The oil droplets are also trapped at the "dead ends" of the flow channels or in microscopic pores of a heterogeneous reservoir (Sheng, 2011). Thus, the reservoir rock and fluid properties such as pore shape and size, reservoir heterogeneities, IFT of the reservoir fluids, and the wetting state of the reservoir rock are major factors causing low oil recovery efficiency of 30%–40% of the original oil in place (OOIP) after primary and secondary recovery stages (Austad et al., 1996). Thus, the manipulation and tuning of reservoir conditions, which include rock properties, fluid properties, and reservoir temperature, are applied in the tertiary oil recovery stage to recover the oil trapped in the reservoir, and this oil recovery stage is termed as enhanced oil recovery. The stages and classifications of oil recovery methods have been shown in Figure 1.5.

1.5   ENHANCED OIL RECOVERY

Enhanced oil recovery (EOR) is oil recovery by the injection of materials not normally present in the reservoir. This definition embraces all modes of oil recovery processes (drive, push-pull, and well treatments) and covers many oil recovery agents. A typical EOR process involves the injection of the EOR solution through the injection well, followed by chase waterflooding that pushes the EOR solution toward the production well, leading to the recovery of trapped oil after primary and secondary recovery stages. The general schematic of the EOR process has been illustrated in Figure 1.6.

1.5.1   PRINCIPLES OF EOR

EOR techniques primarily aim to recover the crude oil trapped in the reservoir after the primary and secondary production stages. However, EOR techniques may be implemented at the primary stage also for better reservoir management and ultimate recovery. The unrecovered crude oil is

Enhanced Oil Recovery

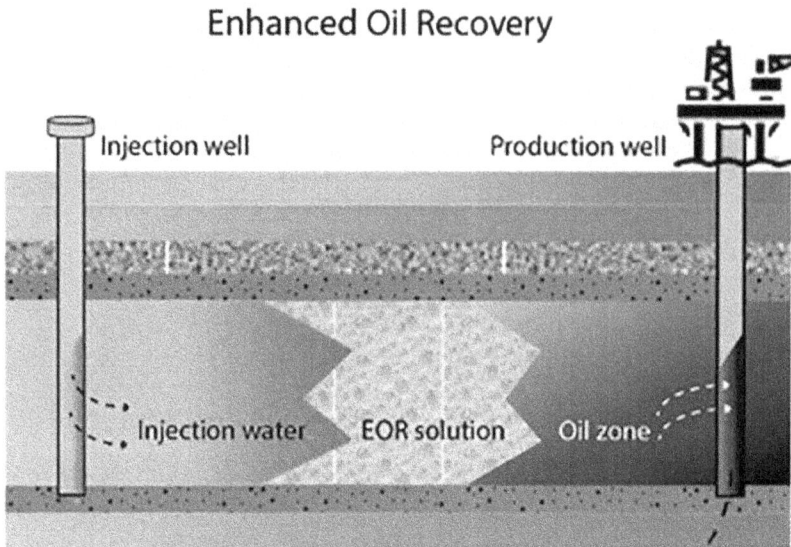

FIGURE 1.6   Schematic illustration of enhanced oil recovery.

Source: Aadland et al., 2019.

trapped in the pores of the reservoir due to various reservoir rock and fluid properties, such as high IFT between the oleic phase and aqueous phase present in the reservoir, oil-wetting state of reservoir rock surface, and high viscosity of crude oil. These properties lead to lower microscopic and macroscopic efficiencies of the crude oil production by the natural and external energies provided to the reservoir. Thus, EOR application focuses on oil production by principles such as IFT reduction, wettability alteration, and improvement of mobility ratio by injection fluid. The reduction of IFT between the oleic phase and aqueous phase present in the reservoir leads to the reduction of capillary forces causing the mobilization of trapped crude oil droplets from the narrow pores of the reservoir. The reduction of capillary forces that traps the crude oil in the reservoir pores is also caused by the wettability alteration of reservoir rock surface from an oil-wetting state to, preferentially, a water-wetting state. The reduction of capillary forces by IFT reduction and wettability alteration leads to the mobilization of crude oil at the pressure differential present in the reservoir. Thus, IFT reduction and wettability alteration cause an increase in the microscopic sweep efficiency of the applied EOR technique. The mobilization of crude oil from the microscopic pores of the reservoir also leads to the formation of an oil bank, which is pushed toward the production well by the injection of chase fluid after EOR application. However, the efficient mobilization of the oil bank toward the production well is also a factor of the macroscopic sweep of EOR techniques. The macroscopic efficiency of the EOR method is improved by controlling the mobility ratio of the displacing fluid and displaced fluid. A lower mobility ratio is preferred for an efficient volumetric sweep of the crude oil. This lower mobility ratio of the crude oil to the applied displacing fluid is achieved either by reducing the viscosity of crude oil or by increasing the viscosity of displacing fluid. Lower mobility ratio improves macroscopic efficiency of EOR by preventing early breakthrough of injected fluid and improving the effective swept volume of reservoir. The macroscopic efficiency, also known as volumetric sweep efficiency (E_v), and

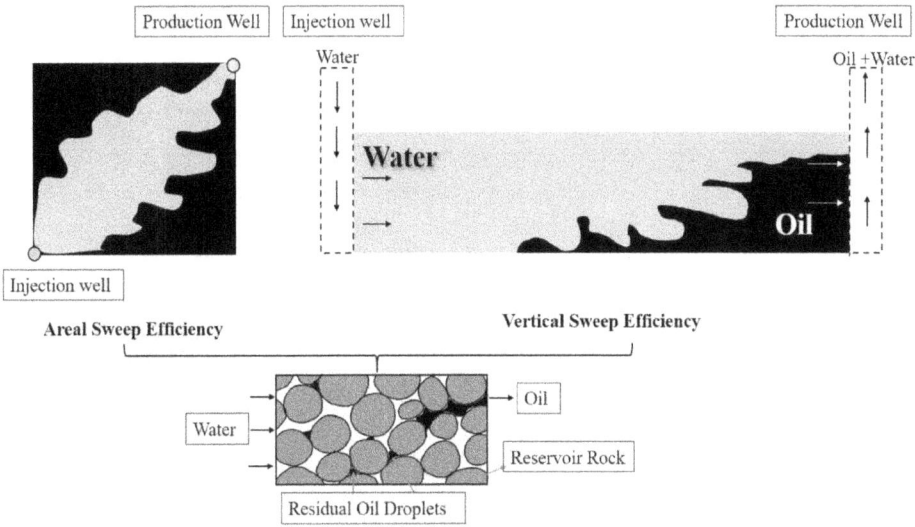

FIGURE 1.7 Schematics of microscopic and macroscopic sweep efficiencies.

microscopic efficiency, also known as displacement efficiency (E_D), is related to the overall efficiency (E) of the EOR method as follows:

$$E = E_D \times E_V = E_D \times E_A \times E_{VI} \tag{1.1}$$

where E_A and E_{VI} are the areal and vertical sweep efficiencies. These respective sweep efficiencies are shown in Figure 1.7.

1.5.2 CLASSIFICATIONS OF EOR

The various oil recovery mechanisms of EOR are achieved by different techniques, which include the use of chemicals in injected fluid, increasing the temperature of the reservoir by steam, hot water injection or in situ combustion, and injecting gases miscible or immiscible into the crude oil (Elgaghah et al., 2007; Srivastava et al., 1999). The classification of these different EOR techniques is shown in Figure 1.8. The increase in reservoir temperature is achieved by the application of thermal EOR, where the methodologies of injection of steam or hot water, in situ combustion, cyclic steam injection, and steam-assisted gravity drainage are employed. The increase in temperature achieved by these techniques causes a reduction in crude oil viscosity, leading to the enhancement of the mobility of the trapped crude oil droplets. Thus, thermal EOR is mostly applicable to reservoirs with heavy oil (Taber et al., 1997). The injection of gases such as natural gas, CO_2, and N_2 are used in the gas EOR process to enhance the oil recovery. These gases are either miscible or immiscible with the reservoir fluids. The immiscible gases push the oil to the production well by expansion in reservoir, whereas the miscible gases reduce the viscosity of the crude oil and IFT after getting dissolved in the trapped crude oil, leading to improvement in oil flowability. Thus, the application of gas EOR is widely implemented in reservoirs with light to moderate oil (Mahdavi and Zebarjad, 2018). Chemical EOR includes the injection of chemicals such as polymers, surfactants, and alkalis

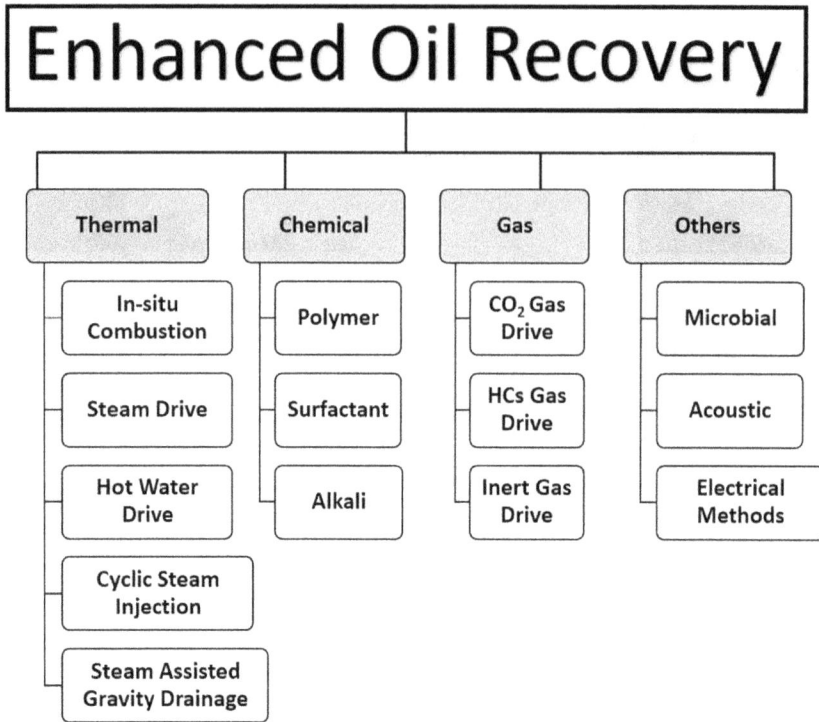

FIGURE 1.8 Conventional EOR classification.

into the reservoir to improve the oil recovery. Polymer injection leads to increased viscosity of the injected fluid, causing viscous forces to act on the trapped crude oil. The injection of surfactants and alkalis causes the reduction of IFT between the displaced and displacing fluids, leading to a decrease of the capillary forces that trap the crude oil in the pores of the reservoir. Thus, a combination of different chemicals is used in chemical EOR for recovery of oil from the reservoir. Other EOR methodologies, which include using microbial and acoustic applications, are mostly in an experimental stage and lack commercial large-scale implementation in reservoirs (Yernazarova et al., 2016).

1.5.3 PARAMETERS FOR THE IMPLEMENTATION OF AN EOR PROJECT

There exist numerous practical and technical as well as economic challenges that need to be addressed before EOR technologies can be deployed in the field. The parameters that significantly impact the selection and implementation of EOR are as follows:

1. Anticipated oil recovery: Before implementing any EOR project, it is important to know the projected additional recovery. An intensive study on the laboratory analysis and simulation is essential to predicting the recovery.
2. Fluid production rates: EOR methods require the injection of chemicals, solvents, heat, etc., which leads to enhanced production of reservoir fluids. Thus, necessary installation for the proper treatment and handling of those fluids is of utmost importance.
3. Monetary investment: Two costs are considered, operational costs (OPEX) and capital investment costs (CAPEX). The capital investment cost is the cost to invest in the infrastructure required to perform the EOR method, including additional costs for facility maintenance, engineering, supervision, construction, and contingency.

4. Cost of water treatment and pumping equipment: Enhanced oil production means increase is always associated with increased water production. This requires additional cost for water treatment and pumping.
5. Cost of maintenance and operation of the water installation facilities: The operational costs are associated with the cost of the chemical/solvent/heat injected and the cost of using, maintaining, and repairing the required surface facilities.
6. Cost of drilling new injection wells or converting existing production wells into injectors: To improve recovery, it is important to improve the sweep efficiency. The sweep efficiency can be increased by drilling new wells or converting existing production wells into injectors. The drilling of new wells requires additional costs, which must be considered for any EOR project.

1.5.4 TIME FOR DESIGN AND IMPLEMENTATION OF AN EOR PROJECT

Design and implementation of an EOR project are time-consuming processes. Even after implementation (especially as a tertiary project) the production response does not occur immediately. Hence, planning has to be made in advance.

Field Selection: Primary and secondary recovery techniques are the most economically viable methods; however, due to the poor efficiencies of these methods, more than 60% of the oil remains trapped in the reservoir. As global demand for oil increases, so does its value, and this makes more expensive oil extraction techniques increasingly viable. However, the production history of the field and the residual oil saturation play an important role in considering a field for a suitable EOR project.

Process Selection: The EOR criteria published by Taber et al. in 1997 and the updated criteria are useful in selecting the suitable process for any target reservoir. The criteria tabulate a range of oil and reservoir properties for the various EOR methods.

Geological Studies: Geologic and permeability heterogeneities are the most probable causes for the low recovery efficiency realized (Goodrich, 1980). The injected recovery medium simply traversed the more permeable layers of reservoir rock, bypassing the less permeable layers because of the heterogeneity of the reservoir. Thus, the geological characterization of a reservoir is as important as the process characterization for a better understanding and planning of an EOR project.

Design Parameters: Laboratory tests are considered the cornerstone for proof of concept during any proposed EOR implementation. Input data from laboratories are used not only to de-risk the EOR process by giving an insight into the potential pore-level incremental displacement efficiency but also to measure key EOR agent-rock interactions, indispensable for the numerical screening of any EOR flooding (Moreno et al., 2018).

Economic Evaluation: Recent advances in computer technology have incorporated advanced screening criteria and allowed the risk and economic analysis of EOR projects. The economic evaluation can be run using optimistic production profiles and oil prices for the purposes of screening (Manrique and Wright, 2005).

Pilot Testing: Defining clear pilot objectives is the first step in designing and executing a successful pilot. Pilots are conducted to address key technical and business uncertainties in addition to the risks associated with applying an EOR technology in a specific field. Pilots are a scaled-down version of the full commercial implementation of an EOR process. The specific piloting objectives may include the following (Teletzke et al., 2010): (i) evaluate the EOR process recovery efficiency in the field of interest; (ii) assess the effects of reservoir geology on process performance, particularly sweep efficiency; (iii) improve field production forecasts to reduce technical and economic risk; (iv) obtain data to calibrate reservoir-simulation models for full-field predictions; (v) identify operational issues and concerns for full-field development; (vi) assess the effect of development options on recovery (e.g., well spacing, processing rate, and completion strategy); and (vii) guide improvements in current operating strategy to improve economics/recovery.

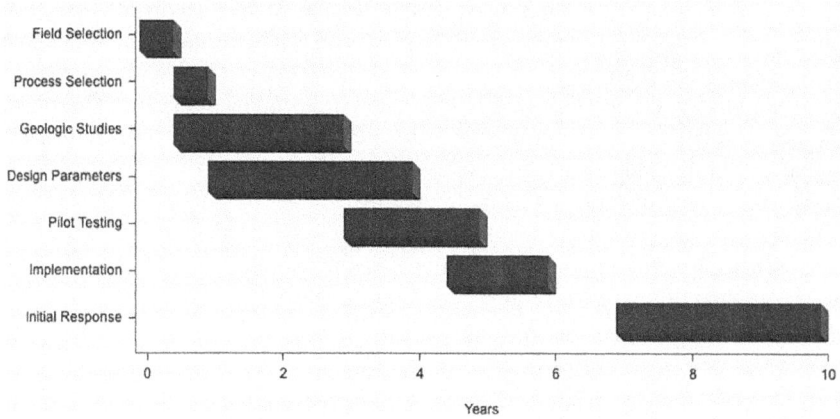

FIGURE 1.9   Milestones for an EOR project.

Implementation: Implementation of the EOR project is a complex and lengthy process. To achieve a greater level of efficiency and reduce the risks and uncertainties of the project outcome, it is necessary to define a comprehensive management of the EOR process, starting from EOR method selection, process design, and performing the pilot test and implementation on a full-scale reservoir (Karović-Maričić et al. 2014).

Initial Response: To assess the efficacy and efficiency of any demonstration project, it is imperative to have a robust monitoring and surveillance plan to understand, track, monitor, and predict the flow paths, injectant plume evolution, and breakthrough times of the injected fluids. Furthermore, a good understanding of the remaining oil saturation (*Sor*) and its distribution in the reservoir after recovery processes is important for reserves assessment (Al-Qasim et al., 2021). As shown in Figure 1.9, after successful completion, the initial response may be observed after seven years (in general). So, a long-term plan is required for any EOR project.

1.5.5   SCREENING CRITERIA FOR THE APPLICATION OF EOR METHODS

Most EOR projects are costly and time consuming. The high risk and technical complexity expose these projects to failure risk (Mashayekhizadeh et al., 2014; Gharbi, 2005; Bourdarot and Ghedan, 2011). Screening criteria usually comprise the first tool for a reservoir engineer in selecting a proper EOR method. It is accomplished by collecting data from a number of successful projects and analyzing them to find out important and effective parameters. These criteria determine the application of the selected parameters for an EOR process (Saleh et al., 2014).

With the utilization of modern reservoir management practices, engineers consider the various improved/enhanced oil recovery options at a much earlier stage in the productive life of the field. For many fields, the decision is not whether, but when, to start an EOR project. Obviously, economics also plays a major part in the "go/no go" decisions for expensive injection projects, but a cursory examination of the technical criteria is helpful to rule out less likely candidates. The criteria are also useful for surveys of a large number of fields to determine whether specific gases or liquids can be used for oil recovery subject to their availability and economic viability. The exploitation strategy through EOR methods is considered primarily based on fluid characteristics followed by reservoir setting and drive mechanism. Other aspects like economics and logistics also play vital roles in screening an EOR method for its implementation. The International Database Projection Map, created in PDVSA-INTEVEP, is used to identify the likely EOR processes in the candidate reservoir. The database is generated with the information of 290 EOR cases of thermal, chemical, and gas injection methods implemented in different countries across the world.

1.6 CURRENT EOR SCENARIO

EOR has been regarded as a technological effort that can reverse the declining trend of mature fields and improve the overall recovery percentage. Currently, 2% of global oil production is the result of implementation of EOR techniques in mature fields, which is expected to increase to 4% of the global petroleum production in 2040. EOR projects have already been started in North America, followed by implementation in other countries such as Malaysia, the United Arab Emirates, Kuwait, Saudi Arabia, India, Columbia, Ecuador, and Oman (Manrique et al., 2010). The number of EOR projects implemented globally till 2022 is around 950, where thermal and CO_2 EOR projects constitute 44% and 21% of the total implemented projects, respectively (Worldwide EOR Projects Distribution, 2022).

The implementation of EOR projects in India started with the pilot testing of an in situ combustion project in Balol in 1990 and later commercialization in 1997 (Chattopadhyay et al., 2004). Other ongoing EOR projects include thermal EOR projects at Santhal and Lanwa, gas injection projects at Gandhar and Borholla, and polymer flooding at the Sanand and Mangala fields (Baskaran et al., 2014; Tiwari et al., 2008). Future EOR projects include the commercial implementation of ASP flooding at Viraj and pilot testing at Bechraji, Lanwa, and Sobhasan. These EOR projects contribute to 3.3% of the total onshore production. Offshore EOR projects include the implementation of low-salinity waterflooding in the Mumbai High and South Heera fields, where the high salinity and complex rock mineralogy pose major challenges for EOR implementation.

1.7 CHALLENGES OF EOR

Worldwide, oil industries are urgently realizing that they must focus on improving ultimate oil recovery instead of immediate oil recovery, which is driven solely by short-term profit. This pivot to a long-term view ensures the optimal utilization of oil resources by improving secondary oil recovery through sustainable development, keeping lower depletion rates, and converging on long-term profit. Suitable EOR techniques can then be employed to optimize ultimate oil recovery. As the conventional light crude oil is beginning to get depleted, a move toward more difficult hydrocarbon resources like heavy crude oil, shale oil, bitumen, and oil sands is already well under way. Usually, the conventional oil recovery for these resources is low. An EOR technique has to be applied comparatively early in these reservoirs. The incentive toward exploiting recovery, rather than thinking about short-term profits, helps in better resource exploitation. Life cycle planning includes thinking about EOR early enough to conduct relevant R&D studies, feasibility tests, and pilots to enable key decisions to be made at the right time (Archibald, 2003). Investment in R&D is essential to generate the right option for field development. Sometimes, incorrect reservoir development strategies have been adopted to recover crude oil as fast as possible or at a higher production rate. This can eventually lower the overall recovery from the reservoir considerably. EOR professionals also ensure better improved oil recovery (IOR) implementation strategies due to their capability development. A stepwise implementation involves moving from laboratory-scale tests, single well tests, and pilot tests to full field implementation. EOR implementation may be aided by a company's or country's need for energy security. The United States is a prime example of this need and has taken a true leadership role in EOR implementation in its fields, despite being a free economy.

1.8 ENVIRONMENTAL IMPACTS

In recent years, a strong boost to EOR has come from environmental concerns. Due to limitations in overboard disposal of produced water, deepwater projects and disposal wells are needed, but they hamper the economics of the project. The chemicals in the water may contain toxic heavy metals and radioactive substances as well as be very salty. This can be very damaging to drinking water sources and the environment, in general, if the disposal is not properly controlled. Therefore, reinjection of produced water is an attractive option here, as it minimizes the material (including the cost of the chemicals) and process costs as well as environmental impact. Advanced electro dialysis reversal technologies use ion exchange membranes to treat EOR-produced waters. These systems desalinate the produced water for reinjection and allow for the recycling of EOR polymers. A more severe impact on global climate is during hybrid chemical EOR, where the production of gases like CO_2 can increase the greenhouse effect. There are incentives to sequester this CO_2. In many ways, this is a win-win situation: sequestering CO_2 at the same time as producing incremental oil. In this case, CO_2 is injected into the well, which is good for environment.

1.9 ARTIFICIAL INTELLIGENCE (AI) IN EOR

Another growing area in EOR is the applications of AI encompassing data mining, data analytics, machine learning, deep learning, fuzzy logic, expert systems, and robotics. The vast availability of real-time data has paced up the quest for the industry and research personnel to develop efficient data management and decision-making solutions. AI can continuously refine production strategies, adapting to reservoir complexities and optimizing injection and extraction rates to achieve the highest possible recovery factor. EOR is aimed at minimizing the residual oil saturation from the reservoir with techno-economic feasibility. AI offers solutions to EOR screening, reservoir characterization, well placement optimization, pressure maintenance, rock-fluid interactions, and improving the ultimate recovery.

The data-driven and/or physics-driven predictive modeling techniques facilitate the performance optimization of the conventional EOR methods such as chemical, thermal, immiscible/miscible fluid injection methods, and unconventional microbial enhanced oil recovery (MEOR) techniques. AI aids in identifying the most effective chemical agents and their optimal concentrations for mobilizing reservoir fluid, reducing the risks of incorrect chemical dosing. AI can optimize thermal injection patterns to ensure efficient heat distribution in the reservoir to enhance the mobility of the oil. AI-driven simulations can determine the ideal gas injection rate and composition to boost overall recovery.

EOR operations generate vast amounts of complex data from various sources, including seismic surveys, well logs, production rates, and reservoir simulations. AI-powered algorithms can learn higher-level abstractions from these data, identifying patterns and relationships for better performance on complex operations. These algorithms provide valuable insights into reservoir properties, fluid behavior, and connectivity, leading to improved reservoir characterization and understanding of fluid behavior.

Precise placement of injector and producer wells is critical to improve macroscopic and microscopic recovery efficiency for optimizing oil recovery. AI algorithms can analyze geological and reservoir data in real time and assist in designing the operational ecosystem. The propagation of the injection fluid front is monitored and controlled using AI tools during the flooding operations.

AI enables real-time monitoring of reservoir behavior, including reservoir pressure, reservoir temperature, and fluid properties by integrating soft sensors and AI algorithms. Consequently, reservoir models can be updated continuously, leading to informed decisions on production strategies. Moreover, AI-based production optimization techniques are designed to respond dynamically to changing reservoir conditions. These algorithms leverage historical and real-time data to predict future production scenarios. By incorporating machine learning, AI can continuously refine production strategies, adapting to reservoir complexities and optimizing injection and extraction rates to achieve the highest possible recovery factor.

EOR operations often raise environmental concerns due to the injection of chemicals and water into the reservoir. Smart AI algorithms can determine the most suitable injection scenarios, reducing the carbon footprint of EOR operations while maximizing oil recovery. However, the challenges related to data integrity, AI model robustness, and human-machine collaboration emphasize the need for research and development to unlock the full potential of AI in EOR.

REFERENCES

Aadland, R.C., Jakobsen, T.D., Heggset, E.B., Long-Sanouiller, H., Simon, S., Paso, K.G., Syverud, K., Torsæter, O., 2019. High-temperature core flood investigation of nanocellulose as a green additive for enhanced oil recovery. Nanomaterials 9, 665.

Al-Qasim, A.S., Kokal, S.L., Al-Ghamdi, M.S., 2021, April. The State of the Art in Monitoring and Surveillance Technologies for IOR, EOR and CCUS Projects. In SPE Western Regional Meeting. OnePetro.

Aminzadeh, F., Dasgupta, S.N., 2013. Fundamentals of Petroleum Geology, in: Developments in Petroleum Science (Vol. 60). Elsevier, pp. 15–36.

Archibald, R.D., 2003. Managing High-Technology Programs and Projects. John Wiley & Sons.

Austad, T., Hodne, H., Strand, S., Veggeland, K., 1996. Chemical flooding of oil reservoirs 5. The multiphase behavior of oil/brine/surfactant systems in relation to changes in pressure, temperature, and oil composition. Colloids and Surfaces A: Physicochemical and Engineering Aspects 108, 253–262.

Baskaran, V.K., Dani, K.C., Kumar, K.P., Urkude, A.M., 2014. Implementation of Enhanced Oil Recovery Techniques in India: New Challenges and Technologies. In SPE EOR Conference at Oil and Gas West Asia, pp. 227–242.

Bourdarot, G., Ghedan, S., 2011, October. Modified EOR Screening Criteria as Applied to a Group of Offshore Carbonate Oil Reservoirs. In SPE Reservoir Characterisation and Simulation Conference and Exhibition. OnePetro.

Campbell, C.J., Laherrère, J.H., 1998. The end of cheap oil. Scientific American 278(3), 78–83.

Chapman, I., 2014. The end of Peak Oil? Why this topic is still relevant despite recent denials. Energy Policy 64, 93–101.

Chattopadhyay, S.K., Binay, R., Bhattacharya, R.N., Das, T.K., 2004. Enhanced Oil Recovery by In-Situ Combustion Process in Santhal Field of Cambay Basin, Mehsana, Gujarat, India—A Case Study. In SPE/DOE Symposium on Improved Oil Recovery. Society of Petroleum Engineers.

Elgaghah, S., Zekri, A.Y., Almehaideb, R.A., Shedid, S.A., 2007. Laboratory Investigation of Influences of Initial Oil Saturation and Oil Viscosity on Oil Recovery by CO_2 Miscible Flooding. In EUROPEC/EAGE Conference and Exhibition. Society of Petroleum Engineers.

Gharbi, R., 2005. Application of an expert system to optimize reservoir performance. Journal of Petroleum Science and Engineering 49(3–4), 261–273.

Goodrich, J.H., 1980. Target Reservoirs for CO2 Miscible Flooding (Vol. 8341, No. 17). US Government Printing Office.

Guo, Y., Zhang, L., Zhu, G., Yao, J., Sun, H., Song, W., Yang, Y., Zhao, J., 2019. A pore-scale investigation of residual oil distributions and enhanced oil recovery methods. Energies 12(19), 3732.

Hubbert, M.K., 1956. Nuclear Energy and the Fossil Fuels: American Petroleum Institute Drilling and Production Practice. In Proceedings of the Spring Meeting, San Antonio, TX (pp. 7–25). American Petroleum Institute.

Iglauer, S., Wu, Y., Shuler, P., Tang, Y., Goddard, W.A., 2010. New surfactant classes for enhanced oil recovery and their tertiary oil recovery potential. Journal of Petroleum Science and Engineering 71, 23–29.

Karović-Maričić, V., Leković, B., Danilović, D., 2014. Factors influencing successful implementation of enhanced oil recovery projects. Podzemni Radovi 22(25), 41–50.

Kutcherov, V.G., Krayushkin, V.A., 2010. Deep-seated abiogenic origin of petroleum: From geological assessment to physical theory. Reviews of Geophysics 48, RG1001.

Maggio, G., Cacciola, G., 2012. When will oil, natural gas, and coal peak? Fuel 98, 111–123.

Mahdavi, E., Zebarjad, F.S., 2018. Screening Criteria of Enhanced Oil Recovery Methods, in: Fundamentals of Enhanced Oil and Gas Recovery from Conventional and Unconventional Reservoirs. Elsevier, pp. 41–59.

Manrique, E.J., Thomas, C.P., Ravikiran, R., Izadi Kamouei, M., Lantz, M., Romero, J.L., Alvarado, V., 2010. EOR: Current Status and Opportunities. In SPE Improved Oil Recovery Symposium. Society of Petroleum Engineers, pp. 1584–1604.

Manrique, E.J., Wright, J.D., 2005, April. Identifying Technical and Economic EOR Potential Under Conditions of Limited Information and Time Constraints. In SPE Hydrocarbon Economics and Evaluation Symposium. OnePetro.

Mashayekhizadeh, V., Kord, S., Dejam, M., 2014. EOR potential within Iran. Special Topics & Reviews in Porous Media: An International Journal 5(4).

Moreno, J.E., Flew, S., Gurpinar, O., Liu, Y., Gossuin, J., 2018, April. Effective Use of Laboratory Measurements on EOR Planning. In Offshore Technology Conference, D021S024R003. OTC.

Saleh, L.D., Wei, M., Bai, B., 2014. Data analysis and updated screening criteria for polymer flooding based on oilfield data. SPE Reservoir Evaluation & Engineering 17(1), 15–25.

Satter, A., Iqbal, G.M., 2016. Primary Recovery Mechanisms and Recovery Efficiencies, in: Reservoir Engineering. Elsevier, pp. 185–193.

Sheng, J.J., 2011. Modern Chemical Enhanced Oil Recovery. Elsevier.

Speight, J.G., 2015. Occurrence and Formation of Crude Oil and Natural Gas, in: Subsea and Deepwater Oil and Gas Science and Technology. Elsevier, pp. 1–43.

Speight, J.G., 2016. General Methods of Oil Recovery, in: Introduction to Enhanced Recovery Methods for Heavy Oil and Tar Sands. Elsevier, pp. 253–322.

Srivastava, R.K., Huang, S.S., Dong, M., 1999. Comparative effectiveness of CO_2, produced gas, and flue gas for enhanced heavy-oil recovery. SPE Reservoir Evaluation & Engineering 2, 238–247.

Taber, J.J., Martin, F.D., Seright, R.S., 1997. EOR screening criteria revisited – Part 1: Introduction to screening criteria and enhanced recovery field projects. SPE Reservoir Engineering 12, 189–198.

Teletzke, G.F., Wattenbarger, R.C., Wilkinson, J.R., 2010. Enhanced oil recovery pilot testing best practices. SPE Reservoir Evaluation & Engineering 13(1), 143–154.

Tissot, B.P., Welte, D.H., 1978. An Introduction to Migration and Accumulation of Oil and Gas, in: Petroleum Formation and Occurrence. Springer, pp. 257–259.

Tiwari, D., Marathe, R.V., Patel, N.K., Ramachandran, K.P., Maurya, C.R., Tewari, P.K., 2008. Performance Of Polymer Flood in Sanand Field, India—A Case Study. In SPE Asia Pacific Oil and Gas Conference and Exhibition. Society of Petroleum Engineers.

Vishnyakov, V., Suleimanov, B., Salmanov, A., Zeynalov, E., 2020. Oil Recovery Stages and Methods, in: Primer on Enhanced Oil Recovery. Elsevier, pp. 53–63.

Walters, C.C., 2006. The Origin of Petroleum, in: Practical Advances in Petroleum Processing. Springer, pp. 79–101.

Worldwide EOR Projects Distribution, 2022. https://public.tableau.com/app/profile/na.zhang/viz/Worldwide EOR/WorldwideEORDistribution

Yernazarova, A., Kayirmanova, G., Baubekova, A., Zhubanova, A., 2016. Microbial Enhanced Oil Recovery, in: Chemical Enhanced Oil Recovery (CEOR) – A Practical Overview. InTech, pp. 295–304.

Zhao, J., Cao, Q., Bai, Y., Er, C., Li, J., Wu, W., Shen, W., 2017. Petroleum accumulation: From the continuous to discontinuous. Petroleum Research 2, 131–145.

2 Fundamentals of Immiscible Fluid Displacement Processes

2.1 INTRODUCTION

Petroleum formations are naturally complex systems owing to the existence of heterogeneity resulting from variation in lithology, minerology, structure, pore size variations, etc. Besides, the presence of different fluid phases—i.e., oil/water/gas as individual or in combinations—with unique interactions with rock enhances the complexity of the petroleum reservoir system and makes it very difficult to describe.

Proper understanding of the physical laws governing the equilibrium and flow of several fluids in a porous medium, at the pore level, is very much important to establish the mechanisms of primary and enhanced oil recovery (EOR). To know the mechanisms of fluid flow in a multiphase system, one has to first understand the displacement mechanisms. The system becomes more complicated, when continuous phase transfer occurs among existing fluids like those in case of volatile oil, gas condensate, or those during miscible flooding/thermal recovery. Various models are used to describe the systems differing in rock heterogeneity as well as the complexity of the flowing fluids and their respective interactions. The most simplified model is applied for one-dimensional (1D), single-phase flow, whereas all the formation heterogeneity and three-phase fluid interactions are included in three-dimensional (3D), compositional models. Individual fluid saturation terms are important for simultaneous flow of more than one fluid phase.

The fundamental principles of the flow of fluids through porous media and the important factors dominating the fluid flow and corresponding equations, specifically the fractional flow and conventionally used frontal advance theories, are discussed in this chapter.

2.2 FACTORS INFLUENCING THE FLUID FLOW

The fundamental principle that is used by almost everyone to describe the flow of a fluid in porous media was developed by Darcy in 1856 (Darcy, 1856). This basic equation is a simplified form of Poiseuille's equation of fluid flow through pipes by incorporating the complexities of porous media like pore size distributions, pore throat sizes and shapes, and pore interconnectivity, all of which are lumped into a single-valued parameter in Darcy's single-phase flow equation. The lumped parameter is designated as the absolute permeability, K. In case of multiphase flow, the individual phase velocity is represented by a simple extension of Darcy's single-phase flow equations, considering the individual phase velocity at their respective saturations, with the absolute permeability replacing the effective one at the prevailing saturation. The individual phase equations at their respective saturation in a multiphase system are solved simultaneously to get the solution. All rock-fluid and fluid-fluid interactions are included by making the permeability of each phase a function of its phase saturation. Wetting and nonwetting phase pressures are related by the capillary pressure, which is also a function of saturation.

The common parameters that affect the flow of fluids through the porous medium of the reservoirs are as follows:

- Porosity
- Permeabilities: absolute and effective
- Rock mineralogy
- Reservoir geometry

DOI: 10.1201/9781003098850-2

- Viscosity of each fluid
- Density of each fluid
- Saturation of each fluid
- Filtration velocity of each fluid
- Phase pressure

The relevant equations for the flow of fluids are discussed in subsequent sections.

2.3   FLOW EQUATIONS

Various formation fluids—i.e., oil, gas, and water—as well as injection fluids (water, gas, or various other liquids) continuously flow through the permeable rock of a reservoir once a pressure gradient prevails within it. Because of the complexity and heterogeneity in the porous media, the flow through porous reservoir rock is very much different from that through simple pipes or conduits. Unlike simple pipeline flow, we need to go for predictive analysis, instead of a direct-solution approach, owing to the fact that we never get the actual data from the reservoir; in contrast, we need to simulate the reservoir data from various correlations indirectly. Determination of the actual rate and volume of fluids flowing through the reservoir and through a production or injection well is very important in estimating the production of hydrocarbons. Accurate prediction of the flow of these fluids is only possible through various modeled equations, developed by physical principles as well as boundary conditions. The equations may be developed for steady state, unsteady state, or a combination of both, depending on the flow conditions. The number of phases present within the reservoir and their behavioral characteristics decide the nature of the flow models—from basic single phase to black oil or compositional models. Again, the reservoir heterogeneity dictates whether to follow the simple 1D, two-dimensional (2D), or 3D flow models. Reservoir flow modeling is the most sophisticated methodology available for achieving the primary reservoir management objective for hydrocarbon reservoirs.

In general, the incorporation of all the different factors to form a complex flow model should result in more realistic results. However, the necessity of huge volumes of accurate data and their calculation efficiency may eventually lead to less accurate outcomes. Hence, the models are simplified to the extent that the relevant equations can effectively describe the flow of the fluids within tolerance limits.

Besides detailed numerical modeling that incorporates most of the reservoir rock and fluid characteristics, some other 1D or 2D models are used in reservoir engineering for fast prediction of fluid flow under primary recovery, improved recovery, or enhanced recovery techniques. The commonly used flow equations for immiscible displacement of fluids include frontal advancement and the Buckley-Leverett equation, Welge method, Stile method, Dykstra-Parsons method, and reverse Welge method.

These model equations are applied for quick estimation of (i) the effective recovery; (ii) breakthrough time; (iii) effectiveness of water or chemical injection in displacing oil; and (iv) volume of fluid to be injected.

These models are frequently used to describe the displacement of oil by water and may be used for oil displacement by polymer also. The application of these models to the displacement of oil by immiscible gas needs some modifications to the model equations.

2.3.1   PHYSICAL ASSUMPTIONS

In general, the applications of these model equations in displacement of oil by immiscible fluids are restricted to the fulfillment of certain conditions as described in the following:

(i)   Reservoir is preferably water wet,
(ii)   Displacement occurs under diffuse flow,

(iii) Displacement is incompressible ($q_t = q_o + q_w = q_i$),
(iv) Linear displacement is considered 1D,
(v) Reservoir is homogeneous and may be layered.

2.3.1.1 Water-Wet Reservoir

Wettability determines the preferential adherence of a fluid on a solid surface in the presence of another immiscible fluid. In other terms, the interaction between solid and fluid can be explained through the characterization of wettability in rock. When two immiscible fluids flow through a porous media, their wettability—i.e., preferential adherence to the surface—may differ. The wettability is measured by their contact angle (θ). A rock surface is generally called water wet if $\theta < 90°$ and oil wet if $\theta \geq 90°$. The distribution of the immiscible phases flowing through the rock depends on their wettability. The wetting phase generally adheres to the rock surface and fills the smaller capillaries, whereas the nonwetting phase occupies the larger pores within the reservoir. Thus, the relative permeability at a particular saturation differs from that at a different saturation. In addition, the saturation history plays an important role in variation of the relative permeabilities to the formation fluids and thereby the GOR/water cut of the production streams at the surface. The process at which wetting phase saturation is increased is known as imbibition, and draining out of wetting phase (decreasing saturation) is called drainage process. It is observed from various experiments that the contact angle is larger when rock is first saturated with the wetting phase fluid and gradually decreases with progressive retreatment (Dake, 2001). This difference is described as hysteresis of the contact angle.

Most of our petroleum reservoirs are considered to be water wet. It is geologically assumed that before the migration of oil/gas to the reservoir, they were filled with water, and this makes the formation water wet. The migration of oil/gas into the reservoir (drainage process) gradually displaced the water to the immovable saturation level—i.e., interstitial water saturation. The wettability of rock may vary with its aging in presence of oil. During water injection in a water-wet reservoir, displacement of oil by water is an imbibition process. The capillary pressure and relative permeability data required for the evaluation process must be measured under imbibition conditions.

However, if the formation becomes oil wet, then the displacement of oil by water will be considered as the drainage process, and the capillary pressure and relative permeability data required for the process should be measured under drainage conditions to eliminate the error incurred because of hysteresis.

2.3.1.2 Diffuse Flow

Immiscible flow is assumed to occur under diffuse flow conditions, which is valid under two extreme conditions: i.e., vertical equilibrium and total lack of vertical equilibrium. The flow under any condition between these two extreme conditions should be solved by numerical simulation—i.e., simplification for 1D flow will not be applicable.

Vertical Equilibrium: The immiscible fluids present in the reservoir are distributed under hydrostatic equilibrium—i.e., the capillary force and the gravity balance each other at this condition. The saturation distribution under vertical equilibrium may be determined as a function of capillary pressure (or the capillary height) as (Dake, 1998):

$$P_C\left(S_W\right) = \frac{\Delta\rho\, g y \cos\theta}{1.0133\times10^6} \tag{2.1}$$

where the capillary pressure P_C is expressed in atm;
y is the vertical distance, as shown in Figure 2.2;
θ is the dip angle; and
$\Delta\rho$ or $(\rho_w - \rho_o)$ is the density difference between the two fluids.

The vertical equilibrium condition is interpreted by the following:

- The vertical velocities of two immiscible fluids—in most cases, oil and water—under capillary-gravity equilibrium appear to be infinite in comparison to the velocities of the fluid parallel to the bedding planes.
- During the displacement of oil by water, when the saturation of water increases by a small amount at any point, the new water saturation is redistributed instantaneously as per Equation 2.1.

The following conditions favor the vertical equilibrium condition:

- Very large vertical permeability compared to other directional permeabilities
- Small reservoir thickness
- Large density difference between the moving fluids
- High capillary forces
- Low fluid viscosity
- Low injection rates

As a rule of thumb, vertical equilibrium exists when the capillary transition zone is very large compared to the reservoir thickness.

Absence of Vertical Equilibrium: Under diffuse flow, the capillary pressure zone may be negligible compared to the total reservoir conditions—i.e., absence of vertical equilibrium. At this condition, the injection rate may be so high that the fluid flow (oil and water) rates are much higher in the bedding plane compared to those at dip-normal direction (vertical direction). The impacts of the capillary force and gravity force on the fluid flow are considered negligible at this condition.

These two opposite conditions (vertical equilibrium or total lack of vertical equilibrium) represent the uniform saturation distribution of fluid in the dip-normal direction. Hence, the saturation is mathematically differentiable as a function of time and displacement distance.

Though diffuse flow may occur theoretically under the two opposite conditions as described earlier, the vertical equilibrium condition mostly satisfies the actual reservoir system.

(i) Incompressible displacement: It implies that the pressure-dependent variables—i.e., density and viscosity of fluid—will remain unchanged, or the variation will be negligible throughout the fluid flow path, from injection well to production well under the prevailing differential pressure conditions. The condition will be valid only if

$$q_t = q_o + q_w = q_i \tag{2.2}$$

where q_t is the total production rate; q_i is the injection rate; q_o and q_w represent the oil and water flow rate. All the flow rates are described at the reservoir condition in vol/time.

The assumption is quite realistic in the sense that at stabilized flooding conditions, the pressure variation within the reservoir is very less, so the variations in density along the flow path becomes insignificant.

(ii) Linear displacement: The flow of the fluids is considered to be unidirectional from the injection well to the production well. The simultaneous flow of two immiscible fluids within the reservoir, with the variation in saturations along the distance from the injection well to the production well at different times may be represented as in Figure 2.1.

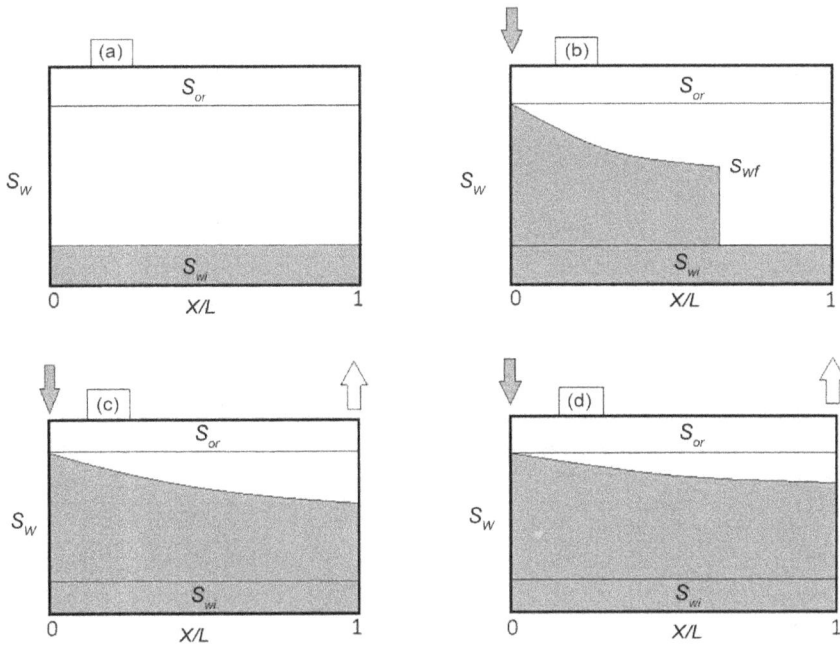

FIGURE 2.1 Saturation variations in the reservoir during waterflood: (a) initial condition of the reservoir, prior to water injection, (b) saturation condition before breakthrough, (c) Condition at the breakthrough, (d) end of the waterflood project with residual oil saturation, S_{or}.

2.4 FRACTIONAL FLOW

The classical mathematical models used for multiphase flows are based on a straightforward generalization of Darcy's law for a single-phase flow, as proposed in Muskat (1946). Buckley and Leverett in 1942 (Buckley and Leverett, 1942) introduced the fractional flow and computed the sharp saturation front position to model the displacement behavior of fluids through porous media. Welge proposed his graphic method (Welge, 1952) to describe the evolution of the saturation front.

The fractional flow equation is used to describe the flow of two immiscible fluids under a steady-state, diffuse flow condition. Diffuse flow means that the viscous, or dynamic, forces are more dominant than the gravity forces, so the vertical variation in saturations may be neglected. The typical characteristics for this type of flow condition occur in a low vertical permeability reservoir combined with a high horizontal pressure gradient.

The converse to diffuse flow condition is the more common condition of segregated flow. Segregated flow is where the gravity forces are more dominant than the viscous, or dynamic, forces, so the vertical variations in saturations are significant. The typical characteristics for this type of flow condition are high vertical permeability, combined with a low horizontal pressure gradient.

Both of the aforementioned flow conditions are implemented and discussed in greater detail in the following sections.

2.4.1 PRINCIPLE AND DERIVATION

Frontal advance theory is an application of the law of conservation of mass—i.e., Mass entering into the element per unit time − Mass leaving the element per unit time = Rate of accumulation of the mass (for a particular phase).

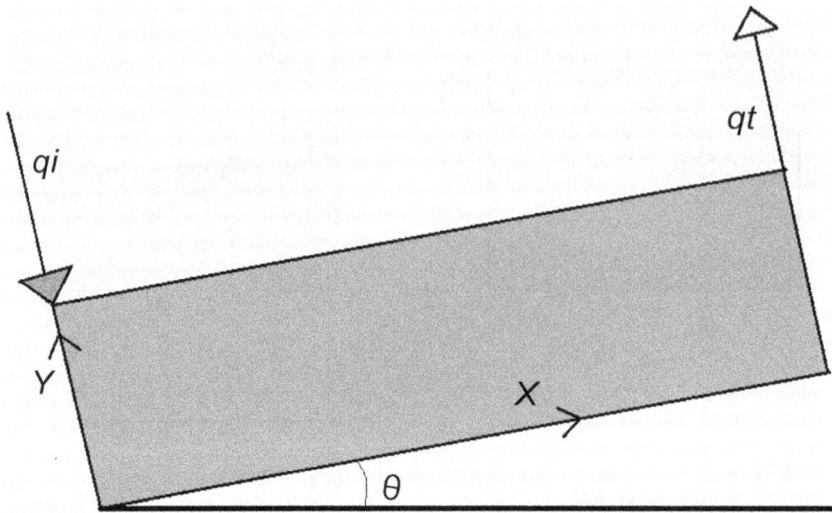

FIGURE 2.2 Direction of flow from injection well to production well in X direction, dipping angle of reservoir is θ.

Consider a part of a reservoir (Fig. 2.2) through which fluids are flowing linearly in the x-direction from the injection well to the production well. The water is injected at a rate of q_i to displace the oil present in the reservoir; q_t is the total production rate that includes either only the oil phase or the oil and water mixture, depending on the breakthrough.

At steady state, the total injection rate (q_t, rb) = production rate (q_t, rb) = Water flow rate (q_w, rb)+oil flow rate (q_o, rb).

$$\text{Mathematically, } q_t = q_o + q_w \tag{2.2}$$

$$q_w = q_t \times f_w \tag{2.3}$$

$$\text{and } q_o = q_t \times \left(1 - f_w\right) \tag{2.4}$$

where f_w is the fraction of water cut. As per Darcy's law, the water flow rate is given by

$$q_w = -\frac{k_{rw}kA}{\mu_w}\left(\frac{\partial p_w}{\partial x} + \frac{\rho_w g \text{Sin}\theta}{1.1033 \times 10^6}\right) \tag{2.5}$$

Similarly, the flow rate of oil is represented as

$$q_o = -\frac{k_{ro}kA}{\mu_o}\left(\frac{\partial p_o}{\partial x} + \frac{\rho_o g \text{Sin}\theta}{1.1033 \times 10^6}\right) \tag{2.6}$$

where p represents the individual phase pressure and k is the absolute permeability of the rock, with k_{ro} and k_{rw} being the relative permeability of the oil and water phase, respectively.
By subtracting Equation 2.6 from Equation 2.5 and rearranging it, we have

$$\frac{q_w \mu_w}{k_{rw}kA} = \frac{q_o \mu_o}{k_{ro}kA} + \left(\frac{\partial p_o}{\partial x} - \frac{\partial p_w}{\partial x} + \frac{\left(\rho_o - \rho_w\right)g\text{Sin}\theta}{1.1033 \times 10^6}\right) \tag{2.7}$$

By rearranging the equation and putting the capillary pressure term

$$\frac{q_w\mu_w}{k_{rw}kA} + \frac{q_w\mu_o}{k_{ro}kA} = \frac{q_w\mu_o}{k_{ro}kA} + \frac{q_o\mu_o}{k_{ro}kA} + \left(\frac{\partial P_c}{\partial x} - \frac{\Delta\rho g\mathrm{Sin}\theta}{1.1033\times10^6}\right) \tag{2.8}$$

where, $P_C = P_o - P_W$ considering water-wet rock and density difference $\Delta\rho = \rho_w - \rho_o$.

Considering steady-state conditions, with total flow ate $q_t = q_o + q_W$

$$q_w\left(\frac{\mu_w}{k_{rw}kA} + \frac{\mu_o}{k_{ro}kA}\right) = \frac{q_t\mu_o}{k_{ro}kA} + \left(\frac{\partial P_c}{\partial x} - \frac{\Delta\rho g\mathrm{Sin}\theta}{1.1033\times10^6}\right) \tag{2.9}$$

Fraction of water cut at a particular time and location with constant water saturation,

$$f_w = q_w/q_t$$

or

$$f_w = \frac{1 + \dfrac{kk_{ro}A}{\mu_o q_t}\left(\dfrac{\partial P_c}{\partial x} - \dfrac{\Delta\rho g\mathrm{Sin}\theta}{1.1033\times10^6}\right)}{\left(1 + \dfrac{\mu_w}{\mu_o}\times\dfrac{k_{ro}}{k_{rw}}\right)} \tag{2.10}$$

Equation 2.10 is called the fractional flow equation. The fraction of water cut inside the reservoir is a function of capillary pressure, dipping angle (defines the gravitational impact on the fractional flow) of the reservoir, density difference between the oil and displacing water, and the mobility ratio. Now, let us consider the effects of each and every factor on the fractional recovery.

2.4.2 Effect of Gravity

The gravity term $\dfrac{\Delta\rho g\mathrm{Sin}\theta}{1.1033\times10^6}$ will be positive—i.e., upward movement of oil with the dipping angle lying between $0°$ and $180°$. The heavier the oil (lesser the density difference), lesser will be the impact of the gravitational term on the fluid flow.

2.4.3 Impact of Capillary Pressure

It is a well-known fact that, mathematically, the capillary pressure is expressed as a sole function of saturation. Thus, a quantitative or qualitative description of capillary pressure gradient with respect to length (x) from the injection well cannot be determined easily. To understand the same, the gradient may be expressed as a product of the two gradients as expressed in Equation 2.11.

$$\frac{\partial P_c}{\partial x} = \frac{dP_C}{dS_W}\cdot\frac{\partial S_W}{\partial x} \tag{2.11}$$

Both the terms—i.e., $\dfrac{dP_C}{dS_W}$ (Fig. 2.3a) and $\dfrac{\partial S_W}{\partial x}$ (Fig. 2.3b)— of the aforementioned equation are negative, thus their product becomes positive. As may be observed from the fractional flow expression (Eq. 2.10), the existence of capillary pressure will contribute to a higher f_w, as $\dfrac{\partial P_c}{\partial x} > 0$, thus resulting in a less efficient displacement. However, this argument alone is not truly valid, as the Buckley-Leverett solution assumes a discontinuous water-oil displacement front (i.e., discontinuity

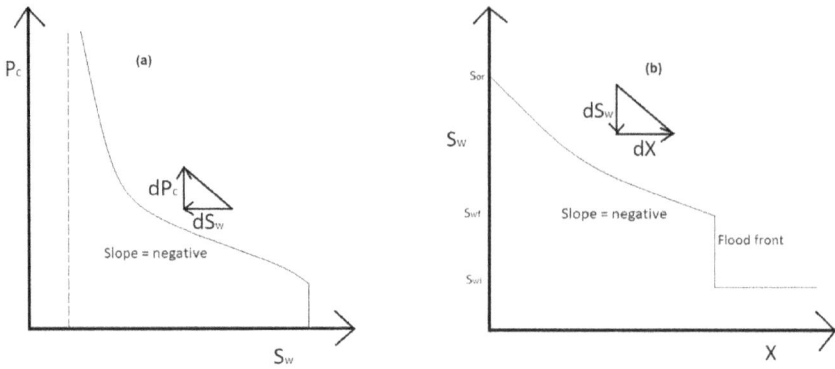

FIGURE 2.3 Capillary pressure and water saturation variations.

of saturation at the flood front, sudden drop from S_{wf} to S_{wc}, as shown in Fig. 2.3b), whereas the capillary pressure is assumed to be a continuous function of water saturation in ideal cases. If capillary pressure gradient is included in the analysis, such a front will not exist, as capillary dispersion (i.e., imbibition) will take place at the front. Thus, in addition to a less favorable fractional flow curve, the dispersion will also lead to an earlier water breakthrough at the production well.

As the presence of capillary pressure gradient will increase the fractional water flow, it is not desirable to have a capillary pressure gradient during the water injection to displace the oil. Practically, it may be observed from the water saturation profile after injecting a specific volume of water into the reservoir that there exists a distinct flood front/shock at ideal conditions. A discontinuity exists at this shock front, and behind this front, the change is gradual with the maximum value of gradient at the shock front. The values of both the gradients are very small, and therefore, the product of these two components becomes insignificant and, in most cases negligible, in comparison with the other terms of the fractional flow equation.

Thus, for a horizontal reservoir with $\theta = 0$, the fractional flow equation is simplified as follows:

$$f_w = \frac{1}{1 + \dfrac{\mu_w}{\mu_o} \times \dfrac{k_{ro}}{k_{rw}}} \tag{2.12}$$

The fractional flow as defined by Equation 2.12 will only depend on the saturation, with the assumption that the temperature remains constant throughout the process—i.e., $f_w = \varnothing(S_w)$, as relative permeability of the most-wetting/least-wetting phase is a function of saturation only.

Figure 2.4 is a plot of the relative permeability ratio, k_o/k_w, versus water saturation. Because of the wide range of k_o/k_w values, the relative permeability ratio is usually plotted on the semilog plot. Most of the cases, the central or main portion of the curve is quite linear. As a straight line on semilog paper, the relative permeability ratio may be expressed as a function of the water saturation by:

$$\frac{k_{ro}}{k_{rw}} = ae^{-bS_w} \tag{2.13}$$

The constants a and b may be determined from the plot of water saturation versus relative permeability as shown in Figure 2.4 or determined through simultaneous equations from known data of saturation and relative permeability.

With the known relative permeability data from the special core analysis (SCAL), the resultant plot of f_w vs. S_w will be of S shape, as shown in Figure 2.5, with the saturation limits between interstitial water saturation (S_{wir}) and $1 - S_{or}$.

Substituting Equation 2.13 into Equation 2.12 will result in

$$f_w = \frac{1}{1 + \dfrac{\mu_w}{\mu_o} \times ae^{-bS_w}} \qquad (2.14)$$

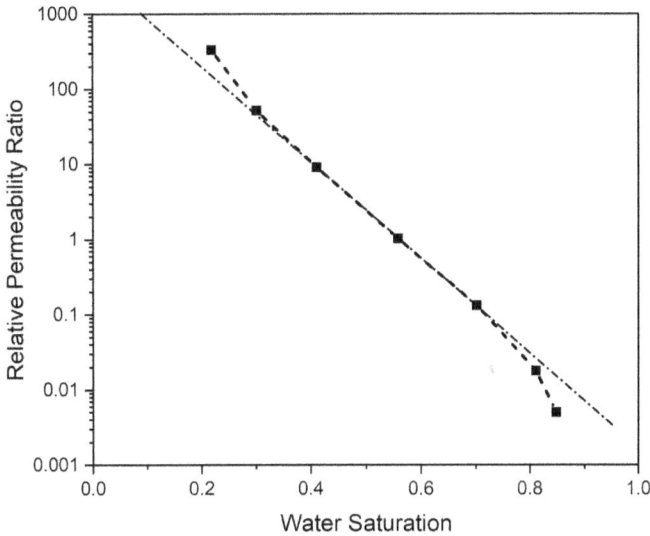

FIGURE 2.4   Semilog plot of relative permeability ratio versus saturation.

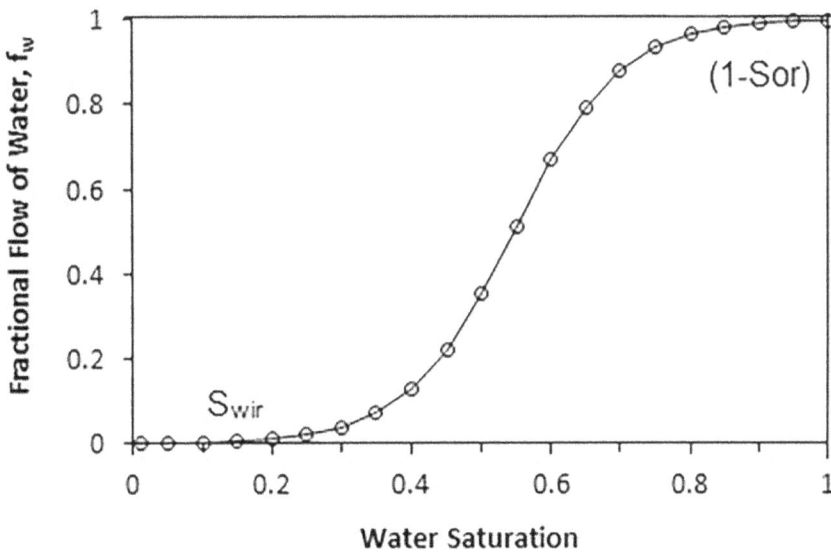

FIGURE 2.5   Fraction of water cut versus saturation.

If the water fractional flow is plotted versus water saturation, an S-shaped curve (Fig. 2.5) will result, which is named fractional flow curve. The fractional flow equation is used to calculate the fraction of the total flow that is water, at any point in the reservoir, assuming the water saturation at that point is known.

2.5 BUCKLEY-LEVERETTE THEOREM FOR ONE-DIMENSIONAL IMMISCIBLE DISPLACEMENT

The fractional flow equation, as described earlier, can provide quantitative information about the fraction of water cut at a particular saturation, without indicating its location or velocity within a reservoir. However, we will not be able locate the flood front or its velocity from this fractional flow equation. The Buckley-Leverette theorem, based on the immiscible displacement theory for 1D flow can help us to track the flood front movement as a function of time and location.

The Buckley-Leverette equation, developed in 1942 (Buckley and Leverett, 1942), for displacement of oil by water in 1D systems is considered to be the basic equation for describing immiscible displacement systems in one dimension. It is applied to determine the velocity of a plane of constant water saturation traveling through a homogeneous, linear system. The equation is derived using the principle of conservation of mass of water.

Let us consider a cylindrical system as shown in the Figure 2.6, through which oil and water are flowing in the "x-direction" only and following the diffuse flow condition. Initially, the system was saturated with oil and interstitial water. Then, water at a rate of q_i is injected at $x = 0$. At a certain time, water starts to displace oil. Only the oil phase is being produced before breakthrough; thereafter, both oil and water (rate q_t) are being produced at the other end once breakthrough occurs. Figure 2.6 represents an ideal, linear system containing mobile oil and water phases.

The area perpendicular to the flow direction X is A, the porosity of the system is Φ, and the length is L. Let us consider an element of volume $A\varnothing\Delta X$ at a distance of X from the reference injection

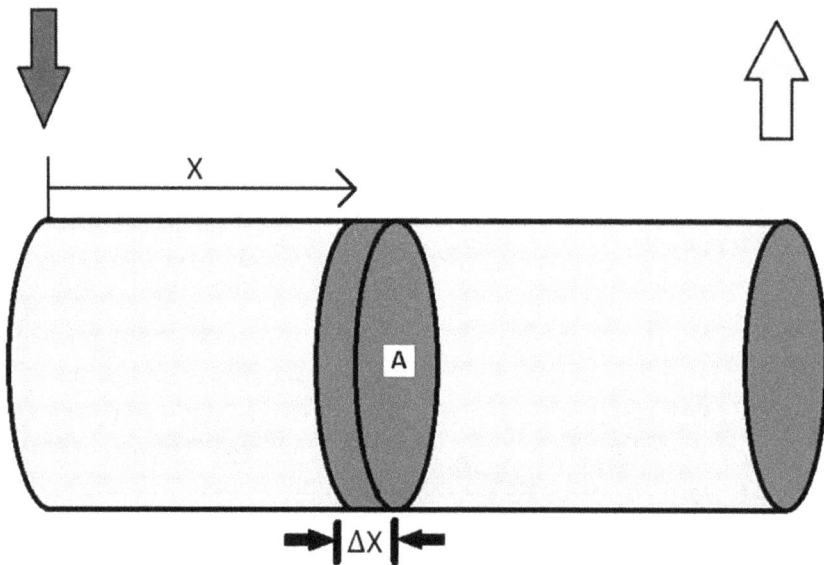

FIGURE 2.6   1-D cylindrical system containing oil and water.

point through which fluid (water + oil) is moving with a constant saturation of S_w. The water density is ρ_w, and its volumetric flow rate is q_w.

Principle of the conservation of mass is expressed as

$$\text{Rate of mass in} - \text{rate of mass out} = \text{Rate of increase of mass}$$

Applying the principle of conservation of mass to the water phase, we get

$$\left(q_w\rho_w\right)_X - \left(q_w\rho_w\right)_{X+\Delta X} = A\varnothing\Delta X \frac{\left\{\left|\rho_w S_w\right|_{t+\Delta t} - \left|\rho_w S_w\right|_t\right\}}{\Delta t} \tag{2.15}$$

Putting the limits,

Limit $\Delta X \to 0$ and $\Delta t \to 0$, Equation 2.15 becomes

$$\frac{\partial\left(q_w\rho_w\right)}{\partial x} = -A\varnothing\frac{\partial\left(\rho_w S_w\right)}{\partial t} \tag{2.16}$$

Incompressible system: During the injection process, the temperature is assumed to be constant— i.e., isothermal process prevails throughout; moreover, for the steady-state condition, total pressure of the reservoir remains constant. Even if there is a change in the pressure, the decline rate is very slow. Water compressibility with respect to pressure is also very less, which results in almost negligible or no change in water density.

So, Equation 2.16 is reduced to

$$\frac{\partial\left(q_w\right)}{\partial x} = -A\varnothing\frac{\partial\left(S_w\right)}{\partial t} \tag{2.17}$$

During the flooding process, it is easier to track the rate of the waterflood as a function of its saturation rather than time or distance. Moreover, it is more important to track the change of the flood front—i.e., the velocity of the flood front—which is not possible from Equation 2.17. This may be possible by reorientation of the mathematical expression of the equation in terms of the measurable variables in the following way.

It is established that the water saturation (S_w) is a continuous function of both space (X) and time (t) behind the flood front.

$$S_w = f\left(X, t\right)$$

So we can express the change in water saturation as

$$dS_w = \left(\frac{\partial S_w}{\partial x}\right)_t dX + \left(\frac{\partial S_w}{\partial t}\right)_x dt \tag{2.18}$$

As per the assumption that flood front is moving with a constant saturation, $dS_w = 0$. So

$$\left(\frac{\partial S_w}{\partial x}\right)_t dX = -\left(\frac{\partial S_w}{\partial t}\right)_x dt \tag{2.19}$$

So

$$\left(\frac{\partial S_w}{\partial t}\right)_x = -\left(\frac{\partial S_w}{\partial x}\right)_t \left(\frac{dX}{dt}\right)_{S_w} \tag{2.20}$$

Moreover, expressing the rate of change of q_w w.r.t. S_w

$$\left(\frac{\partial q_w}{\partial X}\right)_t = \left[\left(\frac{\partial q_w}{\partial S_w}\right)\left(\frac{\partial S_w}{\partial X}\right)\right]_t \tag{2.21}$$

Combining Equations 2.20 and 2.21, we get

$$\left(\frac{\partial q_w}{\partial S_w}\right)_t\left(\frac{\partial S_w}{\partial X}\right)_t = -A\varnothing\left(\frac{\partial S_w}{\partial X}\right)_t\left(\frac{dX}{dt}\right)_{S_w}$$

So

$$\left(\frac{\partial q_w}{\partial S_w}\right)_t = A\varnothing\left(\frac{dX}{dt}\right)_{S_w} \tag{2.22}$$

$\left(\dfrac{dX}{dt}\right)_{S_w}$ defines the linear velocity (V_{sw}) of the flood front moving with a constant water saturation

of S_w. In addition, the water flow rate may be expressed in terms of the fraction of total flow rate q_t (oil + water) as $q_w = f_w \times q_t$ and incompressible water.

2.5.1   DETERMINATION OF FLOOD FRONT VELOCITY

Equation 2.22 may be rewritten to determine the flood front velocity with a constant saturation of S_w.

$$V_{S_w} = \left(\frac{dX}{dt}\right)_{S_w} = \frac{q_t}{A\varnothing}\left(\frac{df_w}{dS_w}\right)_{S_w} \tag{2.23}$$

Equation 2.23 is known as the Buckley-Leverette equation, applicable to constant water injection rate $q_i = q_t$. This equation is used to determine the velocity of a plane of constant water saturation that is directly proportional to the derivative of the fractional flow at saturation—i.e.:

$$V_{S_w} \propto \left(\frac{df_w}{dS_w}\right)_{S_w} \text{, for constant values of } q_t, A, \text{ and } \phi$$

It is derived that under diffuse flow conditions and negligible capillary pressure, the fraction of water cut is the sole function of water saturation, irrespective of whether the gravity term is included or not. The derivative $\left(\dfrac{df_w}{dS_w}\right)_{S_w}$, is the slope of the fractional flow curve and constant at a particular saturation. The derivative dx/dt is the velocity of the moving plane with water saturation S_w. Thus, the derivative dx/dt is sole function of the saturation and is constant at a particular saturation.

2.5.2   TRACKING OF FLOOD FRONT AS A FUNCTION OF TIME AND INJECTION VOLUME

The integral of Equation 2.23 may be used to determine the distance to which the flood front may travel at a time on injection of a cumulative volume of water W_i from Equation 2.24.

$$X_{S_w} = \frac{W_i}{A\varnothing}\left(\frac{df_w}{dS_w}\right)_{S_w} \tag{2.24}$$

where W_i is the cumulative volume of water injected for time t with an injection rate q_i and $W_i = 0$ at time $t = 0$.

The representative reservoir condition is shown in Figure 2.7.

2.5.3 LIMITATION OF THE BUCKLEY-LEVERETTE EQUATION

Equations 2.23 and 2.24 represent the variation in the velocity and location of flood front as continuous functions of water saturation in the reservoir. The lower limit of the water saturation is S_{wc} whereas the upper limit is $1 - S_{or}$—i.e., $S_{wc} \leq S_w \leq (1 - S_{or})$. The typical plots of df_w/ds_w and f_w as functions of S_w are presented in Figure 2.8.

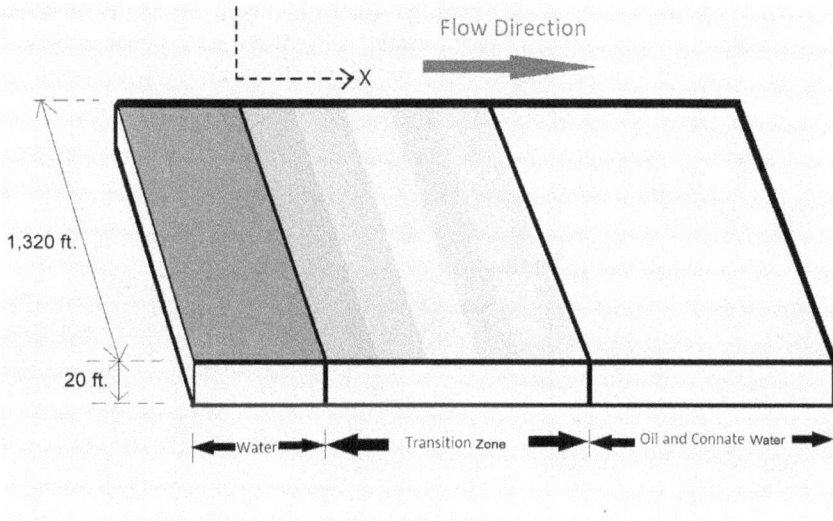

FIGURE 2.7 Occurrence of various phases in reservoir.

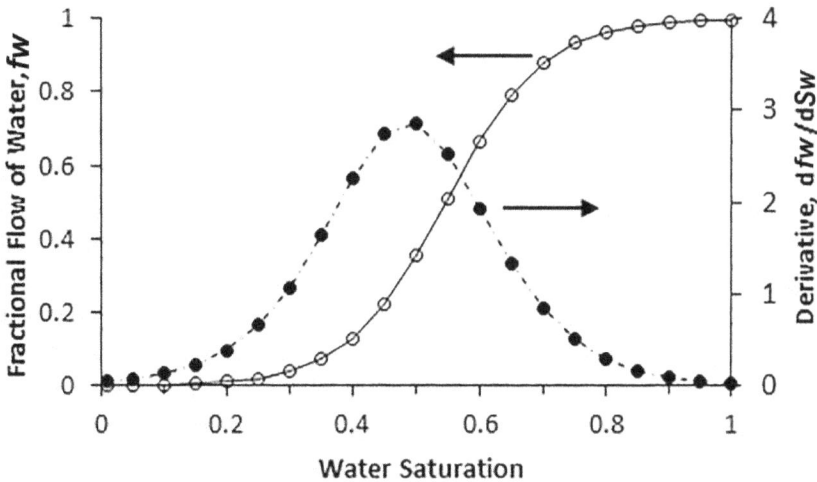

FIGURE 2.8 Variation of water cut and its derivative with water saturation.

It may be observed from the Figure 2.8 that at a particular velocity, represented by $\left(\dfrac{df_w}{dS_w}\right)_{S_w}$, two saturation values exist. However, there should be only one saturation value at a particular place and time. In the actual system, fluid will move from the higher saturation point to even higher saturations—i.e., the intermediate values of the water saturation have the maximum velocity and will initially tend to overtake the lower saturations resulting in one saturation at a particular point behind the shock front. But at the shock/flood front (X_f), because of pistonlike behavior, there should be a sudden drop in the water saturation from its front saturation to its initial value—i.e., S_{wf} to S_{wc}, the connate water or initial water saturation of the reservoir. If we plot the water saturation distribution using Equation 2.24, it can be represented by Figure 2.9. This discontinuity of the water saturation cannot be described by the Buckley-Leverett equation.

Behind the flood front, the water saturation ranges between $S_{wf} < S_w < 1 - S_{or}$, where S_{or} is the residual oil saturation; ahead of the flood front, the water saturation will be the same as its initial value. The actual water saturation profile that prevails w.r.t. distance is shown in Figure 2.10. $\overline{\overline{S}}_w$ indicates the average water saturation of the reservoir flooded by water.

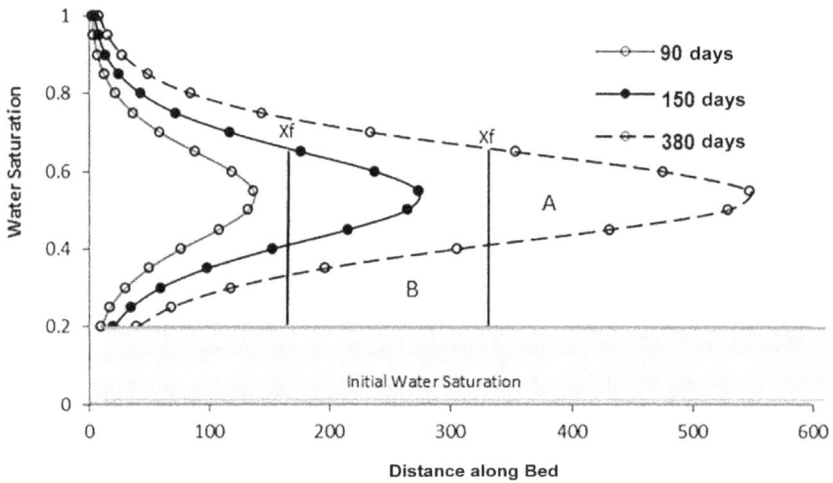

FIGURE 2.9 Water saturation distribution with time and distance, determined from the Buckley-Leverett equation.

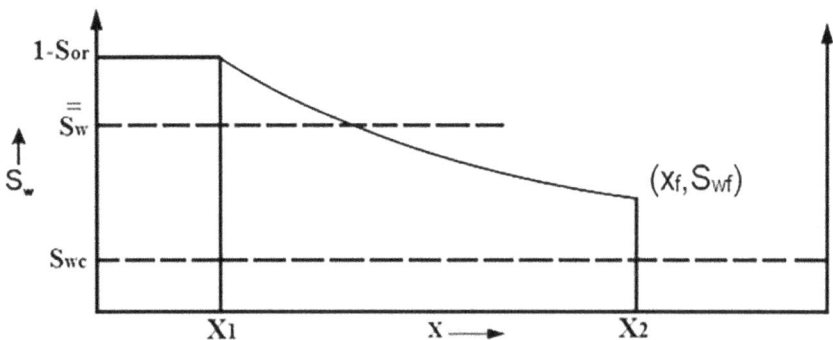

FIGURE 2.10 Water saturation distribution with time and distance before breakthrough, modified from the Welge method.

2.6 WELGE METHOD

The mathematical problem regarding the coexistence of two saturations at a particular point of the reservoir, as presented in Figure 2.9, was mitigated by Welge (1952). The solution consists of integrating the water saturation over the distance from the injection well to the flood front. Thus, it helps in determining the average saturation behind the front $\bar{\bar{S}}_w$ as shown in Figure 2.10. The situation described here prevails for the injection of W_i volume of water over a period of time t with an injection rate of q_i before breakthrough occurs in the production well. At the condition shown in the figure, the water saturation reaches its maximum $(1 - S_{or})$ at a distance X_1. In between the points X_1 and X_2 or X_f, the saturation is a continuous function of distance (X). The discontinuity occurs just at the flood front (X_f or X_2 here). The point X_2 represents a discontinuity, at which there is a sudden drop in saturation from S_{wf} to initial water saturation S_{wc} under the ideal pistonlike movement condition of the flood front.

Using the simple principle of material balance, the total volume of water to be injected at this condition could be correlated with the reservoir parameters by the following equations.

$$W_i = A\emptyset X_2 \left(\bar{\bar{S}}_w - S_{wc} \right) \tag{2.25}$$

or

$$\left(\bar{\bar{S}}_w - S_{wc} \right) = W_i / A\emptyset X_2 \tag{2.26}$$

Correlating Equation 2.24 with Equation 2.26, the average saturation may be correlated with the slope of the fractional flow curve (f_w vs. S_w) at S_{wf} as given in the following:

$$\left(\bar{\bar{S}}_w - S_{wc} \right) = \frac{1}{\left(\dfrac{df_w}{dS_w} \right)_{S_{wf}}} \tag{2.27}$$

The average saturation behind the flood front also can be determined by integrating the saturation profile within the limits between the injection well and the location of the flood front, X_2 from the definition of the saturation—i.e., (Total volume of water present in the porous medium)/(Total pore volume) under consideration (Fig. 2.10).

$$\bar{\bar{S}}_w = \frac{\int_0^{X_2} A\emptyset S_w dX}{\int_0^{X_2} A\emptyset dX}$$

For constant values of porosity and cross-sectional area, the average saturation becomes

$$\bar{\bar{S}}_w = \frac{(1 - S_{0r}) X_1 + \int_{X_1}^{X_2} S_w dX}{X_2} \tag{2.28}$$

Considering Equation 2.25—i.e., $X_{sw} \propto \dfrac{df_w}{dS_w}$ —for a given volume of water injection and $S_w \geq S_{wf}$, Equation 2.28 can be expressed as

$$\bar{\bar{S}}_w = \frac{(1-S_{0r})\left(\dfrac{df_w}{dS_w}\right)_{1-S_{0r}} + \displaystyle\int_{1-S_{or}}^{S_{wf}} S_w \, d\left(\dfrac{df_w}{dS_w}\right)}{\left(\dfrac{df_w}{dS_w}\right)_{S_{wf}}} \tag{2.29}$$

Using the formulas of integration by parts, the second part of the numerator of Equation 2.29 may be expressed as

$$\int_{1-S_{or}}^{S_{wf}} S_w \, d\left(\frac{df_w}{dS_w}\right) = \left[S_w \left(\frac{df_w}{dS_w}\right)\right]_{1-S_{or}}^{S_{wf}} - \left[f_w \right]_{1-S_{or}}^{S_{wf}} \tag{2.30}$$

Incorporation of Equation 2.30 into Equation 2.29 results in

$$\bar{\bar{S}}_w = S_{wf} + \frac{(1-f_w)_{Swf}}{\left(\dfrac{df_w}{dS_w}\right)_{S_{wf}}}$$

or

$$\frac{\left(\bar{\bar{S}}_w - S_{wf}\right)}{(1-f_w)_{Swf}} = \frac{1}{\left(\dfrac{df_w}{dS_w}\right)_{S_{wf}}} \tag{2.31}$$

In Equation 2.31, both f_w and its derivative are derived at the shock saturation S_{wf}, similar to in Equation 2.27.

Thus, combining Equations 2.27 and 2.31 results in

$$\frac{(1-f_w)_{Swf}}{\left(\bar{\bar{S}}_w - S_{wf}\right)} = \frac{(1-f_w)_{Swc}}{\left(\bar{\bar{S}}_w - S_{wc}\right)} = \left(\frac{df_w}{dS_w}\right)_{S_{wf}} \tag{2.32}$$

$$\left(f_w = 0 \text{ at } S_{wc}\right)$$

Equation 2.32 describes an equation of a straight line joining the points (S_{wc}, 0) and (S_{wf}, f_w at the shock front) with a tangent at the point (S_{wf}, f_{wf}). The extrapolation of the line to $f_w = 1$ gives the average saturation behind the flood front, $\bar{\bar{S}}_w$. This equation is known as the Welge equation.

This method enables us to determine the average water saturation behind the flood front only if we have the related permeability data from the SCAL analysis and the fractional flow curve plotted from the data (Fig. 2.11). In addition, this helps in fast calculation of the oil recovery up to breakthrough. Even a minor modification in the calculation allows us to determine the oil recovery after breakthrough occurs in the reservoir under ideal conditions as mentioned before. Equation 2.32 is valid for the saturation behind the flood front where the assumptions of diffuse flow conditions are valid. Ahead of the flood front, the saturation is not continuous (sudden drop from S_{wf} to S_{wc}) as described before and, hence, is plotted as a dotted line in Figure 2.11.

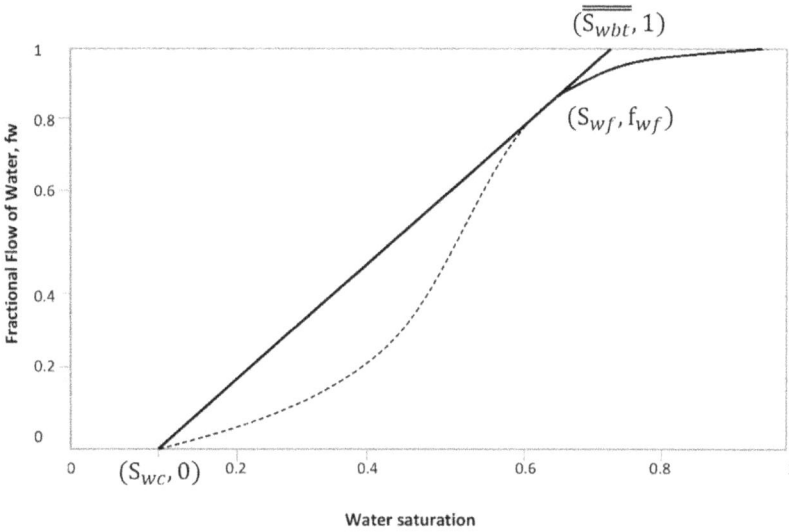

FIGURE 2.11 Tangent to the fractional flow curve at the flood front (S_{wf}, f_{wf}) from $S_w = S_{wc}$.

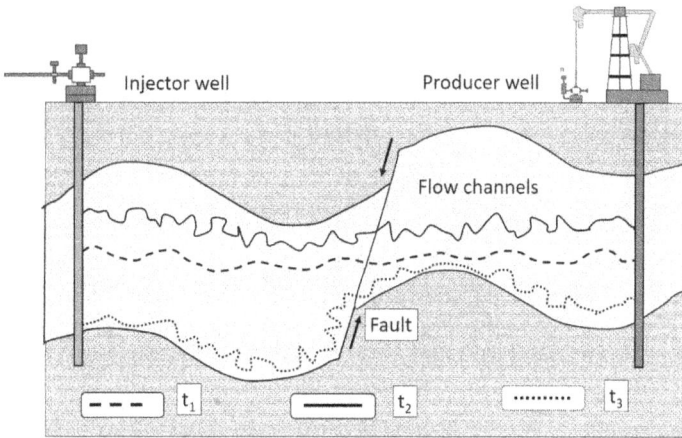

FIGURE 2.12 Schematic representation of breakthrough.

2.6.1 Oil Recovery Calculation

For oil recovery calculations by immiscible flooding, it is necessary to understand its condition, location of the displacing fluid (whether it reaches the production well or not), ideality/nonideality of the flood movements, heterogeneity of the reservoir, etc. The equations/principles applied for the recovery calculations will differ greatly depending on the location of the displacing fluid, whether at the producer or in between the injector and producer.

2.6.1.1 Concept of Breakthrough

Breakthrough condition defines the time at which the injected fluid reaches the production well—i.e., $x = L$, where L is the length of the reservoir block between the injector and producer. At the ideal breakthrough condition, without any deviation in its direction, the injected fluid follows the tortuous path in the porous media with a time $t1$, whereas in the presence of some high-permeability path or heterogeneity (straight line, $t2$), the fluid reaches faster than the ideal one as presented in Figure 2.12.

The longest time is $t3$, in which the fluid follows a diverted path in the presence of some fault or barrier. Even so, it may never reach the production well. Thus, the breakthrough time also gives us an idea about the connectivity between the injection and production wells.

2.6.1.2 Recovery Prior to Breakthrough

Prior to breakthrough (bt), ideally, only oil should be produced, assuming that there exists no mobile water before water injection. Thus, the volume of produced oil will be equal to the injected fluid volume, for incompressible flow with maintenance of the constant pore volume. Here, the voidage replacement ratio is 1.0. The fraction of water cut (f_w) remains zero. The water saturation distribution at different instances before and after breakthrough may be visualized from Figure 2.13.

Mathematically, the total volume of injected water at breakthrough, $W_{i,bt}$ equals $N_{p,bt}$, the cumulative volume of oil produced during this period.

Before breakthrough: Equation 2.24 $\left(X_{S_w} = \dfrac{W_i}{A\varnothing} \left(\dfrac{df_w}{dS_w} \right)_{S_w} \right)$ can be applied to determine the locations

of the planes with constant saturation, for $S_{wf} < S_w < 1 - S_{or}$. It is considered that the flood front moves through the reservoir with a change in water saturation profile; X_{sw} at a particular location varies with time. At the moment when the flood just touches the production well—i.e., $X = L$ (but oil is yet to be produced)—Equation 2.24 may be rewritten as

$$\frac{W_i}{A\varnothing L} = W_{id} = \frac{1}{\left(\dfrac{df_w}{dS_w} \right)_{S_w}} \qquad (2.33)$$

where S_w is the present value of water saturation at the production well and W_{id} is the dimensionless number of the injected pore volume (*PV*) of water ($1PV = A\varnothing L$).

2.6.1.3 Recovery at the Instance of Breakthrough

The flood front reaches the producing well with $S_{wf} = S_{wbt}$, and the water cut suddenly reaches to f_{wbt} from zero. This phenomenon is frequently observed in the field, which also proves the existence of the shock front. At the instance of breakthrough,

$$\text{Volume of oil produced}, N_p = \text{Volume of water injected} \left(W_i \right)$$

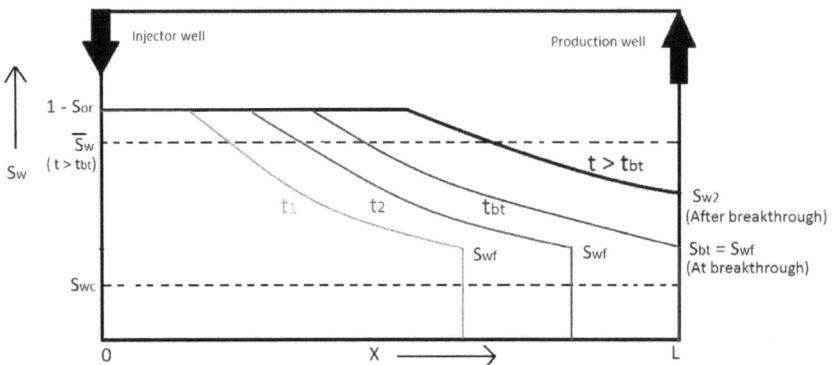

FIGURE 2.13 Saturation distribution before and after breakthrough.

This, in terms of dimensionless parameters, may be expressed as

$$N_{pd} = W_{id} = \left(\bar{\bar{S}}_w - S_{wc}\right) = \frac{1}{\left(\dfrac{df_w}{dS_w}\right)_{S_{wbt}}}$$

(2.34)

where N_{pd} = dimensionless volume of cumulative oil production = $\dfrac{N_p}{A\varnothing L}$

Similarly, with the known injection rate per unit pore volume (q_{id}), the breakthrough time (t_{bt}) may be determined from the following equation:

$$t_{bt} = W_{idbt} / q_{id}$$

(2.35)

2.6.1.4   Recovery after Breakthrough

- Flood front already reaches the production well—i.e. $X = L$ (constant).
- Water saturation gradient still exists and its average value ($\bar{\bar{S}}_w$) increases gradually, leading to an increase in the fractional water cut.
- Total production includes oil as well as water, unlike only oil before breakthrough:

$$W_{id} = N_{pd} + W_{pd}$$

where W_{pd} is the cumulative produced water per unit pore volume.

At this condition, using the Welge equation, the average water saturation of the reservoir may be expressed as

$$\bar{\bar{S}}_w = S_{we} + \left(1 - f_{we}\right) \times \frac{1}{\left(\dfrac{df_w}{dS_w}\right)_{swe}}$$

$$= S_{we} + \left(1 - f_{we}\right) \times W_{id}$$

(2.36)

Where S_{we} and f_{we} are water saturation and fractional flow at producing well. The cumulative oil recovery now may be determined from the principle of material balance,

$$N_{pd} = \left(\bar{\bar{S}}_w - S_{wc}\right)$$

$$= \left(S_{we} - S_{wc}\right) + \left(1 - f_{we}\right) \times W_{id}$$

(2.37)

The average saturation between the two cross sections at x_1 and x_2, after breakthrough (difference in change in water volume in reservoir − water produced per unit pore volume) is expressed as

$$\bar{\bar{S}}_w = \frac{\left(x_2 S_{w2} - x_1 S_{w1}\right)}{x_2 - x_1} - \frac{q_t t\left(f_{w2} - f_{w1}\right)}{A\varnothing\left(x_2 - x_1\right)}$$

(2.38)

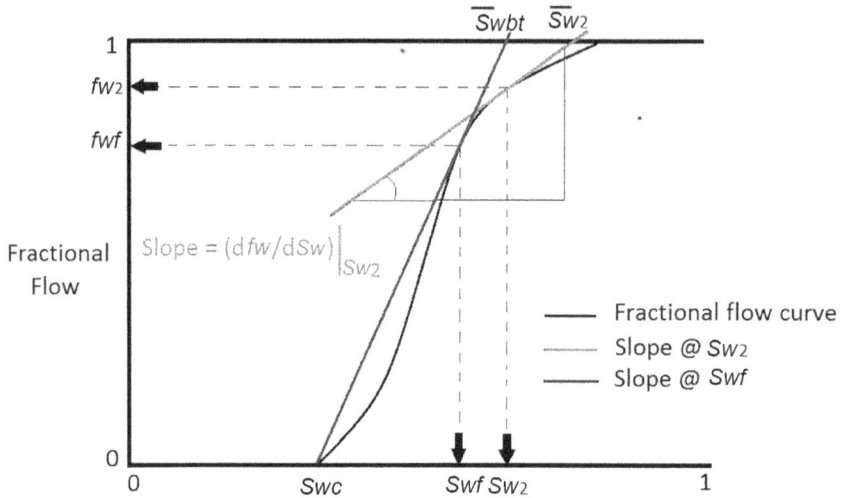

FIGURE 2.14 Determination of oil recovery beyond breakthrough.

2.6.2 GRAPHICAL METHOD DETERMINATION OF OIL RECOVERY

Oil recovery at different conditions can be determined graphically in an easier way from the fractional flow curve. The steps are described in the following.

2.6.2.1 Oil Recovery Calculation before/at Breakthrough

1. Draw the fractional flow curve from the relative permeability versus saturation data as shown in Figure 2.11, and consider gravity effect, if any.
2. Draw a tangent to the curve from point S_{wc}, where $f_w = 0$. The coordinate of the point of tangency will be $(S_{wf}, f_{wf}) = (S_{wbt}, f_{wbt})$.
3. Extend the tangent to $f_w = 1$. This point of intersection gives the value of average water saturation at the breakthrough, $\bar{\bar{S}}_w = \bar{\bar{S}}_{w,bt}$.

Use Equation 2.34 to determine the oil recovery at this breakthrough.

2.6.2.2 Oil Recovery Calculation after Breakthrough

1. Follow steps 1–3 given in Section 2.6.2.1.
2. Select an arbitrary saturation value above S_{wbt}. In general, the saturation values are chosen at 5% incremental saturation (e.g., at S_{w2} at the first increment) for each stage of the calculation as shown in Figure 2.14.
3. Draw a tangent at this selected point, and extend it to $f_w = 1$ to get the new average saturation.
4. Use Equation 2.37 to determine the oil recovery at each stage.

2.7 STILE'S METHOD

2.7.1 BASIC PRINCIPLE

Stile's method is used for predicting immiscible floods for stratified reservoir. Stiles (1949) proposed an approach that considers the effect of permeability variations in predicting the performance of waterfloods. The derivation of the water cut and the recovery equations are based on two principal assumptions: (1) fluid flow is linear and (2) the distance of penetration of the flood

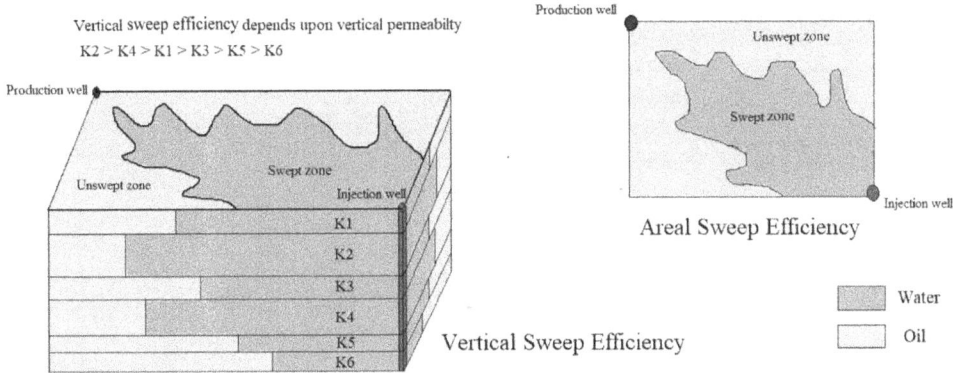

FIGURE 2.15 Flood front movement in stratified reservoir: (a) 2D areal view, (b) 3D.

front is proportional to permeability, in addition to the other basic principles of the Buckley-Leverette theory.

The water breakthrough occurs in a sequence that starts in the layer, as depicted in Figure 2.15, with the highest permeability.

Assuming that the reservoir is divided into n layers that are arranged in a descending permeability order with breakthrough occurring in a layer i, all layers from 1 to i have already been swept by water. The remaining layers obviously have not reached breakthrough. The criteria to be followed to apply Stile's method in a multilayer system are given here:

- Gravity plays no part in Stile's type of displacement, on account of the vertical separation of the layers.
- This results in the displacement efficiency being entirely dictated by the mobility ratio and heterogeneity.
- The method is restricted to reservoirs with no vertical pressure communication between layers and where the mobility ratio is close to a value of one.
- The mobility ratio assumption ensures that the velocity of frontal advance of water in each layer remains constant during the flood.
- The velocities will be different in each layer, as dictated by the following equation, but as the flood progresses, the differences between layers will remain constant: i.e., there is no velocity dispersion. The velocity of water in the horizontal direction in the layer i is denoted by Equation 2.39:

$$v_i = \frac{\Delta P K k_{rwi}}{\varnothing \left(1 - S_{ori} - S_{wci}\right) \mu_w L} \tag{2.39}$$

where K_{rwi} is the relative permeability of water, S_{or} is the residual oil saturation, and S_{wci} is connate water saturation of the ith layer.

2.7.2 Steps to Predict Water Cut and Oil Recovery

The following steps are used to determine the fraction of water cut and oil recovery using the Stile's method:

- Inspect the core and log data, and divide the section into a total of N separate layers.
- Order the N layers in the sequence in which they will successively flood out with water, by applying the aforementioned velocity equation.

- Generate pseudo-relative permeabilities by applying Equations 2.40–2.42:

$$\bar{S}_{wn} = \frac{\sum_{i=1}^{n} h_i \varnothing_i \left(1 - S_{ori}\right) + \sum_{i=n+1}^{N} h_i \varnothing_i S_{wci}}{\sum_{i=1}^{N} h_i \varnothing_i} \tag{2.40}$$

$$\bar{k}_{rwn} = \frac{\sum_{i=1}^{n} h_i k_i k_{rwi}}{\sum_{i=1}^{N} h_i k_i} \tag{2.41}$$

$$\bar{k}_{ron} = \frac{\sum_{i=1}^{n} h_i k_i k_{roi}}{\sum_{i=1}^{N} h_i k_i} \tag{2.42}$$

where N is the total number of layers where n number of layers are flooded.
- Oil recovery equations: Once the average relative permeability to oil and water is determined from the individual layer data, the fraction of water cut, oil recovery, and cumulative volume of injected water are determined similar to the Welge method using the following equations:

The fraction of water cut in at breakthrough where n layers are flooded is given in Equation 2.43:

$$f_{wen} = \frac{1}{1 + \frac{\mu_w}{\bar{k}_{rwn}} \times \frac{\bar{k}_{ron}}{\mu_o}} \tag{2.43}$$

Cumulative water injection (dimensionless) at breakthrough point may be determined as

$$W_{id} = \frac{1}{\left(\frac{\Delta \bar{f}_{we}}{\Delta \bar{S}_{we}}\right)} \tag{2.44}$$

Dimensionless oil recovery from the reservoir after breakthrough is

$$N_{pd} = (\bar{S}_{we} - \bar{S}_{wc}) + \left(1 - \bar{f}_{we}\right) W_{id} \tag{2.45}$$

Oil recovery per unit hydrocarbon pore volume, N_{pdo} is determined by Equation 2.46:

$$N_{pdo} = \frac{(\bar{S}_{we} - \bar{S}_{wc}) + \left(1 - \bar{f}_{we}\right) W_{id}}{1 - S_{wc}} \tag{2.46}$$

The water-oil ratio at surface (WOR$_s$) may be determined simply from the following equation

$$WOR_s = \frac{\bar{k}_{rw} \mu_o B_0}{\bar{k}_{ro} \mu_w B_w} \times \left[\frac{\sum_{i=1}^{n} (Kh)_i}{\sum_{i=n+1}^{N} (Kh)_i}\right] \tag{2.47}$$

n = number of layers through which breakthrough occurs, N = total number of layers.

2.7.3 LIMITATION OF STILE'S METHOD

Stile's method should not be applied where a gas or water zone is present immediately above or below the oil zone under consideration, because there would be a bypassing of the oil zone by the injected water, or there would be coning. In such cases, the oil recovery to any given water cut would be less than the calculated recovery. However, in the case of gas or water zones of known permeability, certain modifications can be made in the basic equations to adjust for those conditions.

2.8 OTHER ONE-DIMENSIONAL METHODS TO PREDICT RECOVERY

There are some other methods like Dykstra-Parsons method and reverse Welge method, which are used to determine the water cut for heterogeneous reservoirs from the rock data and production data, respectively.

The reverse Welge method is used to verify the water cut from the production data itself, to determine the average water saturation from the fractional water cut or cumulative recovery data collected from the field. This model simply assumes the reservoir as a Buckley-Leverett "black-box," no matter how complex it may be, containing numerous vertical/horizontal wells, fractures or heterogeneity, in which the injection q_{wi} and production ($q_o + q_w$) are linked together by the fractional flow relationship.

2.9 APPLICATIONS AND LIMITATIONS OF IMMISCIBLE DISPLACEMENT THEORY

Limitations

- Fluids are considered immiscible and incompressible. For most cases, the situation deviates from ideality.
- Porous media was assumed to be isotropic and homogeneous, with uniform saturation distribution, which is also not true. Actual reservoir parameters vary from layer to layer and even from point to point.
- Only 1D, linear flow is assumed, which is also rarely fulfilled. Because of permeability and saturation variations in different directions, flow occurs mostly in all three directions.
- Stabilized displacement process, irrespective of injection rate and length of the sample/reservoir. Mostly, unsteady state flow is observed, at least at the initial stage of the flooding.

Applications

- Under stabilized field conditions, stable displacement may be considered.
- From the lab test, operating conditions could be determined for a stabilized flow.
- Determination of the maximum possible oil recovery by immiscible displacement at ideal condition.

2.10 UNSTEADY-STATE MULTIPHASE FLOW EQUATIONS FOR RECOVERY PREDICTION

All the methods described so far assume immiscible, 1D flow for incompressible fluids without any mass transfer between the phases, in addition to their limitation to describe laminar flow under steady-state conditions. In the actual reservoir, these idealities are rarely observed. However, in many cases, the methods described earlier can closely and quickly predict the oil recovery without the need for a large data requirement.

For detailed and accurate analyses and prediction, 2D or 3D black oil or compositional models in the presence of all the mobile phases (oil, gas, and water) are applied depending on the type of fluids and reservoir flow characteristics. The detailed description of the models and their solving methods are beyond the scope of the book. Only the basic equations are described in the following as reference.

2.10.1 Multiphase Flow Compositional Model

The equation derived for single-phase flow can be converted to multiphase flow (multicomponent) with the incorporation of saturation and concentration terms of individual components and phases, with multidirectional flow. The phases present in a petroleum reservoir are nothing but oil (o), water (w), and gas (g). To develop the equation, let us consider the flow of a single component (*i*th component) present in all three phases within the reservoir.

Incorporating Darcy's equation in the continuity equation and considering the continuous interphase mass transfer, the diffusivity equation for the compositional model can be written as follows (Eq. 2.48):

$$-\nabla\left(\sum_{p=1}^{3}\frac{C_{ip}\rho_{ip}kk_{rp}}{\mu_p}\left(\nabla P_p - \rho_p g \nabla D\right)\right) \pm C_{ip}q_p = -\frac{\partial}{\partial t}\left(\Phi\sum_{p=1}^{3}C_{ip}\rho_p S_p\right) \qquad (2.48)$$

Because of complexity of the compositional model, it is only used for the simulation of volatile oil reservoirs and gas condensate reservoirs. In other cases, mostly the black oil model is used.

2.10.2 Black Oil Model

The black oil model assumes the presence of three pseudo components: i.e., oil, gas, and water. It is further simplified that there is only a one-way phase transfer of gas into or out of the oil phase. Mass transfer between water-oil, water-gas, and oil-gas is assumed to be nil.

Incorporating the assumptions mentioned earlier and equations of states into the compositional model, the black oil model is obtained as follows:

Water:

$$\nabla\left[\frac{kk_{rw}}{\mu_w B_w}\left(\nabla P_w - \rho_w g \nabla D\right)\right] \pm q_{ws} = \frac{\partial}{\partial t}\left(\frac{\Phi S_w}{B_w}\right) \qquad (2.49)$$

Oil:

$$\nabla\left[\frac{kk_{ro}}{\mu_o B_o}\left(\nabla P_o - \rho_o g \nabla D\right)\right] \pm q_{os} = \frac{\partial}{\partial t}\left(\frac{\Phi S_o}{B_o}\right) \qquad (2.50)$$

Gas:

$$\nabla\left[\frac{kk_{ro}R_s}{\mu_o B_o}\left(\nabla P_o - \rho_o g \nabla D\right)\right] + \nabla\left[\frac{kk_{rg}}{\mu_g B_g}\left(\nabla P_g - \rho_g g \nabla D\right)\right] \pm q_{gs} = \frac{\partial}{\partial t}\left(\frac{\Phi S_g}{B_g} + \frac{\Phi S_o R_s}{B_o}\right) \qquad (2.51)$$

Equations 2.49–2.51 can be incorporated into a single equation using the capillary pressure and saturation relationships.

The unsteady-state equations for the compositional and black oil models are solved using numerical methods.

2.11 WELL PATTERNS FOR EXTERNAL FLUID INJECTION

2.11.1 NEED OF EXTERNAL FLUID INJECTION

During production, reservoir pressure gradually decreases for all three types of primary drive mechanisms:

a) water drive;
b) solution gas or depletion drive; and
c) gas cap drive.

The decline in pressure adversely affects the oil production, particularly, when the reservoir pressure decreases below the saturation pressure. The different problems associated with the oil production as pressure declines are given in the following:

- Decline in pressure reduces the driving force—i.e., the force that pushes the oil from the reservoir drainage boundary to the wellbore.
- Decline in pressure causes solution gas to be evolved out of oil.
- These gas bubbles occupy an increasingly greater space of reservoir rocks, resulting in an increase in gas saturation. Increase in gas saturation above a certain value, known as critical gas saturation, makes gas mobile, which, in turn, decreases the oil production, as the relative permeability to oil decreases. As the mobility of gas is much higher than that of oil, more gas is produced, leaving the oil behind.
- As the gas liberates from the oil, due to depletion, the oil viscosity increases, thus making it less mobile.
- Also due to gravity difference, the part of segregated gas migrates to the crestal area of the reservoir and forms a secondary gas cap.
- Due to formation of a secondary gas cap, many wells at the crestal area become high GOR wells, thereby, rendering these oil wells non-contributing. All these factors finally result in the reduction of recovery from the reservoir.

Therefore, it is necessary to maintain reservoir pressure above a certain value by injecting water, gas, or other immiscible fluid from the surface to the reservoir. As a normal practice, the reservoir pressure is required to be maintained above bubble-point or saturation pressure to prevent segregation of solution gas.

In addition, there is a need for implementing EOR technology to improve the mobility of the residual oil beyond a certain period of primary/secondary recovery processes. Various external fluids—i.e., alkali, polymer, chemicals, various gases (miscible flooding)—are injected. The incremental recovery by injection of external fluids is a function of the well location and pattern, besides being strongly influenced by the rock-fluid properties, temperature, pressure, and injection-fluid parameters. A brief description about the injection well locations and the patterns are given in the following.

2.12 INJECTION WELL LOCATIONS

In large reservoirs, the fluid is injected in a predefined pattern at the reservoir depth mainly to:

- achieve uniform and balanced displacement of oil throughout the field;
- achieve higher overall sweep efficiency;
- obtain higher recovery;
- supplement pressure depletion throughout the field; and
- compensate the voidage created by produced fluids.

There are two types of injection well locations:

(a) central and peripheral and
(b) pattern flooding.

2.12.1 CENTRAL AND PERIPHERAL

In central and peripheral flooding, the injections wells are grouped together. In a reservoir with gas cap, the injection wells are grouped in the gas cap zone, at the top of an anticline as shown in Figure 2.16a. For the anticlinal reservoir with an underlying aquifer, the injection wells form a ring around the oil pay zone as shown in Figure 2.16b. For the monoclinal reservoir with a gas cap or water aquifer, the injection wells are located one side, while the production wells are on other side as shown in Figure 2.16c.

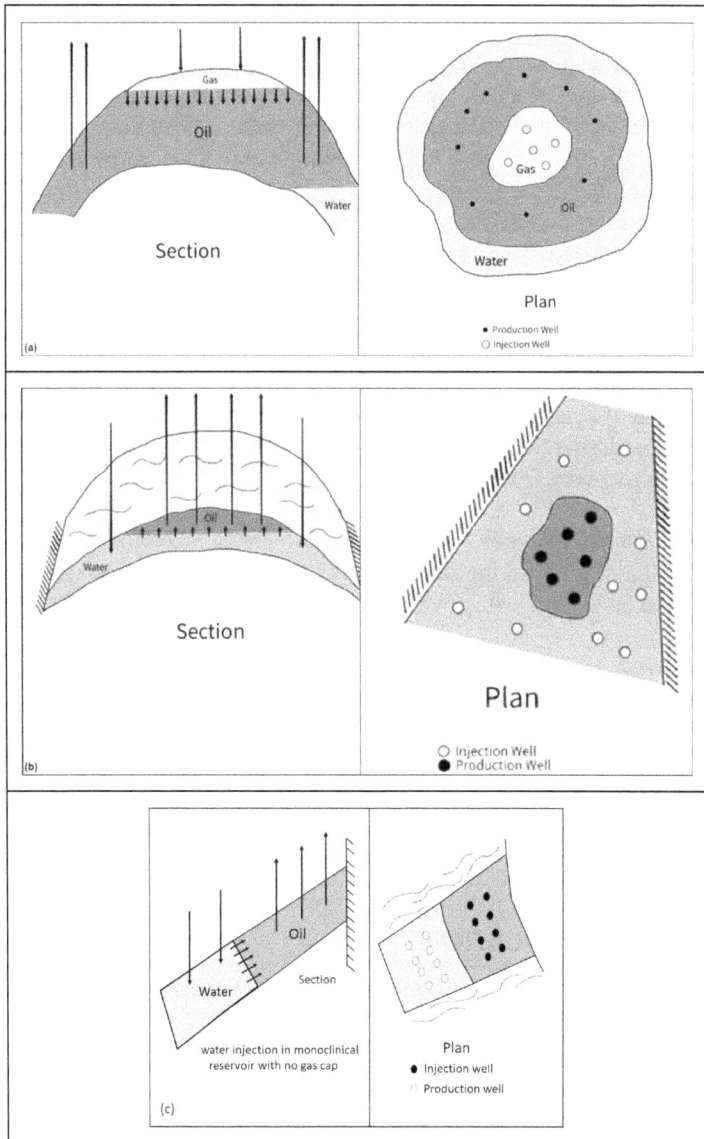

FIGURE 2.16 Injection and production wells in central and peripheral flooding.

2.12.2 PATTERN FLOODING

Pattern flooding is generally effective for reservoirs having a small dip and large surface area. To ensure a better sweep efficiency, the injection wells must be distributed among the production wells in a definite pattern, which depends on the properties of the reservoir rock and fluids. One of the first steps in designing a waterflooding project is flood pattern selection. The objective is to select the proper pattern that will provide the injection fluid with the maximum possible contact with the crude oil system. This selection can be achieved by (1) converting existing production wells into injectors or (2) drilling infill injection wells. When making the selection, the following factors must be considered:

i)   reservoir heterogeneity and directional permeability;
ii) the direction of formation fractures;
iii) availability of the injection fluid (gas or water);
iv) desired and anticipated flood life;
v)   maximum oil recovery; and
vi) well spacing, productivity, and infectivity.

The most common patterns include direct line drive, staggered line drive, five-spot, seven-spot, nine-spot, etc. The ratio of the number of injection wells to that of the production well is shown in Table 2.1. Schematics of the different patterns are discussed in the following.

(a) Direct line drive: The lines of the injection and production well are directly opposed as shown in Figure 2.17a.
(b) Staggered line drive: It is similar to the direct line drive but laterally displaced as shown in Figure 2.17b.
(c) Four-spot pattern: An injection pattern in which three input or injection wells are located at the corners of an equilateral triangle and the production well sits in the center (Fig. 2.17c).
(d) Five-spot pattern: An injection pattern in which four input or injection wells are located at the corners of a square and the production well sits in the center (Fig. 2.17d). The injection fluid, which is normally water, steam, or gas, is injected simultaneously through the four injection wells to displace the oil toward the central production well.
(e) Seven-spot pattern: The injection wells are located at the corners of a hexagon with the production well at the center (Fig. 2.17e).
(f) Nine-spot pattern: It is similar to the five-spot with one extra injection well drilled at the middle of each side of the square (Fig. 2.17f).

TABLE 2.1
Ratio of the Injection Well and Production Wells for Different Flooding Patterns

Pattern	Ratio of Injection Wells to Production Wells	Drilling Pattern Required
Line drive	1:1	Rectangle
Staggered line drive	1:1	Offset lines of wells
Four-spot	1:2	Equilateral triangle
Five-spot	1:1	Square
Seven-spot	2:1	Equilateral triangle
Nine-spot	3:1	Square
Inverted seven-spot	1:2	Equilateral triangle
Inverted nine-spot	1:3	Square

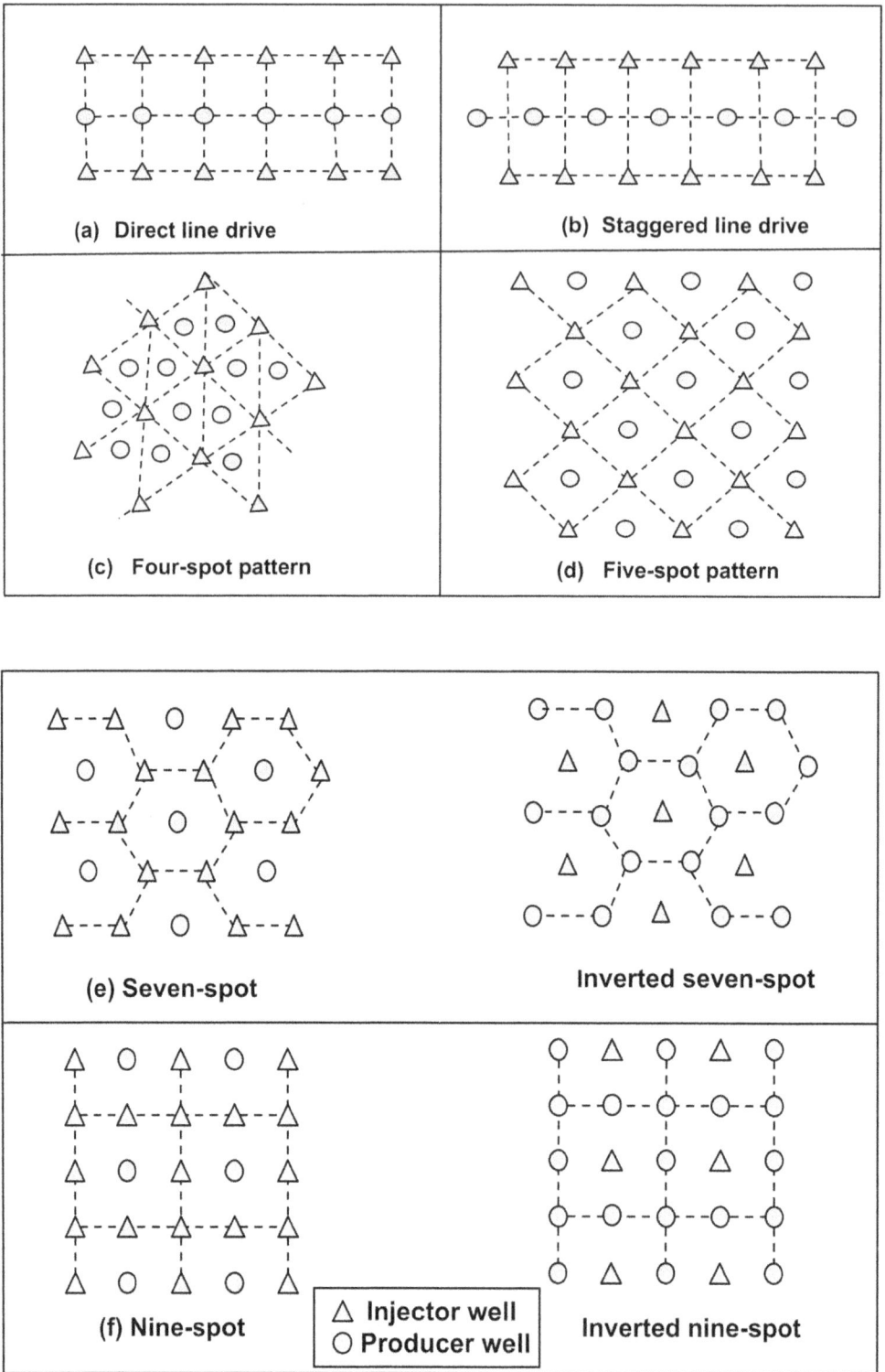

FIGURE 2.17   Different types of pattern flooding.

2.13 MACROSCOPIC EFFICIENCY

Oil recovery in a displacement process depends on the volume of the oil reservoir contacted by the injected fluid. The overall oil recovery efficiency (E) can be mathematically represented as follows:

$$E = E_D \times E_V \tag{2.52}$$

Where E_D and E_V represent the displacement efficiency (microscopic), volumetric sweep efficiency (macroscopic), respectively. The macroscopic sweep efficiency is a measure of how effectively the displacing fluid contacts the volume of a reservoir both areally (E_A) and vertically (E_{VI}). Mathematically, E_V can be represented by the equation:

$$E_V = E_A \times E_{VI} \tag{2.53}$$

Macroscopic displacement efficiency describes the effectiveness of the displacing fluid in contacting the reservoir in a volumetric sense. It is a measure of how effectively the displacing fluid sweeps out the volume of a reservoir, both areally and vertically, as well as how effectively the displacing fluid moves the displaced oil toward production wells (Jin, 2016).

The depending parameters of different efficiency terms are given here:

$E_A = f$ (mobility ratio, pattern, directional permeability, pressure distribution, cumulative injection, and operations)

$E_{VI} = f$ (rock property variation between different flow units)

$E_D = f$ (primary depletion; K_{rw} and K_{ro}; μ_o and μ_w)

E_A is the areal sweep efficiency for a given portion of a reservoir or pattern. Areal sweep efficiency is the fractional area of a pattern that is swept by water (Fig. 2.18).

E_{VI} is the vertical sweep efficiency, defined as the pore space invaded by the injected fluid divided by the pore space enclosed in all layers behind the location of the leading edge of the flood front. It is defined as the cross-sectional area enclosed in all layers behind the injected fluid front. The vertical sweep efficiency is a measure of the 2D (i.e., vertical cross section) effect of reservoir nonuniformities (Fig. 2.15).

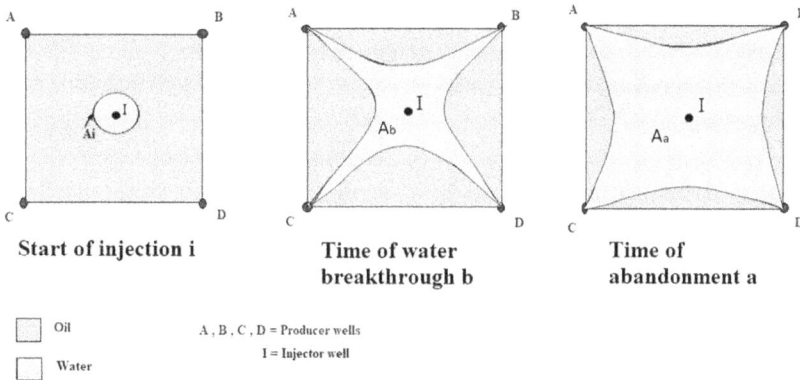

Start of injection i	**Time of water breakthrough b**	**Time of abandonment a**

Oil
Water

A , B , C , D = Producer wells
I = Injector well

FIGURE 2.18 Swept area at the start of injection, time of breakthrough, and time of abandonment.

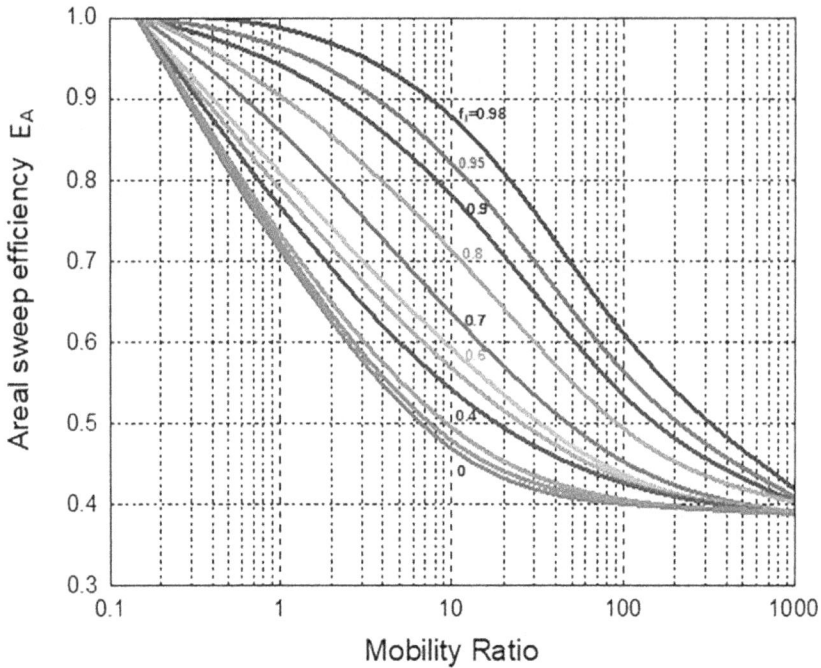

FIGURE 2.19 Areal sweep efficiency versus mobility ratio for various values of f_D.

The sweep efficiency is related to the mobility ratio of the injected fluid (water) to the displaced fluid (oil). The mobility ratio is defined as follows:

$$M = \frac{Mobility\,of\,injected\;fluid}{Mobility\,of\,displaced\;fluid}$$

$$M = \frac{K_W / \mu_W}{K_o / \mu_o} \tag{2.54}$$

A value of one for M means that injected water and produced oil can flow through the reservoir with equal ability. When $M = 10$, the ability of water to flow is ten times greater than that of the oil.

The mobility ratio has a significant impact on areal sweep efficiency. The curves in Figure 2.19 indicate the fraction of a typical five-spot that will be swept at water breakthrough and at increasing water cuts (f_D) of the produced fluid for different values of M. The sweep efficiency gradually increases with a continuous injection of water. The water cut at the abandonment of a waterflood is generally in the range of 90%–98%.

2.13.1 FACTORS AFFECTING AREAL SWEEP EFFICIENCY

- *Injector-producer well pattern*: It is one of the factors that controls the areal sweep efficiency. Areal sweep efficiencies at breakthrough, for different patterns and a mobility ratio of one, are summarized in Table 2.2.
- *Mobility ratio:* As the mobility ratio increases, the areal sweep efficiency decreases because the displacing fluid (water) travels faster than the displaced fluid (oil). This causes an inequality in velocities to develop between the two fluids, giving rise to an unstable front and viscous fingering.

TABLE 2.2
Areal Sweep Efficiency at Breakthrough (*M* = 1) (Craig, 1971)

Type of Pattern	Areal Sweep Efficiency, %	Ep, %
Direct line drive (*d/a* = 1)	57.0	57
Direct line drive (*d/a* = 1.5)	70.6	–
Staggered line drive	80.0	75 (*d/a* = 1)
Five-spot	72.3	68–72
Seven-spot	74.0	74–82
Nine-spot diagonal/directional rate	–	–
0.5	–	49
1	–	54
5	–	69
10	–	78

- *Permeability anisotropy:* It is one of the factors that affect areal sweep efficiency, particularly when rectilinear impermeable barriers and highly permeable channels are present between the injectors and producers. An impermeable barrier has an interference modulus greater than zero; thus, the conductivity, as well as sweep efficiency of the pattern, decreases. Permeability channels, when located along the streamlines, shorten the distance between the injector and producer, causing an early breakthrough.
- *Continued water injection after breakthrough:* Craig et al. (1955) studied the effects of continued injection after breakthrough and concluded that such an injection would result in a substantial increase in recovery, even in the case of an adverse mobility ratio.
- *The effects of fractures:* Both horizontal and vertical fractures, play important roles in the areal sweep efficiency. The effective wellbore diameter is increased by horizontal and vertical fractures either at the injectors or at the producer.

2.13.2 FACTORS AFFECTING VERTICAL SWEEP EFFICIENCY

The affecting vertical sweep efficiency are the following:

- gravity segregation due to density difference;
- mobility ratio;
- vertical to horizontal permeability variation; and
- capillary forces.

2.14 MICROSCOPIC DISPLACEMENT

The microscopic displacement efficiency (E_D) is the fraction of movable oil that has been displaced from the swept zone at any given time or pore volume injected. If we assume constant oil density, the definition of displacement efficiency for oil becomes:

$$E_D = \frac{Amount\ of\ oil\ displaced}{Amount\ of\ oil\ contacted\ by\ displaceing\ agent}$$

$$= \frac{S_{oi} - S_{or}}{S_{oi}} \tag{2.55}$$

where S_{oi} is initial oil saturation and S_{or} is residual oil saturation.

The microscopic displacement efficiency (E_D) relates to the displacement or mobilization of oil at the pore scale.

2.14.1 Factors Affecting the Displacement Efficiency

The displacement efficiency of the oil recovery process is influenced by the following:

- Pore Geometry
- Capillary Pressure
- Reservoir Continuity
- Fluid Saturation
- Wettability

2.14.1.1 Pore Geometry

Trapping of oil and gas on a microscopic scale in a petroleum reservoir rock is affected by the geometric and topologic properties of the pores and fluid properties. Pore properties of importance are pore to throat size ratio; the types and degrees of heterogeneity in the arrangement of pores; throat to pore coordination number or connectivity; and the properties of the pore surfaces, which include composition and degree of roughness.

2.14.1.2 Capillary Pressure

Capillary pressure in a porous media can be explained as the pressure difference existing between two immiscible fluids, both of which are static, in a capillary system. In general, capillary pressure can be expressed as the pressure difference between the nonwetting phase and the wetting phase, which is a positive value in the case of water-wet rocks and a negative value in the case of oil-wet rocks.

Capillary pressure in a reservoir determines the saturation distribution and, hence, the total in situ volumes of fluids (oil/water/gas). Accurate knowledge of the capillary pressure distribution on drainage is one of the primary factors in the reliable estimation of hydrocarbon reserves. Imbibition capillary pressure is used to describe reservoir fluid displacement processes on production and interpret the results of relative permeability experiments.

Capillary pressure is the pressure difference across the curved interface between two immiscible fluids in contact in a small capillary tube. The pressure difference is expressed in terms of wetting and nonwetting phase pressures; thus:

$$P_c = P_{nw} - P_w \tag{2.56}$$

where P_c: capillary pressure [psi]
P_{nw}: pressure in the nonwetting phase [psi]
P_w: pressure in the wetting phase [psi]

Capillary pressure is related to IFT, contact angle, and pore radius by:

$$P_c = \frac{2\sigma \cos\theta}{r} \tag{2.57}$$

where
P_c: capillary pressure [dynes/cm^2]
r: pore radius [cm]
σ: interfacial (or surface) tension [mN/m or dynes/cm]
θ: contact angle [degrees]

The aforementioned expression shows that capillary pressure in reservoirs depends on σ, the IFT between two immiscible fluids; θ, the contact angle between rock and fluid, which is a function of wettability; and r, pore radius, a microscopic rock property. An increase in pore radius leads to a decrease in P_c. Thus, high-permeability rocks with relatively large pore radii have lower P_c than lower-permeability rocks containing the same fluids.

Capillary forces explain why water is retained in oil and gas zones. In water-wet reservoirs, water coats rock surfaces and is preferentially held in smaller pores. Nonwetting hydrocarbon phases occupy the central space of larger pores (Fanchi, 2002).

Consider a sample of reservoir rock or similar porous solid that is fully saturated with an aqueous phase, and assume that the condition $\theta = 0$ holds. The quasi-static displacement of this phase by a nonwetting phase is then defined by the set of increasing values of the capillary pressure, P_c, and a corresponding set of decreasing values of the wetting phase saturation, S_w (Fig. 2.20). The relationship defined by such data is called the drainage capillary pressure curve. The pressure versus

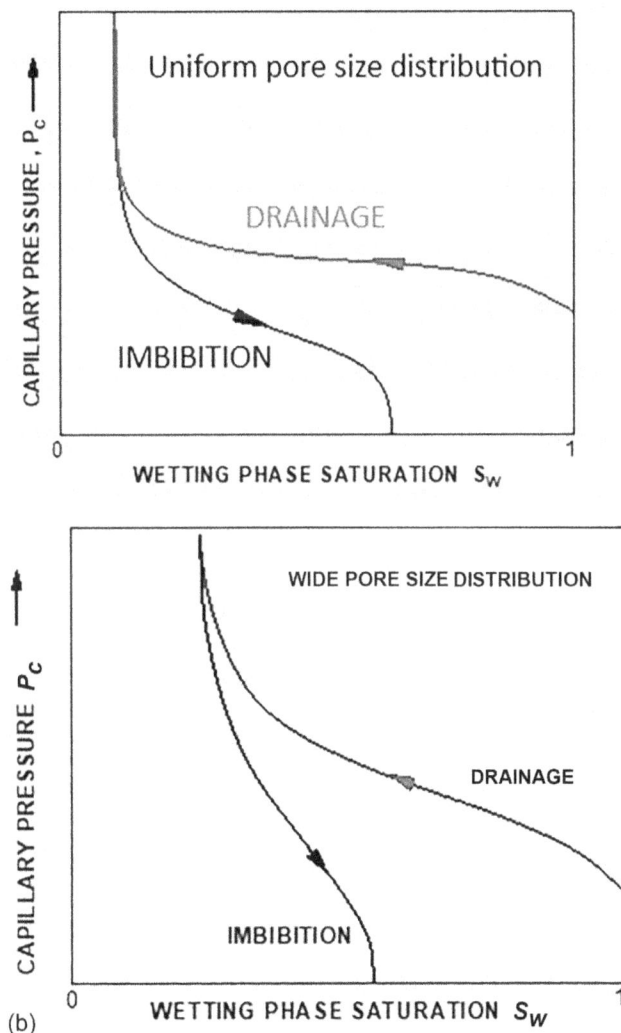

FIGURE 2.20 Typical capillary pressure versus wetting phase saturation: (a) narrow pore size distribution, (b) wide pore size distribution.

saturation relationship for the reverse process, in which the wetting phase displaces the nonwetting phase, is known as the imbibition capillary pressure curve. Because, for a typical pore configuration, the imbibition capillary pressure is of the order of half the drainage pressure, the curves will display considerable hysteresis (Melrose, 1974).

2.14.1.3   Reservoir Continuity

A reservoir needs to be continuous between injectors and producers if an injector is to displace oil to a producer. The injector and the producer need to be in the same zone to produce oil. Faults and other no-flow barriers between injector/producer are a determent to the success of a waterflood.

A combination of analytical and interpretive techniques is used to test whether the injector and producer are in the same zone. If the samples taken from the injector and producer have the same composition, then they are considered to be of the same reservoir.

2.14.1.4   Fluid Saturation

While considering a waterflood, high initial oil saturation is desirable, because a high recovery efficiency, as well as high ultimate recovery, is expected. If the water saturation is greater than the irreducible water saturation ($S_w > S_{wirr}$), then a sharp front may not develop.

Frick and Taylor (1962) reported that if the initial water saturation exceeds a critical value, then the formation of an oil bank may not take place. Above the critical value, water will be more mobile than the oil and create viscous fingering, which will yield lower displacement efficiency. This critical water saturation can be identified on a fractional flow curve as the tangent that cannot be formed while drawing the tangent from the irreducible water saturation.

2.14.1.5   Wettability

Wettability can be considered an important factor that influences the residual saturation, capillary pressure, and relative permeability, which in turn control the fluid flow in porous rocks (Alnili et al., 2018). Preferentially, the oil recovery from a water-wet reservoir is more than that of an oil-wet reservoir. The wettability of a reservoir rock is determined by contact angle measurement. For a reservoir rock containing only oil and water, a contact angle of less than 90° indicates that the reservoir rock is water wet, but when the contact angle is more than 90°, it denotes the reservoir rock is oil wet. For a strong water-wet rock, the angle is close to 0°, while a strong oil-wet rock can be presumed when the contact angle approaches 180°.

Wettability is a function of the reservoir rock's petrophysical properties: for instance, residual saturation, capillary pressure, relative permeability, and end point of relative permeability curves. The crossover points and the relative permeability end points of the wetting and nonwetting phases are related to wettability as shown in Figure 2.21. For a strong water-wet rock, the relative permeability curves will cross over at a wetting phase saturation point greater than 0.5, while for a strong oil-wet rock, the relative permeability curves will cross over at a wetting phase saturation point lesser than 0.5. A crossover point, at a saturation of 0.5 and equal end points of relative permeability curve, implies that the reservoir rock is in a neutral-wet condition.

Example 2.1: For waterflood recoveries from a five-spot well pattern, the following data are given:

Well spacing	= 50 acres	μ_o	= 7 cp
Pay thickness	= 20 ft.	μ_w	= 0.8 cp
Porosity	= 25%	k_{ro}	= 0.75
S_{oi}	= 70%	k_{rw}	= 0.25
S_{or}	= 30%	B_o	= 1.25

If the injection rate is 200 reservoir barrels per day, calculate the oil recovery at breakthrough and the time needed to break through.

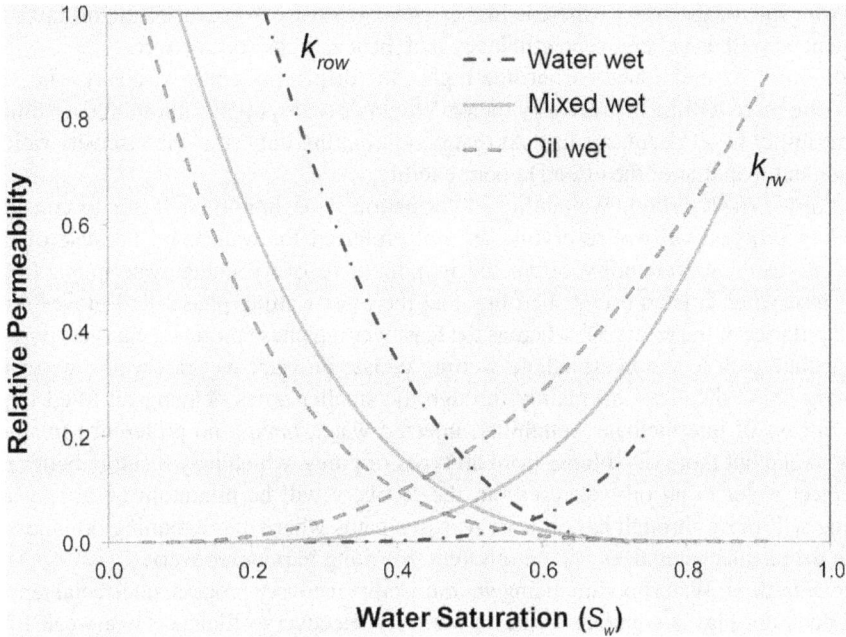

FIGURE 2.21   Effect of wettability on relative permeability curve.

Source: Agada et al., 2016.

Solution:

$$\text{Displaceable oil } = V_D = V \times \phi \times \left(S_{oi} - S_{or}\right)$$

$$= 7758 \times 50 \times 20 \times 0.25 \times \left(0.70 - 0.30\right)$$

$$= 775,800 \text{ RB}$$

$$M = \frac{K_W / \mu_W}{K_o / \mu_o}$$

Here, $M = 2.92$. From Figure 2.19, E_A at breakthrough = 0.57.
   Therefore, oil recovery at breakthrough = $0.57 \times 775,800/1.25 = 353,765$ STB.
   Time needed to break through = $0.57 \times 775,800/200 = 2,211$ days.

2.15   EFFECTS OF VARIOUS PARAMETERS ON RECOVERY BY WATERFLOOD

The following parameters play important roles in the variation in efficacy of the waterflood:

Mobility ratio $\left(M, \dfrac{k_w \mu_o}{k_o \mu_w}\right)$.

It is the most important deciding factor in the success of a waterflood project. An unfavorable mobility ratio will lead to immature, early breakthrough, and lower recovery of oil. For effective sweep efficiency, the displacing phase fluid must move behind the displaced phase—i.e., the oil phase. Thus, a mobility ratio of 1.0 or less is considered to be favorable. Higher effective permeability of oil is preferred over that of the water, whereas higher water viscosity over that of the oil favors uniform displacement as well as better sweep efficiency and, hence, higher oil recovery.

Viscosity ratio: As mentioned earlier, the higher the displacing phase viscosity—i.e., the water viscosity—the more will be the recovery factor. With a decrease in oil saturation over time, the relative permeability to oil decreases fast. At that condition, the impact of the viscosity ratio is much more prominent than that of the effective permeability.

Contact angle (wettability): Wettability of formation is an important factor to consider in the waterflooding process. Oil-wet reservoirs are not preferred for waterflood because of their poor recovery. Intermediate wettability offers the maximum recovery under waterflood, followed by water-wet reservoirs. It is an established fact that the most wetting phase fluid moves through the smallest capillaries of the reservoir, whereas the least wetting phase prefers the largest pores, leaving the intermediate path to the intermediate wetting phase. In water-wet reservoirs, water being the most wetting phase displaces oil mainly through the smaller pores, leaving oil-filled larger pores unswept. In case of intermediate wettability, injected water, having no preference for the smaller pores, can sweep out more oil volume from all types of pores, which may result in better recovery.

If we inject water in an oil-wet reservoir, the recovery will be minimum because water being less wetting will move through larger, least resistive paths where the remaining oil is less. Oil that remains in the smaller interstices will be unswept, resulting in poor recovery.

Interfacial tension: Waterflooding being an immiscible recovery process, interfacial tension (IFT) generally does not play a significant role in recovery. Recovery efficiency is a weak function of interfacial tension for waterflooding. Capillary pressure is not a dominating factor here. This is the reason why we neglect the capillary pressure gradient in recovery calculation.

Gravity effect: Gravity forces are very important factor in oil displacement for a dipping reservoir. By virtue of the density difference, oil tries to move up to achieve equilibrium. Thus, water moving updip will reduce the fractional flow of water, resulting in better displacement efficiency. However, if water is injected at the crest of the structure, it will move faster under gravity, resulting in reduction of displacement efficiency and lower oil recovery.

2.15.1   RESIDUAL OIL SATURATION

It is the initial saturation at the start of EOR processes in regions of a reservoir previously swept by a waterflood—i.e., the oil saturation remains at the end of the waterflood. The economics of the waterflood process depend on the oil saturation at the start and the end of any recovery process. Implementation of a waterflood may not be considered economical if sufficient oil does not remain in the reservoir, and if the residual oil saturation (ROS, S_{or}) does not fall to a sufficiently low value. The displacement efficiency of a waterflood project depends on the ROS (S_{or}), given as follows:

$$\text{Displacement efficiency, } E_D = 1 - (S_{or} / B_o) / (S_{oi} / B_{oi}) \tag{2.58}$$

ROS depends on (i) wettability (ii) pore size distribution (the broader the pore-size distribution, the larger is the residual nonwetting phase saturation, and probability of bypass increases) (iii) heterogeneity (probability of high ROS due to bypass), and (iv) type of fluid (oil/gas, viscosity variations). Thus, recovery can be improved by reducing the ROS. The volume of oil that may be displaced from a total pore volume of V_P is represented by the following Equation 2.59

$$\text{Volume of oil displaced, } N_p = E_D \times V_p \times S_{oi} / B_{oi} \tag{2.59}$$

2.15.1.1   Injection Rate

It is generally accepted that for a completely water-wet system, the ROS is independent of the flood rate. This implies that, under ordinary flooding conditions, capillary forces dominate the macroscopic displacement process and that the microscopic distribution of the oil and water phases is determined by the conditions for hydrostatic equilibrium. The recovery of oil phase during the immiscible displacement by water is primarily controlled by the relative effect of two forces—i.e., viscous force (depends on the viscosity and velocity) and the interfacial force (function of IFT) besides the gravity force. Where the first force tries to move the fluids, the second one holds the wetting fluid against the surface. The dominating force between the two decides the relative movement of the reservoir fluids. In this context, capillary number, a dimensionless parameter, may be used to explain the impact of the velocity on the recovery. The capillary number is a dimensionless parameter describing the ratio of viscous to capillary forces as in Equation 2.60 (Melrose, 1974; Abrams, 1975; Morrow, 1979).

Mathematically,

Capillary Number = Velocity × Viscosity/Interfacial Tension

$$N_C = \frac{v\mu_w}{\sigma_{ow}\cos\theta} \tag{2.60}$$

In the petroleum industry, it has been generally recognized that capillary pressure versus saturation curve reflects the character and arrangements of the pores within the media and the distribution of fluids within the pores, while capillary pressure holds the fluid in the pore and the viscous force drags it out (Guo et al., 2017). Thus, the capillary number (N_{pc}) is crucial in determining the remaining oil saturation as well as its recovery under waterflood or EOR processes. Thus, the higher the capillary number, the lower the effect of capillary pressure in the flow behavior.

There is a critical value of the capillary number beyond which the ROS may be enhanced. In typical reservoir conditions, the capillary number varies from 10^{-8} to 10^{-2}. Where the value of capillary number is higher, viscous forces dominate, and the effect of interfacial tension between fluids in the rock pores is reduced, thereby augmenting recovery. It is generally accepted that to reduce the ROS by a factor of about half, it is necessary to increase the capillary number by a factor of 10–100. The capillary number may be improved for a particular rock by increasing the velocity of displacing fluid, which may not be technically/economically feasible.

The facts regarding the capillary number and recovery under waterflood are as follows:

- At lower capillary numbers, capillary forces dominate.
- With increasing values of the capillary number, viscous forces dominate.
- Importance: A guide to ensure ROS from lab tests on small cores at high rates are representative of the ROS in the field.

2.15.1.2   Free Gas Saturation on Recovery under Waterflood

Free gas may exist within many fields. It is generally assumed that presence of gas in an oil reservoir reduces the oil recovery; mobile gas prohibits the effective displacement of oil, causing reduction in the displacement efficiency. However, the presence of immobile trapped gas will improve the displacement efficiency by reducing the ROS. The reasons are as follows:

(i) Immobile gas phase occupies a certain part of the reservoir pore, resulting in reduction of the effective pore volume available for the mobile oil phase. Thus, there is an enhancement of the pseudo-oil saturation (i.e., the volume of oil/effective pore volume) resulting in improved relative permeability (Kro) as shown in Figure 2.22.

(ii) Sometimes, the immobile gas phase may be enclosed inside oil droplets, resulted in the appearance of larger oil droplets, which can move with lesser difficulty than smaller ones. In addition, the effective permeability to oil phase increases due to an increase in the apparent oil volume (the total volume of oil and trapped gas). The phenomenon helps in improving the oil recovery.

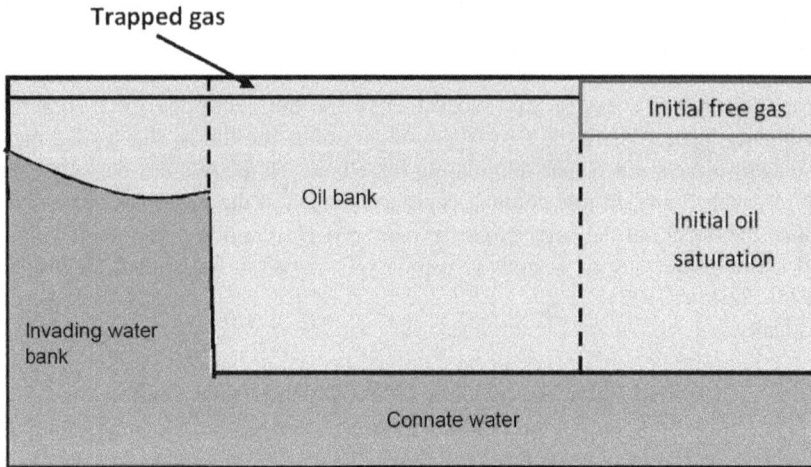

FIGURE 2.22   Change in effective pore volume in the presence of trapped immobile gas.

REFERENCES

Abrams, A., 1975. Influence of fluid viscosity, interfacial tension, and flow velocity on residual oil saturation left by waterflood. Society of Petroleum Engineers Journal 15(5), 437–447.

Agada, S., Geiger, S., Doster, F., 2016. Wettability, hysteresis and fracture–Matrix interaction during CO_2 EOR and storage in fractured carbonate reservoirs. International Journal of Greenhouse Gas Control 46, 57–75.

Alnili, F., Al-Yaseri, A., Roshan, H., Rahman, T., Verall, M., Lebedev, M., Sarmadivaleh, M., Iglauer, S., Barifcani, A., 2018. Carbon dioxide/brine wettability of porous sandstone versus solid quartz: An experimental and theoretical investigation. Journal of Colloid and Interface Science 524, 188–194.

Buckley, S.E., Leverett, M.C., 1942. Mechanism of fluid displacement in sands. Transactions of the AIME 146, 107–116.

Craig, F.C., 1971. The Reservoir Engineering Aspects of Waterflooding. Monograph Series. Society of Petroleum Engineers of AIME.

Craig, F.F., Geffen, T.M., Morse, R.A., 1955. Oil recovery performance of pattern gas or water injection operations from model tests. Transactions of the AIME 204(1), 7–15.

Dake, L.P., 1998. Fundamentals of Reservoir Engineering. Elsevier.

Dake, L.P., 2001. The Practice of Reservoir Engineering (Revised Edition). Elsevier.

Darcy, H., 1856. Les Fontaines Publiques De La Ville De Dijon. Dalmont.

Fanchi, J.R., 2002. Shared Earth Modeling: Methodologies for Integrated Reservoir Simulations. Gulf Professional Publishing.

Frick, T.C., Taylor, R.W., 1962. Petroleum Production Handbook (Vol 1). Society of Petroleum Engineers of AIME.

Guo, H., Dou, M., Hanqing, W., Wang, F., Yuanyuan, G., Yu, Z., Yansheng, W., Li, Y., 2017. Proper use of capillary number in chemical flooding. Journal of Chemistry 2017.

Jin, F., 2016. Principles of Enhanced Oil Recovery, in: Physics of Petroleum Reservoirs. pp. 465–506. Berlin, Heidelberg: Springer Berlin Heidelberg.

Melrose, J., 1974. Role of capillary forces in detennining microscopic displacement efficiency for oil recovery by waterflooding. Journal of Canadian Petroleum Technology 13(4).

Morrow, N.R., 1979. Interplay of capillary, viscous and buoyancy forces in the mobilization of residual oil. Journal of Canadian Petroleum Technology 18(3), 35–46.

Muskat, M., 1946. The Flow of Homogeneous Fluids through Porous Media. The Maple Press Company.

Stiles, W., 1949. Use of permeability distribution in waterflood calculations. Transactions of the AIME 186, 9.

Welge, H.J., 1952. A simplified method for computing oil recovery by gas or water drive. Journal of Petroleum Technology 4, 91–98.

3 Chemical Flooding

3.1 INTRODUCTION

Enhanced oil recovery by chemical flooding improves oil recovery efficiency significantly by improvement in macroscopic sweep efficiency and microscopic displacement of oil trapped in the fine pores of the reservoir rock after the conventional waterflooding process (Mandal, 2015). During waterflooding, the oil phase eventually disintegrates into blobs of residual oil, which are trapped in the pores by capillary forces. This entrapped oil can be recovered if the capillary forces, whose strength is set by oil/water interfacial tension (IFT), are reduced or if the viscous forces are increased sufficiently (Larson, 1982).

The effect of capillary forces on the trapping of oil within the pores of reservoir rock is normally generalized by the use of a dimensionless number called capillary number. The capillary number (N_C) is defined as the ratio of viscous to capillary forces.

$$N_C = \frac{Viscous\ forces}{Capillary\ forces} = \frac{v\mu}{\gamma} \tag{3.1}$$

where v and μ are the velocity and viscosity of the displacing fluid, respectively, and γ is the oil-water IFT. The higher the value of the capillary number, the better is the oil recovery. Wettability alteration of rock surface is another important issue in oil recovery mechanism by the application of surfactants.

Oil recovery by the application of chemical EOR methods is accomplished by the injection of chemicals such as polymers, surfactants, and alkalis into the reservoir. These chemicals are mixed with water and injected through the injection well, which displaces the reservoir fluids toward the production well. The displacement efficiency of the injected slug depends on the chemicals used in the injected slug. The chemicals injected into the reservoir lead to changes in the properties of the aqueous phase of the reservoir, causing the displacement of the trapped crude oil. These altered properties include viscosity of the displacing fluid, IFT between aqueous and oleic phases, wettability alteration capacity of aqueous phase, and miscibility of oleic phase in the displacing fluid (Mohammadzadeh et al., 2015). However, the process has some limitations when applied in the field. Subsequently, different modes of chemical slug injections have been devised, studied, and applied for EOR processes. These include the binary mix of alkali-surfactant (AS), surfactant-polymer (SP), alkali-polymer (AP), alkali-surfactant-polymer (ASP) slug, micellar, and foam flood, as shown in Figure 3.1.

The injection of polymers plays an important role in chemical flooding. It increases the viscosity of the injected water and reduces the permeability of the porous media, allowing for an increase in the vertical and areal sweep efficiencies and, consequently, higher oil recovery (Needham and Doe, 1987). The main objective of polymer injection is mobility control, by reducing the mobility ratio between water and oil. The reduction of the mobility ratio is achieved by increasing the viscosity of the aqueous phase. Another important accepted mechanism of displacing the residual oil after waterflooding is that there must be a rather large viscous force perpendicular to the oil-water interface to push the residual oil. This force must overcome the capillary forces retaining the residual oil, to mobilize it and recover it.

Surfactants have been considered as potential chemicals in EOR since the 1970s because they can significantly reduce the IFT and alter the wetting properties of reservoir rocks, lower the capillary forces, facilitate oil mobilization, and hence, enhance oil recovery (Kumar and Mandal, 2016). Surfactants are preferably adsorbed at the oil-water interface resulting an ultralow interfacial tension, which in turn, increases the displacement efficiency. The wettability alteration is mainly influenced

DOI: 10.1201/9781003098850-3

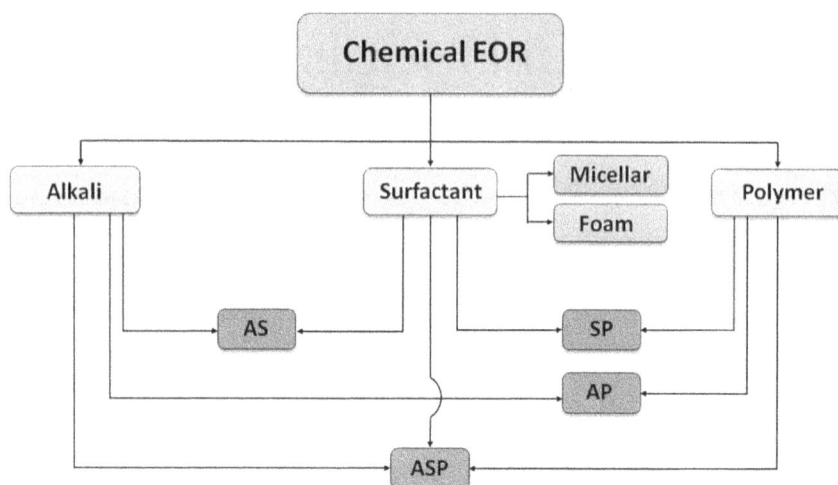

FIGURE 3.1 Classification of chemical EOR technologies.

by the polar interactions between different phases inside the reservoir and the tendency of the surfactants having different charges to spread over the rock surface (Standnes and Austad, 2000). Pore-size distribution and pore surface roughness also control the wettability of reservoir rock. Both ionic and nonionic surfactants have been widely used in EOR processes per their compatibility with the reservoir environment. The mechanism of wettability alteration of reservoir rock in the presence of different surfactants has been reported in literature (Bera et al., 2012) concerning their distinct surface chemistry and chemical interaction between different phases inside the reservoir. Many authors have considered the combined effects of wettability alteration and IFT reduction on the imbibition process. Alteration of rock wettability using surfactants is still a challenge in the EOR field. Adsorption of surfactant on the reservoir rock surface impairs the effectiveness of surfactant flooding and makes the process uneconomical.

Alkalis are also used as chemical flooding agents, as they react with the acidic components of crude oil to form in situ surfactants (Samanta et al., 2011). Alkalis not only reduce the interfacial tension but depending on the rock and crude properties, they also enhance the oil recovery by spontaneous emulsification and wettability alteration of reservoir rock. Thus, the combination of surfactant and alkali shows a synergistic effect on IFT reduction and wettability alteration. In comparison to surfactants, alkalis are much cheaper, which can greatly reduce the cost of an AS flooding project. The commonly used alkalis in EOR application are sodium carbonate, sodium or potassium hydroxide, sodium metaborate, diethanol amine, etc.

The presence of salt can alter the distribution of surface-active components present in oil phase to aqueous phase due to a salting-out effect. Salts can also accelerate the diffusion of surface-active constituents from bulk solution to the interface and, hence, enhance the adsorption of surfactant at the interface, which leads to a reduction in the interfacial tension.

Mixtures of surfactants, due to their superior performance, have been a topic of tremendous interest in recent years (Karaborni et al., 1994; Borse et al., 2006). Surfactant mixtures are gaining considerable attention for use in EOR. In EOR applications, the formation of micelles is an important parameter, as micelles act as a surfactant bank. Almost all anionic-nonionic mixtures exhibit lower values of CMC as compared to individual surfactants. This phenomenon could be attributed to the synergistic interaction between the anionic and nonionic species along with the blanketing effect of nonionic species on the charged head group. The formation of a micelle is favored with reduction in the electrostatic repulsive forces between the charged head groups and with increase in the hydrophobic attraction between the hydrophobic tail groups. The presence of nonionic species blankets the electrostatic repulsive forces, while increasing the hydrophobic attractive forces favors the formation of micelles.

The injection of gas along with surfactant solution can produce a foam that exhibits an apparent viscosity much higher than its constituent phases: liquid and gas. Because of its high apparent viscosity, foam has a mobility much lower than the mobility of gas alone and, therefore, provides an effective means of reducing the mobility of the displacing fluid (Schramm and Wassmuth, 1994; Kumar and Mandal, 2017). Foam can also reduce the gravity override and viscous fingering caused by the low density and viscosity of the gas. Surface tension is one of the most important physical characteristics of the liquid phase, which should be lower for effective foaming. This can be achieved by adding different types of surfactants in the aqueous phase, which stabilizes the dispersions of gas in water to form metastable foam. Co-injection of surfactant solution and gas leads to the generation of foam inside the reservoir. Along with surfactants, the use of other chemicals such as salt, nanoparticles, alcohols, polymer, alkalis, etc. have some notable effects on foam behavior.

The major considerations that must be dealt with before the application of a chemical EOR technique are as follows:

- Gas injection or waterflood provides useful information pertaining to reservoir characterization, which serves a useful impetus for effective EOR planning.
- Chemical flooding study requires detailed laboratory studies, pilot test facilities, and strategic analysis from a functional viewpoint.
- Designing a chemical EOR project requires a long preparation time, simulation and prediction studies, technical planning, and facility installment.
- A greater degree of technical skill, competence, and training is required for engaging oil field managers, engineers, and other workers.
- An EOR project requires more time for approval from decision-makers, which depends on the proposed technical specifications, increasing oil demand, ease of implementation, feasibility, and cost-profitability.

The capability of an EOR route to mobilize and recover oil from rock pores largely determines the success of a project. Therefore, an EOR method is considered beneficial when it can be implemented as a concept-to-pilot approach and boosts production during the process at a rate faster than traditional systems (Teletzke et al., 2010). It is known that EOR processes cannot produce all the remaining crude oil, even if all oleic phases are exposed to injected chemicals. This is largely due to the magnitude of the localized capillary forces, which cause minimal extent of oil entrapment within pore throats. The capillary number measures the chemical fluid's ability to displace the trapped crude oil within a porous medium, such as a petroleum-bearing reservoir. During EOR, a high value of N_C is desirable to achieve favorable oil displacement by decreasing the strength of capillary forces and increasing the magnitude of viscous forces (Andersen, 2020).

3.2 POLYMER FLOODING AND MOBILITY CONTROL PROCESSES

3.2.1 CHEMISTRY OF POLYMERS

Polymers are a class of natural or synthetic materials comprising very large molecules—i.e., macromolecules jointly formed by simpler chemical units called monomers.

3.2.1.1 Types of Polymers Used in EOR

The majority of polymers employed in EOR can be classified into two sets: synthetic polymers and natural polymers or biopolymers or polysaccharide. The most commonly used synthetic polymers in EOR application are polyacrylamide (PAM) and partially hydrolyzed polyacrylamide (PHPA). The natural class of polysaccharide includes xanthan and derivatives of natural polymers, including HEC (hydroxyl ethyl cellulose), guar gum, sodium carboxy methyl cellulose, and carboxyethoxyhydroxyethylcellulose (Olajire, 2014). For EOR application, the polymers must be water soluble. The selection of an appropriate polymer based on its own properties is a very crucial job for its application in a specific reservoir. The details of commonly used polymers are discussed in the following.

3.2.1.2 Polyacrylamide

Polyacrylamides (PAMs) are synthetic chemicals with an approximate molecular weight >1.0 × 10^6 g/mol. The polymer can be tailored to fit for a reservoir with a specific temperature and salinity. The molecule is unique in its strong hydrogen bonding, linearity, and a high degree of non-Newtonian viscosity. The polyacrylamide molecule can be modified by copolymerization with ionic substitutes or by partial hydrolysis of the amide side chain to the carboxylic acid group for EOR applications. Polyacrylamide can be hydrolyzed partially, where some amide groups are converted to the carboxylic acid group. The hydrolysis increases the solubility of the polymer in water with improved properties and is commonly known as partially hydrolyzed polyacrylamide (PHPA). Different grades of polyacrylamides are hydrolyzed to different degrees. Generally, the highest degree of hydrolysis is ~35%. The advantages of PHPA include its tolerance to high mechanical forces during the polymer flooding in a reservoir, cost-effectiveness, and its resistance to microbial attack. PHPA can be used at high temperatures of up to 99°C depending on the salinity of the reservoir.

Based on the specific requirement for EOR application, an increasing number of copolymers of polyacrylamide are becoming available to the industry, and more will be available in the future. With few modifications, like HPAMAMPS, copolymers and sulfonated polyacrylamide can tolerate temperatures up to 104°C and 120°C, respectively (Abidin et al., 2012).

Because of typical straight long-chain structure, polyacrylamide is highly susceptible to mechanical degradation but relatively immune to bacterial action. PAM is thermostable up to 90°C at normal salinity, and thermal stability decrease to 62°C at seawater salinity. Therefore, it is restricted to onshore operations only. High salt content can intensely lower the viscous properties of PAM. The structures of polyacrylamide and partially hydrolyzed polyacrylamide are shown in Figure 3.2.

3.2.1.3 Polysaccharide (Xanthan Gum)

Polysaccharide is greatly accepted in the oil field. It is a high molecular weight, natural carbohydrate. Polysaccharide is a biopolymer with a higher degree of rigidity, as it contains side chains that hold the molecule in a rigid, helical structure and, thus, is immune to mechanical degradation. However, it is highly susceptible to bacterial action.

Xanthan gum is a natural polysaccharide that is produced by different bacterial strains (one of which is *Xanthomonas campestris*) via fermentation of glucose or fructose. The gum so formed generally has a high molecular weight of 2–50 × 10^6 g/mol and has very rigid and long polymer chains, making xanthan gum fairly insensitive to high-salinity conditions and hardness. This polymer is expected to be compatible with the majority of surfactants and other injected fluids and additives

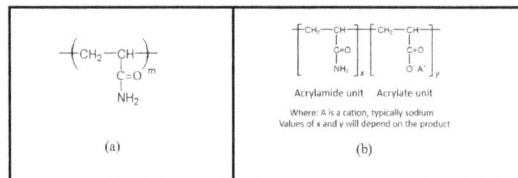

FIGURE 3.2 Structure of (a) polyacrylamide and (b) partially hydrolyzed polyacrylamide.

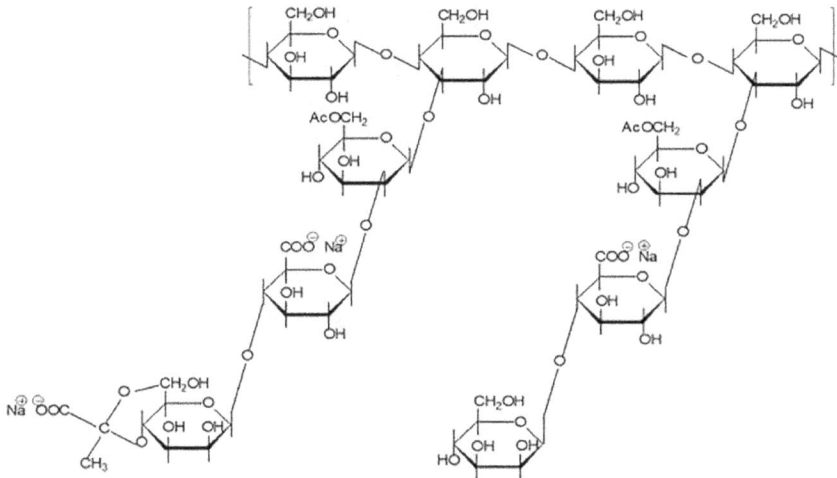

FIGURE 3.3 A schematic molecular structure for xanthan gum.

used in EOR formulations. Xanthan gum is generally formed as broth in concentrated form, and it can be easily diluted to various desired concentrations without any mechanical equipment. Xanthan gum is found to be thermally stable from 70°C to 90°C depending on reservoir salinity. However, xanthan gum is highly susceptible to bacterial degradation, when employed in the field, at low-temperature regions in the reservoir. Also, it has been stated in a few reports that xanthan can possess some cellular debris responsible for plugging the pores. A schematic molecular structure for xanthan gum is shown in Figure 3.3.

3.2.1.4 Needs of New Advanced Polymers

There are many polymer flooding projects reported that are unprofitable for EOR because of improper reservoir description or problems associated with polymer selection. In the past few years, polymer flooding has been field tested extensively and has been emerging as a proven technology. However, there are some limitations with the polymer available in the market such as degradation (thermal, physical, bacterial, and chemical) or rheology of polymer. PAM, the commonly used polymer, suffers from strict salinity and temperature limitations.

It is a prerequisite to select the right polymer for a particular field of application. Reservoir properties like permeability and oil viscosity are used to select polymers with desired molecular weight polymer. Composition of rock and polymer adsorption on the rock surface are used to evaluate the suitability of a polymer for a specific reservoir. A new class of polymer has to be designed that can tolerate high salinity and temperature conditions so that there is no limitation in a high-salinity reservoir (Needham and Doe, 1987). Thomas et al. (2012) provide some critical features for polymers based on acrylamides for EOR applications at high temperatures, high total dissolved solids, low permeabilities, and high oil viscosities.

3.2.2 RHEOLOGY OF POLYMER SOLUTIONS

The rheological behavior of a polymer solution is very important for its application in EOR. Polymer solutions are generally classified as pseudoplastic fluids under most conditions. A pseudoplastic material is one that exhibits a smaller resistance to flow as the shear rate increases. Mathematically, the formula is known as the power-law model (Barnes et al., 1999) (Eq. 3.2):

$$\tau = k\gamma^n \qquad (3.2)$$

$$n = 1,\ K = \mu,\ \text{Newtonian fluids}$$

$$n < 1.0,\ \text{shear thinning, pseudoplastic fluids}$$

$$n > 1.0,\ \text{shear thickening, dilatant}$$

where τ is the shear stress (Pa) and γ is the shear rate (s^{-1}), K is the consistency index (Pa.s^n), and n is the flow behavior index (dimensionless). Note that if $n = 1$, the equation reduces to the Newtonian case with K equivalent to dynamic viscosity, μ. As shown in Figure 3.4, the flow of these non-Newtonian fluids may follow one of the following complex fluid models.

Equation 3.2 may be converted to the logarithmic form as follows:

$$\log\tau = \log K + n\log\gamma \tag{3.3}$$

Thus, a plot of log shear stress against log shear rate will produce a straight line, the slope of which will equal the flow behavior index, and the intercept on the shear stress axis will be equal to the logarithm of the consistency coefficient.

The rheological behavior of the flow of polymer solution through porous media can be divided into four flow regions as shown in Figure 3.5. At low velocities, the apparent viscosity of the polymer solution will approach a maximum limiting value. For a larger flow range, the polymer solutions exhibit pseudoplastic behavior with decreasing apparent viscosity with increasing velocity. At a higher velocity, the viscosity of the polymer solution approaches a minimum value that is equal to or greater than the solvent viscosity. At these high rates of shear, the polymer solution loses its pseudoplastic nature and displays an increasing apparent viscosity with increasing shear rate. Such a fluid is normally classified as dilatant.

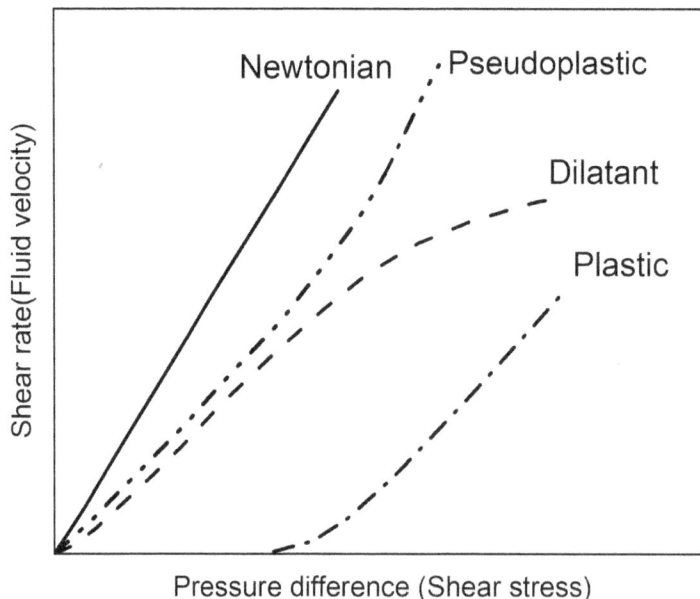

FIGURE 3.4 Newtonian and non-Newtonian flow.

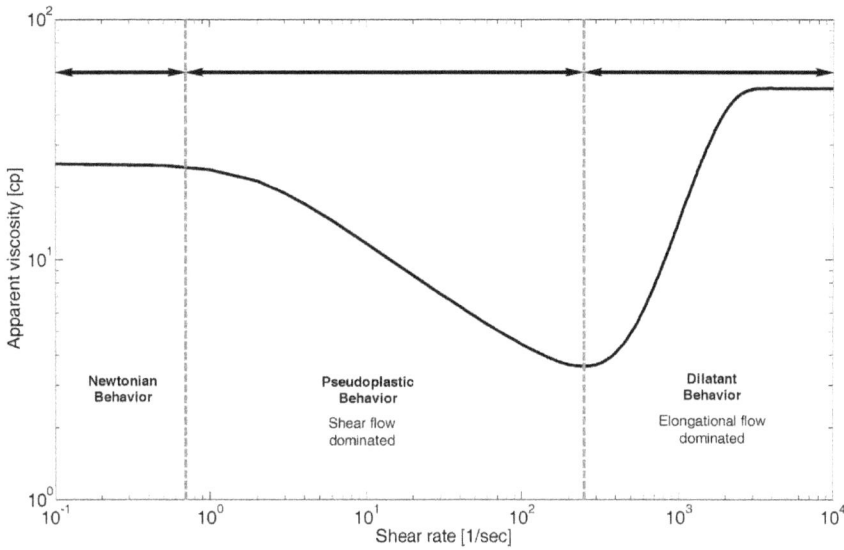

FIGURE 3.5 Rheological behavior of polymer solution in porous media.

Source: Skauge et al., 2018.

3.2.2.1 Apparent Viscosity

Equation 3.2 can be rewritten as follows:

$$\tau = \left(K\gamma^{n-1}\right)\gamma$$

or

$$\tau = \mu_{app}\gamma \qquad (3.4)$$

where $\mu_{app} = K\gamma^{n-1}$ is apparent viscosity, which is the ratio of shear stress to shear rate.

Inspection of the aforementioned formula shows that the apparent viscosity for a pseudoplastic fluid decreases as the shear rate increases, though it is true for a certain shear rate range.

To predict the viscosity-enhancing power of a polymer solution, the empirical Mark-Houwink-Sakurada (MHS) equation is used:

$$[\eta] = K'M_p^{\alpha} \qquad (3.5)$$

where $[\eta]$ is the polymer's intrinsic viscosity, the parameters K' and α are dependent on the polymer-solvent system and temperature, M_P is the polymer molecular weight. For solvents, a value of $\alpha=0.5$ is indicative of a theta solvent. A value of $\alpha=0.8$ is typical for good solvents. For most flexible polymers, $\alpha \geq 0.5$. The intrinsic viscosity, also known as the limiting viscosity number or the Staudinger index, is defined as the ratio of the increase in the relative viscosity by the polymeric solute to its concentration in the limit of infinite dilution:

$$[\eta] = \lim_{c \to 0} \frac{\eta - \eta_0}{\eta_0 c} \qquad (3.6)$$

where η_0 is the viscosity of the solvent, η is the viscosity of the polymer solution.

The empirical Flory equation (Eq. 3.7) can be used to estimate the mean end-to-end distance of a polymer in solution. The Flory equation is given as follows:

$$d_p = 8\left(M_p\left[\eta\right]\right)^{1/3} \tag{3.7}$$

where d_p is in angstroms (10^{-10} m), and $[\eta]$ is in dl/g.

The typical flow behavior of a polymer solution along with that of oil and water is shown in Figure 3.6 (Skelland, 1967). It is observed that the apparent viscosities of a polymer solution are significantly higher than the viscosity of water, even at high shear rates.

In a good solvent, the polymer molecules extend fully to maximize contact with the solvent. For chemical EOR application, the polymer must be water soluble. In water, polymers with polar functional groups are extended to form H-bonds, which gives a flexible and gel-like appearance to the solution. With the polymer molecules extended, the polymer-polymer entanglements are maximized, which leads to increase in apparent viscosity of the polymer solution.

As reported by Samanta (2011), the viscosity of solutions increased with increasing polymer concentration due to intermolecular entanglements (Fig. 3.7). This is a mass effect, as more polymer molecules are dissolved. However, after a certain concentration, the increase in viscosity is marginal. On the other hand, with increase in concentration, the cost of the project increases. Thus, it is very important to use an optimized concentration of the polymer. Also, higher molecular weight polymers exhibit a greater apparent viscosity than their low molecular weight counterparts under similar conditions. The rheology of certain polymers significantly changes with the hydrolysis of those polymers. It has been reported (Martin and Sherwood, 1975) that the apparent viscosity of polyacrylamide increases with the degree of hydrolysis, although this difference is quite small compared to the effect of the initial hydrolysis (Fig. 3.8).

Temperature has a substantial impact on viscosity and the viscous behavior of polymer solutions. With increase in temperature, the apparent viscosity of the polymer solution diminishes. As temperature increases, the average speed of the molecules in a liquid increase, and the time they spend in contact with their nearest neighbors decreases. Thus, as temperature increases, the average intermolecular forces decrease. So, the selection of a suitable polymer and its concentration

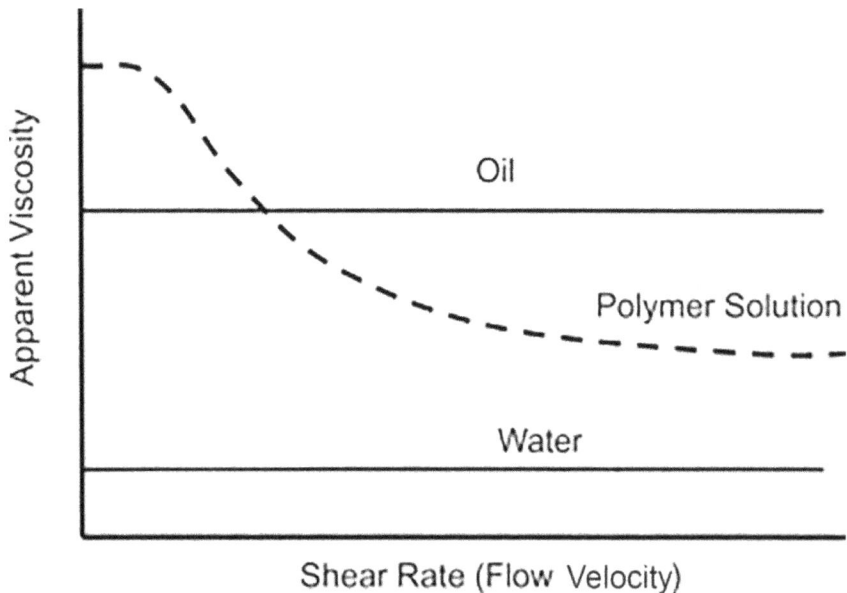

FIGURE 3.6 Apparent viscosity versus shear rate (flow velocity).

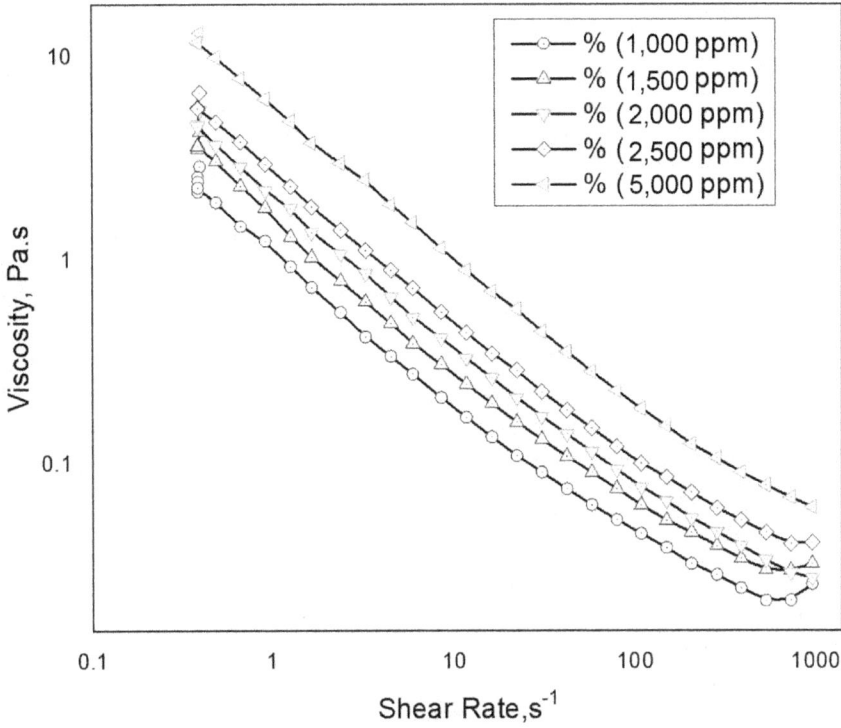

FIGURE 3.7   Effects of concentration of PHPA polymer on viscosity.

Source: Samanta, 2011.

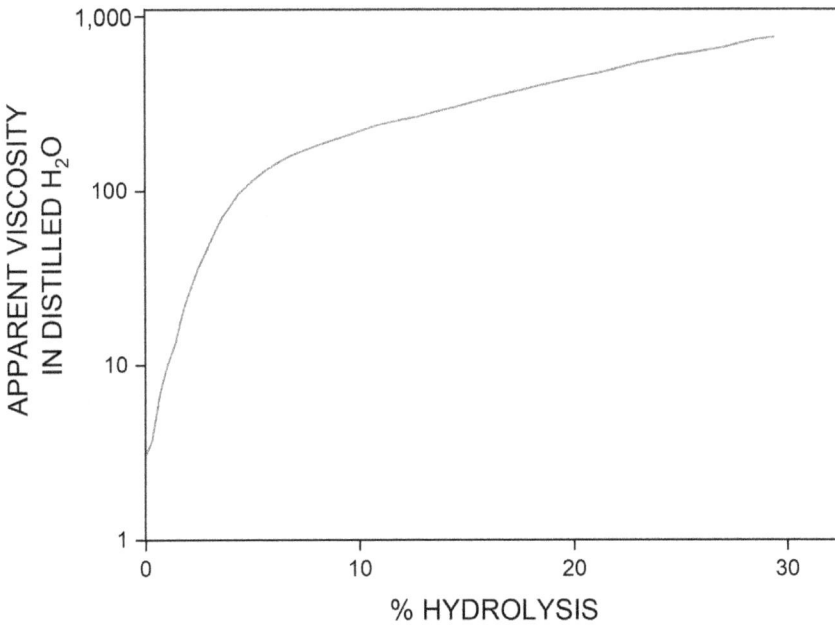

FIGURE 3.8   Typical apparent viscosity in distilled water versus hydrolysis.

is highly challenging considering the decrease in viscosity of polymer solutions and thermal degradation at high reservoir temperatures. A high reservoir temperature means not only a technical restriction but also an economic one. PHPA suffers excessive hydrolysis at high temperatures and precipitates in brine solutions present in reservoirs. Polyacrylamides demonstrate a stronger tendency to flocculate at high temperatures due to the fact that the decomposition reactions are also accelerated at higher temperatures (Hashmet et al., 2014). Depending upon the nature of polymer, the effect of temperature on the degree of hydrolysis, concentration, and its molecular weight, the rheological behavior of each polymeric solution is unique from the others (Nouri and Root, 1971). Obviously, reservoir temperature is a factor affecting solution viscosity and must be certainly weighted among other factors such as polymer concentration, molecular weight, degree of hydrolyzation, and salt concentration.

The salinity of the formation water has a significant impact on the rheology of the injected polymer solution. The addition of Na^+ can effectively neutralize the negative charge of the polymer molecule, which brings about the shrinkage of the molecular chains and the decrease of the hydrodynamic radius. Because of the increase of ionic strength by adding NaCl aqueous solution, the double electrical layer on PHPA molecular chains is compressed, and electrostatic repulsion among the anions is shielded. The electrolytes NaCl and $CaCl_2$ are reported to suppress the viscosity and pseudoplasticity, with the latter suppressing it to a greater extent because of the doubly charged cations (Fig. 3.9). In presence of high concentrations of Ca^{2+} and Mg^{2+}, the polymers get precipitated from the solution and lose their viscous properties. The mechanism is complex, but precipitation occurs when the divalent cations interact with carboxylate groups along the polymer backbone.

The partially hydrolyzed polyacrylamide chain is stretched in deionized water because of the repulsive forces between the negative charges (carboxylate groups) on the chain. This means that

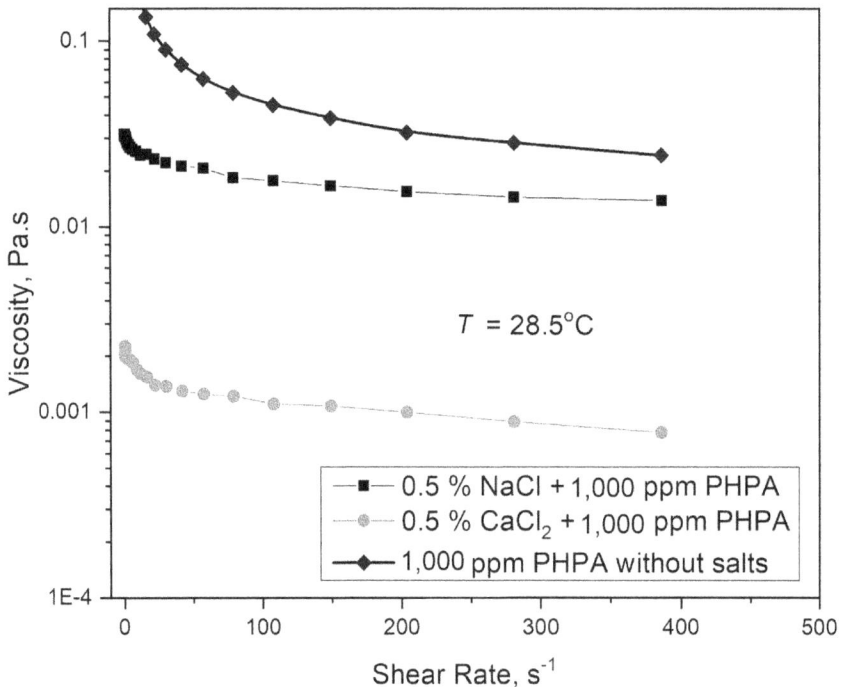

FIGURE 3.9 Effects of monovalent and divalent salts on viscosity of PHPA.

Source: Samanta, 2011.

the hydraulic radius of the polymer chain is larger in deionized water, and consequently, the polymer solution viscosity is high. As the concentration of sodium ions in the solution is increased, the repulsive forces within the polymer chain decrease, due to charge screening effects, and the chain coils up. This change in the polymer conformation causes the hydraulic radius of the chain to decrease and the degree of polymer chain entanglement to diminish. Both factors cause the polymer solution viscosity to decrease. Also, the reduction in the polymer chain size, due to the charge shielding, increases the critical shear rate.

3.2.2.2   Viscoelasticity of Polymer Solutions and EOR

Viscoelasticity is defined as the ability of particular polymer solutions to behave as a solid and liquid simultaneously in certain flow conditions. Complex viscosity basically represents the overall resistance to deformation of a fluid, irrespective of whether that deformation is elastic (recoverable) or viscous (unrecoverable). It is measured as a function of angular frequency under forced (oscillatory) shear conditions. This represents the ability of the injection fluid to flow and ultimately exceed capillary pressure within a pore throat. Initially, a chemical fluid enters a rock pore section with variable throat diameter but is unable to exceed its capillary pressure (force). However, the fluid possesses an internal energy, which undergoes gradual oscillation (under dynamic flow) under the injected force to ultimately exceed the capillary pressure and displace crude oil from rock pores.

The EOR mechanism of viscoelastic polymer flooding is twofold. On the one hand, additional viscosity further prohibits viscous fingering so that volumetric sweeping is expanded macroscopically. At the same time, the oil displacement efficiency is enhanced due to deformation of the long-chained molecular structure microscopically so that the residual oil can be hauled out from dead ends or pore throats and onto the rock surface.

During polymer flooding, the polymeric chain entanglements formed within the injection fluid gradually improve their elastic and viscous characteristics during the displacement process and are gradually able to block/plug high-permeability pores (where capillary effect is low) and forcibly mobilize trapped oil in low-permeability zones (wherein high capillary forces exist). This is the effect of the complex viscosity/rheology of fluid on EOR in porous media.

When polymer solutions flow through porous media, polymer molecules are distorted by the flow through tortuous and converging or diverging pore channels. The curly molecules need enough time to relax and change their configurations, so the polymer solution will show an elastic response in addition to the viscous one. Especially in high shear rate regions, such as in the vicinity of injectors and producers, the "elastic viscosity" cannot be neglected. A polymer solution with high elasticity yields a significantly high pressure drop during flow through porous media. The elasticity can also maintain the stability of the propagating front and minimize fingers.

Compared with viscous polymer solutions, the required injection pressure of viscoelastic polymer flooding is higher. Hence, the injection of displacement fluid becomes more difficult with the increasing elastic effect. This is a challenging issue in chemical flooding techniques.

The relaxation time is a characteristic parameter representing the elastic behavior of polymers and plays a decisive role in selecting a viscoelastic polymer for chemical flooding. The longer the relaxation time, the higher the oil recovery; however, the injection pressures increase with relaxation time.

Figure 3.10 depicts the complex viscosity η^*, storage modulus G', and loss modulus G'', for a linear polymer. There is a decrease in the complex viscosity at high frequency, which is a typical shear thinning behavior. The shear thinning behavior indicates the pseudoplastic behavior of the polymer. The storage modulus G' is a measure of the elasticity of materials, at low frequencies. However, at a high frequency, there is a slight elastic behavior. On the other hand, the loss modulus G'' reveals the viscous nature of the material, which dominates and varies linearly with frequency. The G'' indicates the energy lost to the viscous deformation in the course of the deformation of materials; it reveals the viscosity of materials. The higher the G'' value, the higher the viscosity of the polymer.

Usually in polymer flooding, the polymer solution is not assumed to be a viscoelastic fluid but a non-Newtonian or pseudoplastic one. This assumption is valid when the flow rate in the pore space

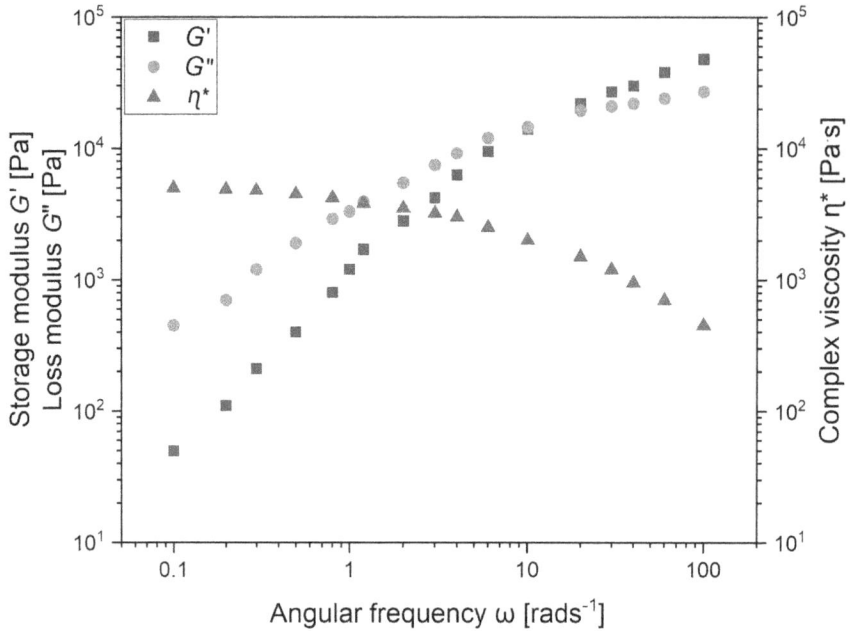

FIGURE 3.10 Complex viscosity η^*, storage modulus G', and loss modulus G'', for a linear polymer at 70°C.
Source: Habibu et al., 2018.

is so slow that the viscoelastic effect of the polymer solution can be neglected. Polymer viscoelasticity must be considered in high shear rate regions, such as in the vicinity of injectors and producers, and in reservoirs having strongly heterogeneous pore geometry.

3.2.3 Mobility Control Process

That the mobility of injected fluids affects waterflood performance, was first recognized by Muskat (1949) in 1949. Later, Stiles (1949) showed the effect of mobility ratio and permeability variations on oil recovery. Dyes et al. (1954) reported the influence of mobility ratio on oil production after breakthrough. Polymer flooding has been evaluated at field scale for more than five decades and is becoming a technology of great interest within the oil and gas community based on recent laboratory studies and field trials that are widely documented in the literature (Needham and Doe, 1987). Due to decreased water production and enhanced oil production, the total cost of using the polymer flooding technique is less than that of waterflooding. The polymer flooding efficiency ranges from 0.7 to 1.75 lb. of polymer per barrel of incremental oil production (Abidin et al., 2012). Polymers added to water increase its viscosity and reduce water permeability due to mechanical entrapment, thus decreasing its mobility.

3.2.3.1 Why Polymer Flooding Is Required?

Reservoirs are porous media with heterogeneity in permeability and porosity. When water or other fluids are injected under pressure, they follow the path of least resistance. High-permeability zones and fractures offer the least resistance to flow. Most of the oil remaining in the low-permeable zones is bypassed. The residual oil saturation in high-permeable zones decreases drastically, and hence, the relative permeability to water increases in high-permeable zones, resulting in ever-increasing water cut. Polymers can reduce the detrimental effect of permeability variations and fractures and, thereby, improve both the vertical and areal sweep efficiency.

3.2.3.1.1 Mobility Control

Polymers are chemicals with high molecular masses and have chains or rings of repeating subunits, known as monomers, linked together. The networked structure of polymers causes the increased viscosity of their aqueous solutions (Yang et al., 2019). The higher viscosity of the injected fluid leads to the reduction of its mobility and improves the mobility ratio of displacing fluid to displaced fluid (Shaker Shiran and Skauge, 2013).

The mobility ratio is defined as the mobility of the displacing fluid at the average water saturation behind the advancing front of displaced oil (S_{wa}), divided by the mobility of the oil at the average saturation in the advancing oil bank ($S_{w\beta}$) (Fig. 3.11).

There is an inverse relation between the volumetric sweep efficiency and the mobility ratio. The value of M greater than unity is unfavorable, as this will cause instability of the displacement process and the so-called viscous fingering effect. Waterfloods in secondary recovery stages have a lower volumetric sweep efficiency due to the lower viscosity and higher mobility of water that causes fingering and early breakthrough of the injected fluid from the production well through the high-permeable streaks in heterogeneous reservoirs (Kumar et al., 2008). Under the condition of a large viscosity difference between the displacing (water, lower viscosity) and displaced (oil, higher viscosity) fluid, the mobility ratio will become larger than one, and thus, poor recovery will be reached. The fingering effect is highly undesirable, as it reduces production as soon as the finger reaches the production well site. Polymers act as mobility control agents, and their application is highly favored in heterogeneous reservoirs, as the polymer solution reduces the mobility of the aqueous phase through the high-permeability region of the reservoir, allowing the aqueous solution to flow through the less permeable region, leading to an increase in the vertical and areal sweep efficiencies of the injected slug (Carrero et al., 2007; Hernandez et al., 2001). This improved sweep efficiency of polymer flooding in comparison to conventional waterflooding has been depicted in Figure 3.12. It can be seen from the figure that conventional waterflooding has fingering of flood front that causes early breakthrough, whereas polymer flooding has a stable

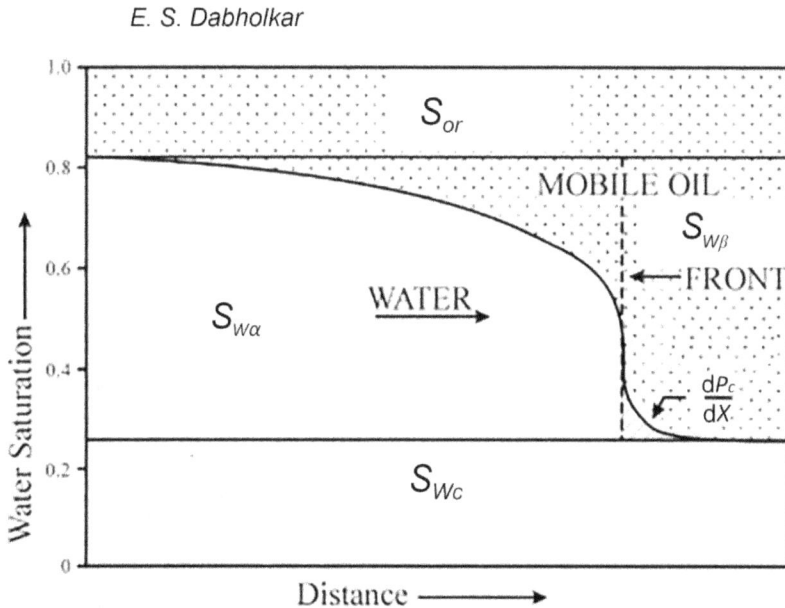

E. S. Dabholkar

FIGURE 3.11   Displacement of oil by water in a water-wet system.

Source: Donaldson and Alam, 2013.

flood front that leads to its higher volumetric sweep efficiency. A schematic of polymer flooding is shown in Figure 3.13.

3.2.4 POLYMER ADSORPTION AND PERMEABILITY REDUCTION

Polymer molecules are favorably adsorbed on solid surfaces during the injection of polymer solutions through reservoir rock in polymer flooding for enhanced oil recovery. Polymer adsorption in the flow of polymer solutions through porous media is usually accompanied by a variety of additional complex phenomena. Viscoelasticity has been shown to contribute to increased flow resistance, particularly at high flow velocities. Consequently, the reduction of the permeability of the reservoir rock by polymer adsorption is a well-established mechanism that improves the mobility ratio. The determination of flow reduction of the mobility control fluid requires knowledge of the degree of adsorptive retention of the polymer in porous media.

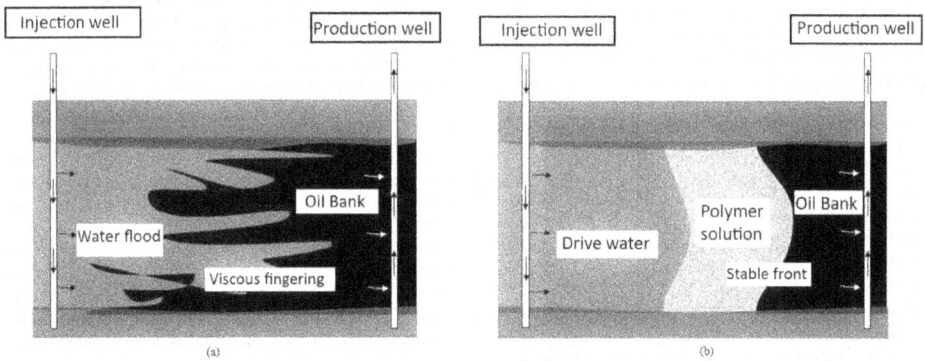

FIGURE 3.12 Typical flow of flood front of (a) waterflooding process with viscous fingering and (b) polymer flooding process.

Source: Gbadamosi et al., 2019.

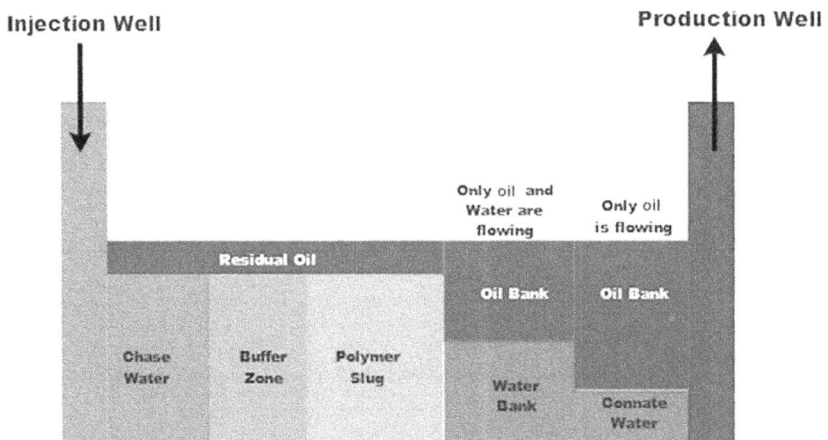

FIGURE 3.13 Polymer flooding as a secondary recovery process.

The equilibrium or residual concentrations of polymer solutions are determined by a UV spectrophotometer. The equilibrium concentrations (C_e) are calculated using Equation 3.7:

$$A = \varepsilon * L * C \tag{3.7}$$

where A is absorbance, ε is molar extinction coefficient (L/mole-cm), L is the path length (cm), and C is concentration (mole/liter) of the solution.

The adsorption capacity of polymer on the adsorbent Γ_i (mg/g) is calculated by a mass-balance relation (Eq. 3.8):

$$\Gamma_i = \left(C_0 - C_e \right) \frac{V}{m} \tag{3.8}$$

where C_0 and C_e are the initial and equilibrium concentrations of the polymer solutions (mg/L), respectively; V is the volume of the polymer solution (L); and m is the weight of the sand particles (g) (adsorbent) used. The time-dependent polymer adsorption on sand surface at different concentrations is depicted in Figure 3.14. With time, the polymer adsorption increases, and after a certain period, the adsorption remains constant with time. With time, the number of adsorption sites decreases due to the formation of an adsorbed polymer layer on the sand surface. When all the sites are covered with the polymer, then further adsorption does not take place, and time-independent adsorption occurs.

3.2.4.1 Permeability Reduction

Polymer adsorption on the reservoir rock surface leads to a decrease in the permeability of the formation to water, which also improves the mobility ratio and, hence, the oil recovery. Once the polymer is adsorbed on the rock surface, as oil passes through the pores with the adsorbed polymer, the polymer remains at shrinking stage, as there is no ionic interaction due to the nonpolar nature of the oil. Thus, it allows the oil to flow through the pores. On the other hand, when water tries to flow,

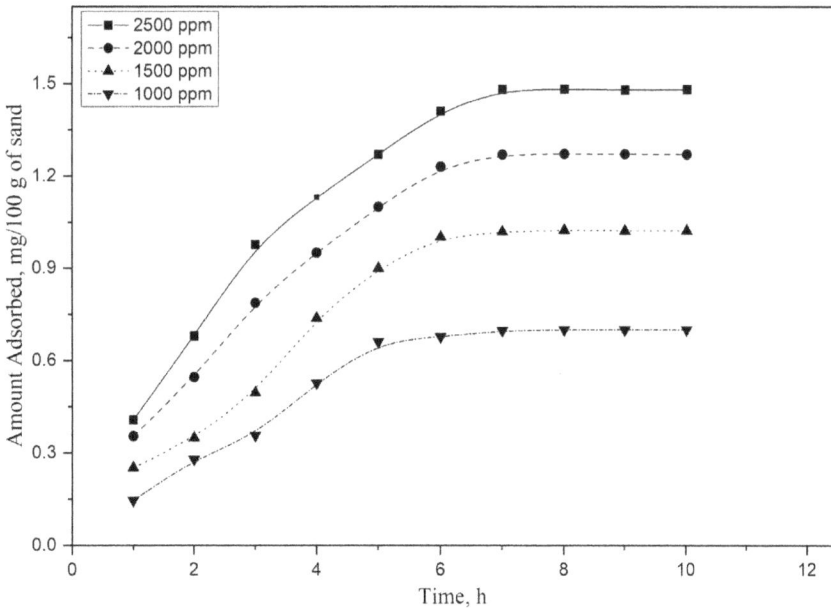

FIGURE 3.14 Effect of polymer (PHPA) concentration on polymer adsorption.

Source: Samanta, 2011.

the polymer molecules get extended due to hydrophilic interaction because of the polar nature of the water. Thus, it resists the flow of water through the pores. This leads to higher relative permeability of oil compared to water and, hence, recovery increases. The mechanism is depicted in Figure 3.15.

3.2.4.2   Polymer Retention

Polymer retention has a major influence on the rate of propagation of a polymer solution through porous media during polymer flooding, and hence, this affects oil recovery. It has two opposite effects on additional oil recovery. The retention of polymer by adsorption on reservoir rock improves relative permeabilities as discussed in the last section. On the other hand, it has a significant effect in porous media on the transport and performance of the polymer solution. High polymer retention can dramatically delay oil movement and recovery during polymer flooding. With very high polymer retention of 200 μg/g and above, polymer retention might have a tremendous impact on the rates of oil displacement and the economic feasibility of polymer flooding (Al-Hajri et al., 2018). Polymers can be retained in the porous media through different mechanisms—i.e., (1) physical adsorption on the rock surface; (2) mechanical entrapment; or (3) due to hydrodynamic retention (Glasbergen et al., 2015). These are depicted in Figure 3.16.

3.2.4.3   Physical Adsorption

Physical adsorption is bound by the number of available adsorption sides on the rock, and the process of adsorption can be described by an adsorption isotherm. The adsorption isotherm is shown in Figure 3.14.

3.2.4.4   Mechanical Entrapment

Polymer molecules are large relative to the size of the narrow pores. The entrapment of polymers occurs in porous media, as the polymer molecules are large relative to the size of the pores. Some polymer molecules entering the pores may not move further due to their hydrodynamic size because the open pore throats for exit are too narrow to allow further propagation and, therefore, are entrapped.

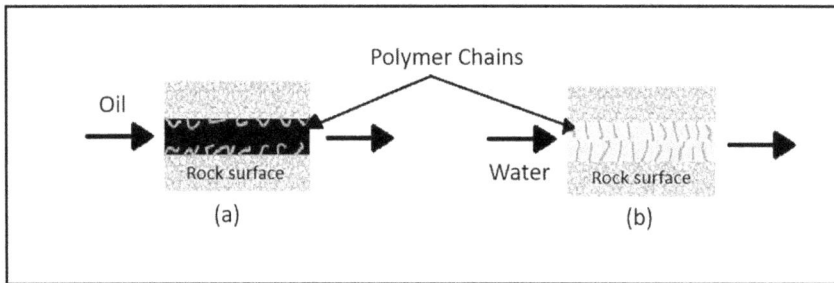

FIGURE 3.15   (a) Flow of oil through the pores with polymer adsorbed on the rock surface and (b) resistance to the flow of water through the pores due to hydrophilic interaction of polymer and polar water molecules.

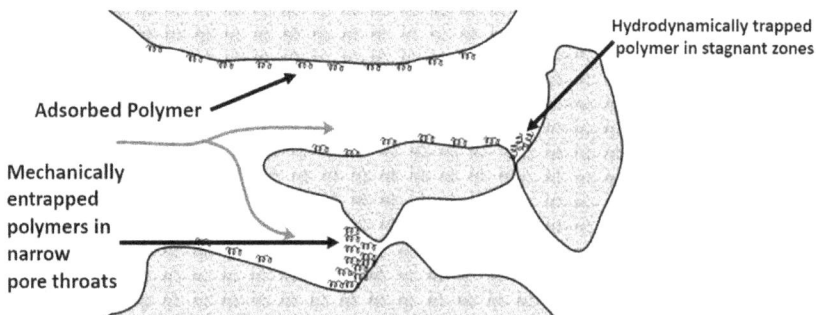

FIGURE 3.16   Schematic diagram of polymer retention mechanisms in a porous medium.

3.2.4.5   Hydrodynamic Retention

Another type of polymer retention is hydrodynamic retention. Some polymer molecules enter in areas (e.g., corners or dead ends) where they are shielded from the driving forces due to being outside the flow streamlines. Therefore, they cease to flow and get temporarily trapped until the streamlines are redistributed. This phenomenon is often referred to as flow-induced hydrodynamic retention (Akbari et al., 2019).

3.2.4.6   Inaccessible Pore Volume

Inaccessible pore volume (IPV) has been widely reported in the literature (Dominguez and Willhite, 1977; Shah et al., 1978) as a phenomenon that needs to be considered in polymer flooding applications. When the polymer molecular size is larger than some pores in a porous medium, polymer molecules cannot flow through those pores. The volume of those pores that cannot be accessed by the polymer molecules is the IPV.

3.2.4.7   Resistance Factor, RF

When a polymer in solution flows through the porous reservoir rock, its mobility reduces significantly compared to water. The mobility reduction can be imparted by an increase in the viscosity of the brine being flooded through the porous media and reduction of the permeability of the reservoir matrix rock due to polymer adsorption. Mobility reduction is the primary conformance improvement benefit of polymer waterflooding. The resistance factor (RF) describes the reduction in water mobility and is defined as the ratio of the brine mobility to that of the polymer solution, with both mobilities measured under the same conditions—i.e.:

$$RF = \lambda_w / \lambda_p \tag{3.9}$$

where λ_w is the mobility of water and λ_p is the mobility of polymer solution.

3.2.4.8   Residual Resistance Factor, RRF

During polymer flooding, only a certain pore volume of polymer solution is injected followed by chase waterflooding. Adsorption of polymer on the rock surface of the porous reservoir causes reduction of permeability as discussed earlier. Due to the retention of the polymer, the effect of permeability reduction is observed even after a significant amount of chase waterflooding after the polymer injection. The residual resistance factor (RRF) is defined as the ratio of the mobility of the injected brine before and after the injection of the polymer solution.

$$RRF = \frac{k_b}{k_a} \tag{3.10}$$

where k_b is brine permeability measured before polymer flooding and k_a is brine permeability measured after polymer flooding. The ability of the polymer solution to reduce the permeability is measured in the laboratory and expressed in a property called RRF.

The change of RF and RRF with variation of pore volume injected at different polymer concentrations is depicted in Figure 3.17. Referring to the definition of RF (Equation 3.10) at a particular concentration of polymer, the effective permeability to water decreases with increasing injected polymer volume because of adsorption of polymer on the rock surface as described in section 3.2.4.1. This results in improvement in RF with injected pore volume. Initially, when the polymer is injected, it prefers high-permeability zones due to low resistance in flow. Results show that an increment in the polymer concentration brings higher values of RF and RRF. The interaction between polymer and rock matrix leads to profile modification. This profile modification depends on how much polymer is adsorbed on the wall of rock surface in high-permeability zones. Due to polymer adsorption, the pore throat radii reduce, and this results in unexpected decrement in the mobility of the polymer.

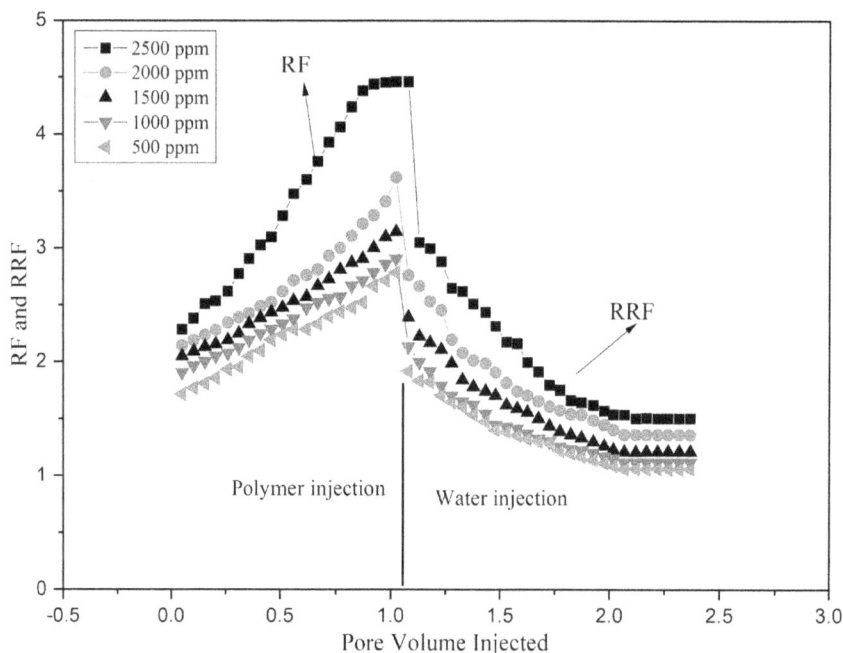

FIGURE 3.17 Plot of RF and RRF with different concentrations of polymer solution.

The thickness of adsorbed layer of polymer (χ_p) can be calculated from the permeability reduction data. The calculation is based on the assumption of Poiseuille's fluid flow through a capillary having its cross section reduced by a uniform layer (Masuda et al., 1992). The thickness of the adsorbed layer of polymer is calculated as follows:

$$\chi_p = r_p \left(1 - \text{RRF}^{-1/4}\right) \qquad (3.11)$$

where RRF is the residual resistance factor and r_p is the average pore diameter, which can be calculated as follows:

$$r_p = (8k_w / \varphi)^{1/2} \qquad (3.12)$$

where k_w is the effective permeability to water and φ is the porosity of the sand pack or core as applicable.

3.2.5 OIL RECOVERY BY POLYMER FLOODING

Polymer flooding offers a higher sweep efficiency and higher oil recovery per pore volume (PV) of displacing fluid injected, enhances oil flow, decreases water cut, and delays the breakthrough of the displacing fluid relative to waterflooding. It also improves recovery by modification of relative permeabilities. During polymer flooding, a dilute aqueous solution of polymer is injected into the reservoir, which displaces the mobile oil toward production oil (Fig. 3.18). Polymer flood design includes laboratory tests to screen commercially available polymers, computer simulation to forecast potential reserves, and field injectivity tests to verify laboratory polymer characteristics. The concentration of polymer and the volume of polymer slug injected depend on the reservoir rock

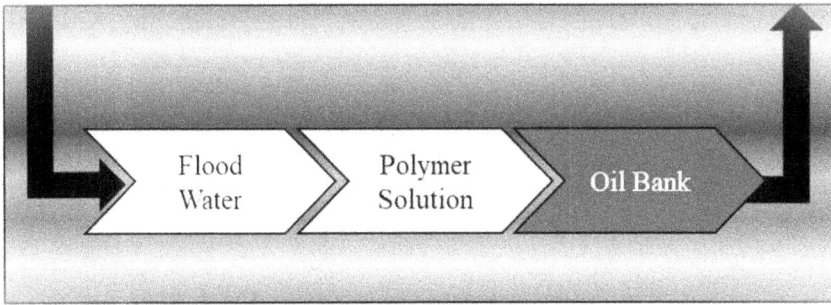

FIGURE 3.18 Simplified schematic of the polymer flooding process for enhanced oil recovery.

Source: Scott et al., 2020.

and fluid properties. Manning (1983) reported a median injected polymer concentration of 250–260 ppm, though the concentration of 1,000–1,500 ppm of polymer is very common. They suggested median polymer bank size is minimum of 17% PV for field-wide projects and 5% PV for pilot projects.

The flow behavior of polymers through reservoir rocks is determined by the type of polymer, flow rate (time dependent), and the porous media's topology and morphology. In addition, the change of the permeability of the porous media during polymer flooding due to polymer retention has an important effect on polymer flow. The economics of EOR polymer flooding are determined by the polymer solution's injectivity (defined as the ratio between the volumetric injection rate and the pressure drop), which should allow using practical injection rates in field applications. To establish whether or not a polymer solution has suitable injectivity, one can compare the injectivity of the polymer solution to the injectivity of water (or brine) in the same reservoir. If the volumetric injection rate and the differential pressure are known (for both the primary brine injection and the polymer solution injection), the injectivity loss (I_{loss}) can be calculated as per Equation 3.13 (Scott et al., 2020):

$$I_{loss} = 1 - \frac{I_p}{I_b} = 1 - \frac{i_p/\Delta P_p}{i_b/\Delta P_b} \tag{3.13}$$

where I represents the injectivity, i represents the volumetric injection rate, and ΔP represents the pressure drop for the polymer solution (p) and brine (b). Han et al. (2012) suggested that a range of injectivity loss from 0.5 to 0.9 is acceptable but that adjustments should be made if the injectivity loss is higher than 0.9.

The additional oil recovery by polymer flooding and payout time depends on the reservoir rock and oil properties. Generally, the initial production response to polymer injection is observed after about 0.1 PV injection, showing increased oil rate and reduced water cut. A typical oil recovery by polymer flooding in a lab-scale study is shown in Figure 3.19. Approximately 20% incremental recovery may be observed by polymer injection over conventional water flooding process. As the polymer slug is injected, the pressure differential increases rapidly as the polymer increases the viscosity of the injected slug and, thus, reducing its mobility. The pressure differential again decreases with the injection of chase water; however, due to significant adsorption and retention of polymer in the core sample, the pressure differential does not fall to the value as measured after the initial brine flooding. In addition, the water cut is reduced significantly during polymer injection.

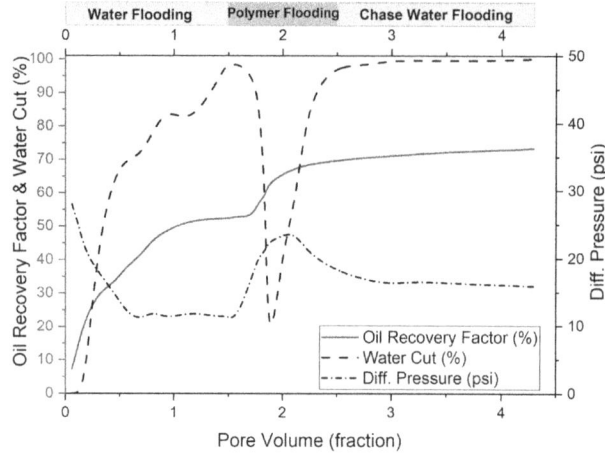

FIGURE 3.19 Oil recovery factor (%) and differential pressure as a function of PV injected during oil recovery.

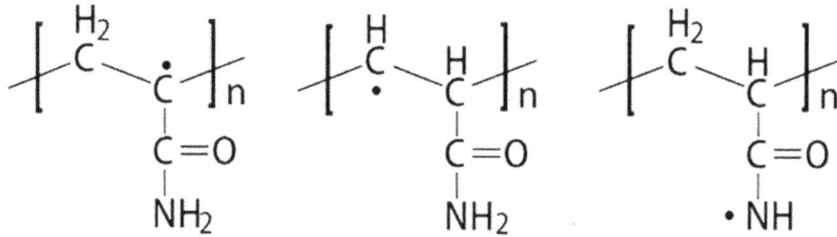

FIGURE 3.20 Three possible forms of polymer radicals of polyacrylamide.

Source: Grollmann and Schnabel, 1982.

3.2.6 Degradation of Polymers

The in-situ viscosity of polymer solutions can be adversely affected by the degradation of polymers. Degradation refers to processes that break down the molecular structure of the macromolecule (Sorbie, 2013). The degradation of polymer can be categorized as follows:

1. Chemical degradation
2. Thermal degradation
3. Mechanical degradation
4. Biological degradation

3.2.6.1 Chemical Degradation

The mechanism of chemical degradation is the formation of free radicals that cut the polymer chain (Fig. 3.20). These free radicals are formed by the reaction of an oxidizer with a reducer (redox). Hydroxyl radicals are the most common radicals generated in the environment from the Fenton reaction, interactions between oxygen and dissolved Fe^{2+} or other transition metals, and sulfate radicals through persulfate activation (Shupe, 1981; Lu et al., 2012; Carman et al., 2007). Some oxidizers and reducers responsible for the chemical degradation of polymers are given in Table 3.1. The high molecular weight (MW) chains are more sensitive to chemical degradation, especially those above 15 million MW. The best polyacrylamide stability is found in a reducing media.

3.2.6.2 Thermal Degradation

During the polymer thermal degradation process, the polymer molecular chains are broken down, and the average MW decreases. The thickening effect is reduced, while the polymer mass concentration remains unchanged. Thermal degradation varies with the type of polymer and reservoir conditions. For typical PHPA polymers, a temperature increase will lead to an increase in the hydrolysis of acrylamide moieties, generating a higher charge density of anionic functionalities along the polymer backbone (Thomas, 2016). An elevated temperature acts as an initiator by providing the activation energy to promote the hydrolysis of the amide groups into carboxylate groups. The mechanism of chemical transformation during thermal degradation is shown in Figure 3.21. Needham and Doe (1987) reported that the biopolymers thermally degrade too fast at temperatures above 200°F. At temperatures above 170°F, polyacrylamides may precipitate in waters containing too much calcium.

3.2.6.3 Mechanical Degradation

During the flow of a polymer solution through the porous media of the reservoir, high shear cuts the polymer chains into pieces. A high MW long-chain polymer with a certain viscosity will be more sensitive to shear than a low MW polymer. This shear degradation is amplified by the formation of free radicals as a mechanism of degradation. Shearing time has a significant impact on the degradation of the polymer. During injection time, solids come from the injected water or are precipitated as $CaCO_3$, $Mg(OH)_2$, FeS, S_2, or biological molds. The solids increase the shearing time by forming channels, where the polymer solution can be under high degradation conditions for a length of time. Morris and Jackson (1978) tested polyacrylamide in a porous medium and reported that mechanical degradation depends on polymer concentration (where mechanical degradation increases as the concentration decreases) and the degree of degradation depends on polymer stretch rate, flow path, and MW.

3.2.6.4 Biological Degradation

Biodegradation utilizes the functions of microbial species to convert organic substrates (polymers) into small MW fragments that can be further degraded to carbon dioxide and water.

TABLE 3.1

Oxidizers and Reducers Responsible for the Chemical Degradation of Polymers

Oxydizer	Reducer
O_2	H_2S
Fe^{2+}, Fe^{3+}	Scavenger
Hydrocarbon	Fe^{2+}
Peroxydes	Sulfato reducing bacteria
	NH_3

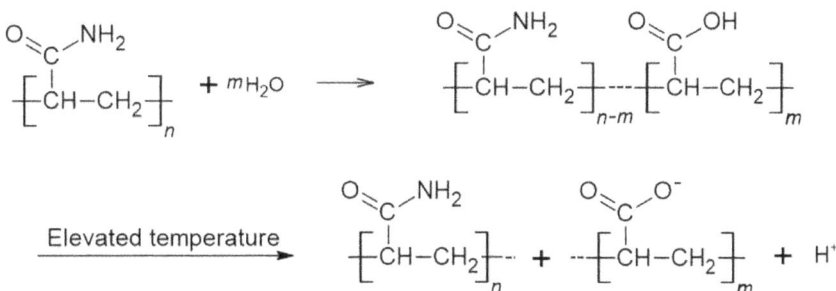

FIGURE 3.21   Effect of elevated temperature in hydrolyzing PAM from amide group to carboxylate.

Polysaccharide is highly susceptible to bacterial action, while polyacrylamide is relatively immune to bacterial action.

In polymer flooding, as the injected polymers move through the reservoir formation, they undergo the aforementioned degradation process depending upon reservoir conditions (Fig. 3.22). So, ultimately, the final viscosity of the polymer is much less than its viscosity prior to injection. Thus, long-term stability tests are useful to ensure the actual polymer degradation within the reservoir during the transit time between the injection and production wells. This information is very important when planning for any polymer-based EOR project.

3.2.6.5   New Types of Polymers

Fine-tuning of the polymer chemical structure (i.e., composition and molecular weight) is crucial to optimize the polymer stability for application at specific reservoir conditions. The selection of the best reservoir for a polymer flooding application must take into consideration the analysis of the following three reservoir parameters: reservoir temperature, brine composition (salinity, divalent cations, dissolved oxygen, iron, and hydrogen sulfide), and permeability. Some new types of polymers with specific characteristics are mentioned here:

- Thermostable polymers—which increase the stability of polyacrylamides from 75°C to 90°C, with new monomers, FLOPAAM AN 125–132.
- Associative polymers, with a main polyacrylamide chain and statistic repartition of hydrophobic groups. There is an association of these hydrophobic groups in a specific brine to give a high viscosity, SUPERPUSHER.
- Star polymers, with three or more branches on a central polymer group. These polymers are normally associated with a high viscosity, ST5030.
- Comb and T-shape polymers, with a main hydrophobic chain and end hydrophobic chain.
- Block associative polymers, with multiple hydrophobic groups inside a hydrophylic chain.
- Structured polymers with hydrophilic branches in a main hydrophilic chain.
- Soft or movable gels are totally insoluble yet injectable gels mainly used in profile modification but with high potential in EOR, FLOPERM 2000.

3.2.7   SCREENING CRITERIA FOR POLYMER APPLICATION

As most commercially available polymers suitable for EOR application are anionic in nature, sandstones are preferred over carbonates when considering polymer injection. Anionic polymers have a high viscosifying power and very high molecular weights and are cheap to produce by opposition

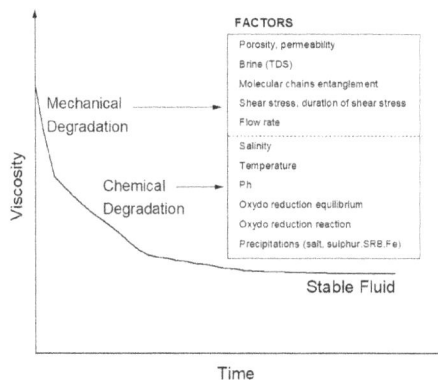

FIGURE 3.22   Apparent viscosity versus degradation of polymer.

to synthetic cationic polymers that are expensive to produce and highly shear sensitive and display lower molecular weights on average. For sandstone and clayey reservoirs, which are negatively charged, the injection of anionic macromolecules is obviously preferred to limit ionic interactions. The selection of a reservoir suitable for polymer flooding depends on the temperature, pressure, and properties of reservoir rock and fluids. Generally, the reservoirs with good waterflooding recovery are also suitable for polymer flooding. Some criteria described by different authors are summarized in Table 3.2.

3.2.8 DESIGN AND IMPLEMENTATION PLANNING OF POLYMER FLOOD

The design and implementation planning of a polymer flood for EOR consists of the following steps (Thomas, 2016):

- Preliminary screening: It includes gathering of reservoir description, including temperature, salinity, and permeability. There is a database available to identify the likely EOR processes in the candidate reservoir. The database is generated with the information of a large number of EOR cases implemented in different countries of the world. Based on the thorough study, potential polymer candidates are selected.
- Preliminary laboratory test: For polymer flooding, it is of the utmost importance to test the rheological behavior of polymer solutions at different environments of pressure, temperature, salinity, etc., to select the appropriate polymer. Then, based on the reservoir properties, the target concentration and viscosity are determined. The compatibility of the polymer solutions with other chemicals is checked. A basic simulation study at this stage is necessary. Based on the laboratory data, it is also important to check the economic feasibility of the process.
- Detailed laboratory tests and equipment design: At this stage, it is important to test the long-term stability of the polymer solutions. Oil recovery with the injection of polymers by core flooding should be performed to check the technical feasibility. Other activities include improved reservoir simulation, field test, equipment design, and feasibility study.

TABLE 3.2
General Screening Guide for Polymer Flooding (Saleh et al., 2014)

Author	Published Year	Gravity °API	Oil Viscosity, cp	Oil Saturation, %	Average Permeability, md	Temperature, °F	Depth, ft	SPE Paper No.
Brashear and Kuuskraa	1978	15	< 200	50	> 20	< 200	NC	6,350
Carcoana	1982	–	50–80	> 50	> 50	< 180	< 6,562	10,699
Goodlett et al.	1986	> 25	< 100	> 10	> 20	< 200	< 9,000	15,172
Taber et al.	1997	> 15	10–100	> 50	> 10	< 200	< 9,000	35,385
Al-Bahar et al.	2004	-	< 150	> 60	> 50	< 158	-	88,716
Dickson et al.	2010	> 15	10–1,000	> 30	> 100 if ($10 < \mu < 100$) > 1,000 if ($100 < \mu < 1,000$)	< 170	800–9,000	129,768
Aladasani and Bai	2010	13–42.5	0.4–4,000	34–82	1.8–5,500	74–237.2	700–9,460	130,726

- Field test and pilot: It includes the testing of polymer mixing facilities; assessing injectivity, maximum rates, and viscosity; evaluation of oil recovery; assessing back-produced water treatment; and updating the reservoir model and economic analysis.
- Full-field deployment: This is the final application stage. The activities include dissolution and deployment of injection facilities, logistics, monitoring and surveillance plans, and updating the reservoir model.
- Breakthrough time of injected polymer: Considering the typical distance of 200m between the injector and producer and the velocity of polymer through the formation ~0.0275m/h, the breakthrough time will be 303.3 days.

3.3 SURFACTANT FLOODING

The main purpose of the EOR process is to decrease the oil saturation within the rock strata. The oil saturation decreases until the oil production becomes discontinuous or the capillary forces become exceedingly high to restrict oil movement. It is necessary to have a clear understanding of static and dynamic rock/fluid behavior in subsurface petroleum systems on various scales (from microscopic/ pore scale to field scale) to improve hydrocarbon recovery. Surfactant flooding is a proven technique to produce a significant quantity of trapped/bypassed residual oil. Surfactant flooding improves pore-scale displacement efficiency through the mechanism of interfacial tension reduction, wettability alteration, or a combination of both and emulsification. Surfactants are amphiphilic organic compounds comprising hydrophobic as well as hydrophilic components in the same structure. Each surfactant possesses unique structural and rheological properties suitable for application in varying reservoir conditions.

3.3.1 CHEMISTRY OF SURFACTANTS

The application of surfactants for recovery of trapped oil was introduced in the early 1900s (Negin et al., 2017). Since then, significant research in the development and understanding of surfactant EOR has been carried out, and reducing IFT to an ultralow value and altering the wettability of a reservoir rock surface to, preferably, a water-wet condition have been regarded as the main mechanisms of oil recovery by surfactant flooding. However, these mechanisms are affected by various factors such as structure of surfactant, type of reservoir rock, salinity of reservoir fluid, and temperature of reservoir. Structurally, the surfactants are categorized as nonionic, anionic, cationic, and zwitterionic surfactants, depending on the charge of the hydrophilic head group of the surfactants. Nonionic surfactants have a neutral charge, and their surface behavior is the result of ethylene oxide or propylene oxide groups present in their molecular structure. Anionic surfactants have a negatively charged hydrophilic head group such as carbonate, sulfate and sulfonate ions, whereas cationic surfactants have positively charged hydrophilic head groups such as ammonium and imidazolium ions. Zwitterionic surfactants are a specific type of surfactants containing both positively and negatively charged groups in their hydrophilic head. Nonionic surfactants have a cloud point temperature, above which their solubility decreases and the surfactant molecules start to phase out, causing the solution to become cloudy (Karnanda and Benzagouta, 2013). However, ionic surfactants have a Krafft temperature, below which the surfactant molecules precipitate out of solution. Thus, reservoir temperature is an important parameter that is considered for the selection of surfactants for chemical EOR, as a high temperature also decreases the time required to reach equilibrium IFT (Negin et al., 2017). The IFT reduction capability of a surfactant is also affected by the salt content of the reservoir fluid. Higher salinity of reservoir can have an unfavorable effect on the surfactant's properties; thus, maintaining optimal salinity in the reservoir has been suggested to obtain maximum oil recovery (Chou and Shah, 1981). Optimal salinity is defined as the salinity at which solubilization of both oil and water is equal in the microemulsion, leading to achievement of minimum IFT (Hirasaki et al., 2008). Furthermore, the compatibility of surfactant to reservoir rock type is significantly important, as similar charge of rock surface and surfactant head group can lead to significant adsorption loss, leading to ineffectiveness of the injected surfactant

TABLE 3.3

Categories of Surfactants

Category	Charge on Hydrophilic Head	Functional Groups	Examples
Nonionic	No charge	Ethoxylated alcohol/alkylphenol and polysorbate	Tween 20, Span 40, Brij 35
Anionic	Negative	Sulfate, sulfonate, phosphate, carboxylate, sulfosuccinate, isethionate, taurate, and sarcosinate	Sodium dodecyl sulfate, sodium stearate, sodium dodecylbenzene sufonate
Cationic	Positive	Quaternary ammonium, pyridinium, and guanidium	Benzyldodecyldimethylammonium bromide, cetylpyridinium chloride, cetrimonium bromide
Zwitterionic	Positive and negative	Carboxybetaine, sulfobetaine, and hydroxysulfobetaine	Cocamidopropyl betaine, lauryl sultaine, lauryl hydroxysultaine

slug. Thus, the type of surfactant is an important selection criterion for a particular type of reservoir. The categories and examples of surfactants are shown in Table 3.3.

3.3.1.1 Nonionic Surfactants

Nonionic surfactants are neutrally charged amphiphiles that have ethylene oxide (EO) and propylene oxide (PO) molecular groups imparting their surface-active properties. The EO group act as a hydrophilic group, whereas the PO group leads to the hydrophobic properties of the surfactant. The different positions of EO and PO and their respective number in the surfactant structure have a significant effect on the critical micelle concentration (CMC), viscosity, IFT, and salt tolerance of the surfactant solution (Hirasaki et al., 2008; Levitt et al., 2012). Nonionic surfactants also include ethoxylates of alcohol, alkylphenol, and polysorbates. These nonionic surfactants have a lower CMC in comparison to ionic surfactants and possess consistent surface properties in the presence of salt (Muherei and Junin, 2008). Thus, nonionic surfactants have better applicability in high-salinity reservoirs, but their dominating oil recovery mechanism is wettability alteration (Negin et al., 2017). The lower adsorption of nonionic surfactants has also led to the use of mixed surfactant systems of nonionic and anionic surfactants for oil recovery from reservoirs with positively charged rock surfaces (Wang and Kwak, 1999). A mixed surfactant system of anionic-nonionic and cationic-nonionic surfactants are also applied as a hydrophilic group of nonionic surfactants inserts itself between the hydrophilic head groups of ionic surfactants, leading to a decrease in repulsion between the polar head groups and reduction in the precipitation of surfactants in aqueous solution. However, the reduction of IFT to the desired values are not achievable by nonionic surfactants, and their low cloud point leads to limited applicability of nonionic surfactants as an efficient chemical EOR agent. Some examples of nonionic surfactants are shown in Figure 3.23.

3.3.1.2 Anionic Surfactants

Anionic surfactants are compounds that have negatively charged ions such as, sulfate, sulfonate, phosphate, carboxylate, sulfosuccinate, isethionate, taurate, and sarcosinate ion as the hydrophilic head group in their molecular structure. These surfactants are used the most widely due to their lower cost in comparison to other classes of surfactants. These surfactants have low solubility in water at low temperatures and have high sensitivity to the presence of divalent ions. The presence of divalent ions leads to the formation of a surfactant divalent cation complex that has low solubility in the aqueous phase and causes the precipitation of the surfactant in aqueous phase with divalent ions (Olajire, 2014). However, sulfonate surfactants have been found to have stability in high-saline aqueous phase with a high amount of divalent ions, but they have low effectiveness in high-temperature reservoirs (Shupe and Maddox, 1978). Alkylbenzene sulfonates are the most commercially used anionic surfactant, and

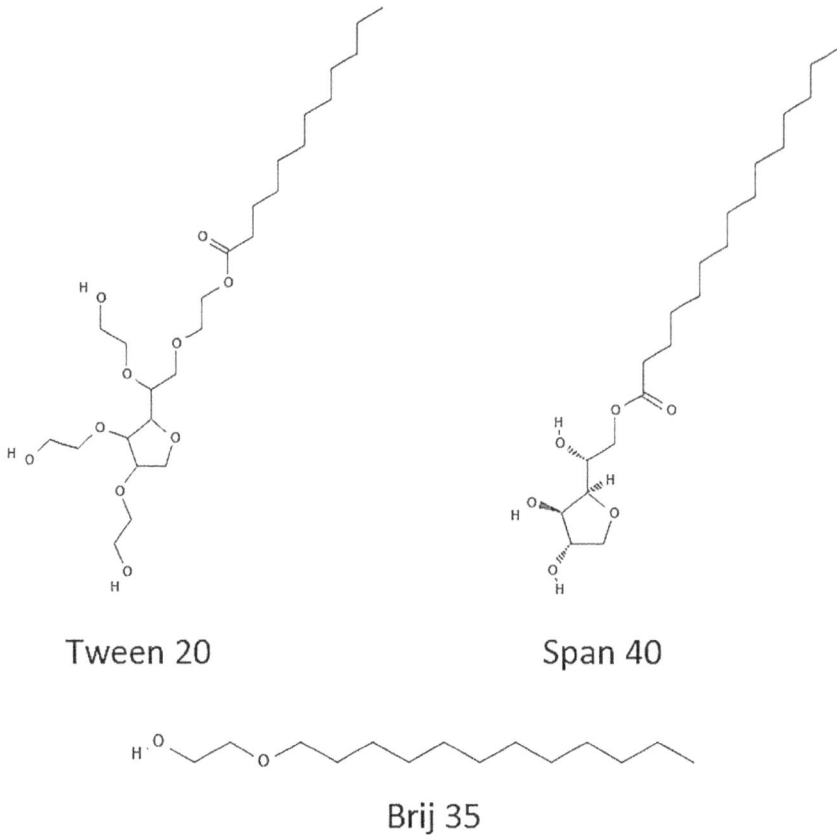

Tween 20   Span 40

Brij 35

FIGURE 3.23 Nonionic surfactants.

they have slow biodegradation in comparison to linear alkylbenzene sulfonates. Petroleum sulfonates produce low IFT, but its aqueous solution forms a precipitate at salinity of 2.5% NaCl (Bansal et al., 1980). Adding an EO group with the sulfonate group increases its applicability in high-salinity and high-temperature reservoirs. Carboxylate surfactants also have better applicability at high temperatures and high salinity, along with having a higher resistance to surfactant loss due to adsorption on the rock surface. However, carboxylate surfactants have poor solubility in the presence of multivalent cations, such as calcium and magnesium. Salt tolerance of surfactant slug is also improved by using a mixture of anionic and nonionic surfactants. Anionic surfactants are applied in sandstone reservoirs, as they have low adsorption on negatively charged clay in comparison to cationic surfactants. However, anionic surfactants cannot be used in carbonate reservoirs due to the excess loss of surfactant due to adsorption. Most surfactants applied in EOR are alcohol propoxylate sulfonate and alcohol propxylate sulfate. Some examples of anionic surfactants are shown in Figure 3.24.

3.3.1.3 Cationic Surfactants

Cationic surfactants are amphiphiles with a positively charged hydrophilic head. The positively charged polar head group of cationic surfactants is mostly a quaternary ammonium ion (Hayes and Smith, 2019). Other positively charged head groups that are present in cationic surfactants are pyridinium and guanidium cations (Miyake and Yamashita, 2017). Cationic surfactants have excellent antibacterial properties and significant applications as antiseptics. Other applications of cationic surfactants include their use as foam depressants, textile softeners, and flotation chemicals. They have lower solubility in high pH solutions due to the formation of uncharged insoluble compounds. In

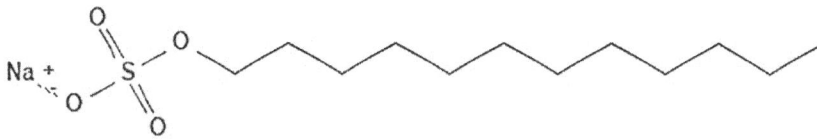

Sodium dodecyl sulfate

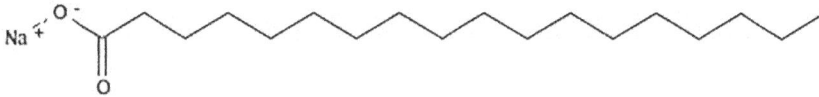

Sodium stearate

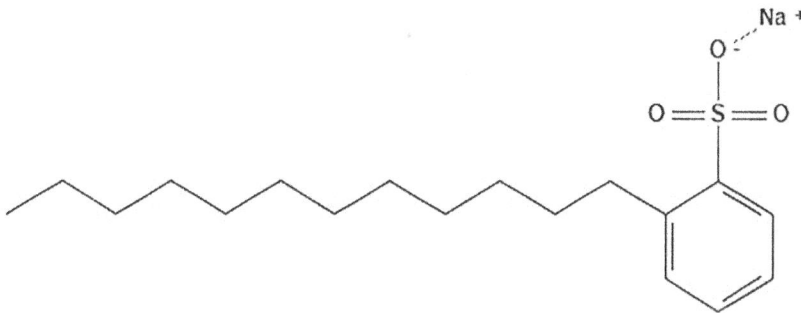

Sodium dodecylbenzene sulfonate

FIGURE 3.24 Anionic surfactants.

EOR applications, cationic surfactants are most widely used in carbonate reservoirs, and their main mechanism is wettability alteration of carbonate reservoirs. Cationic surfactants interact with the carbonate surface and the strong and irreversible ion-pair formation led to a higher recovery of oil in comparison to oil recovery by anionic surfactant (Standnes and Austad, 2000). Cationic surfactants also have low loss due to adsorption on carbonate reservoirs due to the presence of a repulsive electrostatic force; however, the presence of negatively charged impurities in carbonate rock can cause significant loss of cationic surfactants while surfactant flooding. Cationic surfactants also cannot be applied in sandstone reservoirs due to significant higher surfactant adsorption in comparison to anionic surfactants. Increase of pH also increases the loss of cationic surfactant on sandstone reservoirs. Some examples of cationic surfactants are shown in Figure 3.25.

3.3.1.4 Zwitterionic Surfactants

Zwitterionic surfactants have electrically neutral surfactants as their hydrophilic head group, which possesses both positively and negatively charged moieties. The cationic moiety can be quaternary ammonium or imidazolium ion, and the anionic moiety can be phosphate, carboxylate, sulfonate, or sulfate ion. The mildness of the zwitterionic surfactant has led to its significant application in cosmetics and household products (Chu et al., 2010). The properties of zwitterionic surfactants are dependent on the charge separation between the polar groups in their hydrophilic head. The increase in separation between the charged molecular groups of the zwitterionic surfactant leads to

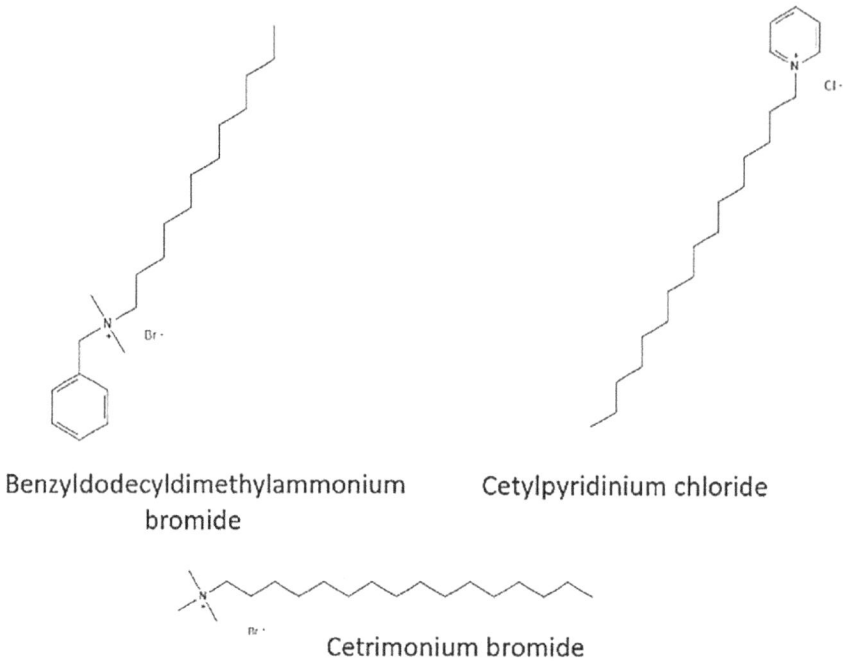

Benzyldodecyldimethylammonium   Cetylpyridinium chloride
bromide

Cetrimonium bromide

FIGURE 3.25 Cationic surfactants.

increase in area per molecule and the CMC of the surfactant with the formation of smaller micelles (Chevalier et al., 1992). The solubility and Krafft temperature of ionic surfactants depend on their counterions; however, for the zwitterionic surfactant, the molecular head group plays a vital role in the solubility of zwitterionic surfactants in aqueous media. Carboxybetaine-based zwitterionic surfactants have better solubility in comparison to sulfobetaine-based zwitterionic surfactants, due to the higher hydrophilicity of the carboxylate ion than the sulfate group. Similarly, imidazolium ions have lower solubility than ammonium ions; thus, carboxybetaines are the most suitable zwitterionic surfactants (Tondo et al., 2010). Zwitterionic surfactants have excellent compatibility with anionic and cationic surfactants (McLachlan and Marangoni, 2006). These surfactants have better electrolyte tolerance, temperature resistance, wettability, and foaming performance. Zwitterionic surfactants can also form wormlike micelles that imparts viscoelasticity to surfactant solutions. Some examples of zwitterionic surfactants are shown in Figure 3.26.

3.3.1.5 Gemini Surfactants

Surfactant design and molecular architecture are pivotal parameters that can greatly influence a chemical's potential in EOR. Menger and Littau (1991, 1993) developed an efficient class of surfactants, referred to as gemini (or dimeric) surfactants. Unlike conventional monomeric surfactants that consist of a hydrophilic head and a hydrophobic tail, the gemini surfactant has a certain unique structural attribute with two amphiphilic components attached with a spacer chain (Fig. 3.27). The spacers may be short or long methylene groups with rigid/flexible entities. Therefore, the presence of two heads and tails renders the gemini surfactant with enhanced hydrophilicity/hydrophobicity. Due to the ability of the gemini surfactant molecules to self-aggregate at low concentrations, their CMC values are measured to be lower than that of their corresponding monomeric counterparts (Zana, 2002). Gemini surfactants offer an ultralow IFT and exhibit favorable tolerance to high temperature and salinity conditions (Sharma and Gao, 2014). The retention of

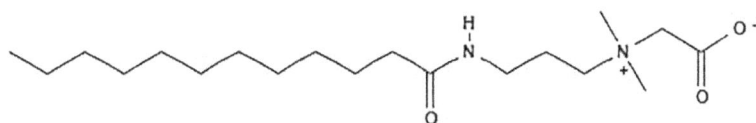

Cocamidopropyl betaine

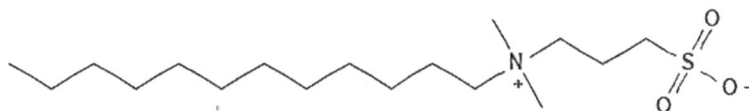

Lauryl sultaine

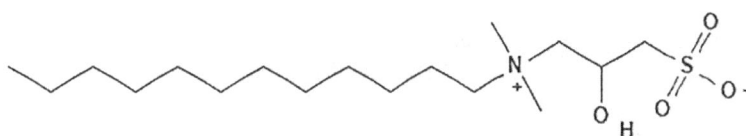

Lauryl hydroxysultaine

FIGURE 3.26 Zwitterionic surfactants.

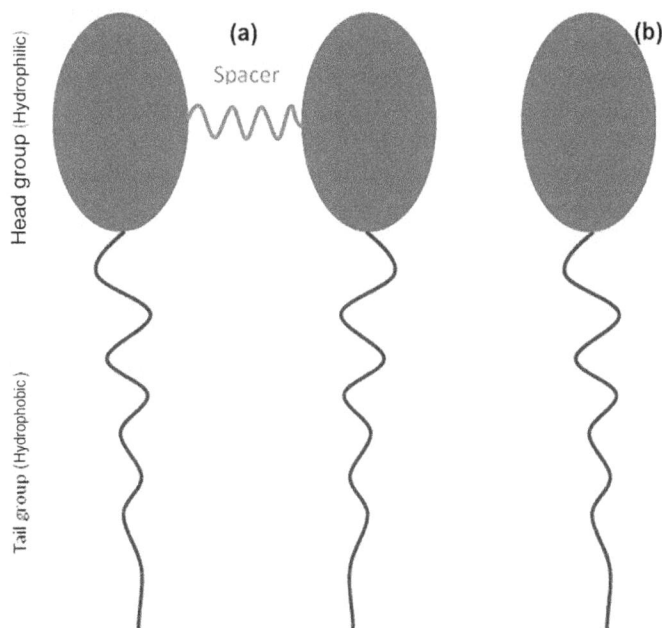

FIGURE 3.27 Basic molecular structure of (a) a gemini surfactant and (b) a monomeric surfactant.

GS molecules onto rock surfaces is lesser as compared to that of monomeric units. Gemini surfactants have been shown to achieve enhanced interfacial and wetting properties, responsible for improved microscopic sweeping of crude oil. Therefore, these surfactants may be introduced in existing oil fields with significant deposits of tertiary hydrocarbons, wherein standard EOR agents are not economically profitable.

3.3.1.6   Bio-based Surfactants

Natural surfactants have gained attention in the petroleum industry as an alternative to nonbiodegradable surfactants (Negin et al., 2017). "Green" surfactants are derived from natural oils, resins, amino acids, and a wide variety of other plant extracts. One of the first attempts to introduce natural surfactants into oil recovery was made by an investigation on a *Zyziphus spina-christi* extract by Pordel Shahri et al. (2012). Thereafter, many research teams have successfully worked on the design, formulation, and application of surfactants from vegetable oils and other bio-resources (Elraies et al., 2010; Babu et al., 2015; Pal et al., 2018). Mulberry leaf-based surfactant act as an effective stimulator in petroleum reservoirs due to low oil-aqueous IFT (Ahmadi et al., 2014). Nowadays, waste cooking oil is employed as a viable source of surfactant for oil recovery applications (Zhang et al., 2015; Rabiu et al., 2016). Another group of bio-based surfactants is derived from natural bacteria such as *Bacillus* sp. (Surfactin, Rhamnolipid, Lichenysin); Acinetobacter sp. (Emulsan, Alasan); *Pseudomonas* (Rhamnolipid) (Sen, 2008; Zhang et al., 2015). This biotechnological technique imparts specific metabolic events that favor oil recovery. Torres et al. (2011) studied the relative efficacies of natural surfactants of vegetal and bacterial origins. Natural surfactants, in most cases, provide similar and even better results than commercial synthetic surfactants. These fluids are characterized by their biodegradability, low environmental footprint, and cost-profitability.

3.3.1.7   Ionic Liquids

Ionic liquids are a relatively new family of chemical EOR agents, which if applied properly, can be a potent alternative to traditional surfactants (Fig. 3.28). Generally referred to as organic salts, ionic liquids have a melting point below 373 K and often exist in liquid state at ambient temperature (Xiao and Malhotra, 2005; Domanska, 2005; Simoni et al., 2008). The molecular compositions of ionic liquids include organic cations such as N-dialkylimidazolium, N-alkylpyridinium, alkylammonium, alkyl-phosphonium, alkylsulfonium and tiazolium; inorganic anions such as halide (e.g., Br^-, Cl^-), tetrachloroaluminate $[AlCl_4]^-$, tetrafluoroborate $[BF_4]^-$, hexafluorophosphate $[PF_6]^-$, bis(trifluoro-methyl-sulfonyl)imide $[(CF_3SO_2)2N^-]$ and acetate $[CH_3CO_2^-]$; and organic anions such as alkylsulfate $[R\text{-}O\text{-}SO_3^-]$, tosylate $[C_7H_7O_3S]$ and methanesulfonate $[R_3C\text{-}S\text{-}O_3^-]$.

Designation	Structure	Molecular weight (g/mol)	Density ($\rho/g \cdot cm^{-3}$)
1-methyl-3-octylimidazolium tetrafluoroborate (C_8mimBF_4)	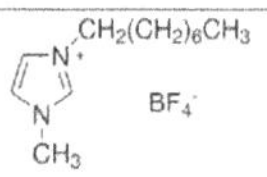	282.129	1.09
1-methyl-3-decylimidazolium tetrafluoroborate ($C_{10}mimBF_4$)	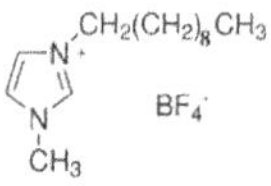	310.18	1.05
1-methyl-3-dodecylimidazolium tetrafluoroborate ($C_{12}mimBF_4$)	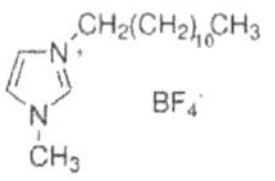	338.38	0.99

FIGURE 3.28   Ionic liquids.

In addition to negligible vapor pressure and nonflammability under ambient conditions, ionic liquids possess high values of polarity, density, heat capacity, and thermal conductivity; and low degrees of volatility and thermal stability (Domanska, 2005; Johnson et al., 2007). Ionic liquids attain better interfacial tension, wettability, and salt tolerance properties over organic surfactants, and hence, it is considered an important chemical for enhanced oil recovery (Hezave et al., 2013; Somasundaran and Zhang, 2006).

3.3.1.8 Polymeric Surfactants

Nowadays, the field of polymeric surfactants is being explored by petroleum researchers as a promising EOR source, though problems associated with cost and technical favorabilities exist. Significant strides in the field of polymeric surfactants were made by a number of researchers (Ezell and McCormick, 2007; Elraies et al., 2011; Raffa et al., 2015; and Pal et al., 2016). The basic idea behind the application of polymeric surfactants in EOR is to graft hydrophobic groups of surfactants into a water-soluble polymer backbone. Wever et al. (2011) and Taylor and Nasr-El-Din (1998) developed several polymer structures based on hydrophobically modified polyacrylamides and polysaccharides. The rheology of SP fluids is influenced by the presence of hydrophobic groups and the resultant strength of grafting between associating SP molecules. Though the overall effect of polymeric surfactants may be less pronounced as compared to surfactants, this approach may be considered when polymer systems are assessed for use in oil recovery systems. The structure of a typical polymeric surfactant is reported in Figure 3.29.

3.3.2 CRITERIA FOR SURFACTANT SELECTION

Most of the chemical flooding processes use petroleum sulfonates or mixtures of them with property-modifying additives. The choice of surfactant(s), which has been surveyed by various authors (Iglauer et al., 2010; Jamaloei et al., 2011), is made based on several criteria that may be classified into two groups: performance criteria versus economic aspect.

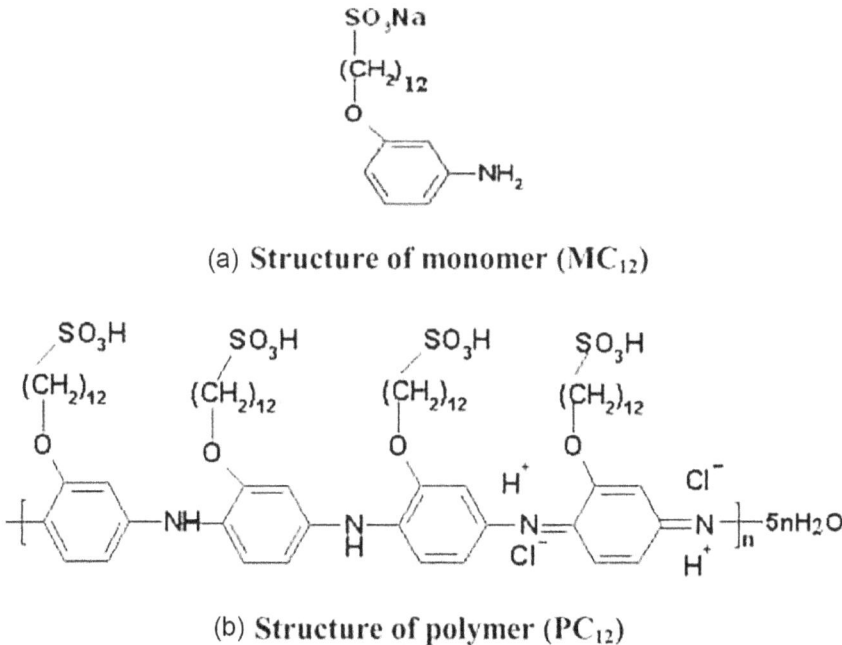

(a) Structure of monomer (MC$_{12}$)

(b) Structure of polymer (PC$_{12}$)

FIGURE 3.29 Structure of monomeric and polymeric surfactants.

3.3.2.1 Performance Criteria

A chemical flooding surfactant formulation must exhibit the following properties:

1. Ability of the surfactant to reduce the surface tension of its solution and IFT between oil and surfactant solution system.
2. Hydrophilic-lipophilic balance of surfactants: The hydrophilic-lipophilic balance (HLB), often used to describe surfactants, is calculated from the weight percentage of the hydrophilic groups to the hydrophobic groups in a molecule, with values ranging from 1 to 20 (Kralova and Sjöblom, 2009). The HLB value of a surfactant should match the HLB value of the oil phase based on the notion of "like dissolves like." In the range of 3.5–6.0, surfactants are more suitable for use in W/O emulsions. Surfactants with HLB values in the 8–18 range are most commonly used in O/W emulsions (Griffin, 1949).
3. The CMC, thermodynamic and surface-active properties, and solubilization parameters.
4. Good displacement efficiency—i.e., it must produce a low interfacial tension against crude oil.
5. Low adsorption on reservoir rocks and clays to reduce surfactant losses.
6. Good compatibility with reservoir fluids, especially tolerance to divalent cations such as Ca^{2+} and Mg^{2+}.

3.3.2.2 Economical Aspects

The current research effort may soon lead to formulations having excellent performances. However, economic considerations might restrict their use. Economic criteria include:

1. The availability of raw materials may be a problem, as enormous quantities of surfactants and additives will be required for extensive application of chemical flooding.
2. Efficient manufacturing and low cost of production of surfactants.

Careful planning is required for the application of chemical EOR because of the shortage of useful raw materials in the market. Owing to the gap between supply and demand, the raw material price may hike abruptly, thereby affecting the economics of the project.

3.3.3 CHARACTERIZATION OF SURFACTANTS

It is necessary to characterize surfactants for technical and economic reasons. Before using a surfactant in any application, it is important to ensure its effectiveness and stability over its period of use. In this section, the general characterization and thermodynamic parameters of the surfactant solution are discussed.

3.3.3.1 Surface Tension and Critical Micelle Concentration

Surfactants display distinct behavior when interacting with water. The polar part of the molecule seeks to interact with water while the nonpolar part shuns interaction with water. There are two ways in which such a molecule achieves both these states. An amphiphilic molecule can arrange itself at the surface of the water such that the polar part interacts with the water and the nonpolar part is held above the surface, either in the air or in a nonpolar liquid. The presence of these molecules on the surface disrupts the cohesive energy at the surface and thus lowers the surface tension. These molecules can allow each component to interact with its favored environment. Molecules can form aggregates in which the hydrophobic portions are oriented within the cluster and the hydrophilic portions are exposed to the solvent. Such aggregates are called micelles. The proportion of molecules present at the surface or as micelles in the bulk of the liquid depends on the concentration of the amphiphile. At low concentrations, surfactants will favor arrangement on the surface. As the surface becomes crowded with surfactant, more molecules will arrange into micelles. At some concentration, the surface becomes completely loaded with surfactant, and any further additions must arrange as micelles. This concentration is called CMC. Figure 3.30 shows the schematic diagram of micelle formation with increasing surfactant concentration. It follows that measurement of surface tension and conductance may be used to find CMC. At very low concentrations of surfactant, only a slight change in surface tension is detected. The additional surfactant decreases surface tension, and as the surface becomes fully saturated, no further change in surface tension is observed.

The surface tension measurements impart the amount of cohesive forces acting at the air-liquid interface (Paul and Moulik, 2015). The interfacial tension-concentration isotherm curves (IFCV) give the CMC values of the surfactant solutions, which are crucial in adsorption behavior study (Fig. 3.31a, b, c). A decrease in surface tension values with increase in surfactant concentration is observed due to adsorption of surfactant molecules at the air-liquid interface. The surface tension values become constant for a wide range of surfactant concentrations when surfactant molecules start to saturate at the interface and form micelles. The minimum surfactant concentration above which the values of surface tension tend to cease is taken as the CMC value of the surfactant. The surface tension values decrease with increase of temperature due to the lowering of the hydrophilic nature and increase in molecular activity of the surfactant molecules at the air-liquid interface. Initially, the CMC value of each surfactant decreases with rise in temperature.

3.3.3.2 Surface-Active Parameters

The CMC values obtained through IFCV curves are vital for determination of the surface-active parameters at the air-liquid interface. The Gibbs adsorption equation gives the amount of surfactant adsorbed per unit area at the interface. The surfactant adsorbed at the interface reduces the surface tension at equilibrium condition per Gibbs law. Table 3.4 presents the values of surface-active parameters for surfactants at varying temperatures.

3.3.3.3 Surface Excess Concentration (Γ_{max})

Γ_{max} is the measure of effectiveness of surfactant adsorption at the air-liquid or liquid-liquid interface and defines the change in the interface by surfactant adsorption. It is the maximum amount

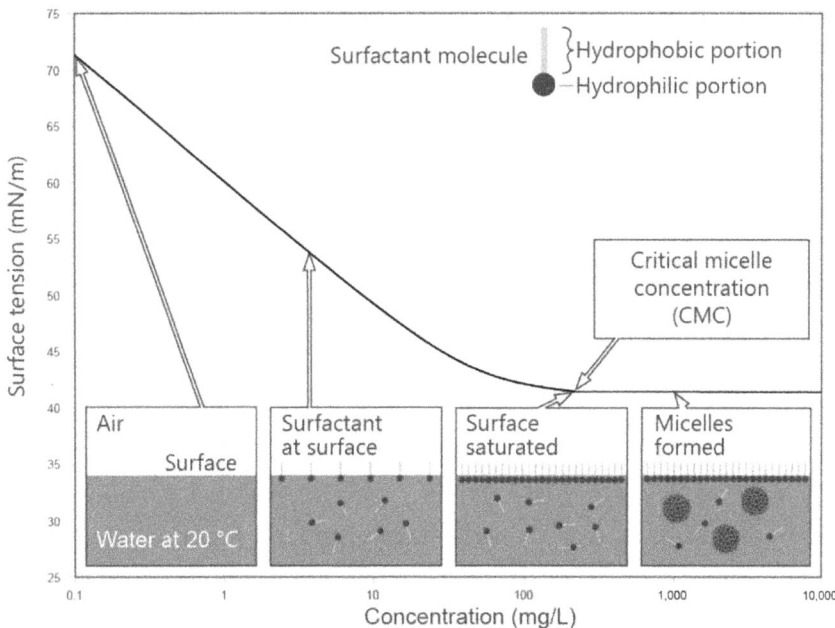

FIGURE 3.30 Formation of micelle as surfactant concentration increases.

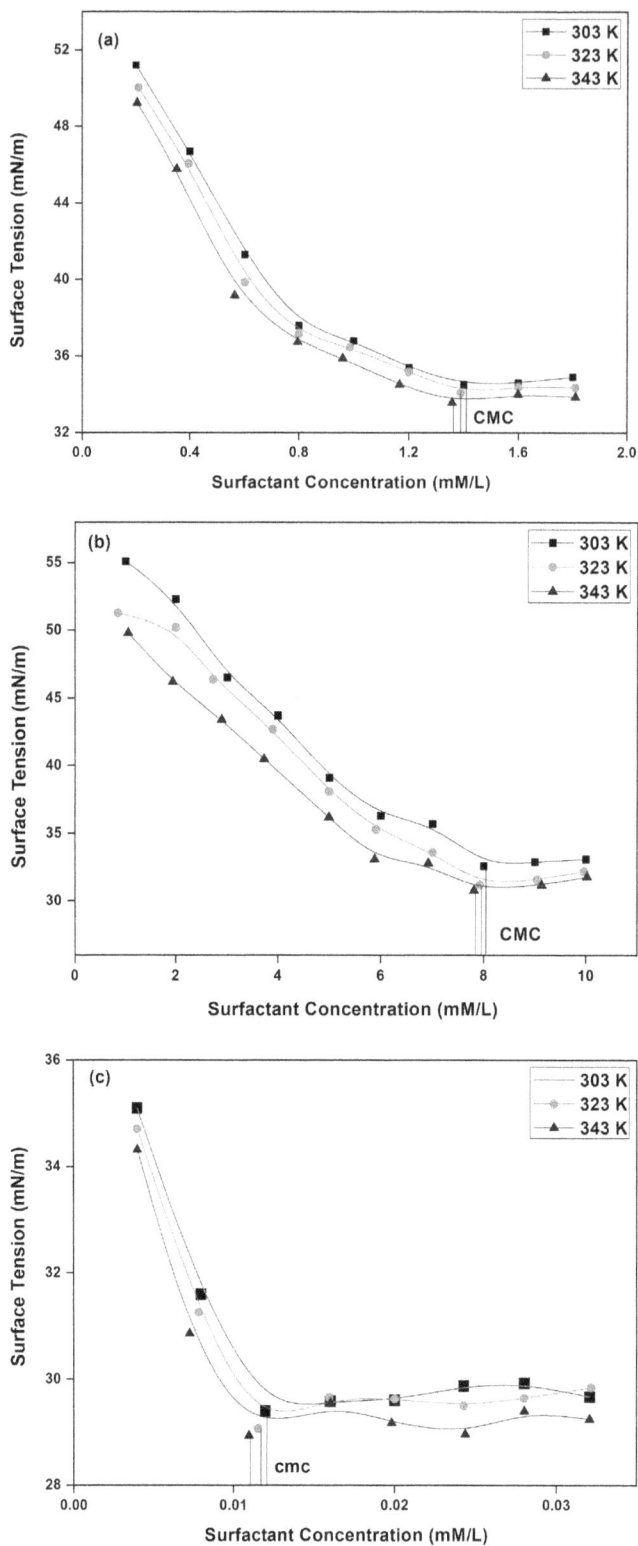

FIGURE 3.31 IFCV curves for surfactants: (a) Cetyltrimethyl ammonium bromide (CTAB), (b) Sodium dodecyl sulfate (SDS), and (c) Tween 80, at 303, 323, and 343 K.

Source: Kumar, 2017.

TABLE 3.4

Surface-Active Properties of CTAB, SDS, and Tween 80 Solutions at 303, 323, and 343 K (Kumar, 2017)

Surfactant	Temp. (K)	CMC (mmol/l)	γ_{CMC} (mN/m)	Γ_{max} (x10^6 mol./m^2)	A_{min} (nm^2/ molecule)	Π_{CMC} (mN/m)	pC_{20}	CMC/C_{20}
CTAB	303	1.40	34.50	2.451	6.773	37.50	0.698	7
	323	1.38	34.07	2.860	5.803	37.92	0.721	7.305
	343	1.35	33.57	2.907	5.711	38.42	0.744	7.533
SDS	303	8.00	32.60	3.205	5.180	39.40	0.045	8.888
	323	7.92	31.20	3.484	4.764	40.80	0.070	9.322
	343	7.80	30.80	3.855	4.306	41.20	0.136	10.696
Tween 80	303	0.012	29.40	7.978	20.809	42.60	3.036	13.043
	323	0.011	29.07	8.032	20.067	42.92	3.060	13.252
	343	0.010	28.93	9.089	18.267	43.06	3.119	14.421

up to which surfactant adsorption can be obtained depending upon the molecular structure of the interacting components. The degree of adsorption is calculated by the Gibbs isotherm equation, Equation 3.14:

$$\left(\Gamma_{max} \right) = -\left(\frac{1}{RT} \right)\left(\frac{\partial \gamma}{\partial \ln C} \right)$$

(3.14)

where Γ_{max} is the surface excess concentration (mol/m^2), T is absolute temperature (K), R is universal gas constant ($R = 8.314$ J mol^{-1} K^{-1}, and $\frac{\partial \gamma}{\partial \ln C}$ is surface activity. Since, surface excess concentration (Γ_{max}) is the measure of the amount of surfactant adsorbed at the interface as calculated by Equation 3.14, the larger the value Γ_{max}, the lower the interfacial tension.

A_{min}, Π_{CMC}, and pC_{20} are important parameters that help in the determination of the adsorption ability of the surfactant, as it cannot be decided alone by the surface excess concentration (Γ_{max}) parameter. The packing density and orientation of the adsorbed surfactant molecules at the interface, as derived from area per molecule, are considered while determining their surface activity.

3.3.3.4 Minimum Area per Molecule (A_{min})

The minimum area per molecule in m^2/molecule is known as A_{min}. It determines the average area occupied by each adsorbed surfactant molecule at the air-liquid interface at surface saturation and expresses the surface activity of surfactants (Rosen, 1978) and is given by Equation 3.15:

$$A_{min} = 10^{18} / \left(N_A \times \Gamma_{max} \right)$$

(3.15)

where N_A is Avogadro's number (6.023×10^{23} molecule/mol). The molecular arrangement of surfactant molecules at the interface is reflected by the A_{min} and Γ_{max} values obtained from the Gibbs isotherm equation.

3.3.3.5 Effectiveness of Surfactants (Π_{CMC})

The difference between the surface tension values of the prepared surfactant solutions at their CMC and that of pure water is known as their effectiveness (Π_{CMC}). The surface pressure or the effectiveness of surface tension reduction of the surfactant is calculated by Equation 3.16. Π_{CMC} defines the minimum surface tension value and, hence, signifies the capability of the surfactant in lowering the surface tension of the solution.

$$\Pi_{CMC} = \gamma_o - \gamma_{\text{CMC}} \tag{3.16}$$

where γ_o and γ_{CMC} are the surface tension values for a pure solvent and solutions at CMC, respectively, at the air-liquid interface. A large value of Π_{CMC} or lesser value of γ_{CMC} denotes a better ability of the surfactant to reduce the surface tension of solutions. Π_{CMC} defines the ability of the surfactant to reduce the surface tension of the solution. With a rise in temperature, there is an increase in Π_{CMC} value (Table 3.4). This signifies the enhanced capability of the surfactant in surface tension reduction with increase in temperature.

3.3.3.6 Adsorption Efficiency (pC_{20})

The negative log of the surfactant concentration that reduces the surface tension of the pure solvent by 20 mN/m gives the adsorption efficiency. The adsorption efficiency is given by Equation 3.17:

$$pC_{20} = -\log C_{20} \tag{3.17}$$

where C is surfactant molar concentration and C_{20} is amount of surfactant required for surface tension reduction of pure solvent by 20 mN/m; that means C_{20} is the minimum surfactant concentration that denotes the saturation of the surface adsorption. Therefore, C_{20} is the measure of efficiency of the surfactant molecules' adsorption at the air-water interface (Rosen, 1989). The greater the pC_{20} value, the higher is the surfactant adsorption efficiency.

3.3.3.7 Occurrence of Adsorption or Micellization (CMC/C_{20})

To investigate the effect of structural factors on the adsorption and micellization process, the CMC/C_{20} ratio is determined. The CMC/C_{20} value defines the occurrence of either the adsorption process or the micellization process. A high value of CMC/C_{20} ratio denotes the spontaneity of the adsorption process than micellization and vice versa. The CMC/C_{20} ratio increases with a rise in temperature, indicating the adsorption process is favored with a drop in surface tension values (Table 3.4). Thus, an examination of the CMC/C_{20} ratio imparts useful knowledge about the adsorption and micellization phenomenon.

3.3.3.8 Thermodynamics of Micellization and Adsorption

The potential of the micellization process depends upon the thermodynamic parameter, Gibbs free energy of micellization (ΔG_{mic}). The pseudo-phase separation model is used to determine the thermodynamic parameter of ionic and nonionic micelles. The CMC values of surfactants play an important role in the calculation of ΔG_{mic}. The Gibbs free energy of micellization of nonionic surfactants is approximated by Equation 3.18:

$$\Delta G_{micelle} = RT \times \ln(cmc) \tag{3.18}$$

where ΔG_{mic} is the molar Gibbs energy of micellization in kJ/mol; cmc is the CMC (cmc = CMC/55.4) in molar fraction unit; since CMC is in mol/L, a factor of 55.4 is used for water as 1 L of water accounts for 55.4 mol of water at 298 K. CMC gives the exact surfactant concentration that yields the aggregate to be thermodynamically soluble. In ionic (cationic and anionic) surfactants, the ionic micelles of the surfactant monomers are fully ionized; therefore an understanding of counterion binding by the micelles is required to comprehend the aggregation and micellization of ionic solutions. The average degree of counterion binding to the micelle (β) is determined by Equation 3.19:

$$\beta = 1 - \alpha \tag{3.19}$$

where α is the degree of ionization of the micelle, determined from the ratio of S_2/S_1. S_1 and S_2 are the slopes below and above the CMC, obtained from the IFCV curves. For nonionic surfactants $\alpha = 0$.

Usually, ΔG_{mic} for ionic surfactant is approximated when micelles turn to microscopic phase and is given by Equation 3.20:

$$\Delta G_{micelle} = (1+\beta)RT \times \ln(cmc) \tag{3.20}$$

The Gibbs free energy change due to adsorption (ΔG_{ads}), is calculated from the adsorption data and is given by Equation 3.21:

$$\Delta G_{adsorption} = \Delta G_{micelle} - [\pi_{CMC} / \Gamma_{max}] A_{min} \tag{3.21}$$

The $\Delta G_{adsorption}$ and $\Delta G_{micelle}$ values indicate the tendency of the surfactant to adsorb at the air-liquid interface and to form micelles in solution. If $\Delta G_{adsorption}$ is observed to be higher in magnitude, then adsorption at the air-liquid interface is more spontaneous than micellization in an aqueous system and vice versa. Negative values of ΔG_{mic} and ΔG_{ads} denote the formation of micelles in the solution and spontaneous adsorption of surfactant at the air-liquid interface, respectively. A high value of ΔG_{ads} indicates better adsorption than micellization.

Table 3.5 presents the values of Gibbs free energy of micellization (ΔG_{mic}) and Gibbs free energy change due to adsorption (ΔG_{ads}) for ionic and nonionic surfactants at different temperatures determined from the Gibbs isotherm equation. The negative values obtained for ΔG_{mic} and ΔG_{ads} signify the occurrence of both phenomena: the adsorption of surfactant molecules at the air-liquid interface and the micelle formation in the bulk phase. The negative value of ΔG_{mic} indicates the aggregation of surfactant molecules. With a rise in temperature, the value of the Gibbs aggregation energy becomes more negative. The high magnitude of ΔG_{ads} than ΔG_{mic} for each surfactant denotes the spontaneous manifestation of the adsorption process than micellization. Thus, surfactant molecules impulsively adsorb at the air-liquid interface because of the increment in their curvature. The adsorption phenomenon is also supported by high magnitudes of Π_{cmc} and pC_{20}.

3.3.3.9   Krafft and Cloud Points

The Krafft point indicates the temperature at which micelles are first formed in aqueous solution. The surface tension values initially decrease with increase in temperature but eventually reach a

---

TABLE 3.5

Micellization and Adsorption Parameters of CTAB, SDS, and Tween 80 at 303, 323, and 343 K (Kumar, 2017)

Surfactant	Temp. (K)	CMC (mmol/l)	Degree of Ionization (α)	Binding Degree (β)	ΔG_{mic} (kJ/mol)	ΔG_{ads} (kJ/mol)
CTAB	303	1.408	0.40	0.60	−42.64	−57.94
	323	1.384	0.41	0.59	−45.25	−58.51
	343	1.353	0.42	0.58	−47.85	−61.07
SDS	303	8.000	0.41	0.59	−35.42	−47.71
	323	7.921	0.44	0.56	−37.09	−48.80
	343	7.801	0.45	0.55	−39.20	−49.89
Tween 80	303	0.012	0	1.0	−77.31	−82.65
	323	0.011	0	1	−82.88	−88.29
	343	0.010	0	1.0	−88.56	−93.29

critical point. This indicates that the minimum temperature at which micelles are formed initially increases with concentration, up to a specified point. Beyond this critical point referred to as Krafft temperature, tension values are observed to remain constant or increase. This is attributed to a dramatic increase in the solubility of the surfactant in the aqueous phase because of micelle formation (Gaudin et al., 2018). With increase in surfactant concentration, the Krafft temperature is generally found to increase due to decreased surfactant solubilization ability in the aqueous phase. Studies on changes in cloud temperature are pivotal at controlling the solubilization properties of the surfactant. To measure the cloud point of a surfactant, the solution is heated with a gradually increasing temperature in step and allowed to equilibrate at each temperature for five minutes. The temperature values at which turbidity appears as well as disappears is noted for each sample solution. The average value of the temperatures indicating the initial appearance and subsequent disappearance of the cloudy state of aqueous system is identified as cloud point (Batıgöç and Akbaş, 2017; Gaudin et al., 2018). Figure 3.32 depicts the variation of Krafft point and cloud point with the concentration of a typical Gemini surfactant.

3.3.3.10 Hydrophile-Lipophile Balance

An important criterion for surfactant characterization is the HLB. This criterion measures the extent to which a surfactant is lipophilic or hydrophilic (Reham et al., 2015). The HLB is a number on a scale from 0 to 20 that indicates relatively the tendency of a surfactant to dissolve in oil or water (Massarweh and Abushaikha, 2020). A value of 0 indicates that the surfactant is a completely hydrophobic (lipophilic) molecule, while a value of 20 indicates that the surfactant molecule is composed entirely of hydrophilic components. Table 3.6 depicts the property or application of surfactants based on their HLB values. For microemulsion-based EOR at low salinity, a low-HLB surfactant needs to be selected. Whereas, a high-HLB surfactant is better for high-salinity formations (Sheng, 2010).

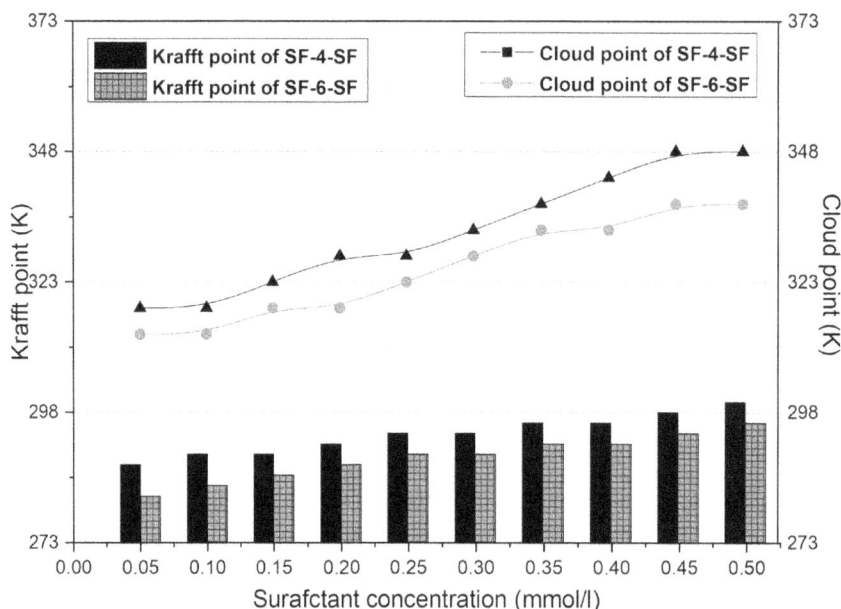

FIGURE 3.32 Krafft temperature and cloud temperature values SF-s-SF GSs with varying concentrations.
Source: Pal, 2020.

TABLE 3.6
Relation between HLB Value and the Expected
Properties/Applications of Surfactants (Sheng, 2010)

HLB Value	Property/Application
0–3	Antifoaming agent
4–6	W/O (water in oil) emulsifier
7–9	Wetting agent
8–18	O/W emulsifier
13–15	detergent
10–18	Hydrotrope or solubilizer

Royer et al. (2018) stated Griffin's equation to calculate the HLB for nonionic ethoxylated surfactants as follows:

$$HLB_{Griffin} = \frac{1}{5}\left(\frac{M_H}{M_T} \times 100 \right) \tag{3.22}$$

where M_H denotes the molecular mass of the hydrophilic part of the surfactant molecule and M_T denotes the total molecular mass of the surfactant molecule. Davies (1957) found that HLB values for surfactants can be calculated from their chemical formulas based on group numbers, as shown in the following equation:

$$HLB_{Davies} = 7 + \sum \left(hydrophilic\ group\ numbers \right) - \sum \left(lipophilic\ group\ numbers \right) \tag{3.23}$$

Later, other researchers developed experimental methods to determine HLB values. Some of these methods are based on the emulsion inversion point (Marszall, 1978), the phase inversion temperature (Kunieda and Ishikawa, 1985), and others.

3.3.4 Mechanisms of Oil Recovery by Surfactant Flooding

The oil recovery potential by the surfactant flooding process is best explained by capillary number calculations. Besides, capillary desaturation curves (CDC) may be employed to express the ratio of residual oil (after waterflood) to that after surfactant flooding as a dependent function of the capillary number. The plot between residual oil saturation and capillary number is termed as the capillary desaturation curve (Fig. 3.33). Note that the CDC curve depends on many parameters such as rock pore structure and rock wettability, among others. The capillary number (N_c) in an oil field is determined with Equation 3.24:

$$N_c = \left[\frac{\mu V}{\sigma} \right] \tag{3.24}$$

For an example, if an injected fluid having a viscosity 10^{-2} Pa-S and IFT of 10^{-2} mN/m moves with a velocity of 7.5×10^{-6} m/s through the formation, then the value of N_c will be ~0.0075. Surfactant selection is dependent on the properties of the crude oil, brine, reservoir, and other factors (temperature, pressure, compatibility, and stability). A successful EOR operation must possess sufficient driving power to mobilize the trapped oil and retain fluid stability during the flooding period. With proper knowledge about oil displacement and reservoir properties, cost-effective flooding technology can be applied with favorable efficacy. Mechanisms associated with surfactant EOR are IFT reduction, wettability alteration, and emulsification. Figure 3.34 illustrates the different types of mechanisms associated with surfactant-induced flooding processes.

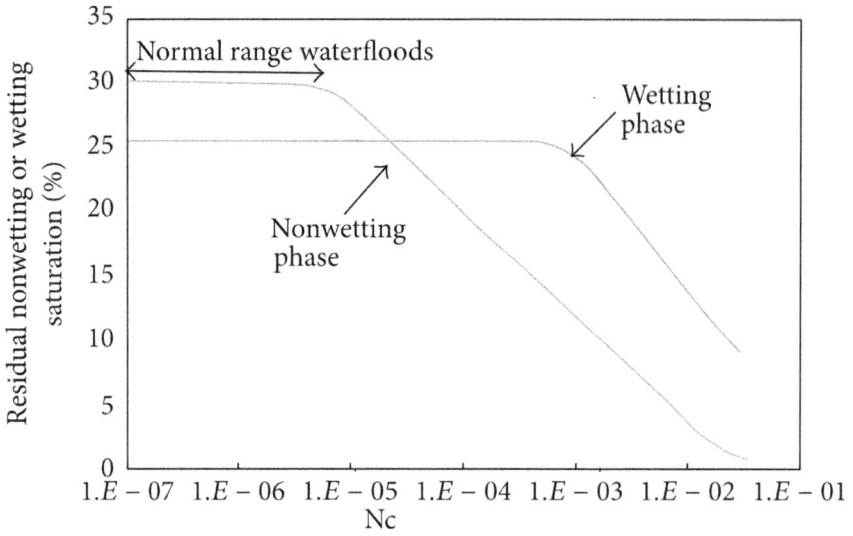

FIGURE 3.33   Schematic of wettability-affected CDC.

Source: Guo et al., 2017a.

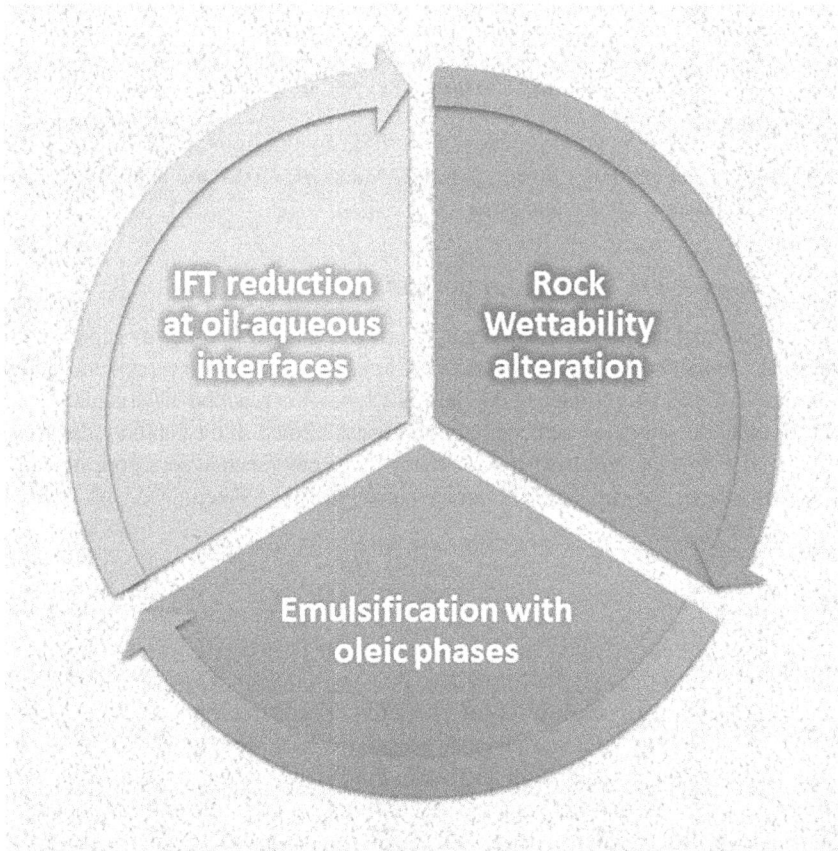

FIGURE 3.34   Different flooding mechanisms during surfactant EOR.

3.3.4.1 Reduction of Oil-Water IFT

A significant amount of oil globules remains trapped or bypassed within pore throats at the end of waterflooding. The capillary number at this stage is observed in the range of 10^{-7} to 10^{-8}, thereby making it difficult to displace all the oil, due to its eventual entrapment by capillary forces (Melrose and Brandner, 1974). Surfactant flooding causes the IFT to decrease, which is desirable for crude oil displacement. With increasing concentration of surfactant, the IFT initially decreases until the CMC value is reached. Thereafter, the IFT increases slightly or remains steady with further increase in concentration. The presence of surfactants increases the value of N_c to the order of 10^{-2}. The reduction of IFT is a practical method of increasing the capillary number (N_c) in rock pores. The hydrophilic head groups of surfactants interact with water molecules, whereas the nonpolar tails interact with crude oil components to form an adsorbed film. This film is responsible for the weakening of capillary forces withholding the trapped/bypassed oil and causing the oleic phases to flow through pore spaces. Figure 3.35 demonstrates how the oil droplet passes through pore throat restrictions at high and low IFT. In the absence of any surfactant, the oil-water interfacial tension remains high, and hence, the trapped oil is quite rigid and cannot pass through the fine pore under the applied viscous force. However, in the presence of a surfactant as it adsorbs at the oil-water interface, the IFT value is reduced to an ultralow level. Under this condition, the oil droplet can be easily deformed under the applied viscous force and passes through the fine throats if the pressure is maintained, which leads to enhancement of oil displacement.

3.3.4.1.1 Effect of Different Surfactants on IFT between Crude Oil and Water

The initial distribution of crude oil in a reservoir is dependent on the shape and size of the pores, the interfacial tension between the oleic and aqueous phase, and the wettability of the reservoir rock surface (Pal, 2020). These factors lead to the capillary forces that act on the oil drop in a reservoir. Thus, the movement of oil through the pores is possible only when the capillary forces holding the drop in the pores are overcome either by the reservoir pressure or by the pressure of the injected flood. The high IFT between the trapped crude oil and the brine injected during the waterflooding process leads to a lower recovery of crude oil from the reservoir. The additional recovery of the residual oil is achieved by the reduction of the IFT of the oil-water interface by the use of surfactants in the injected flood. Figures 3.36 (a)–(d) depict that for each surfactant, the IFT value initially decreases and reaches a minimum value and then increases up to a certain concentration of surfactant and finally attain a constant trend. At a lower concentration range, the surfactant molecules preferably adsorb at the interface between oil and

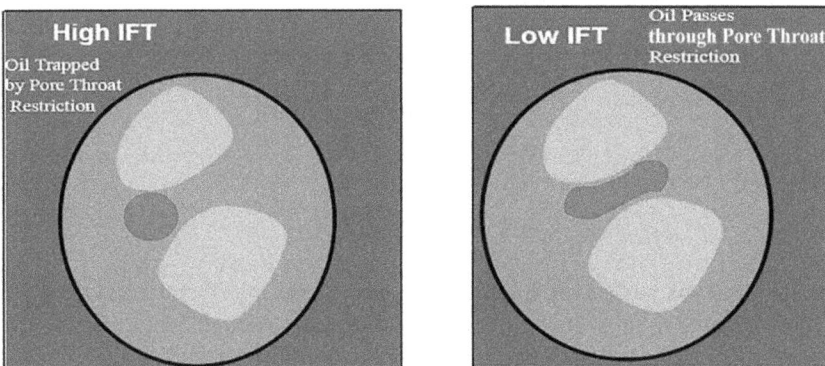

FIGURE 3.35 Oil drop passes through pore throat restriction at high IFT and low IFT.

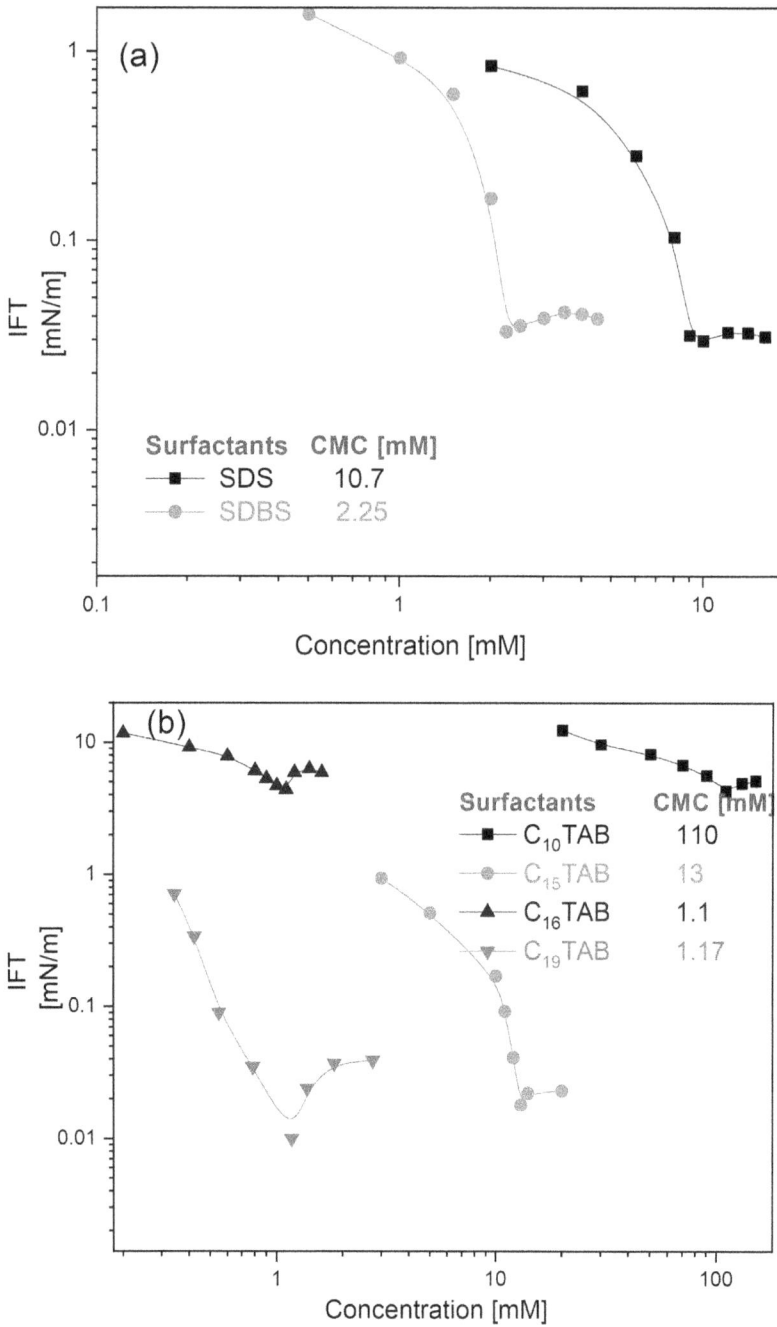

FIGURE 3.36 *(Continued)*

water from the aqueous surfactant solution, and hence, the interfacial tension of the solution starts to decrease rapidly. Subsequently, the IFT values pass through a minimum as the interface gets saturated with surfactant monomers. This concentration of surfactant corresponding to the minimum IFT is the CMC. Upon reaching the CMC, any further addition of surfactant just

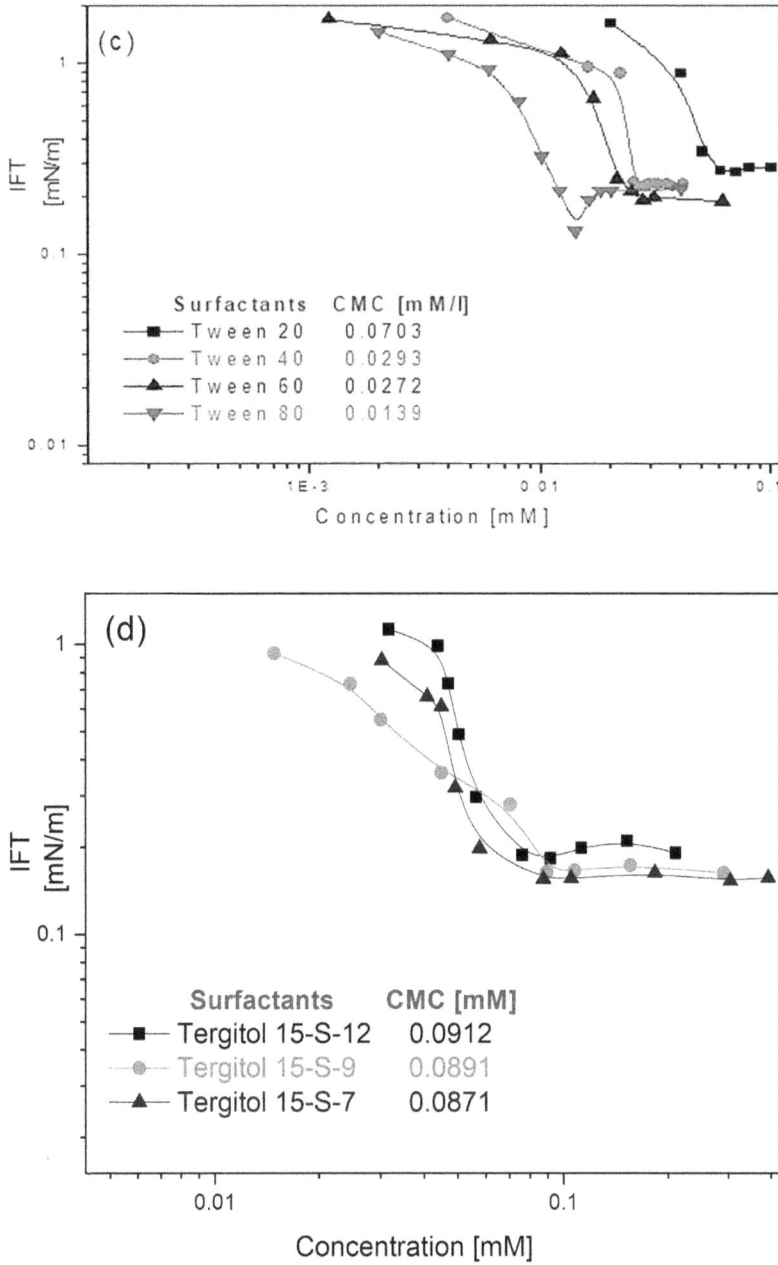

FIGURE 3.36   Interfacial tension of different surfactants: (a) anionic, (b) cationic, (c) tween series, and (d) tergitol series at their varying concentrations at 25°C.

Source: Kumar, 2017.

changes the structure/size of micelles and increases the number of micelles; hence, the effective surfactant concentration reduces a little due to the solubilization of micelles in the aqueous phase and reduction of surfactant molecule at the interface, which leads to an increase in the IFT (Yu et al., 2011).

3.3.4.1.2   Effect of Salinity on IFT between Different Surfactant Solutions and Crude Oil

The interaction between crude oil and brine plays an important role in the measurement of interfacial tension. As in situ surface-active agent formation takes place in the presence of brine from crude oil, therefore, with increasing time, the accumulation of surface-active agent at the crude oil–water interface increases. This phenomenon helps to decrease the IFT of the system. But when the formation of the surface-active agent will be in equilibrium, then no further decrease of IFT takes place; rather, it increases due to the desorption of surface-active agents from the crude oil–water interface. Generally, for most of the surfactants, the IFT first decreases and then increases after passing through a minimum value. The salinity corresponding to the minimum IFT is called optimum salinity.

3.3.4.1.3   IFT between Surfactant Solution and Crude Oil in the Presence of Different Alkalis

Several mechanisms have been critically reviewed to explain the role of alkalis in AS flooding, including reduced surfactant adsorption, balancing of surfactant adsorption by forming new soap and surfactant during the reaction inside the reservoir, low IFT, wettability reversal, emulsification, and entrapment (Samanta et al., 2011; Mohammadi et al., 2009; Pei et al., 2012). The interfacial properties depend upon the rate at which the natural acid migrates from the crude oil to the interface and interacts with the alkaline solution. It is believed that the amount of surface-active species mainly depends upon the acid value of crude oil and basicity of the alkaline solution.

The IFT between oil and water remarkably decreases in the presence of alkalis. At a lower concentration of the alkali, more IFT reduction is observed unlike at higher concentrations. This dramatic behavior is observed for all surfactant-alkali systems, as reported by Kumar (2017), since an optimum alkali concentration can provide a suitable level of -OH at the oil-water interface, which reacts with petroleum acid to produce enough in situ surfactants to decrease the IFT. The effects of different alkalis on the IFT of SDS surfactants in the presence of optimum salinity of 4 wt. % are shown in Figure 3.37. Among these three alkalis, Na_2CO_3 is a better choice due to its favorable properties in EOR (Chan and Yen, 1982; Hirasaki and Zhang, 2004). Diethanolamine (DEA) can be used as a replacement of inorganic alkalis due to its minimum IFT value as well as its organic characteristics, which minimizes scale formation.

3.3.4.1.4   Effect of Temperature on IFT of Crude Oil–Water System

The IFT between two immiscible liquids is also influenced by temperature. The effect of temperature on the IFT between pure water and crude oil has been investigated by many authors, and the results show the linear variation of IFT with temperature (Ye et al., 2008; Flock et al., 1986). In case of pure liquid systems, the IFT decreases nearly linearly with temperature, which means that increasing temperature modifies the mutual solubility of both crude oil and water and favors the reduction of IFT. With increase in temperature, the in situ formation of the surface-active agent also increases, and the solubility of the produced surfactant molecules is increased with temperature; then the hydrophobic interactions among surfactant molecules become stronger, which facilitates the adsorption of surfactant molecules at the oil-water interface and makes the arrangement of surfactant molecules in the interfacial layer closer. Interfacial tensions in the presence of different surfactants at different temperatures are depicted in Figure 3.38. As temperature increases, the surfactant molecules move faster and also the viscosity of the crude oil decreases with temperature, which results in the enhanced migration of the produced surfactant to the interface and causes the reduction of IFT.

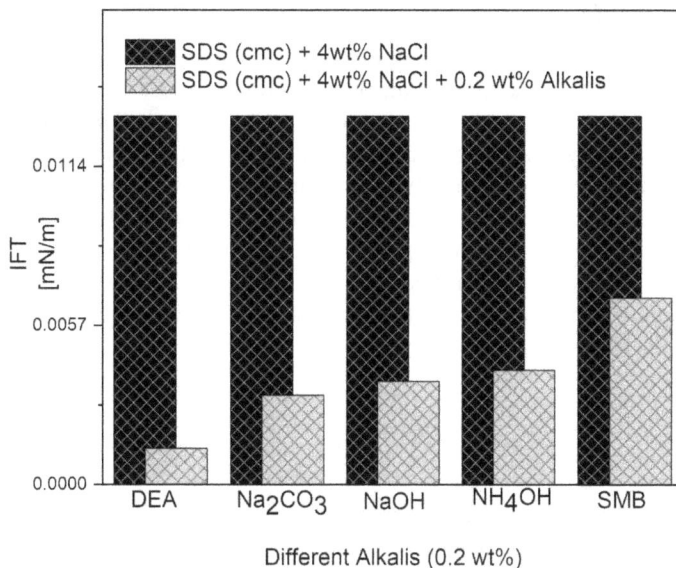

FIGURE 3.37 Variations of IFT between crude oil and surfactant (SDS) solutions (at CMC) at optimum salinity with varying concentrations of different alkalis at 25°C.

Source: Kumar, 2017.

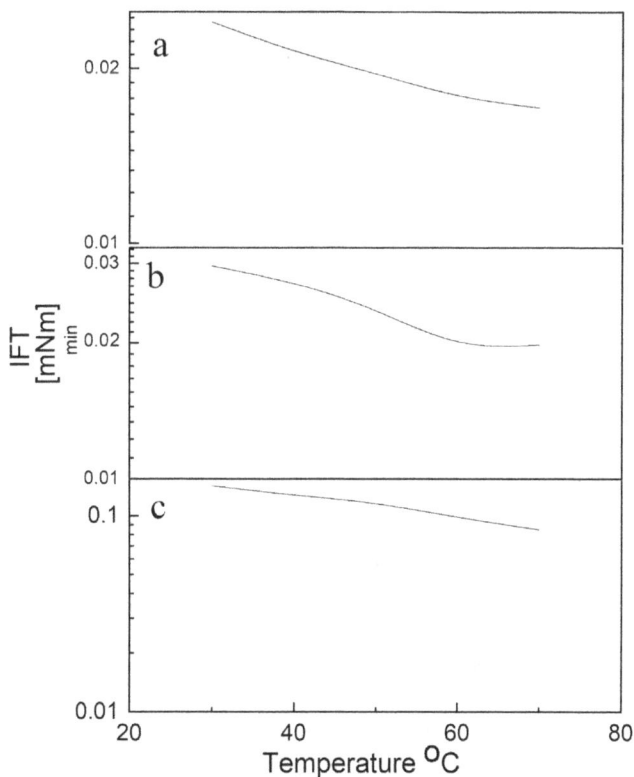

FIGURE 3.38 Effect of temperature on IFT of different surfactants: (a) CTAB, (b) SDS, and (c) Tween 80 at their CMC values.

Source: Kumar, 2017.

3.3.4.2 Rock Wettability Alteration

Wettability describes the preference of a solid to be in contact with one fluid rather than another. The wettability of a reservoir rock has a dominant effect on interface movement and associated displacement of oil through porous media. Changes of wettability from oil wet to water wet increases the oil recovery significantly. The wettability of a solid surface relates directly to the solid-fluid and fluid-fluid interactions. The interaction between two immiscible phases implies interfacial energy. Attraction between the substrates causes lower interfacial energy, and repulsion forces result in a higher energy surface. It is believed that the ions have the capability to change the reservoir rock wettability. At equilibrium, different interfacial forces on a liquid drop residing on a substrate can be mathematically expressed by Young's law (Young, 1805) as represented in Equation 3.25 (Rosen, 2004; Kwok et al., 1998).

$$\cos\theta_C = \frac{\sigma_{sw} - \sigma_{so}}{\sigma_{wo}} \qquad (3.25)$$

where θ_C is contact angle, and σ values indicate the interfacial tensions between solid-water (σ_{sw}), solid-oil (σ_{so}), and water-oil (σ_{wo}) interfaces. Young's equation is valid at equilibrium conditions for the ideal state of a perfectly smooth, chemically homogeneous, rigid, insoluble, and nonreactive surface. Figure 3.39 illustrates the situation behind the oil drop residing on a solid surface in the presence of another immiscible liquid such as water.

According to Craig (1971), wettability is defined as "the tendency of one fluid to spread on or adhere to a solid surface in presence of other immiscible fluids." Wettability of solid surfaces can be modified by introducing surface-active agents or surfactants and ionic substances into the solid-liquid systems and also by regulating thermodynamic parameters such as temperature (Adamson, 1990).

In petroleum reservoirs, the term "wettability alteration" usually refers to the process of restoring the original reservoir wettability, which is presumed to be water wet. The target of this restoration treatment is the unrecoverable oil by conventional waterflooding. The great role of reservoir wettability on primary oil recovery methods such as water drive was recognized by early research (Bobek et al., 1958). Secondary recovery by waterflooding is directly related to the wettability of the oil reservoir as well. Wagner and Leach (1959) stated that oil recovery during waterflooding for an oil-wet reservoir can be less by 15% in comparison to a water-wet reservoir. Reservoir rocks, as a whole, may not show the same wettability. It depends upon the reservoir rock properties, oil

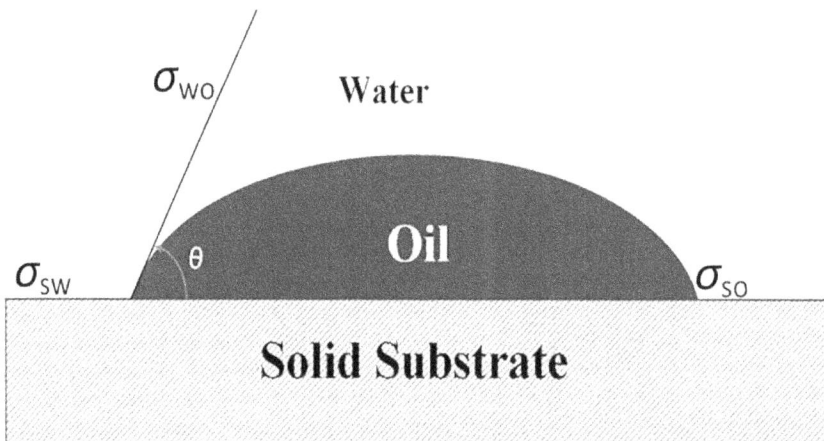

FIGURE 3.39 Illustration of contact angle in three-phase system on solid surface.

chemistry, and water salinity of the system. The rocks may be oil wet, water wet, or mixed wet. This wetting behavior depends upon the interaction between rock and crude oil. Most reservoirs, on the other hand, exhibit some degree of oil-wetness, and it is rare to find a strongly water-wet reservoir. If a reservoir has similar affinity to oil and water, the wettability is defined as neutral, and when some parts of the reservoir exhibit a different wettability than other parts, the term "mixed wet" is used (Salathiel, 1973). The wettability conditions of a rock-brine/oil system are depicted in Figure 3.40.

The wetting nature of a reservoir is usually intermediate to oil wet, and it is dependent on the rock mineralogy, composition of crude oil and brine, and temperature and pressure conditions of the reservoir. The mechanism of wettability alteration of the originally water-wet reservoir to oil wet has been related to the adhesion of the polar components of crude oil that had migrated into the reservoir. After the migration of crude oil into reservoir pores, there exists a layer of brine between the crude oil and rock surface, which prevents the contact of crude oil and rock surface. The rupture of the brine layer occurs due to the presence of attractive forces such as charge transfer, van der Waals forces, and hydrogen bonding between the crude oil and rock surface. Thus, these forces lead to adsorption of crude oil components on the rock surface and the oil-wet state of the reservoir rock. A schematic of wettability alteration of initially oil-wet rock in presence of surfactant is presented in Figure 3.4.1. When the surfactant is injected into the reservoir, the surfactant molecules interact with the adsorbed components of the crude oil like naphthenic acid, asphaltenes, and resins. These adsorbed components of the crude oil, which were responsible for the oil-wet state of the surface, can have either a hydrophilic interaction or a hydrophobic interaction with the injected surfactant molecules. Hydrophilic interaction leads to the formation of an ion-pair between the hydrophilic head of the surfactant and adsorbed crude oil. The bonding of the carbon chain of surfactant molecules with the adsorbed crude oil components leads to a hydrophobic interaction between the crude oil and surfactant molecules. With the movement of the flood front, the crude oil components bonded with the surfactant molecules are desorbed into the bulk of the aqueous phase, and the available sites get occupied by surfactant molecules. As per the DLVO theory, surfactant molecules are preferentially adsorbed onto rock surface along with water molecules to form a wetting-water film, thus altering the nature of rock to water-wet. The extra stability of the detached oil droplet by formation of micelles stabilized by surfactant in the bulk water phase further promotes the wettability alteration process. The adsorption of surfactant molecules leads to an increase in the zeta potential of the surface, as shown in Figure 3.42. This increase in the surface charge is due to the dominance of the negative charge by sulfonate ions and hydroxyl groups. Thus, surfactant molecules adsorbed on the rock surface prevent further interaction of the crude oil and rock surface and change the wettability of the rock to water wet.

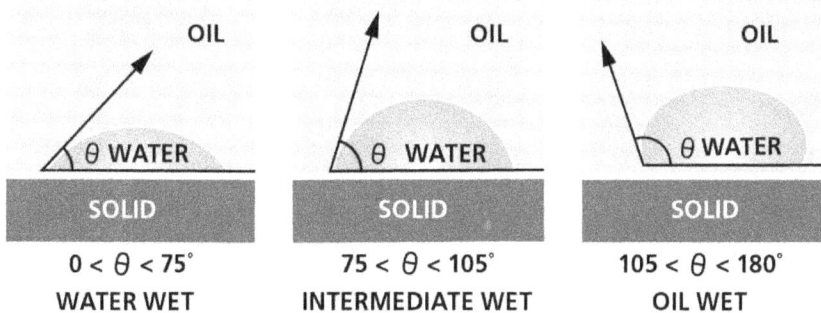

FIGURE 3.40 A schematic diagram of rock wettability conditions of a rock-brine/oil system.

3.3.4.2.1 Measurement of Contact Angle

The contact angle of the droplet of the surfactant solutions on the surface of the rock sample can be measured to determine the wettability of the surfactant solution to the surface of the sample. Contact angle can be measured by the sessile drop technique using different types of goniometers available in market. The oil-aged rock sample is placed in the temperature-humidity chamber to maintain the temperature and humidity of the environment near the sample. To measure contact angle, approximately, 5 μL of surfactant solution is dropped on the rock surface, and images of drop are taken for 15–30 min at the desired temperature for the liquid-air-solid system.

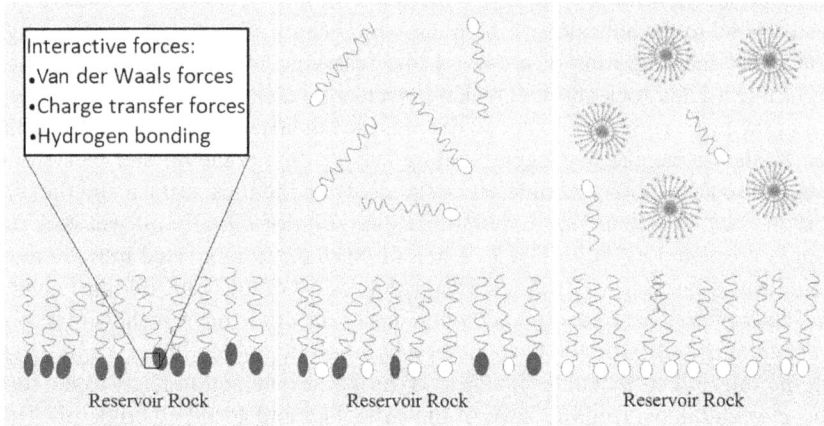

FIGURE 3.41 Schematic of preferential adsorption of surfactant on oil-aged sandstone rock with wettability alteration.

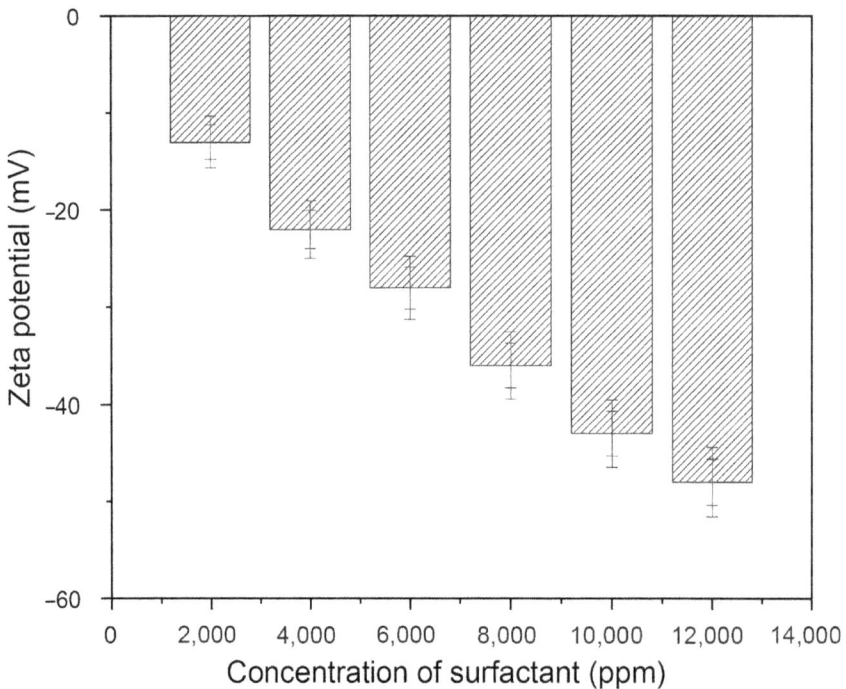

FIGURE 3.42 Change of zeta potential on adsorption of surfactant on oil-aged sandstone rock.

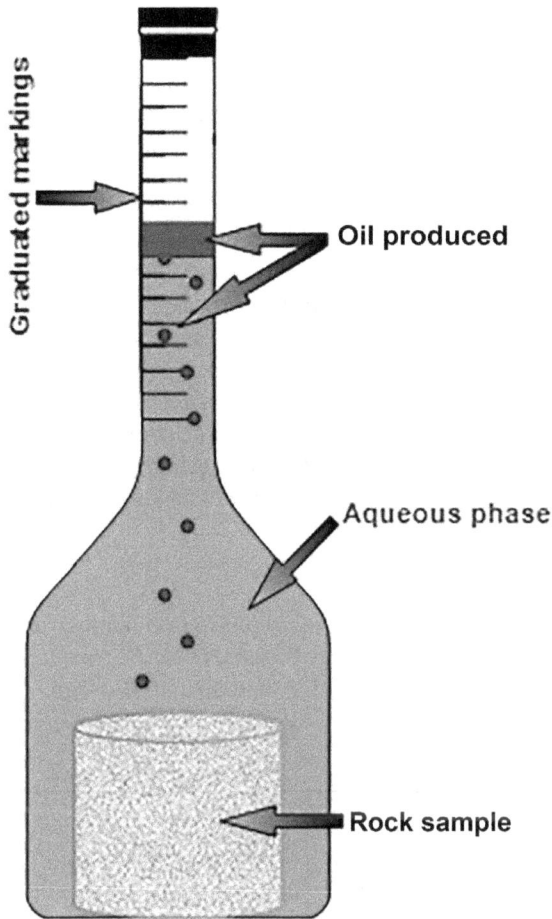

FIGURE 3.43   Schematic of Amott cell.

Spontaneous imbibition experiments are performed using sandstone and carbonate core samples to test the imbibing effectiveness of the surfactant solution. Prior to the imbibition experiment, the core samples are saturated with the brine solution. The drainage of brine from the saturated core samples is done by the flooding of crude oil, and the amount of brine drained is used for the calculation of the initial oil saturation of the respective core samples. The core samples are then placed in Amott cells surrounded by brine or surfactant solutions as imbibing fluids. The schematic of the Amott cell is shown in Figure 3.43. The oil produced is collected at the top of the aqueous solution, and the graduated markings are used to measure the collected oil. The oil produced due to imbibition is plotted as a function of time and compared to evaluate the wettability alteration effectiveness of the surfactant solution.

3.3.4.2.2   Effect of Different Chemicals on Contact Angle Variation in the Presence of Different Surfactants

Rather than the type of surfactants, the contact angle also depends upon the salinity, alkalinity, and viscosity of the chemical solution. The effect of salt, alkali, and polymer on contact angle reduction of surfactant solutions at their CMC on the oil-wet rock surface are depicted in Figure 3.44. It is observed that, in the presence of an optimum concentration of salt, the equilibrium contact angle is much lower than that of lone surfactant solutions at their CMC. The principal effect of NaCl is to partially screen the electrostatic repulsion between the head groups, and so spreading of the droplet results in lowering the contact angle (Bera et al., 2012). In the presence of NaCl, there is a compression of the electric double layer surrounding the ionic heads. The surface potential of the double layer is reduced, so the mutual repulsion is also reduced. A well-known fact is that surfactant solutions can change the wettability of an oil-wet rock toward being more water wet. However, the polymer effect on wettability alteration is not much as discussed in the literature. Injection of surfactant with polymer improves sweep efficiency as a result of improving the mobility ratio of the flood. Partially hydrolyzed polyacrylamide (PHPA) is used to control the mobility of the co-injected surfactant solutions to increase the sweep efficiency. The effects of PHPA on the reduction of contact angle by nonionic, anionic, and cationic surfactants are also shown in Figure 3.44. It is observed that, in the presence of a polymer in surfactant solution, the equilibrium contact angle increases marginally because the polymer adsorbs on the rock and reduces the net polarity of the rock-surfactant system. Alkalis and surfactants show a synergistic effect in producing ultralow IFT between oil and water. As the alkali interacts with the acidic components that are naturally present in crude oil, it forms a petroleum shop, which leads to a reduction in the IFT and the contact angle. The ultimate reduction of the contact angle in the presence of an alkali is higher than the corresponding pure surfactant solution.

3.3.4.3 Emulsification Mechanism

Emulsions are defined as dispersions of a liquid phase into another immiscible solvent phase. The proper choice of emulsifier is necessary to achieve long-term stability as well as rheological properties for effective mobility control. Alvarado and Marsden (1979) and Kumar (2013)

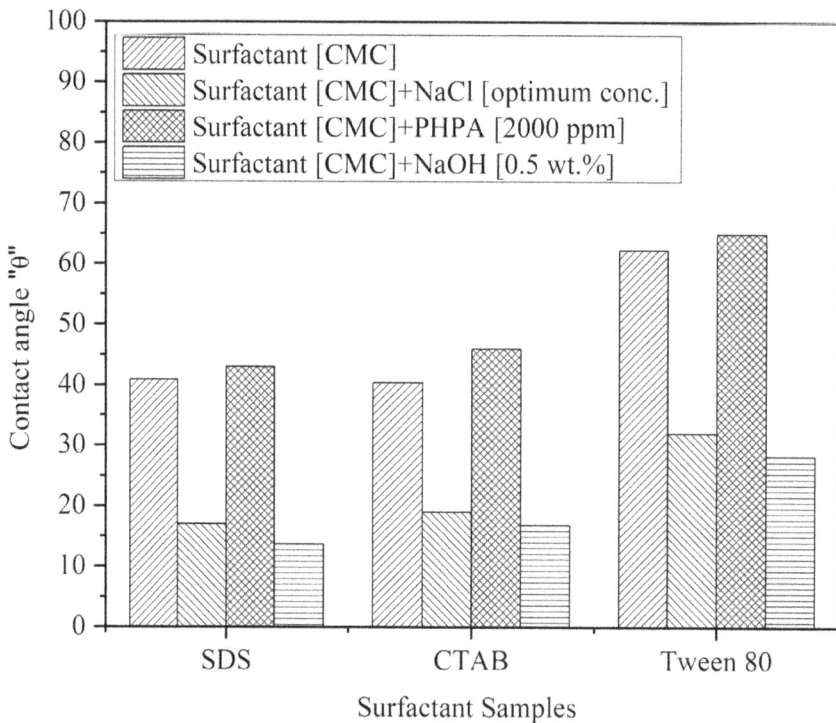

FIGURE 3.44 Contact angle variation in the presence of different additives in the presence of different surfactants.

Source: Kumar, 2017.

studied the role of emulsifier molecule(s) on interfacial and stabilization behavior against the detrimental effects of the impending coalescence of oil droplets. Factors affecting the in situ formation of emulsion include the types of immiscible liquids in contact, emulsifier components, and energy of the injection fluid phase. Surfactant molecules scale off oil from the rock surface to form emulsions, thereby reducing the oil viscosity and improving the mobility ratio (Wasan et al., 1978). The emulsions, so formed, are protected by electrostatic repulsion/steric effects imparted by surfactant entities. This reduces the possibility of oil droplet coalescence, as described by Perazzo et al. (2018). Phase-behavioral studies are important to understand the ability of displaced surfactant fluids to extract crude oil. The emulsification process is a function of surfactant concentration, type of oil, brine salinity, and rock adsorption behavior. The formation of emulsions yields optimal recovery at optimal salinities and, hence, must be properly controlled depending on the type of reservoir and fluid characteristics. A schematic of in situ emulsion formation during surfactant flooding in the oil reservoir is shown in Figure 3.45.

3.3.4.4 Foam Generation

The injection of gas along with surfactant solution can produce a foam that exhibits an apparent viscosity much higher than its constituent phases: liquid and gas. Because of its high apparent viscosity, foam has a mobility much lower than the mobility of gas alone and, therefore, provides an effective means of reducing mobility of the displacing fluid (Schramm and Wassmuth, 1994; Krumova et al., 2015). Foam can also reduce gravity override and viscous fingering caused by the low density and viscosity of the gas. Surface tension is one of the most important physical characteristics of the liquid phase, which should be lower for effective foaming. This can be achieved by adding different types of surfactants in the aqueous phase, which stabilizes the dispersions of gas in water to form metastable foam. Co-injection of surfactant solution and gas leads to the generation of foam inside the reservoir. Along with surfactants, the use of other chemicals such as salt, nanoparticles, alcohols, polymer, alkalis, etc. have some notable effects on foam behavior. Crude oil can be trapped and carried in the foam lamella due to very low IFT of the oil-water-gas interface, which consequently, increases the oil recovery as depicted in Figure 3.46.

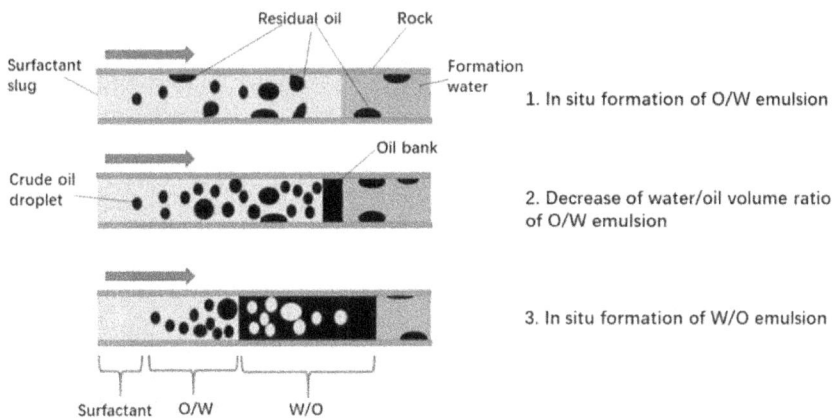

FIGURE 3.45 Schematic diagram of in situ emulsion formation and phase inversion process in porous media.

Source: Yang and Pu, 2020.

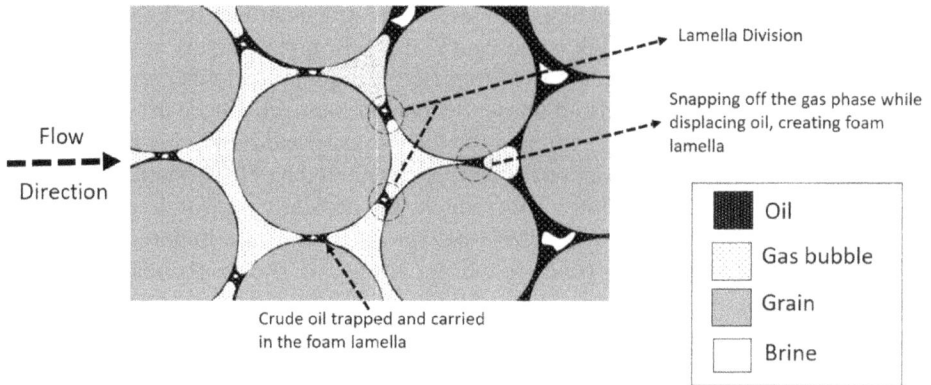

FIGURE 3.46   Crude oil trapping and flow movement through the foam lamella.

Source: Kumar, 2017.

3.3.5   Loss of Surfactant by Adsorption

An excess adsorption of surfactant leads to a reduction in their concentration and limits their efficiency during the surfactant-based EOR process. Thus, a comprehensive knowledge of loss of surfactant by adsorption on the reservoir rock is of utmost importance prior to the injection of surfactant slugs for chemical EOR. Extensive studies in past decades have been carried out to analyze the adsorption behavior of various ionic and nonionic surfactants on rock surfaces. The physiochemical properties that affect the adsorption of a surfactant are pH, temperature, ionic strength, and electrolyte concentration in the reservoir. Any variation in the mentioned physiochemical properties can result in a change in the adsorption pattern of the system. It is widely known that the adsorption of ionic surfactants on the rock surface is governed by electrostatic attraction. The similarity of charges of the anionic surfactant and sandstone reservoir surface leads to a preference for anionic surfactants over nonionic and cationic surfactants for applicability in sandstone reservoirs. Scientists throughout the world have worked in the last few years to reduce surfactant adsorption on the rock surface. Traditionally, sodium carbonate and sodium hydroxide as alkaline reagents are used to maintain the negative charge of the rock surface, causing electrostatic repulsion between the rock surface and surfactant molecules, leading to a significant decrease in surfactant adsorption in sandstone reservoirs. Gogoi (2009) studied the effect of salt concentration and pH on the adsorption of surfactant on the reservoir rock and reported that with an increase in salt concentrations, adsorption of the surfactant increases, and an increase in pH leads to a decrease in surfactant adsorption. Wang et al. (2015) showed that presence of polymers also lowers the adsorption of the surfactant by forming a thick protective layer over the rock surface, thus, reducing the active sites available for adsorption. However, it has been found by some researchers that additives like alkalis and polymers are ineffective at reducing the adsorption in high-salinity and high-temperature reservoirs. Hence, the prevention of adsorption of surfactant on the reservoir rock surface is a challenging task, and it is important to make it an economically viable process for its application in oil recovery from a reservoir. Typical adsorption densities of an anionic surfactant on sandstone, carbonate, and bentonite clay are reported in Figure 3.47. It may be found that because of the negative charges of sandstone rock, the adsorption of negatively charged anionic surfactant on sandstone is less compared to that on positively charged carbonate. In fact, the anionic surfactant is not preferred for carbonate rock because of high adsorption density.

3.3.5.1 Models of Surfactant Adsorption on Rock Surfaces

An adsorption isotherm model is required to predict the loading on the adsorption matrix at a certain concentration of the component. There are different adsorption isotherm models that can be used to describe the equilibrium adsorption relation. These models can be classified into two categories: linearized and nonlinearized. Langmuir, Freundlich, Temkin, Linear, Redlich-Peterson, Sips, and Dubinin-Radushkevich isotherms are very common models for explaining best-fit models (Langmuir, 1916; Freundlich, 1906; Temkin and Pyzhev, 1940; Saxena et al., 2019). Among them, the Langmuir and Freundlich equations are extensively used to describe the equilibrium adsorption. In Table 3.7, the corresponding equations for different models have been depicted.

In Table 3.7, Γ is the amount of adsorbate adsorbed (mg/g); Γ_{max} is the maximum amount adsorbed (mg/g); K_L is the Langmuir equilibrium constant (L/mg); and C_e is the equilibrium aqueous concentration (mg/L). In the Freundlich equation, K_F (mg/g) and n are the Freundlich adsorption constants related to sorption capacity and sorption intensity, respectively. B and K_t are the Temkin constants and equilibrium binding constant, respectively. In case of the linear isotherm model, K_H is a constant in units of L/m^2. K_R (L/mg) and α_R [(L/mg)β] are Redlich-Peterson isotherm constants. K_S [(L/mg)^m_s] is the Sips isotherm constant representing the energy of adsorption, and m_s is the empirical constant. In case of the Dubinin-Radushkevich model D and ϵ are the Dubinin-Radushkevich isotherm constants.

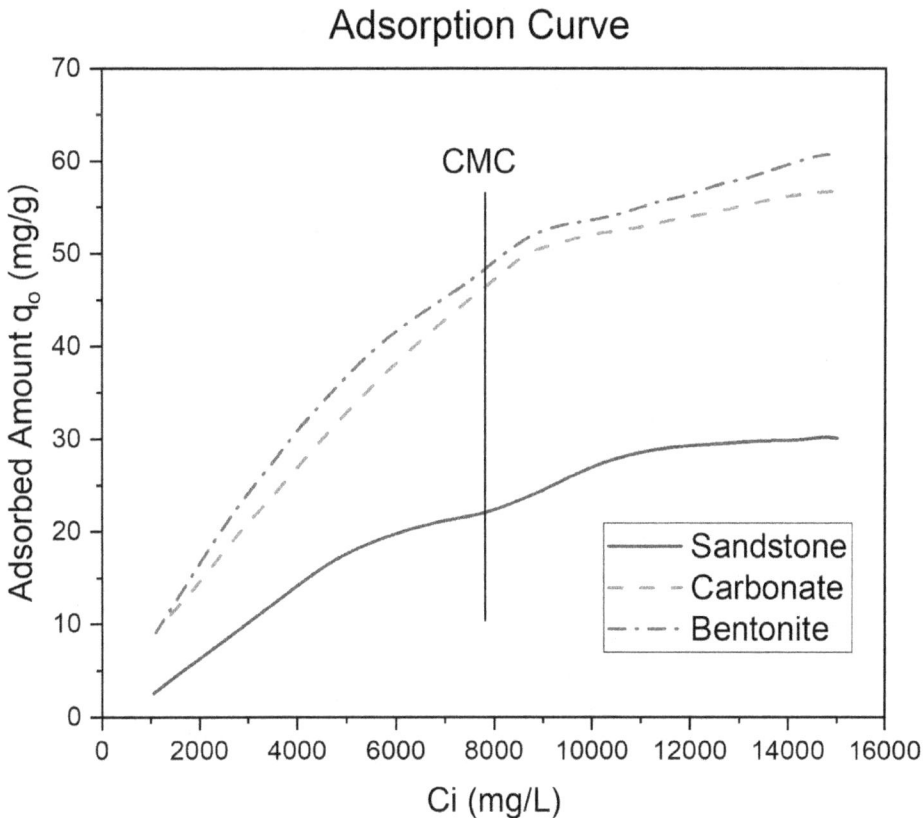

FIGURE 3.47 Equilibrium profiles for adsorption of soap-nut surfactant against the initial concentration.

Source: Saxena, 2020.

TABLE 3.7
Models of Adsorption Isotherms and Their Corresponding Equations

Sl. No.	Models	Equation	Form of the Equations
1.	Langmuir (1916)	$\Gamma = \dfrac{\Gamma_{max} K_L C_e}{1 + K_L C_e}$	Nonlinear
2.	Freundlich (1906)	$\Gamma = K_F C_e^{1/n}$	Nonlinear
3.	Temkin (1940)	$\Gamma = B \ln K_t + B \ln C_e$	Linear
4.	Linear isotherm	$\Gamma = K_H C_e$	Linear
5.	Redlich and Peterson (1959)	$\Gamma = \dfrac{K_R C_e}{1 + \alpha_R C_e^{\beta}}$	Nonlinear
6.	Sips (1950)	$\Gamma = \dfrac{\Gamma_{max} K_S C_e^{m_s}}{1 + K_S C_e^{m_s}}$	Nonlinear
7.	Dubinin (1960)	$\Gamma = \Gamma_{max} \exp\left(D\epsilon^2\right)$	Nonlinear

3.3.6 MICROEMULSIONS

The term "microemulsion" is a misnomer, as pointed out by Winsor in 1954. The concept of microemulsions was introduced by Prof. Jack H. Schulman and his coworker at Columbia University in 1959. A microemulsion is a thermodynamically stable, translucent, micellar solution of oil, water that may contain electrolytes, cosurfactants, and one or more amphiphilic compounds. Cosurfactants are usually used in conjunction with surfactants as most single-chain surfactants are incapable of reducing the interfacial tension of oil and water to form a microemulsion. The most common cosurfactants are medium-chain alcohols, which reduce the tension and increase the fluidity of the oil-water interface, thereby increasing the entropy of the system. These medium-chain alcohols also increase the mobility of the surfactants' nonpolar tail region, allowing greater penetration by oil molecules and, therefore, stabilizing the system and facilitating the formation of a microemulsion. Winsor (1954) pointed out that certain nontransparent but translucent and often opalescent micellar solutions are stable. The degree of translucency is merely a measure of average micelle size and configuration and can be caused to vary continuously from completely transparent to nearly opaque simply by varying, for example, salinity. Although inclusion of the constraint that a microemulsion must be transparent is a matter of choice, it excludes the preponderance of systems that have utility for tertiary oil recovery. It is an experimental fact that when compositions lead to opaque fluids, these fluids are usually unstable, separate on standing, and hence, form macroemulsions; so translucency is an essential aspect.

3.3.6.1 Types and Structures of Microemulsions

Microemulsion structure plays a key role in the different physicochemical parameters of the applied fields. The specific structures of microemulsions have been extensively studied by many researchers (Azouz et al., 1992; Wadle et al., 1993; Li et al., 2010). The three basic types of microemulsions are direct (oil dispersed in water, o/w), reversed (water dispersed in oil, w/o), and bicontinuous. Like multiple emulsions, sometimes, multiple microemulsion are also possible. In this type, another layer is formed outside the o/w or w/o microemulsions. The schematic diagram of the basic three types of microemulsions is shown in Figure 3.48. The structure of a microemulsion depends on salinity, water content, cosurfactant concentration, surfactant concentration, and oil type. At a higher water content, the microemulsion will be a water-external system with oil solubilized in the cores of the micelles. Although the mixtures remain single phase and thermodynamically stable, the microemulsion structure changes through a series of intermediate states (Bourrel and Schechter, 2010).

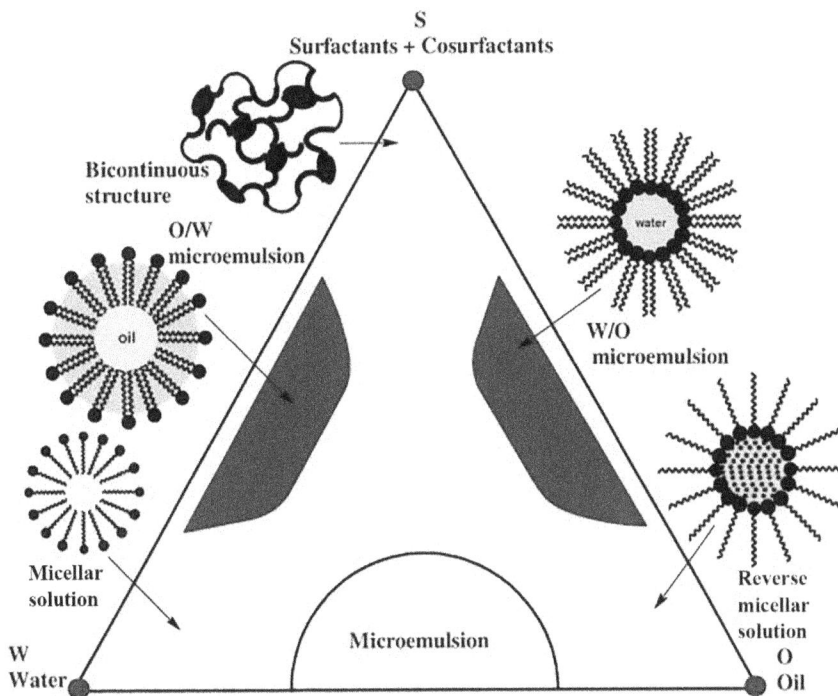

FIGURE 3.48 Hypothetical phase regions of microemulsion systems.

Source: Malik et al., 2012.

The structures of these intermediate states are not well known. However, the solutions are thermodynamically stable and isotropic. Salinity also can reverse the structure of the microemulsion. As salinity increases, the direct microemulsion changes to reverse microemulsion. At low salinity, the system remains in water-external phase, but with increasing salinity, the system separates into an oil-external microemulsion.

Phase diagrams are very useful to describe microemulsion systems. The simplified ternary representation describes the different phases obtained when oil, surfactants, and water are blended. Microemulsions are usually characterized by ternary phase diagrams in which three edges are the components of a microemulsion: oil, water, and surfactant. Any cosurfactant used is usually grouped together with the surfactant at a fixed ratio and treated as a pseudo component. Figure 3.49 shows the formation of micelles at a certain concentration with the help of a hypothetical pseudoternary phase diagram of a microemulsion system. Each corner of the phase diagram represents a pure compound. Each side represents the different compositions of a blend of two components, and a point inside the diagram represents the composition of a blend of the three components.

When the water content of a microemulsion is low (near the oil corner of the phase diagram), then the local structure of the microemulsion is swollen inverse micelle (water in oil). The oil phase is continuous, and the water micellar droplets are the dispersed phase. Bicontinuous microemulsions form spontaneously when the surfactant is in its balanced state, where it has equal affinity for the oil phase as for the aqueous phase. Near the water corner, the oil content is low, and then the microemulsion looks like a direct micellar solution (oil in water).

3.3.6.2 Phase Behavior of Microemulsion System

The phase behavior of the surfactant-oil-brine system is an important, key step in the laboratory to screen proper surfactants for EOR (Engelskirchen et al., 2007; Barnes et al., 2008). The

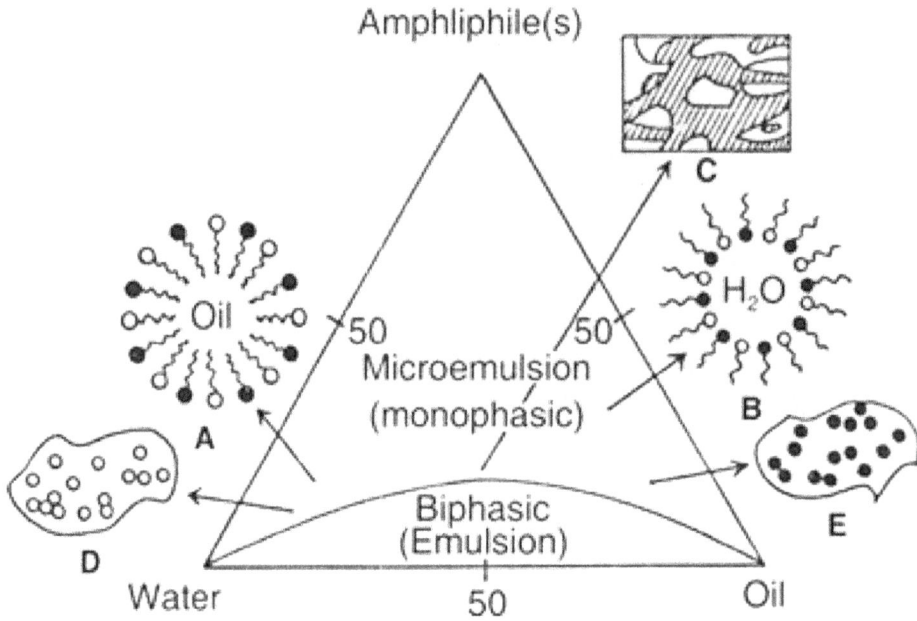

FIGURE 3.49   Micelles formation, pseudoternary phase diagram, and structure inversion of microemulsion system.

Source: Paul et al., 2006.

microemulsion phase behavior changes from Winsor I to Winsor II through Winsor III with the variation of salinity, temperature, and pressure. Surfactant molecules in oil or in water form a variety of structures when structure-assisted parameters such as water content, surfactant concentration, cosurfactant type, cosurfactant concentration, pressure, and/or temperature are varied (Mitra et al., 2006; Ray and Moulik, 1995). The middle-phase microemulsion consists of solubilized oil, brine, surfactant, and cosurfactant. The lower to middle to upper phase transition of the microemulsion phase can be obtained by varying the following factors: (i) increasing salinity; (ii) increasing alcohol concentration (propanol, butanol, pentanol, and hexanol); (iii) decreasing oil chain length; (iv) change in temperature; (v) increasing total surfactant concentration; (vi) increasing surfactant solution/oil ratio; (vii) increasing brine/oil ratio; and (viii) increasing the MW of the surfactant.

For a given overall composition, an oil-in-water (O/W) microemulsion in equilibrium with the excess-oil phase is called Winsor I phase behavior. When the water-in-oil (W/O) microemulsion remains in equilibrium with the excess-water phase, then it is termed as Winsor II phase behavior. Another phase behavior—i.e., Winsor III—can be seen when a microemulsion is in equilibrium with both the water and oil excess phases. Figure 3.50 shows the schematic diagram of the Winsor phase behavior of microemulsion, and Figure 3.51 depicts the corresponding pseudoternary phase diagram. Middle-phase microemulsions are, nevertheless, often favorable for a microemulsion flooding process (Hirasaki et al., 1983; Chen et al., 2007). Hence, it is important to maintain the middle-phase microemulsion system as long as possible during the microemulsion flooding.

3.3.6.3   Solubilization Parameters and IFT

Solubilization parameters express the amount of oil and water that solubilize per unit volume of surfactant and are essential for emulsion formation. For determination of solubilization parameters, different solutions with different compositions of oil, water, and surfactant are prepared at different salinities. The different volumes of oil, emulsion, and brine are then easily measured from the scaled

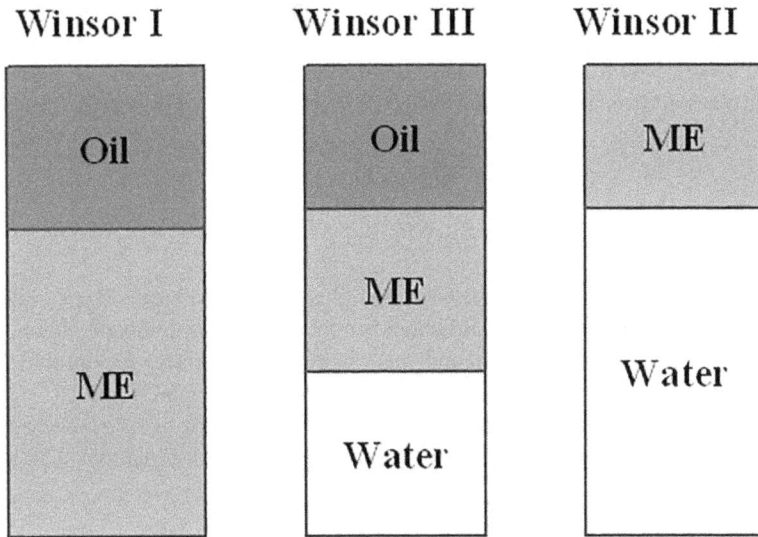

FIGURE 3.50   Schematic diagram of Winsor-type phase behavior of microemulsion (ME).

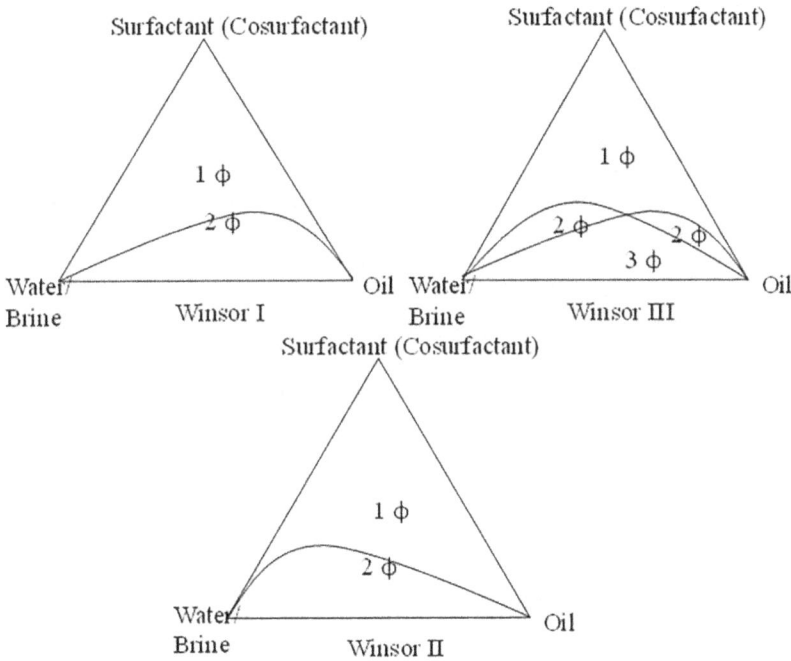

FIGURE 3.51   Pseudoternary phase diagram of oil-water-(brine)-surfactant-(cosurfactant) system (Winsor-type phase behavior of microemulsion).

tubes after settlement. The amount of oil and brine solubilized in the emulsion phase is determined (in volume) by Equations 3.26 and 3.27, respectively:

$$V_{os} = V_{oi} - V_{of} \tag{3.26}$$

$$V_{ws} = V_{wi} - V_{wf} \tag{3.27}$$

where V_{os} and V_{ws} are the volumes of oil and brine solubilized in the emulsion phase, V_{oi} and V_{wi} are the initial volumes of oil and brine in the solution, and V_{of} and V_{wf} are the final volumes of oil in the upper phase and brine in the lower phase respectively. The solubilization parameters (S_{po} and S_{pw}) for oil and water, respectively, are obtained from Equations 3.28 and 3.29, as expressed by Healy et al. (Healy et al., 1975):

$$S_{po} = V_{os}/V_s \tag{3.28}$$

$$S_{pw} = V_{ws}/V_s \tag{3.29}$$

where V_s is the volume of surfactant in the emulsion phase (middle phase). The V_s value is taken to be constant, as it is assumed that all the surfactant is contained in the emulsion phase. The intersection point in the solubilization curve of oil and water helps in determining the optimal solubilization parameter (σ^*), and the salinity at this point is acknowledged as optimal salinity.

The interfacial tension between oil-middle phase and water-middle phase is determined from the solubilization parameters using the Chun-Huh equation (Huh, 1979), Equations 3.30 and 3.31:

$$\sigma_{om} = \frac{c}{\left(V_{os}/V_s\right)^2} \tag{3.30}$$

$$\sigma_{wm} = \frac{c}{\left(V_{ws}/V_s\right)^2} \tag{3.31}$$

where σ_{om} and σ_{wm} are the IFT between oil-middle phase and water-middle phase, respectively. V_{os}/V_s is the ratio of volume of oil solubilized in the total volume of surfactant, and V_{ws}/V_s is the ratio of volume of water solubilized in the total volume of surfactant. The constant c ($c = 0.3$ mN/m) is consistent for colloidal systems (Kanan et al., 2017).

For determination of the amount of oil and water present in emulsion after the equilibrium condition, an examination of the transformation of emulsion phase is done. Table 3.8 presents the values of the solubilization parameters and IFT values at different salinities.

Surfactant molecules are assumed to be fully present in the emulsion phase, and no surfactant molecule exists in the upper oil and lower brine phases. The intensity of the interaction of the surfactant with oil and brine is quantified by the solubilization parameters (S_{po} and S_{pw}). Figure 3.52 shows the image of relative phase volume present in the oil-surfactant-brine system at varying salinities. An increase in the emulsion phase is observed with a rise in the NaCl concentration. This is due

---

TABLE 3.8

Solubilization Ratio and IFT for Tween 80 Surfactant, as Predicted by the Chun-Huh Equation at 303 K, Assuming All the wt. % of the Surfactant Is in the Middle Phase ($c = 0.3$ mN/m)

Salinity (wt. % of NaCl)	Solubilization		IFT (Chun-Huh Equation), mN/m	
	V_{os}/V_s	V_{ws}/V_s	σ_{om}	σ_{wm}
0.1	8	20	4.69E−03	7.50E−04
0.3	12	16	2.08E−03	1.17E−03
0.5	20	12	7.50E−04	2.69E−03
1.0	32	16	2.93E−04	1.17E−03
1.5	60	8	8.33E−05	4.68E−03
2.0	84	4	4.25E−05	1.88E−02

to the increased activity of the surfactant molecules present at the interface in the presence of NaCl, which accumulates on the surface of the surfactant molecules and helps in the reduction of interfacial tension. Figure 3.53 presents the curve of solubilization parameters at varying salinities. The

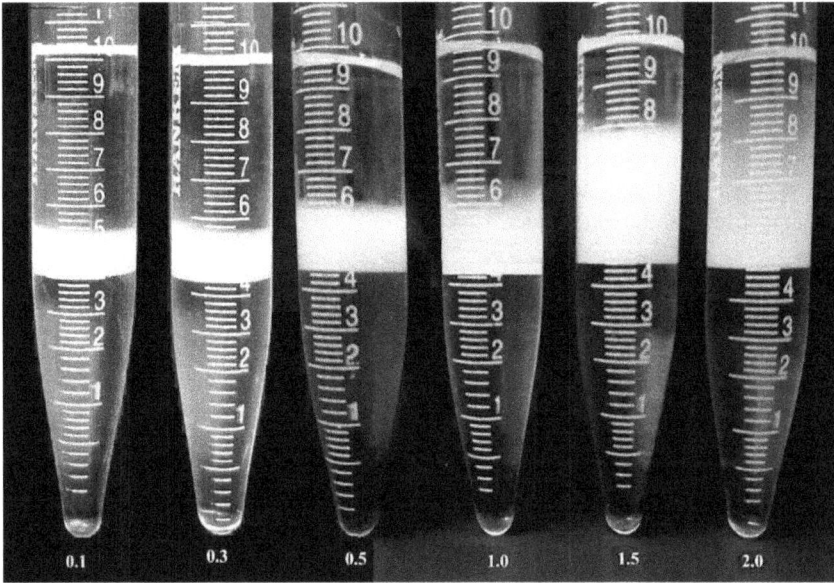

FIGURE 3.52 Salinity scan for oil-surfactant-brine system at 0.5 wt. % of Tween 80 and 1:1 ratio of brine: n-heptane with varying salinity from 0.1 to 2.0 wt. %.

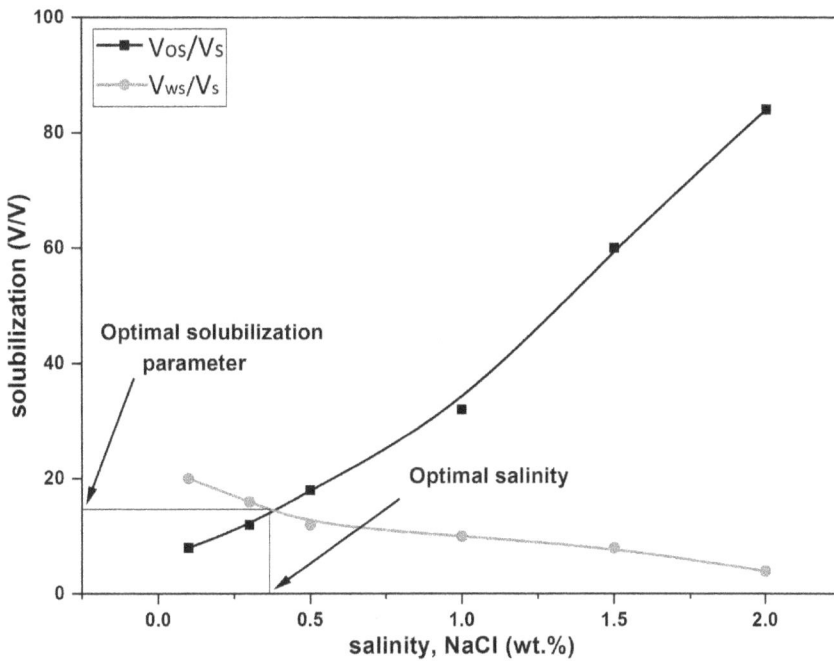

FIGURE 3.53 Solubilization parameters of Tween 80 with varying salt concentrations from 0.1 to 2.0 wt. %.

Source: Kumar, 2017.

salinity value at which the S_{po} and S_{pw} curves mutually intersect each other is accredited as optimal salinity (Pal, 2020). The oil and water solubilization parameters are equal at optimal salinity in the emulsion phase, and the solubilization ratio at this point is known as the optimal solubilization ratio (σ^*). The σ^* strongly depends upon the optimal salinity. Figure 3.53 indicates that for the present oil-surfactant-brine system, optimal salinity and σ^* exist at 0.3 wt. % of NaCl and 15, respectively.

Figure 3.54 shows the relationship of IFT with varying salinities for the oil-middle emulsion phase and water-middle emulsion phase, as predicted by the Chun-Huh equation. A drop in IFT is observed between the oil-middle emulsion phase with an increase in salinity, while the IFT between the water-middle emulsion phase increased, denoting a decline in the solubilization of water. The IFT curves intersect at optimal salinity, designating the presence of an equal amount of oil and water in the middle phase at this point. The ultralow IFT value obtained between the oil-middle emulsion phase and higher solubilization parameter value obtained for oil, indicate better solubilization of oil in surfactant than brine.

3.3.6.4 Viscosity and Density of Microemulsion

The magnitudes of the viscosity and density of displacing fluid relative to the displaced fluid are important design variables that affect volumetric displacement efficiency. The tendency for gravity override and underride to occur is determined by the relative densities of the displaced and displacing fluids. Areal and vertical sweep efficiencies are, in large measure, determined by the mobility ratio in the displacement process. Both viscosity and density are functions of microemulsion composition. Viscosity, in particular, can be varied over a wide range by the proper adjustment of composition and/or by polymer addition. The viscosity of microemulsions depends on the structure of the microemulsion—i.e., whether it is water or oil external. It is well known that at a low water content, the system is oil external, and at high water content, the system shows the reverse—i.e., water external. The viscosity of the microemulsion increases as water content increases, creating

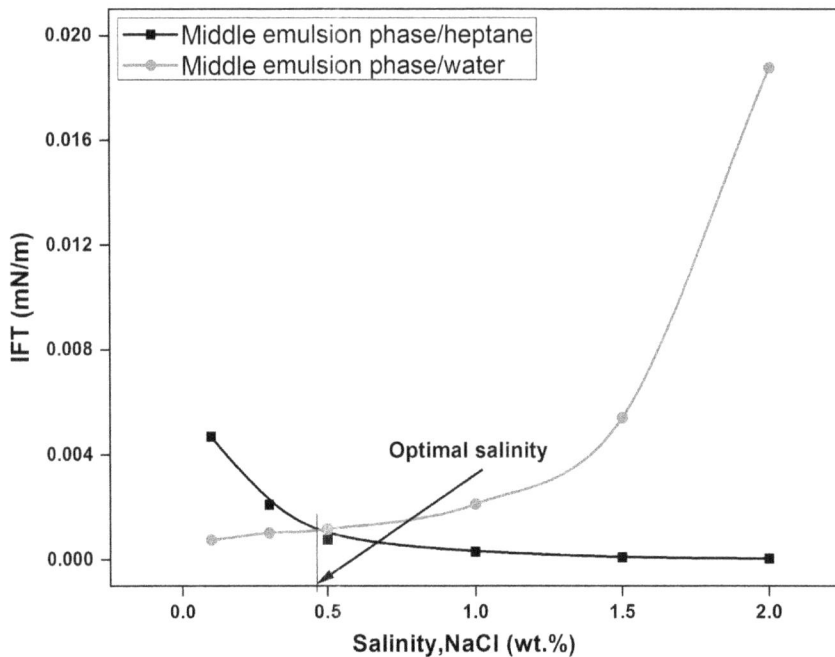

FIGURE 3.54 IFT versus NaCl concentration, as predicted by the Chun-Huh equation, for middle emulsion phase/heptane and middle emulsion phase/water, with water:oil = 1:1.

Source: Kumar, 2017.

swollen micelles. At 50% water content, the viscosity of the microemulsion increases to two orders of the initial value. At a higher water content, after inversion to a water-external system, the viscosity decreases with further addition of water. In general, the viscosity of the displacing slug has been modified by the addition of a polymer, such as polyacrylamide or biopolymer. The viscosity of the microemulsion can be modified by adding a cosurfactant (medium-chain alcohol) and/or polymer to the microemulsion.

3.3.6.5   Factors Affecting Phase Behavior and IFT

A number of factors affect the phase behavior and solubilization parameters of microemulsion and IFT between oil and water systems. The variables are oil type, cosurfactant type, temperature, ions with different valencies, surfactant structure, and pressure. A brief description of the aforementioned factors is given in the following. This discussion provides information to guide the development of design criteria for the selection of surfactant as well as microemulsion slug system for a particular reservoir oil.

(a) Effect of oil type: Oil nature plays an important role in monitoring the phase behavior as well as the IFT between oil and water systems. Healy et al. (1976) reported that as aromaticity decreases, the IFT value increases at a constant salinity, and phase behavior shifts from lower to middle to upper phase microemulsions.

(b) Effect of cosurfactant type: Cosurfactants are originally added to the surfactant system to increase the solubility of certain surfactants and to alter the viscosity of the system. Jones and Dreher (1976) studied the effect of cosurfactants on the phase behavior of microemulsion systems. They reported that water-soluble cosurfactants (lower MW alcohols) make a microemulsion more hydrophilic—i.e., increase in capacity to solubilize water but decrease in capacity to solubilize oil. On the other side, alcohol with low water solubility (higher MW alcohols such as pentanol and hexanol) has the opposite effect—i.e., oil solubilization increases, and water solubilization decreases. The cosurfactant type and concentration influence the IFT and phase behavior of the microemulsion system. As the MW of the cosurfactant decreases (cosurfactant being more water soluble), the optimal salinity is shifted to a higher value and increased IFT value at optimal salinity.

(c) Effect of temperature: Temperature can influence the IFT and phase behavior of microemulsion systems (Skauge and Fotland, 1990). An increase in temperature causes the solubilization parameters of oil and water to decrease at optimal salinity, increasing the IFT and shifting the optimal salinity for a given system to a higher value. Phase transition takes place from upper to middle to lower phase as temperature increases at a specified concentration of surfactant and cosurfactant.

(d) Effect of ions with different valencies: Oil field fluids generally contain different mono, bi, and trivalent ions with other compounds. Mainly divalent ions such as Ca^{2+} and Mg^{2+} contribute to the hardness of the brine, which makes ions to be precipitated and can increase the compatibility with a surfactant. The presence of divalent ions shifts the optimal salinity to a lower value. An experimental study showed that when NaCl was replaced by a mixture of $NaCl/CaCl_2$, the IFT increased, and phase behavior shifted to the upper phase system. This is due to the association of the divalent ions with the surfactant (Glover et al., 1979).

(e) Effect of the molecular structure of the surfactant: A number of considerable research works have been developed focusing on surfactant structures that are tolerant to high salinity and hardness (Graciaa et al., 1982; Carmona et al., 1985). The molecular structure of the surfactant affects phase behavior, solubilization parameters, and consequently, the IFT of the oil and water system. Healy et al. (1976) investigated the effect of chain length of the hydrocarbon tail on IFT. They reported that as chain length increases from 9 to 12 to 15, the optimal salinity gradually reduces IFT values, and an increase in optimal solubilization parameters takes place. Gale and Sandvik (1973) examined the effect of the MW of the surfactant on IFT. They concluded that the high-equivalent-weight molecules dominated the surfactant's properties. They also showed that petroleum sulfonate, with equivalent weights in the vicinity of 400–500, is effective in reducing the IFT to a lower value.

(f) Effect of pressure: In general, the effect of pressure on phase behavior is small. Nelson (1981) studied the phase behavior of live crude oil under pressurized conditions. He concluded that for live crude oils with a significant amount of gas (C_1, C_2, etc.) content, the possible influence of pressure must be considered in the design process. Skauge and Fotland (1990) reported that increasing the pressure caused a shift in phase behavior toward a lower phase microemulsion. For a given system, optimal salinity increased as pressure on the system increased. The effect of pressure on the IFT between oil and water system is also very small.

3.3.7 SURFACTANT SLUG AND MICELLAR SOLUTION FLOOD

Based on the composition of the injection slugs, surfactant floodings are classified as follows:

- Surfactant-polymer (SP) slug
- Micellar-polymer (MP) slug

3.3.7.1 Surfactant-Polymer Flooding

SP flooding is a process where a surface-active agent (surfactant) is injected into a reservoir to improve displacement efficiency by reducing the IFT between the trapped crude oil and injected fluid, followed by the injection of a polymer buffer to improve the sweep efficiency by controlling the mobility ratio. The SP slug injection process consists of the following.

Preflush: Before injecting the surfactant slug, one has to be sure that the formation brine is compatible with the injected surfactant slug. Thus, the objective of the preflush is to condition the reservoir by injecting a brine solution prior to the injection of the surfactant slug so that the mixing of the surfactant with the brine will not cause the loss of interfacial activities. The preflush solution volume typically ranges from 50% to 100% PV.

Surfactant slug: The volume of the surfactant slug ranges between 5% and 20% PVs of the target reservoirs. This is the main oil-recovering agent, and a suitable surfactant slug is selected primarily based on screening studies. Besides the surfactant, some other chemicals are usually needed in actual field applications. Extensive laboratory studies show that the minimum slug size is 5% pore volume to achieve effective oil displacement and recovery.

Mobility buffer: The surfactant slug is displaced by a mobility buffer solution, which is a dilute solution of a water-soluble polymer. It improves sweep efficiency, which increases oil production. The high viscosity of the mobility buffer aids in the displacement of chemicals into the reservoir and also minimizes the channeling of the injected chemical solution. The mobility buffer also protects the surfactant slug and prevents the dilution of the surfactant slug from mixing with chase water.

Mobility buffer taper: This is a volume of brine that contains a polymer at low concentrations, injected after the mobility buffer so that a gradient in the concentration of the polymer is maintained, which mitigates the effect of the adverse mobility ratio between the mobility buffer and the chase water.

Chase water: The purpose of the chase water is simply to reduce the expense of continually injecting the surfactant and polymer. Further, it maintains the viscous force to drive the chemical slugs and to subsequently displace the oil ahead of the slugs.

A schematic of the injection plan of a surfactant flooding is shown in Figure 3.55. The design of the injection volume of different chemical slugs is very important and challenging. It should be small enough that it may vanish before breakthrough, and also, it should not be so large as to

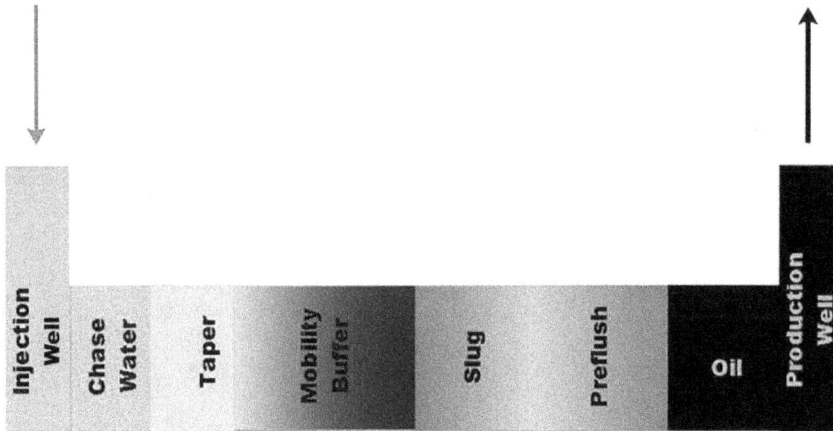

FIGURE 3.55 Schematic of the injection plan of surfactant flooding.

increase the project cost. The actual design of different slugs also depends on the reservoir rock and fluid properties and should be done based on exhaustive laboratory studies.

3.3.7.2 Micellar or Microemulsion-Polymer Flooding

Micellar flooding is variously known as "microemulsion flooding" and "micellar-polymer flooding," with various shades of meaning. Micellar/microemulsion flooding is a very effective way to increase the additional oil recovery than through other methods due to the attainment of ultralow interfacial tension between the oil and water phase. The micellar solutions are composed primarily of a hydrocarbon, surfactant, cosurfactant, and water.

A micellar solution is a dispersion of a surfactant in an oleic or aqueous solvent. Such solutions can solubilize large amounts of water or oil to form either water-in-oil (w/o) or oil-in-water (o/w) microemulsions, respectively. A microemulsion is a stable o/w or w/o emulsion, in which the drop size is less than about one micrometer (micron). A soluble oil is an oleic composition that is capable of solubilizing water (not oil) when mixed with it. A soluble oil may contain water, but the external phase is oleic. MP flooding involves sequential injection of a preflush, micellar slug, mobility buffer, and drive water similar to surfactant flooding as shown in Figure 3.55.

The mechanisms of MP flooding for EOR are (Dang et al., 2014) as follows:

(1) ultralow IFT between chemical bank and residual oil and between the chemical bank and the driving fluid;
(2) small surfactant losses to the reservoir rock (if retention is excessive, IFT will eventually become high enough to retrap the residual oil in the remainder of the reservoir);
(3) brine compatibility and temperature stability;
(4) mobility control; and
(5) economy of the process.

Microemulsion flooding in EOR can be applied over a wide range of reservoir conditions due to its exclusive ultralow interfacial tension property (Santanna et al., 2009; Jeirani et al., 2013). In cases where the pressure exerted by water on the oil phase is not able to overcome capillary forces sufficiently, microemulsions are the key to extracting more than just a minor portion of crude oil. Properly balanced microemulsions are able to do so by drastically reducing the interfacial tension to the magnitude of 0.001 mNm^{-1}.

TABLE 3.9

Screening Criteria of Surfactant/Micellar-Polymer Flooding

Parameters	Values
Oil viscosity, cp	< 35
API gravity, °	> 20
Residual oil saturation, %	> 35
Formation salinity, ppm:	
Chloride	< 20,000
Divalent ions	< 500
Temperature, °F	< 250
Thickness, ft	> 10
Depth, ft	< 9,000
Porosity, %	> 15
Permeability, Md	> 10

3.3.8 SCREENING CRITERIA OF SURFACTANT/MICELLAR-POLYMER FLOODING

The main criteria for SP flooding are salinity and temperature, which dictate the conditions of using surfactants and polymers. It is proven that at high salinity, the surfactant gets precipitated from the solution, which leads to detrimental effects on the efficiency of the surfactant as an EOR agent. Further, for micellar flooding, the salinity plays a very important role on oil/brine/surfactant phase behavior. High temperature also leads to thermal degradation of the surfactant and polymer. However, the selection of a candidate reservoir totally depends on the selection of a surfactant and its interaction with reservoir rock and fluids. Nowadays, tailor-made surfactants are used based on the specific rock/fluid properties of the target reservoir. Based on the reported literature (Sheng, 2013a, 2013b; Mohsenatabar Firozjaii et al., 2019), the screening criteria of surfactant/micellar-polymer flooding are reported in Table 3.9.

3.4 ALKALI FLOODING

Alkaline waterflooding is an EOR process where an alkaline solution is injected within the reservoir through the injection wells. The alkaline solution with a higher pH reacts with the acidic component of the crude oil to form in situ surfactants, leading to lower water-oil interfacial tension; emulsification of oil and water; and solubilization of rigid, interfacial films (Mungan, 1981). The process is relatively cheaper, as the alkaline agents used in this process, such as sodium carbonate, sodium silicate, sodium hydroxide, and potassium hydroxide are relatively inexpensive. Though the mechanisms of alkaline flooding were reported in 1917 by F. Squires, the field applications are not so numerous, and most of them have been unsuccessful. However, presently, the process is drawing attention due to the establishments of other beneficial mechanisms. The alkali also reacts with the reservoir rocks to change the oil-wet rock to water wet, leading to an enhanced recovery of oil. The alkaline flooding process is a relatively simple process as compared to other chemical floods but is still sufficiently complex to warrant careful laboratory investigation and field trials before application (Kumar et al., 1989).

3.4.1 MECHANISM OF ALKALINE FLOODING

Oil recovery by alkaline flooding takes place in a two-stage process (Castor, 1979). The first stage involves the formation of in situ surfactants. The alkali reacts with the acidic components

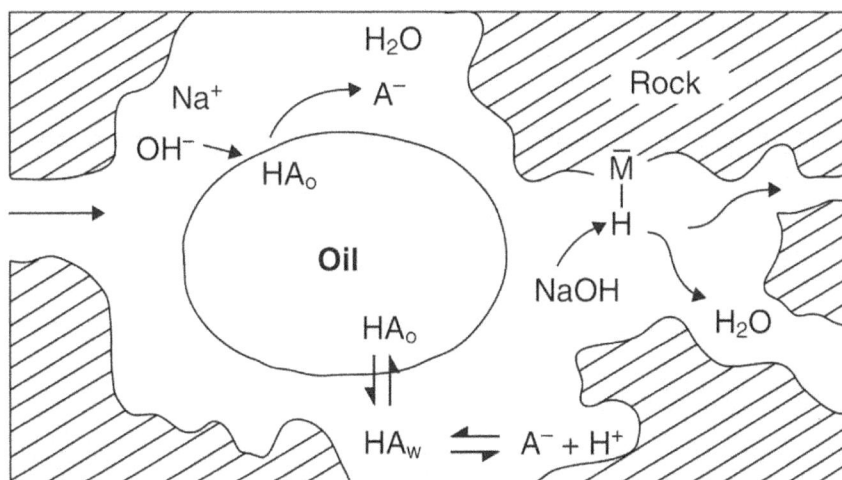

FIGURE 3.56 Schematic diagram of the alkali and petroleum acid interaction.

of crude oil to form in situ surfactants. The surfactants thus generated may (1) adsorb at the oil-water interface to lower the interfacial tension and, in some cases, cause spontaneous emulsification and phase swelling; and (2) react with or adsorb at the rock surface, changing the wettability characteristics of the rock and, hence, the configuration of the residual ganglia of the crude oil (Castor, 1979).

The second stage of the alkaline recovery processes involves macroscopic production of the mobilized oil phase. In this stage, the overall recovery efficiency can be increased by improvement of the displacement efficiency through reduction in the mobility of the floodwater.

Johnson (1976) reviewed the mechanisms by which alkaline flooding may improve the recovery of acidic crudes from partially depleted reservoirs:

(1) Emulsification and entrainment: According to this mechanism, the residual oil is emulsified by alkaline action and entrained into the flowing alkaline solution and, thus, produced as a fine emulsion (Mungan, 1981).

(2) Emulsification and entrapment: In this process, the emulsified oil is trapped again downstream in the porous medium at pore throats that are too small for the emulsion droplets to penetrate, forcing the injection water into pores that have not been previously displaced.

(3) Wettability reversal from water wet to oil wet: Cooke et al. (1974) suggested this mechanism, where the alteration of wettability from water wet to oil wet causes the discontinuous nonwetting residual oil phase to spread out into a continuous wetting phase.

(4) Wettability reversal from oil wet to water wet: In this mechanism, this alkali reacts with the reservoir rocks and displaces the adsorbed oil. Thus, the oil-wet rock is converted to water wet, and the recovery is enhanced. The displaced oil is also emulsified and flows with the injected water.

(5) Lowering of interfacial tension: The in situ surfactant generated by alkaline action with the acidic crude oil gets adsorbed at the oil-water interface and reduces the IFT, which in turn, improves the displacement efficiency.

(6) Solubilization of rigid, interfacial films: Resins and asphaltenes present in the crude oil are responsible for the formation of insoluble films at the crude oil–water interface. These interfacial films may be redissolved in the oil, and thus, the oil is mobilized, but this rate is very slow.

The extent to which each of these mechanisms may contribute to increasing recovery depends on the specific properties of the crude oil, injection water, and the reservoir (Mungan, 1981).

3.4.2 Mechanisms of In Situ Surfactant Generation

The reaction between the injected alkali and crude oil to form in situ surfactant is dependent on the petroleum acid number and also varies with the composition of crude oil (Mayers et al., 1983).

Addition of the alkali in water results in a high pH because of the dissociation in the aqueous phase. NaOH dissociates as

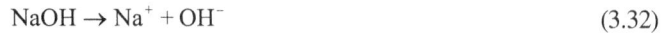

$$NaOH \rightarrow Na^+ + OH^- \tag{3.32}$$

Equilibrium dissociation of water is given as

$$K = \frac{\left[OH^-\right]\left[H^+\right]}{\left[H_2O\right]} \tag{3.33}$$

where $[OH^-]$ and $[H^+]$ is the molar concentration of OH^- and H^+, respectively. An increase in $[OH^-]$ causes a decrease in $[H^+]$. Water concentration is essentially constant.

The hydroxide ion must react with petroleum acids from the crude oil to form a surfactant. A mechanism is shown in Figure 3.56. Some of the petroleum acid in the crude oil partitions into aqueous phase according to the solubility expression:

$$K_D = \left[HA_0\right]/\left[HA_w\right] \tag{3.34}$$

where K_D = distribution or partition coefficient and HA_0 and HA_w denote petroleum acid in the oil and water phase, respectively. This can be also expressed as follows:

$$HA \rightleftharpoons HA_w \tag{3.35}$$

The petroleum acid dissociates in the aqueous phase according to the expression:

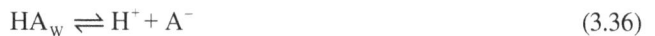

$$HA_w \rightleftharpoons H^+ + A^- \tag{3.36}$$

As governed by the equilibrium relationship:

$$K = \frac{\left[H^+\right]\left[A^-\right]}{\left[HAw\right]} \tag{3.37}$$

The species A^- is an anionic surface-active agent. The effect of caustic soda on petroleum acids is expressed by the following reaction:

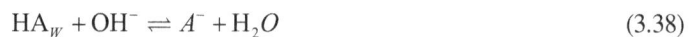

$$HA_w + OH^- \rightleftharpoons A^- + H_2O \tag{3.38}$$

According to Equation 3.38, this results in increases in $[A^-]$. The net effect of all the reaction is shown in Figure 3.56.

Depending on the rock mineralogy, alkalis can interact with reservoir rock in several ways, which include surface exchange and hydrolysis, congruent and incongruent dissolution reactions, and insoluble salt formation by reaction with hardness ions in the fluid and those exchanged from the rock surface (Somerton et al., 1983). Among those alkali-rock (clay) interactions, the reversible sodium/hydrogen-base exchange is a very important mechanism of alkali consumption and cannot be neglected.

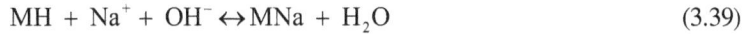

$$MH + Na^+ + OH^- \leftrightarrow MNa + H_2O \tag{3.39}$$

where M denotes a mineral-base exchange site.

The interactions of OH^- ions and alkali molecules with crude oil and rock surface are shown in Figure 3.56. As shown in the figure, the OH^- ions of the alkali molecules interact with the acids present in the crude oil, forming in situ surfactants, which are partitioned at the oil-water interface. The generated surfactant molecules reduce the IFT of the oil-water interface, leading to the mobilization of the trapped crude oil droplets through the pores of the reservoir. However, the reduction of IFT occurs for the alkali concentration in the range of 0.05–0.1 wt. %, and the desired reduction of IFT to ultralow values has not been achieved by alkaline flooding. The molecules of the injected alkali also interact with the minerals present at the rock surface, leading to wettability alteration of the rock surface from an oil-wet state to the preferred water-wet state. Alkalis are also injected as a sacrificial agent that reduces the adsorption of costlier and more effective chemical EOR agents such as surfactants (Donaldson et al., 1989). However, the use of alkali is also associated with corrosion, scaling, and precipitation problems that occurs due to the interaction of the alkali with the divalent ions present in the formation brine. Thus, the use of organic alkali, derived from the sodium salts of weak polymer acids, instead of inorganic alkalis, has been recommended (Berger and Lee, 2006).

3.4.3   ROLE OF ALKALI IN CHEMICAL LOSS

Alkalis like sodium carbonate have been applied as a sacrificial agent to reduce the adsorption of the surfactant in various reservoir rocks. Figure 3.57 shows the adsorption of surfactants on rock and clay surfaces in the presence of alkalis of different concentrations. It has been observed that with an increase in the concentration of sodium carbonate, the adsorption of the surfactant on all the samples decreased. The decrease in the adsorption of surfactants is due to the dissociation of sodium carbonate into weak carbonic acid and the generation of OH^- ions upon interaction with water molecules. This causes an increase in the pH of the system and increases the negativity of the surface of the sample. This leads to electrostatic repulsion between surfactant molecules and the surface of the rock, causing a significant reduction of surfactant adsorption. Sodium carbonate molecules also form in situ surfactants by interacting with the naphthenic acids present in the crude oil, which facilitates the recovery of trapped oil.

3.4.4   RESERVOIR SELECTION

3.4.4.1   Rock Mineralogy

Reservoir rocks with a high content of clays having a large surface area and high ion-exchange capacity are generally undesirable for alkaline flooding as they consume the alkaline slug by transferring hydrogen, calcium, and/or magnesium ions to the alkaline aqueous phase in exchange for sodium. Especially clays of montmorillonite type swell on encountering an injected water of a different salinity (and ionic composition) and reduce the permeability drastically. Thus, the reservoir rock should have a minimal amount of these minerals. For these reasons, carbonate reservoirs enriched in Ca^{2+} and Mg^{2+} are unsuited for alkaline flooding, and only sandstone reservoirs are candidates for this process. The only reported case of alkaline flooding in a carbonate reservoir is in Hungary (Nagylengyel field), where ammonium hydroxide has been used as the alkaline chemical. Because of a high pH, there are some limitations of using only alkalis for EOR, particularly in carbonate reservoirs. However, the use of alkalis with surfactants and polymers are also reported for many carbonate reservoirs (Al-Murayri et al., 2018). Carbonates are positively charged rock

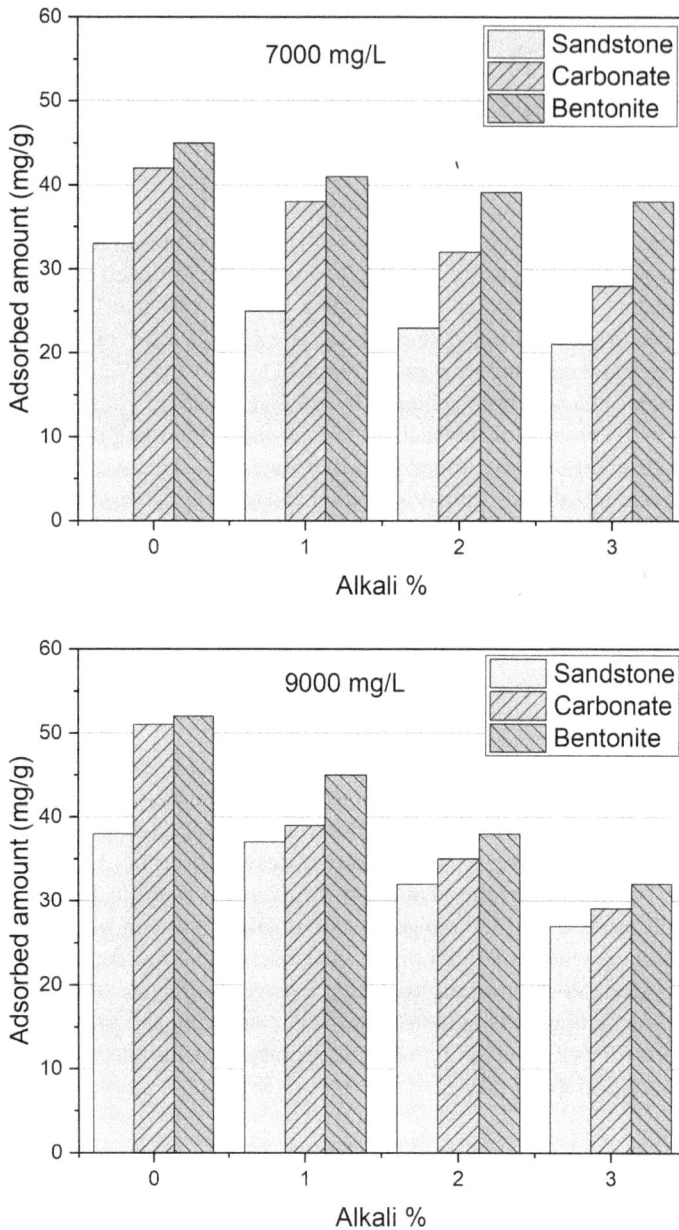

FIGURE 3.57 Effect of alkali on adsorption of surfactant molecules at a concentration of 7,000 mg/L and 9,000 mg/L: (a) soap-nut surfactant and (b) mahua oil surfactant.

Source: Saxena, 2020.

surfaces; thus, anionic surfactants are not preferred there because of high adsorption loss. However, the use of an alkali leads to a change of the positive charge on the surface to negative. When there is an increase in pH by adding an alkali, the calcite surface charge changes from positive to negative. In between, it crosses the point of zero charge (PZC). Once the PZC is crossed, the surface becomes negatively charged, and the adsorption of anionic surfactants is reduced. The PZC for calcite is around 8.5. On the other hand, the use of an alkali improves the efficiency of surfactant flooding in a sandstone reservoir significantly (Volokitin et al., 2018; Hawkins et al., 1994).

3.4.4.2 Crude Oil Properties

The chemical properties of crude oil play a very important role in providing the ultralow IFT value, which is an important mechanism of chemical EOR. For a successful alkaline flooding project, the crude oil must contain a significant amount of acidic components that can react with the injected alkali to form in situ surfactants. An acid number of 0.5 is desirable in alkaline flooding. Ehrlich and Wygal (1976) suggested a minimum acid number of 0.1–0.2 mg KOH/g oil. The oil should preferably have no dissolved acid gases like H_2S and CO_2, which consume much of the injected alkali (Wilson, 1976). The viscosity and API gravity should be ~100 cP and 13°–35° for alkaline flooding, which is the same as for waterflooding.

3.4.4.3 Reservoir Characteristics

The reservoirs that are responsive to waterflooding are favorable for alkaline flooding also. The major requirements and considerations are as follows:

(1) The reservoir should not be very heterogeneous.
(2) Highly faulted and/or fractured reservoirs are unsuitable.
(3) The pay zone must be of a sizable areal extent and sufficiently thick.
(4) An aquifer-connected water-drive reservoir is not suitable for alkaline flooding.
(5) The reservoir permeability and porosity should be well developed.
(6) The reservoir should not have a gas cap.

3.5 ALKALI-SURFACTANT-POLYMER ENHANCED RECOVERY

Currently, the alkali-surfactant-polymer (ASP) enhanced recovery method is considered the most promising chemical method in EOR because it integrates the advantages of alkali surfactant and polymer. There have been several field pilot tests using ASP in the last two decades (Pratap and Gauma, 2004; Guo et al., 2017b; Sheng, 2013a, 2013b). Successful results of pilot tests have been reported from two fields in China. In the Daqing oil field where oil has high paraffin content and low acidity, two tests, the PO and XF pilot tests, were conducted in 1994 through 1996 (Gao et al., 1995; Wang et al., 1997). Wang et al. (1998) also reported another ASP pilot test in the Daqing field, which has been conducted to evaluate the process applicability in the flooded-out zones.

The effect of ASP flooding in EOR depends on many factors, including reservoir geological features, oil-water interfacial tension, reservoir wettability, and rheology of displacing fluid. The behavior of the rheological properties of displacing fluids determines the oil displacement efficiency (Wang et al., 2000, 2001). The oil displacement efficiency of ASP flooding is better than that of individual polymer, surfactant, and alkali, as all the different displacement mechanisms are applicable in the former. The ultralow interfacial tension between oil and water can be obtained using specific surfactants. Alkalis can also reduce the interfacial tension by forming in situ surfactants by reacting with the acidic components of crude oil. The use of a combination of alkali and surfactant not only has a good, synergistic effect on the reduction of interfacial tension, but the alkali also reduces the absorption loss of costly surfactants. A complete knowledge of the rheology of polymers in ASP flooding is essential to examine its ability to improve mobility control during the displacement process (Xia et al., 1999). ASP flooding is a technique developed on the basis of alkali flooding, surfactant flooding, and polymer flooding, and oil recovery is enhanced gently by decreasing interfacial tension, increasing capillary number, enhancing microscopic displacement efficiency, improving mobility ratio, and increasing macroscopic sweep efficiency (Shen and Yu, 2002).

3.5.1 Mechanism of ASP Flooding

The important mechanism of ASP flooding is that the in situ formed soaps—as a result of the alkali reacting with the naturally occurring organic acid in crude oil—interact synergistically

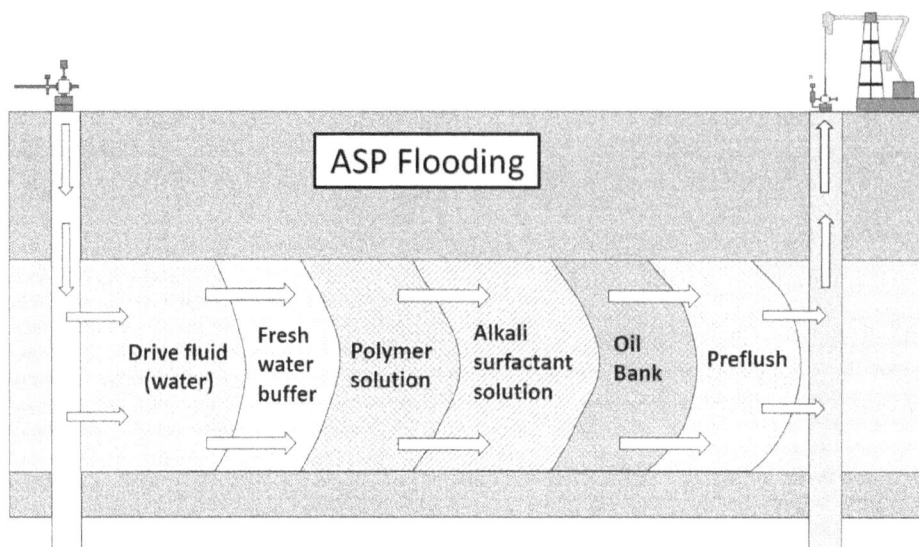

FIGURE 3.58   Schematic of ASP flooding.

Source: www.swiftts.com/Chemical_Injuction.html.

with the added surfactant to produce ultralow interracial tension (IFT). The ultralow IFT of ASP/crude oil may be caused by surfactant distribution between the oil and water phases and surfactant arrangement at interface of oil/water. This is controlled by pH value and ionic strength (Yang et al., 1995). The alkalis injected with surfactants can reduce surfactant adsorption, improve ionic strength, and lower IFT. The surfactant added with the alkalis can raise the salinity requirement of the alkali (Nelson et al., 1984). Under the same displacement efficiency as that of surfactant/polymer flooding, ASP and surfactant-alkali-polymer (SAP) flooding will reduce the concentration of surfactant by more than ten times, as well as the capital cost of the surfactant.

A schematic diagram of ASP flooding is shown in Figure 3.58, which clearly shows the typical stages in an ASP flooding process: a preflush of brine to lower the salinity of the reservoir, an ASP solution used to reduce the interfacial tension between the aqueous and oleic phases, a polymer solution to perform a uniform sweep of the oil and the previous slugs, and chase water to finally drive the oil and the chemicals to the producer wells. Finally, water drive is used to move the front toward the producing well.

3.5.2   LAB-SCALE STUDY

A detailed flow diagram of ASP flooding operation in a laboratory is given in Figure 3.59 (Samanta et al., 2011). The cores were initially flooded with oil to reduce water saturation followed by water (brine) flooding. After injection of 5PV of brine solution, an ASP slug of 0.40 PV was injected followed by 0.8 PV of polymer (2,000 ppm) and 2.8 PV brine solution as chase water.

The additional recovery of oil by ASP flooding and subsequent polymer and chase waterflooding are shown in Figure 3.60. From the figure, it can be seen that the enhanced recoveries by flooding of ASP slug and polymer solution are more than 20% OOIP. This additional recovery of oil is due to significant reduction of IFT between the displacing fluid and crude oil and improvement of viscosity of ASP slug.

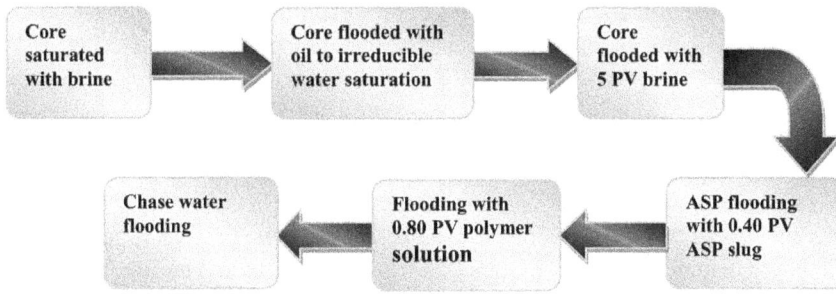

FIGURE 3.59   Flow diagram of ASP flooding.

Source: Samanta, 2011.

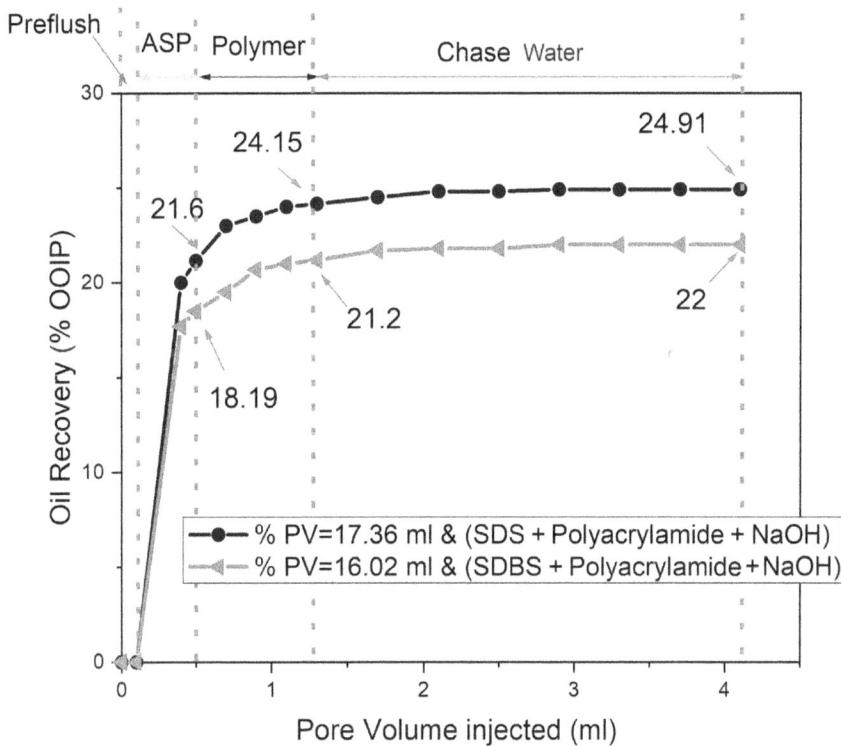

FIGURE 3.60   Enhanced oil recovery by the injected ASP slugs after waterflooding.

Source: Samanta, 2011.

3.6   FOAM FLOODING

Gas flooding is one of the preferred EOR methods for producing oil from a mature field because of its better microscopic sweep through the fine pores of the reservoir compared to a waterflood (Farajzadeh et al., 2012). However, the problem associated with gas flooding is poor volumetric sweep efficiency, viscous fingering, and gravity separation because of low viscosity and density of gas (Yan et al., 2006). Gas channeling through high-permeability streaks in the porous medium further magnifies its poor volumetric sweep efficiency. Water alternating gas (WAG) injection—i.e., the injection of gas slugs alternated by slugs of water—has been successfully applied for partially overcoming the drawbacks of continuous gas injection. Nonetheless, gravity segregation might also occur during WAG flooding, yielding again an early breakthrough of gas (Janssen et al., 2019).

Foam-assisted EOR has gained the increasing attention of oil and gas companies as well as researchers mostly due to its potential advantages over other EOR methods (Guo et al., 2011; Bera et al., 2013). The usage of foam in EOR improves gas mobility by increasing the effective viscosity and decreasing the relative permeability to gas, which results increase in sweep efficiency. Foam involves a discontinuous gas phase—i.e., gas bubbles—within a continuous liquid phase. Foam stability is a strong function of the lamellae thickness—i.e., thin aqueous films that separate gas bubbles within the foam texture, where thinner lamellae tend to rupture more easily. The formation of foam is facilitated through the addition of a foaming agent—for example, a surfactant—to the aqueous phase that inhibits the coalescence of separated gas bubbles, thus promoting foam stability.

The injection of gas along with surfactant solution can produce a foam that exhibits an apparent viscosity much higher than its constituent phases: liquid and gas. Because of its high apparent viscosity, foam has a mobility much lower than the mobility of gas alone and, therefore, provides an effective means of reducing the mobility of the displacing fluid (Schramm and Wassmuth, 1994; Rossen, 2017). Foam can also reduce gravity override and viscous fingering caused by the low density and viscosity of the gas (Li et al., 2010). Surface tension is one of the most important physical characteristics of the liquid phase, which should be low for effective foaming. This can be achieved by adding different types of surfactants in the aqueous phase, which stabilizes the dispersions of gas in water to form metastable foam. Co-injection of surfactant solution and gas leads to the generation of foam inside the reservoir. Along with surfactants, the use of other chemicals such as salt, nanoparticles, alcohols, polymer, alkalis, etc. have some notable effects on foam behavior.

The most common expansion gases are air, carbon dioxide, and low MW hydrocarbons (Turta and Singhal, 2002). The dispersion of these gases can also reduce the surface tension of the liquid system. The implementation of CO_2 flooding is used in mature field because of its capability to reduce oil viscosity and make crude oil swell, hence, improving oil mobility. A CO_2 slug combined with a surfactant slug generally produces foam that significantly reduces the gas mobility and blocks the high-permeability zones in the reservoir. Figure 3.61 shows a schematic of the displacement patterns in continuous gas (CO_2) injection (Fig. 3.61a), conventional WAG injection (Fig. 3.61b), and foam assisted water alternating gas (FAWAG) (in Fig. 3.61c).

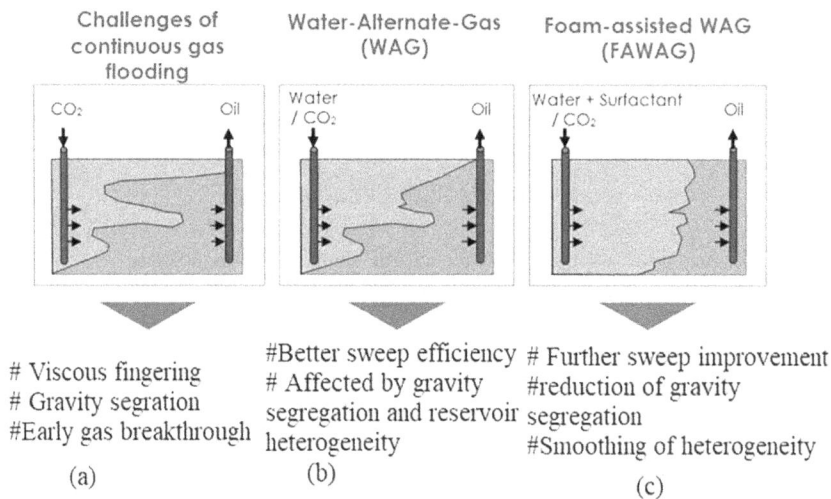

FIGURE 3.61   Schematic of challenges and benefits of (a) continuous gas injection, (b) conventional WAG injection, and (c) FAWAG injection in a reservoir.

Source: Modified after www.eor-alliance.com/solutions/foam.

3.6.1 FOAM PROPERTIES

Several properties important to the characterization of bulk foam, as might exist in a bottle, are foam quality, foam texture, bubble size distribution, foam stability, and foam density. Foam quality is the volume percent gas within foam at a specified pressure and temperature. Foam qualities can exceed 97%. Oil field conformance-improvement foams typically have foam qualities in the range of 75%–90%. When propagated through porous media, the mobility of many foams decreases as foam quality increases up to the upper limit of foam stability in terms of foam quality. Foam texture is a measure of the average gas bubble size. In general, as foam texture becomes finer, the foam will have greater resistance to flow in matrix rock. Bubble size of gas in foam is a measure of the gas bubble size distribution in a foam. When holding all other variables constant, a bulk foam with a broad gas bubble size distribution will be less stable because of gas diffusion from small to large gas bubbles. The imparted resistance to fluid flow in porous media by a foam will be higher when the bubble size is relatively homogeneous (Schramm and Wassmuth, 1994). The stability of an aqueous-based foam is dependent on the chemical and physical properties of the surfactant-stabilized water film separating the foam's gas bubbles. Foams are metastable entities; therefore, all foams will eventually break down. Factors affecting foam lamellae stability include gravity drainage, capillary suction, surface elasticity, viscosity (bulk and surface), electric double-layer repulsion, and steric repulsion.

Interfacial forces that arise in foam films confined between gas bubbles are critical to better understand the drainage properties and stability. The classical Derjaguin-Landau-Verwey-Overbeek (DLVO) theory is commonly used to explain the stability of foams related to the development of the thermodynamics of thin films (Exerowa et al., 1987). In foam films, the hydrophobic force may play a role, in which case, the DLVO theory can be extended for the contributions of hydrophobic force (Yoon and Aksoy, 1999).

$$\Pi_{dis} = \Pi_{vW} + \Pi_{el} + \Pi_{hb} \tag{3.40}$$

where Π_{dis} is the disjoining pressure, Π_{vW} is van der Waals forces, Π_{el} is electric double-layer forces, and Π_{hb} is hydrophobic force.

In foam stability study, the half-decay time ($t_{1/2}$) is defined as the time taken by the foam to reach half of its initial value. It describes the foam stability such that a longer ($t_{1/2}$) corresponds to more stable foam, and it is measured for each sample by a shaker bottle test as well as a bulk foam test. When surfactant solution is mixed or purged with any gaseous phase (air or any individual gas), foam is formed. As aqueous foam is a metastable system, drainage of aqueous phase from plateau border or lamella of foam bubbles starts, which is a kinetic process. The withdrawing of liquid through the lamella is due to gravity, which results in thinning of the top part, and fewer surfactant molecules are occupied; hence, a surface tension gradient is generated inside the bubble (Fig. 3.62a–b).

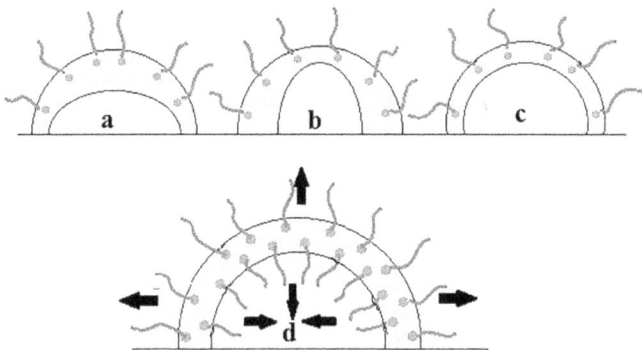

FIGURE 3.62 Stabilization of foam bubble during drainage process.

The surface tension at the top is greater than at other parts of the foam bubble. This creates an unbalance in surface forces, and again, liquid flows back from bottom to the top part of the foam bubble to stabilize the forces acting on it (Fig. 3.62b–c). Figure 3.62 (d) represents a stabilized foam bubble in a static condition. In static condition, the same polar heads start to repel each other and create opposite forces. This balances at some particular value of liquid holdup inside the foam bubble.

3.6.1.1 Foam Stability in the Presence of Different Surfactants at Their Varying Concentrations

Foam stability is significantly dependent on the type of surfactants. It is observed that for stabilization of CO_2 foam, the ionic surfactants—SDS and CTAB—form more stable foams due to the higher strength of the electrostatic double-layer formation, resulting from charge interactions at the film interface, whereas Tween 80 shows poor results on foam stability due to its large surface area per unit molecule (Fig. 3.63) (Kumar and Mandal, 2016). Due to the size and nature of the head group of Tween 80 surfactant, the Gibbs-Marangoni effect is poor; hence, a faster foam decay is observed. It is also observed that a lower concentration of the surfactant is not efficient for a longer life of the foam, whereas foam life at CMC of each surfactant is observed to be quite longer. This is due to the optimization of surface and interfacial properties of the surfactant at CMC, and further addition does not show significant effect on foam stability. SDS shows a larger foam life and slower drainage rate than CTAB and Tween 80. Tween 80 surfactant shows a half-life time of around 7 minutes, whereas SDS and CTAB show a half-life time of around 20 and 11 minutes, respectively, at their corresponding CMCs (Kumar, 2017).

Because of a different rate of drainage for different surfactant systems and the inconvenience in measuring initial drainage, the midrange study is preferred for better understanding (Iglesias et al., 1995). Thus, the time $t_{1/2}$ is taken as the reference at which the foam height is half of the original foam height h_o. From the plot of the dimensionless form of $(t/t_{1/2})$ and (h/h_o) (Fig. 3.64), it is predicted that all the data lines would cross through the central point of the diagram—i.e., $h/h_o = 0.5$ and $\log (t/t_{1/2}) = 0.0$.

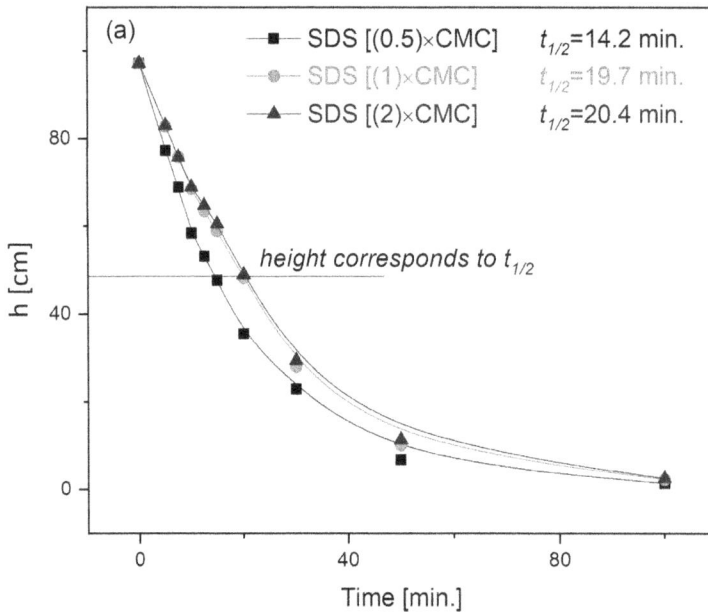

FIGURE 3.63 *(Continued)*

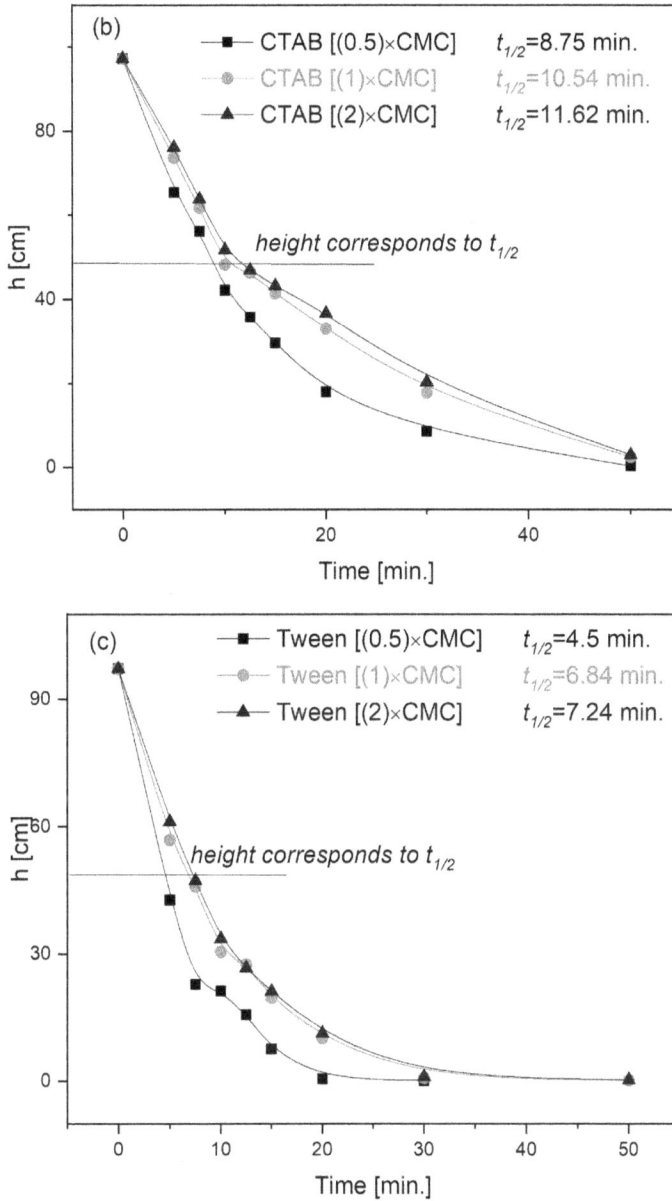

FIGURE 3.63   Variation of foam height with elapse of time of CO_2 gas–based foam in the presence of different surfactants: (a) SDS, (b) CTAB, and (c) Tween 80.

Source: Kumar, 2017.

The general equation for this line is given as follows:

$$h/h_o = 1/2 - \alpha \log \left(t/t_{1/2} \right) \tag{3.41}$$

where h_o is the initial foam height and h is the foam height at any time t. The dimensionless height and time data (or characteristic parameters) for all the three surfactants at their different concentrations fitted in Equation 3.41 pass through the predicted point (0, 0.5). For a particular surfactant system, α does vary only marginally.

3.6.1.2 Effect of Gas Flow Rate on Bubble Size and Foam Generation

The foam stability of liquid depends not only on its chemical properties but also on the bubble size. Jung and Richard et al. investigated the effect of gas flow rate on the bubble diameter (Jung and Fruehan, 2000). They estimated that, at lower gas flow rates, the bubble diameter does not vary, but at higher flow rates, the size of the bubble increases monotonically. It was observed that, as the gas flow rate increases, the foamy region decreases rapidly. This suggests that, as the gas flow rate increases, bubble size increases—i.e., they are less packed, and their frequency is reduced. Therefore, there exists a minimum gas flow rate and a corresponding void fraction for stable foaming beyond which the foam is unstable and instantly drained out. The larger bubble obtained at a higher flow rate may be due to the maximum amount of gas being present inside the bubble.

3.6.2 Foam-Enhanced CO_2-EOR

Bond and Holbrook (1958) were the ones who first claimed that foam generated in oil reservoir by consecutive injection of aqueous surfactant solution and gas, for both miscible and immiscible gas drives, could increase sweep efficiency. In CO_2-based foam injection, co-injection or alternate injection of surfactant solution and CO_2 result in the formation of foam in porous media. Foam-assisted CO_2-WAG can benefit from combined effect of chemical and CO_2-EOR governing mechanisms, known as low-tension gas process. The mechanisms of foam-enhanced CO_2-EOR includes (i) stabilization of the displacement process due to apparent increase in gas viscosity and by reducing the

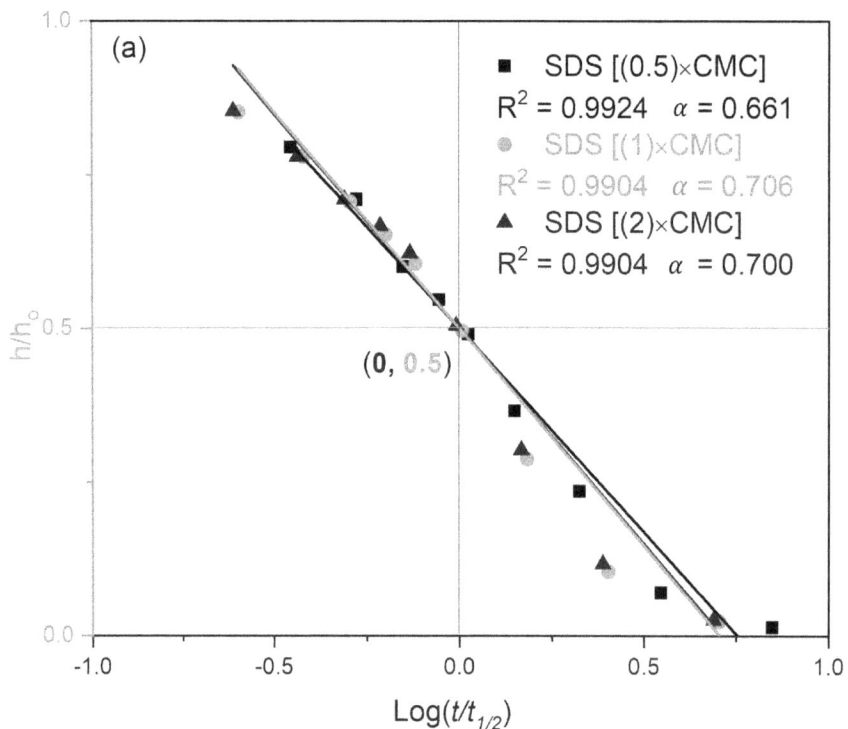

FIGURE 3.64 *(Continued)*

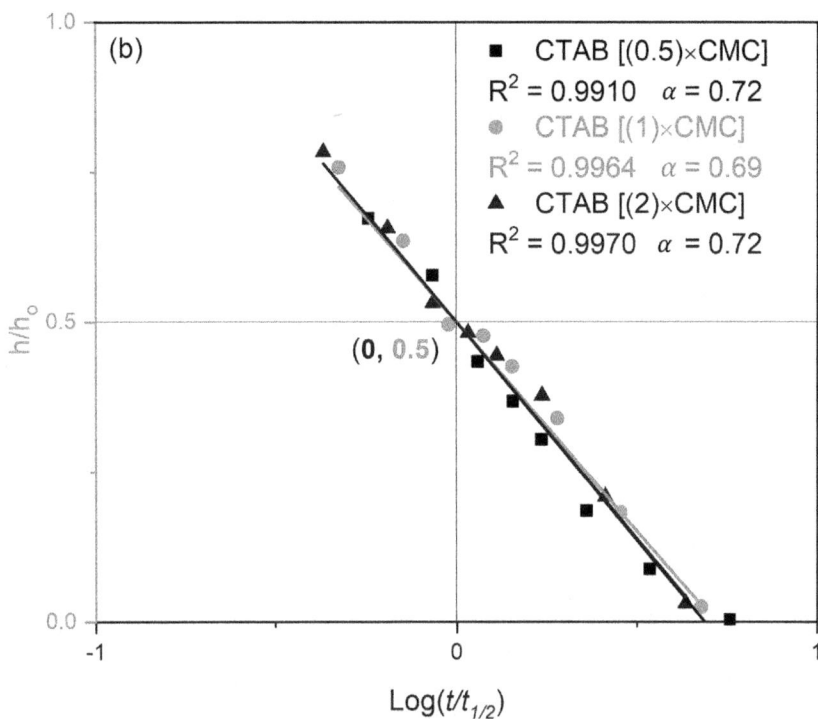

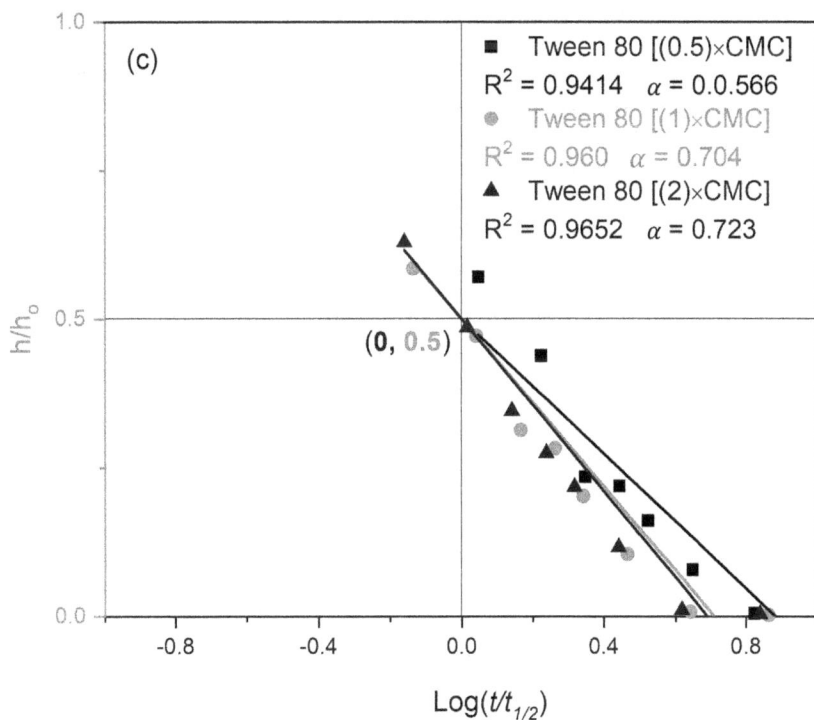

FIGURE 3.64 Dimensionless plot for half-decay time and characteristic time at different surfactant concentrations for (a) SDS, (b) CTAB, and (c) Tween 80.

Source: Kumar, 2017.

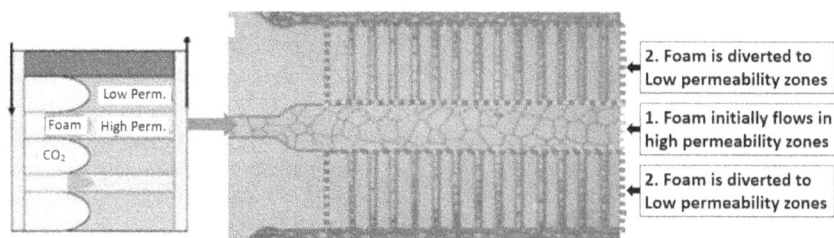

FIGURE 3.65 Foam SMR property at macro and micro scales: Foam efficiently diverts gas bubbles from high- to low-permeability zones.

Source: With courtesy to http://ceor-alliance.com.

capillary forces between oil and water in an ultralow IFT condition due to surfactant presence; (ii) alleviation of the gravity segregation; (iii) IFT reduction; (iv) surfactant-induced wettability alteration; and (v) interfacial mass transfer between the gas and liquid phases.

In addition, foam can create mobility control in a selective manner that can smooth heterogeneities. One of the advantages of foam over polymer is that foam is stronger in high-permeability regions and acts like a more viscous fluid than in low-permeability rocks. The strong foam formed in high-permeability regions diverts the injected fluid to low-permeability regions and, thus, provides mobility and conformance control. This property of foam flow is known as selective mobility reduction (SMR). The SMR function of foam at high gas fraction at both macro (reservoir rock of high-permeability contrast layers) and micro (layered micromodel) scales is illustrated in Figure 3.65.

When the foam eventually ruptures in porous media, the surfactant can cause IFT reduction between trapped oil in unswept channels and water, resulting in a large increase in N_c (capillary number), which is the target of the foam-assisted EOR process. The wettability alteration in foaming processes occurs as consequence of the interaction of surfactants with the solid surface of the porous rock. A surfactant, by adsorbing at the solid surface, reduces interfacial tension and modifies the wetting preference of rock to water or oil. Mass transfer between the gaseous CO_2-rich phase with oil phase has meaningful implications for oil recovery and is of great importance in this field. Foam provides a higher retention time for CO_2 to contact with oil, which results in more CO_2 dissolution in oil. Interfacial mass transfer between gas and oil and gas dissolution leads to oil viscosity reduction and oil swelling, which mobilizes more oil in place (Farajzadeh et al., 2012).

A sufficient amount of effective foam is indispensable for each type of mobility reduction mechanism. Thus, it is practical to understand and apply the necessary conditions to generate a strong and stable foam, with the ability to propagate a considerable distance as well, that can cause the transportation of oil droplets before the rupture. Factors that have significant influences on the propagation of foam through porous media include capillary pressure, surfactant concentration, crude oil and its composition, oil saturation, etc.

3.7 CASE STUDIES

3.7.1 CASE STUDY 1

The West Salym oil field is located in the West Siberian oil province (Russia) with stock tank oil initially in place (STOIIP) of 280 MMm³. It is a sandstone formation with temperatures as high as

83°C. The crude oil is of low viscosities of about 2 cP, and brine salinities in the range of 14,000–16,000 ppm. The other data are reported in Table 3.10 (Volokitin et al., 2018). The field was water-flooded to maintain the reservoir pressure close to its initial level and to optimize oil recovery. Oil production from West Salym peaked in 2011 and since declined with increasing water cuts. The expected ultimate recovery factor due to waterflood, as reported in the field development plan, was between 35% and 40%. To increase the recovery factor, a tertiary oil recovery technique, ASP flooding, was selected.

Laboratory studies were started in 2008 with the surfactant/polymer screening and selection followed by core flooding experiments. In 2009, a successful single-well chemical tracer test was conducted to prove the efficiency of the developed ASP formulation at field conditions. The ASP pilot project was taken by Salym Petroleum Development shareholders at the end of 2012 with two major project objectives: (1) demonstrate that tertiary mode ASP flooding could lead to a significant incremental oil recovery; and (ii) collect enough information for taking decisions on follow-up commercial ASP projects. The pilot operation was started in February 2016 with the start of the ASP mixing and injection plant. Active injection was completed in January 2018.

ASP injection resulted in the mobilization of the remaining oil inside the pilot pattern. Total oil recovery achieved during the operation of the ASP pilot reached 3,400 tons of oil that corresponds to RF = 16%, which was in line with the initial estimate. Low remaining oil saturation prior to ASP injection (pre-ASP recovery factor ~52%) and uneven vertical and areal sweep had a major impact on the oil recovery efficiency by ASP flooding. However, the project demonstrated good long-term injectivity of both ASP slug and polymer-drive solutions into low-permeable (50–200 mD) formation under controlled hydraulic fracturing. The pilot summary of production is reported in Table 3.11.

3.7.2 Case Study 2

Most of the oil fields in the northern part of Oman are tight carbonate oil reservoirs. One such field underwent a special type of polymer injection for EOR (Wu et al., 2019). Initially, the field was produced under natural depletion for almost 15 years until 2005 when a line drive waterflood development with horizontal wells took place and was deployed in the whole field. After more than 10 years of water injection, the water cut reached an average of 75% in the major producing blocks. The reservoir has light oil with viscosity of 0.8 m-Pa-s, a downhole temperature of 87°C, and average permeability of 10 mD. The calcium and magnesium concentration in formation water is high, about 4,000 mg/L.

TABLE 3.10
West Salym Crude and Reservoir Properties

Properties	Values
Reservoir temperature, °C	83
Oil acid number of TAN, mg KOH/g	0.04
Oil viscosity (@ reservoir conditions), cp	2
Reservoir brine salinity, ppm	14,000–16,000
Reservoir brine hardness, ppm	~200
Oil density (@surface conditions), ton/m^3	0.87

Reservoir heterogeneity in tight carbonate reservoirs causes uneven waterflood sweep efficiency and, hence, resulted in plenty of bypassed oil. The conventional polymer was found to be ineffective in terms of additional recovery. However, a new, unique nano-ploymer was found to be a potential EOR method for such tight formation reservoirs. The incremental recovery based on the laboratory and simulation study is reported in Table 3.12. The economic evaluation showed that the best economic scenario is to go for PV injection of 0.3, achievable in ~4 years' time. The pilot tests were also conducted, and the results are encouraging.

3.7.3   CASE STUDY 3

The Viraj field lies in the Ahmedabad-Mehsana tectonic block and is situated at a distance of 45 km north of the Jhalora field in the Cambay basin, India. It has mainly four oil-bearing pay zones: KS-VIII, KS-IX + X (L1 +L2), Chattral member of Kadi formation (L3), and South Kadi sand "C+D." The reservoir is operating under water drive. The field was put on production in July 1980, and till April 1, 2002, the recovery was only 18% of the oil initially in place. The details of the reservoir rock and fluids properties are given in Table 3.13. The initial pressure was 135.5 ksc at 1,265 m datum.

The porosity was of the order of 28%–30%, and initial average oil saturation was in the range of 66%–70%. The pilot scheme was prepared with the use of existing producing wells in which the inter-well spacing is 200–250 m. It was the first reported ASP injection as a chemical EOR

TABLE 3.11
Pilot Production Summary (Volokitin et al., 2018)

Parameters	Values
Pilot pore volume Km³	40
STOIIP, Ktons	21
Initial oil saturation, %	58
Averaged oil saturation before pilot operation, %	28
Recovery factor before pilot operation, %	52
Water cut before ASP injection (producing well), %	98
Minimum water cut during ASP injection (producing well), %	88
Cumulative oil production, Ktons	3.4
Incremental recovery factor, %	16

TABLE 3.12
Incremental Oil Production Nano- to Micro-Sized Particle-Type Polymer SMG (Soft Microgel) (Wu et al., 2019)

Plan No.	SMG Injected, PV	Injection Period, Years	Recovery Incremental, %
1	0.05	0.94	1.2
2	0.1	1.51	3
3	0.2	2.62	5.8
4	0.3	3.73	8.3
5	0.5	5.94	10.4

TABLE 3.13
Reservoir and Fluid Properties (Singh et al., 2017)

Reservoir

Depth	1,300 m
Temperature	81°C
Porosity	31%
Permeability	1–10 Darcy

Oil and Water Properties

Gravity	18.9° API
Viscosity	50 cp
Acid number	1.825 mg KOH/g
Asphaltene (%w/w)	4.48%
Resin (%w/w)	18%
Wax (%w/w)	5.67%

technology implemented in the Viraj field of the Ahmedabad Asset. The total oil production by EOR and mobilized oil obtained by IOR (infill locations) was 156,476 m³, which is about 87% of the envisaged oil production from the pilot.

3.7.4 CASE STUDY 4

The Mangala field, located in the western desert state of Rajasthan, India, contains nearly 1,300 million barrels of oil in place and has been regarded as a very good chemical EOR candidate since its discovery in 2004 (Pandey et al., 2020). Currently, the full field is under polymer flooding, with results largely in line with expectations. The field, which has already produced more than 30% of the STOIIP, is on a declining production trend and large-scale ASP is being considered as the next development initiative to arrest the production decline and increase the overall recovery factor from the field.

The Mangala EOR pilot is a normal five-spot pilot with a central producer surrounded by four injectors with injector-injector spacing of ~100 m and injector-producer spacing of ~70 m. The same wells were used to conduct both the polymer as well as the ASP pilot. The development plan encompasses learning from the successful ASP pilot executed during 2014–2015 and is based on the detailed modeling work using a simulator calibrated to a history-match exercise for the core flood and pilot results. The modeling studies indicate a potential of around 100 million barrels of additional cumulative oil production in a ten-year time frame over the base case of full-field polymer flooding, which has been in execution since 2014.

REFERENCES

Abidin, A.Z., Puspasari, T., Nugroho, W.A., 2012. Polymers for enhanced oil recovery technology. Procedia Chemistry 4, 11–16.
Adamson, A.W., 1990. Physical Chemistry of Surfaces. Wiley-Interscience.
Ahmadi, M.A., Arabsahebi, Y., Shadizadeh, S.R., Behbahani, S.S., 2014. Preliminary evaluation of mulberry leaf-derived surfactant on interfacial tension in an oil-aqueous system: EOR application. Fuel 117, 749–755.

Akbari, S., Mahmood, S.M., Nasr, N.H., Al-Hajri, S., Sabet, M., 2019. A critical review of concept and methods related to accessible pore volume during polymer-enhanced oil recovery. Journal of Petroleum Science and Engineering 182, 106263.

Aladasani, A., Bai, B., 2010, 8–10 June. Recent Developments and Updated Screening Criteria of Enhanced Oil Recovery Techniques. In International Oil and Gas Conference and Exhibition in China, Beijing, China. http://dx.doi.org/10.2118/130726-MS.

Al-Bahar, M.A., Merrill, R., Peake, W., Jumaa, M., Oskui, R., 2004, 10–13 October. Evaluation of IOR Potential Within Kuwait. In Abu Dhabi International Petroleum Exhibition and Conference, SPE-88716. SPE.

Alfazazi, U., AlAmeri, W., Hashmet, M.R., 2019. Experimental investigation of polymer flooding with low-salinity preconditioning of high temperature–High-salinity carbonate reservoir. Journal of Petroleum Exploration and Production Technology 9(2), 1517–1530.

Al-Hajri, S., Mahmood, S.M., Abdulelah, H., Akbari, S., 2018. An overview on polymer retention in porous media. Energies 11(10), 2751.

Al-Murayri, M.T., Al-Kharji, A.A., Kamal, D.S., Al-Ajmi, M.F., Al-Ajmi, R.N., Al-Shammari, M.J., Al-Asfoor, T.H., Badham, S.J., Bouma, C., Brown, J., Suniga, H.P., 2018, March. Successful Implementation of a One-Spot Alkaline-Surfactant-Polymer ASP Pilot in a Giant Carbonate Reservoir. In SPE EOR Conference at Oil and Gas West Asia. OnePetro.

Alvarado, D.A., Marsden Jr., S.S., 1979. Flow of oil-in-water emulsions through tubes and porous media. Society of Petroleum Engineers Journal 19(06), 369–377.

Andersen, P.Ø., 2020. Capillary pressure effects on estimating the enhanced-oil-recovery potential during low-salinity and smart waterflooding. SPE Journal 25(1), 481–496.

Azouz, I.B., Ober, R., Nakache, E., Williams, C.E., 1992. A small angle X-ray scattering investigation of the structure of a ternary water-in-oil microemulsion. Colloids and Surfaces 69(2–3), 87–97.

Babu, K., Pal, N., Bera, A., Saxena, V.K., Mandal, A., 2015. Studies on interfacial tension and contact angle of synthesized surfactant and polymeric from castor oil for enhanced oil recovery. Applied Surface Science 353, 1126–1136.

Bansal, V.K., Shah, D.O., O'Connell, J.P., 1980. Influence of alkyl chain length compatibility on microemulsion structure and solubilization. Journal of Colloid and Interface Science 75(2), 462–475.

Barnes, H.A., 1999. The yield stress—A review or 'παντα ρει'—Everything flows? Journal of Non-Newtonian Fluid Mechanics 81(1–2), 133–178.

Barnes, J.R., Smit, J., Smit, J., Shpakoff, G., Raney, K.H., Puerto, M., 2008, April. Development of Surfactants for Chemical Flooding at Difficult Reservoir Conditions. In SPE Symposium on Improved Oil Recovery. OnePetro.

Batıgöç, C., Akbaş, H., 2017. Thermodynamic parameters of clouding phenomenon in nonionic surfactants: The effect of the electrolytes. Journal of Molecular Liquids 231, 509–513.

Bera, A., Kissmathulla, S., Ojha, K., Kumar, T., Mandal, A., 2012. Mechanistic study of wettability alteration of quartz surface induced by nonionic surfactants and interaction between crude oil and quartz in the presence of sodium chloride salt. Energy & Fuels 26, 3634–3643.

Bera, A., Ojha, K., Mandal, A., 2013. Synergistic effect of mixed surfactant systems on foam behavior and surface tension. Journal of Surfactants and Detergents 16(4), 621–630.

Berger, P.D., Lee, C.H., 2006, April. Improved ASP Process Using Organic Alkali. In SPE/DOE Symposium on Improved Oil Recovery. OnePetro.

Bobek, J.E., Mattax, C.C., Denekas, M.O., 1958. Reservoir rock wettability-its significance and evaluation. Transactions of the AIME 213(1), 155–160.

Bond, D.C., Holbrook, O.C., 1958, 30 December. U.S. Patent No. 2,866,507.

Borse, M., Sharma, V., Aswal, V.K., Goyal, P.S., Devi, S., 2006. Aggregation properties of mixed surfactant systems of dimeric butane-1,4-bis(dodecylhydroxyethylmethylammonium bromide) and its monomeric counterpart. Colloids and Surfaces A: Physicochemical and Engineering Aspects 287(1–3), 163–169.

Bourrel, M., Schechter, R.S., 2010. Microemulsions and Related Systems: Formulation, Solvency, and Physical Properties. Editions Technip.

Brashear, J.P., Kuuskraa, V.A., 1978. The potential and economics of enhanced oil recovery. Journal of Petroleum Technology 30(9), 1231–1239. http://dx.doi.org/10.2118/06350-PA

Carcoana, A.N., 1982, 4–7 April. Enhanced Oil Recovery in Rumania. In Paper SPE 10699 presented at the SPE Enhanced Oil Recovery Symposium, Tulsa, OK. http://dx.doi.org/10.2118/10699-MS

Carman, P.S., Cawiezel, K.E., Co, B.J.S., 2007. Successful Breaker Optimization for Polyacrylamide Friction Reducers Used in Slickwater Fracturing. In SPE Hydraulic Fracturing Technology Conference. Society of Petroleum Engineers, Richardson, TX.

Carmona, I., Schecter, R.S., Wade, W.H., Weerasooriya, U., 1985. Ethoxylated oleyl sulfonates as model compounds for enhanced oil recovery. Society of Petroleum Engineers Journal 25(3), 351–357.

Carrero, E., Queipo, N.V., Pintos, S., Zerpa, L.E., 2007. Global sensitivity analysis of Alkali–Surfactant–Polymer enhanced oil recovery processes. Journal of Petroleum Science and Engineering 58, 30–42.

Castor, T.P., 1979. Enhanced recovery of acidic oils by alkaline agents (Doctoral Dissertation, University of California, Berkeley).

Chan, M., Yen, T.F., 1982. A chemical equilibrium model for interfacial activity of crude oil in aqueous alkaline solution: The effects of pH, alkali and salt. The Canadian Journal of Chemical Engineering 60(2), 305–308.

Chen, H., Mou, D., Du, D., Chang, X., Zhu, D., Liu, J., Xu, H., Yang, X., 2007. Hydrogel-thickened microemulsion for topical administration of drug molecule at an extremely low concentration. International Journal of Pharmaceutics 341(1–2), 78–84.

Chevalier, Y., Melis, F., Dalbiez, J.P., 1992. Structure of zwitterionic surfactant micelles: Micellar size and intermicellar interactions. Journal of Physical Chemistry 96, 8614–8619.

Chou, S.I., Shah, D.O., 1981. The optimal salinity concept for oil displacement by oil-external microemulsions and graded salinity slugs. Journal of Canadian Petroleum Technology 20.

Chu, Z., Feng, Y., Su, X., Han, Y., 2010. Wormlike micelles and solution properties of a C22-tailed amidosulfobetaine surfactant. Langmuir 26, 7783–7791.

Cooke, C.E., Williams, R.E., Kolodzie, P.A., 1974. Oil recovery by alkaline waterflooding. Journal of Petroleum Technology 26(12), 1365–1374.

Craig Jr, F.F., 1971. The Reservoir Engineering Aspects of Waterflooding (Monograph Series 3, Henry L. Doherty Series). SPE, Dallas, TX, Owens and Archie. 3.

Dang, T.Q.C., Chen, Z., Nguyen, T.B.N., Bae, W., 2014. The potential of enhanced oil recovery by micellar/polymer flooding in heterogeneous reservoirs. Energy Sources, Part A: Recovery, Utilization, and Environmental Effects 36(14), 1540–1554.

Davies, J.T., 1957. A Quantitative Kinetic Theory of Emulsion Type, I. Physical Chemistry of the Emulsifying Agent, in: Gas/Liquid and Liquid/Liquid Interface. Proceedings of the International Congress of Surface Activity (Vol. 1). Citeseer. pp. 426–438.

Dickson, J.L., Dios, A.L., Wylie, P.L., 2010, 24–28 April. Development of Improved Hydrocarbon Recovery Screening Methodologies. In Paper SPE 129768 presented at the SPE Improved Oil Recovery Symposium, Tulsa, OK. http://dx.doi.org/10.2118/129768-MS.

Domanska, U., 2005. Solubilities and thermophysical properties of ionic liquids. Pure and Applied Chemistry 77, 543–557.

Dominguez, J.G., Willhite, G.P., 1977. Retention and flow characteristics of polymer solutions in porous media. Society of Petroleum Engineers Journal 17, 111–121.

Donaldson, E.C., Alam, W., 2013, 25 November. Wettability. Elsevier.

Donaldson, E.C., Chilingarian, G.V., Yen, T.F., (eds), 1989. Enhanced Oil Recovery, II: Processes and Operations. Elsevier.

Dubinin, M.M., 1960. The potential theory of adsorption of gases and vapors for adsorbents with energetically nonuniform surfaces. Chemical Reviews 60(2), 235–241.

Dyes, A.B., Caudle, B.H., Erickson, R.A., 1954. Oil production after breakthrough as influenced by mobility ratio. Transactions of the AIME 201, 81–86.

Ehrlich, R., Wygal Jr, R.J., 1976, 22–24 March. Interrelation of Crude Oil and Rock Properties With the Recovery of Oil by Caustic Waterflooding. In SPE-AIME Symposium on Improved Oil Recovery, Tulsa, OK, SPE 5830.

Elraies, K.A., Tan, I.M., Awang, M., Saaid, I., 2010. The synthesis and performance of sodium methyl ester sulfonate for enhanced oil recovery. Petroleum Science and Technology 28, 1799–1806.

Elraies, K.A., Tan, I.M., Fathaddin, M.T., Abo-Jabal, A., 2011. Development of a new polymeric surfactant for chemical enhanced oil recovery. Petroleum Science and Technology 29, 1521–1528.

Engelskirchen, S., Elsner, N., Sottmann, T., Strey, R., 2007. Triacylglycerol microemulsions stabilized by alkyl ethoxylate surfactants—A basic study: Phase behavior, interfacial tension and microstructure. Journal of Colloid and Interface Science 312(1), 114–121.

Exerowa, D., Kolarov, T., Khristov, K., 1987. Direct measurement of disjoining pressure in black foam films. I. Films from an ionic surfactant. Colloids and Surfaces 22(2), 161–169.

Ezell, R.G., McCormick, C.L., 2007. Electrolyte- and pH-responsive polyampholytes with potential as viscosity-control agents in enhanced petroleum recovery. Journal of Applied Polymer Science 104, 2812–2821.

Farajzadeh, R., Andrianov, A., Krastev, R., Hirasaki, G., Rossen, W.R., 2012, 16–18 April. Foam-Oil Interaction in Porous Media: Implications for Foam Assisted Enhanced Oil Recovery. In SPE EOR Conference at Oil and Gas West Asia, Muscat.

Flock, D.L., Le, T.H., Gibeau, J.P., 1986. The effect of temperature on the interfacial tension of heavy crude oils using the pendent drop apparatus. Journal of Canadian Petroleum Technology 25, 72–78.

Freundlich, H.M.F., 1906. Over the adsorption in solution. Journal of Physical Chemistry 57(1906), 1100–1107.

Gale, W.W., Sandvik, E.I., 1973. Tertiary surfactant flooding: Petroleum sulfonate composition-efficacy studies. Society of Petroleum Engineers Journal 13(4), 191–199.

Gao, S., Li, H., Li, H., 1995. Laboratory investigation of combination of alkaline-surfactant-polymer for Daqing EOR. SPE Reservoir Engineering 10(3), 194–197.

Gaudin, T., Rotureau, P., Pezron, I., Fayet, G., 2018. Investigating the impact of sugar-based surfactants structure on surface tension at critical micelle concentration with structure-property relationships. Journal of Colloid and Interface Science 516, 162–171.

Gbadamosi, A.O., Junin, R., Manan, M.A., Agi, A., Yusuff, A.S., 2019. An overview of chemical enhanced oil recovery: Recent advances and prospects. International Nano Letters 9, 171–202.

Glasbergen, G., Wever, D., Keijzer, E., Farajzadeh, R., 2015, October. Injectivity Loss in Polymer Floods: Causes, Preventions and Mitigations. In SPE Kuwait Oil and Gas Show and Conference. OnePetro.

Glover, C.J., Puerto, M.C., Maerker, J.M., Sandvik, E.L., 1979. Surfactant phase behavior and retention in porous media. Society of Petroleum Engineers Journal 19(3), 183–193.

Gogoi, S.B., 2009. Adsorption of non-petroleum base surfactant on reservoir rock. Current Science, 1059–1063.

Goodlett, G.O., Honarpour, M.M., Chung, F.T., Sarathi, P.S., 1986, 19–21 May. The Role of Screening and Laboratory Flow Studies in EOR Process Evaluation. In SPE Rocky Mountain Regional Meeting. OnePetro. http://dx.doi.org/10.2118/15172-MS.

Graciaa, A., Fortney, L.N., Schechter, R.S., Wade, W.H., Yiv, S., 1982. Criteria for structuring surfactants to maximize solubilization of oil and water: Part 1–Commercial nonionics. Society of Petroleum Engineers Journal 22(5), 743–749.

Griffin, W.C., 1949. Classification of surface-active agents by "HLB". Journal of Society of Cosmetic Chemists 1, 311–326.

Grollmann, U., Schnabel, W., 1982. Free radical-induced oxidative degradation of polyacrylamide in aqueous solution. Polymer Degradation and Stability 4, 203–212.

Guo, H., Dou, M., Hanqing, W., Wang, F., Yuanyuan, G., Yu, Z., Yansheng, W., Li, Y., 2017a. Proper use of capillary number in chemical flooding. Journal of Chemistry 2017.

Guo, H., Faber, R., Buijse, M., Zitha, P.L., 2011, July. A Novel Alkaline-Surfactant-Foam EOR Process. In SPE Enhanced Oil Recovery Conference. OnePetro.

Guo, H., Li, Y., Wang, F., Yu, Z., Chen, Z., Wang, Y. and Gao, X., 2017b. ASP flooding: Theory and practice progress in China. Journal of Chemistry 2017.

Habibu, S., Sarih, N.M., Mainal, A., 2018. Synthesis and characterisation of highly branched polyisoprene: Exploiting the "Strathclyde route" in anionic polymerization. RSC Advances 8(21), 11684–11692.

Han, M., Zhou, X., Fuseni, A., Al-Zahrani, B., AlSofi, A., 2012, 16–18 April. Laboratory Investigation of the Injectivity of Sulfonated Polyacrylamide Solutions Into Carbonate Reservoir Rocks. In Proceedings of the SPE EOR Conference at Oil and Gas West Asia, Muscat.

Hashmet, M.R., Onur, M., Tan, I.M., 2014. Empirical correlations for viscosity of polyacrylamide solutions with the effects of temperature and shear rate. II. Journal of Dispersion Science and Technology 35(12), 1685–1690.

Hawkins, B.F., Taylor, K.C., Nasr-El-Din, H.A., 1994. Mechanisms of surfactant and polymer enhanced alkaline flooding: Application to David Lloydminster and wainwright sparky fields. Journal of Canadian Petroleum Technology 33(4).

Hayes, D.G., Smith, G.A., 2019. Biobased Surfactants: Overview and Industrial State of the Art, in: Biobased Surfactants. Elsevier, pp. 3–38.

Healy, R.N., Read, R.L., Stenmark, D.G., 1976. Multiphase microemulsion system. Society of Petroleum Engineers Journal 261, 147–160.

Healy, R.N., Reed, R.L., Carpenter, C.W., 1975. A laboratory study of microemulsion flooding. Society of Petroleum Engineers Journal 259, 87–100.

Hernandez, C., Chacon, L.J., Anselmi, L., 2001, 25–28 March. ASP System Design for an Offshore Application in the La Salina Field, Lake Maracaibo. Paper SPE 69544 Presented at the SPE Latin American and Caribbean Petroleum Engineering Conference, Buenos Aires.

Hezave, A.Z., Dorostkar, S., Ayatollahi, S., Nabipour, M., Hemmateenejad, B., 2013. Effect of different families (imidazolium and pyridinium) of ionic liquids-based surfactants on interfacial tension of water/ crude oil system. Fluid Phase Equilibria 360, 139–145.

Hirasaki, G., Zhang, D.L., 2004. Surface chemistry of oil recovery from fractured, oil-wet, carbonate formations. SPE Journal 9(02), 151–162.

Hirasaki, G.J., Miller, C.A., Puerto, M., 2008. Recent Advances in Surfactant EOR, in: SPE Annual Technical Conference and Exhibition. Society of Petroleum Engineers, pp. 130–164.

Hirasaki, G.J., Van Domselaar, H.R., Nelson, R.C., 1983. Evaluation of the salinity gradient concept in surfactant flooding. Society of Petroleum Engineers Journal 23(3), 486–500.

Huh, C., 1979. Interfacial tensions and solubilizing ability of a microemulsion phase that coexists with oil and brine. Journal of Colloid and Interface Science 71(2), 408–426.

Iglauer, S., Wu, Y., Shuler, P., Tang, Y., Goddard III, W.A., 2010. New surfactant classes for enhanced oil recovery and their tertiary oil recovery potential. Journal of Petroleum Science and Engineering 71(1–2), 23–29.

Iglesias, E., Anderez, J., Forgiarini, A., Salager, J.L., 1995. A new method to estimate the stability of short-life foams. Colloids and Surfaces A: Physicochemical and Engineering Aspects 98(1–2), 167–174.

Jamaloei, B.Y., Kharrat, R., Asghari, K., Torabi, F., 2011. The influence of pore wettability on the microstructure of residual oil in surfactant-enhanced water flooding in heavy oil reservoirs: Implications for pore-scale flow characterization. Journal of Petroleum Science and Engineering 77(1), 121–134.

Janssen, M.T., Mutawa, A.S., Pilus, R.M., Zitha, P.L., 2019. Foam-assisted chemical flooding for enhanced oil recovery: Effects of slug salinity and drive foam strength. Energy & Fuels 33(6), 4951–4963.

Jeirani, Z., Jan, B.M., Ali, B.S., Noor, I.M., See, C.H., Saphanuchart, W., 2013. Formulation, optimization and application of triglyceride microemulsion in enhanced oil recovery. Industrial Crops and Products 43, 6–14.

Johnson, C.E., 1976. Status of caustic and emulsion methods. Journal of Petroleum Technology 28(1), 85–92.

Johnson, K., Pagni, R., Bartmess, J., 2007. Brønsted acids in ionic liquids: Fundamentals, organic reactions, and comparisons. Monatshefte für Chemie—Chemical Monthly 138, 1077–1101.

Jones, S.C., Dreher, K.D., 1976. Cosurfactants in micellar systems used for tertiary oil recovery. Society of Petroleum Engineers Journal 16(3), 161–167.

Jung, S.M., Fruehan, R.J., 2000. Foaming characteristics of BOF slags. ISIJ International 40(4), 348–355.

Kanan, K., Al-Jabari, M., Kayali, I., 2017. Phase behavioral changes in SDS association structures induced by cationic hydrotropes. Arabian Journal of Chemistry 10, S314–S320.

Karaborni, S., Esselink, K., Hilbers, P.A.J., Smit, B., 1994. Simulating surfactant self-assembly. Journal of Physics: Condensed Matter 6(23A), A351.

Karnanda, W., Benzagouta, M.S., 2013. Effect of temperature, pressure, salinity, and surfactant concentration on IFT for surfactant flooding optimization. Arabian Journal of Geosciences, 3535–3544.

Kralova, I., Sjöblom, J., 2009. Surfactants used in food industry: A review. Journal of Dispersion Science and Technology 30(9), 1363–1383.

Krumova, J., Stefanova, D., Deyanov, E., 2015. Drainage of foam films stabilized by nonionic, ionic surfactants and their mixtures. Colloids and Surfaces A: Physicochemical and Engineering Aspects 481, 87–99.

Kumar, M., Hoang, V.T., Satik, C., Rojas, D.H., 2008. High-mobility-ratio waterflood performance prediction: Challenges and new insights. SPE Reservoir Evaluation & Engineering 11, 186–196.

Kumar, R., 2013. Enhanced oil recovery of heavy oils by non-thermal chemical methods (Doctoral Dissertation, The University of Texas at Austin, Austin).

Kumar, S., 2017. Studies on surface and interfacial properties of cationic, anionic and nonionic surfactants for their application in enhanced oil recovery (PhD Thesis).

Kumar, S., Mandal, A., 2016. Studies on interfacial behavior and wettability change phenomena by ionic and nonionic surfactants in presence of alkalis and salt for enhanced oil recovery. Applied Surface Science 372, 42–51.

Kumar, S., Mandal, A., 2017. Investigation on stabilization of CO_2 foam by ionic and nonionic surfactants in presence of different additives for application in enhanced oil recovery. Applied Surface Science 420, 9–20.

Kumar, S., Yen, T.F., Chilingarian, G.V., Donaldson, E.C., 1989. Alkaline Flooding, in: Developments in Petroleum Science (Vol. 17). Elsevier, pp. 219–254.

Kunieda, H., Ishikawa, N., 1985. Evaluation of the hydrophile-lipophile balance (HLB) of nonionic surfactants. II. Commercial-surfactant systems. Journal of Colloid and Interface Science 107, 122–128.

Kwok, D.Y., Lam, C.N.C., Li, A., Leung, A., Wu, R., Mok, E., Neumann, A.W., 1998. Measuring and interpreting contact angles: A complex issue. Colloids and Surfaces A: Physicochemical and Engineering Aspects 142(2–3), 219–235.

Langmuir, I., 1916. The constitution and fundamental properties of solids and liquids. Part I. Solids. Journal of the American Chemical Society 38(11), 2221–2295.

Larson, R.G., Davis, H.T., Scriven, L.E., 1982. Elementary mechanisms of oil recovery by chemical methods. Journal of Petroleum Technology 34(2), 243–258.

Levitt, D.B., Dufour, S., Pope, G., Morel, D., Gauer, P., 2012, February. Design of An ASP Flood in a High-Temperature, High-Salinity, Low-Permeability Carbonate. In IPTC 2012: International Petroleum Technology Conference (pp. cp-280). European Association of Geoscientists & Engineers.

Li, F.T., Liu, Y., Liu, R.H., Sun, Z.M., Zhao, D.S., Kou, C.G., 2010. Preparation of Ca-doped LaFeO$_3$ nanopowders in a reverse microemulsion and their visible light photocatalytic activity. Materials Letters 64(2), 223–225.

Lu, M., Wu, X., Wei, X., 2012. Chemical degradation of polyacrylamide by advanced oxidation processes. Environmental Technology 33, 1021–1028.

Malik, M.A., Wani, M.Y., Hashim, M.A., 2012. Microemulsion method: A novel route to synthesize organic and inorganic nanomaterials: 1st Nano Update. Arabian Journal of Chemistry 5(4), 397–417.

Mandal, A., 2015. Chemical flood enhanced oil recovery: A review. International Journal of Oil Gas and Coal Technology 9, 241.

Manning, R.K., 1983. A technical survey of polymer flooding projects (Doctoral Dissertation). Austin: The University of Texas.

Marszall, L., 1978. Relationship among emulsion type, emulsion stability and the presence of additives. Fette Seifen Anstrichmittel 80, 289–293.

Martin, F.D., Sherwood, N.S., 1975. The Effect of Hydrolysis of Polyacrylamide on Solution Viscosity, Polymer Retention and Flow Resistance Properties. In Paper SPE, 5339.

Massarweh, O., Abushaikha, A.S., 2020. The use of surfactants in enhanced oil recovery: A review of recent advances. Energy Reports 6, 3150–3178.

Masuda, Y., Tang, K.C., Miyazawa, M., Tanaka, S., 1992. 1D simulation of polymer flooding including the viscoelastic effect of polymer solution. SPE Reservoir Engineering 7(2), 247–252.

Mayers, E.H., Berg, R.L., Carrmichael, J.D., Weinbrandt, R.M., 1983. Alkaline Injection for Enhanced Oil Recovery – A Status Report. Journal of Petroleum Technology, 209–221.

McLachlan, A.A., Marangoni, D.G., 2006. Interactions between zwitterionic and conventional anionic and cationic surfactants. Journal of Colloid and Interface Science 295, 243–248.

Melrose, J.C., Brandner, C.F., 1974. Role of capillary forces in determining microscopic displacement efficiency for oil recovery by water flooding. Journal of Canadian Petroleum Technology 13(4), 54–62.

Menger, F.M., Littau, C.A., 1991. Gemini-surfactants: Synthesis and properties. Journal of the American Chemical Society 113(4), 1451–1452.

Menger, F.M., Littau, C.A., 1993. Gemini surfactants: A new class of self-assembling molecules. Journal of the American Chemical Society 115(22), 10083–10090.

Mitra, R.K., Paul, B.K., Moulik, S.P., 2006. Phase behavior, interfacial composition and thermodynamic properties of mixed surfactant (CTAB and Brij-58) derived w/o microemulsions with 1-butanol and 1-pentanol as cosurfactants and n-heptane and n-decane as oils. Journal of Colloid and Interface Science 300(2), 755–764.

Miyake, M., Yamashita, Y., 2017. Molecular Structure and Phase Behavior of Surfactants, in: Cosmetic Science and Technology. Elsevier, pp. 389–414.

Mohammadi, H., Delshad, M., Pope, G.A., 2009. Mechanistic modeling of alkaline/surfactant/polymer floods. Society of Petroleum Engineers 110212, 518–570.

Mohammadzadeh, O., Chatzis, I., Giesy, J.P., 2015. A novel chemical additive for in-situ recovery of heavy oil using waterflooding process. Journal of Petroleum Science and Engineering 135, 484–497.

Mohsenatabar Firozjaii, A., Zargar, G., Kazemzadeh, E., 2019. An investigation into polymer flooding in high temperature and high salinity oil reservoir using acrylamide based cationic co-polymer: Experimental and numerical simulation. Journal of Petroleum Exploration and Production Technology 9, 1485–1494.

Morris, C.W., Jackson, K.M., 1978, 16–18 April. Mechanical Degradation of Polyacrylamide Solutions in Porous Media. In SPE 7064, 5th Symposium on Improved Oil, Tulsa, OK

Muherei, M.A., Junin, R., 2008. Mixing effect of anionic and nonionic surfactants on micellization, adsorption and partitioning of nonionic surfactant. Modern Applied Science 2, 2–12.

Mungan, N., 1981. Enhanced oil recovery using water as a driving fluid-5. alkaline flooding field applications. World Oil 193(1).

Muskat, M., 1949. Physical Principles of Oil Production. McGraw-Hill Book Co.

Needham, R.B., Doe, P.H., 1987. Polymer flooding review. Journal of Petroleum Technology 39(12), 1503–1507.

Negin, C., Ali, S., Xie, Q., 2017. Most common surfactants employed in chemical enhanced oil recovery. Petroleum 3, 197–211.

Nelson, R.C., 1981. Further Studies on Phase Relationships in Chemical Flooding, in: Surface Phenomena in Enhanced Oil Recovery. Springer, pp. 73–104.

Nelson, R.C., Lawson, J.B., Thigpen, D.R., Stegemeier, G.L., 1984. Cosurfactant-Enhanced Alkaline Flooding. In SPE/DOE 12672 Presented at the 4th Symposium on Enhanced Oil Recovery, Tulsa, OK.

Nouri, H.H., Root, P.J., 1971, 1 January. A Study of Polymer Solution Rheology, Flow Behavior, and Oil Displacement Processes. In Fall Meeting of the Society of Petroleum Engineers of AIME. Society of Petroleum Engineers.

Olajire, A.A., 2014. Review of ASP EOR (alkaline surfactant polymer enhanced oil recovery) technology in the petroleum industry: Prospects and challenges. Energy 77, 963–982.

Pal, N., 2020. Gemini surfactant assisted enhanced oil recovery: Synthesis, formulation design and performance assessment (PhD Thesis, IIT (ISM), Dhanbad, India).

Pal, N., Babu, K., Mandal, A., 2016. Surface tension, dynamic light scattering and rheological studies of a new polymeric surfactant for application in enhanced oil recovery. Journal of Petroleum Science and Engineering 146, 591–600.

Pal, N., Saxena, N., Laxmi, K.V.D., Mandal, A., 2018. Interfacial behaviour, wettability alteration and emulsification characteristics of a novel surfactant: Implications for enhanced oil recovery. Chemical Engineering Science 187, 200–212.

Pandey, A., Jain, S., Prasad, D., Koduru, N., Raj, R., 2020, August. Planning for Large Scale ASP Flood Implementation in Mangala Oil Field. In SPE Improved Oil Recovery Conference. OnePetro.

Paul, B.K., Mitra, R.K., Moulik, S.P., 2006. Microemulsions: Percolation of Conduction and Thermodynamics of Droplet Clustering, in: Encyclopedia of Surface and Colloid Science, 2nd ed. Taylor & Francis, p. 3927.

Paul, B.K., Moulik, S.P., 1997. Microemulsions: An overview. Journal of Dispersion Science and Technology 18(4), 301–367.

Paul, B.K., Moulik, S.P., 2015. Ionic Liquid-Based Surfactant Science: Formulation, Characterization, and Applications. John Wiley & Sons.

Pei, H., Zhang, G., Ge, J., 2012. Comparative effectiveness of alkaline flooding and alkaline-surfactant flooding for improved heavy-oil recovery. Energy & Fuels 26, 2911–2919.

Perazzo, A., Preziosi, V., 2018. Catastrophic Phase Inversion Techniques for Nanoemulsification, in: Nanoemulsions. Academic Press, pp. 53–76.

Perazzo, A., Tomaiuolo, G., Preziosi, V., Guido, S., 2018. Emulsions in porousmedia: From single droplet behavior to applications for oil recovery. Advances in Colloid and Interface Science 256, 305–325.

Pordel Shahri, M., Shadizadeh, S.R., Jamialahmadi, M., 2012. A new type of surfactant for enhanced oil recovery. Petroleum Science and Technology 30(6), 585–593.

Pratap, M., Gauma, M.S., 2004, October. Field Implementation of Alkaline-Surfactant-Polymer (ASP) Flooding: A Maiden Effort in India. In SPE Asia Pacific Oil and Gas Conference and Exhibition. OnePetro.

Rabiu, A.M., Elias, S., Oyekola, O., 2016. Evaluation of surfactant synthesized from waste vegetable oil to enhance oil recovery from petroleum reservoirs. Energy Procedia 100, 188–192.

Raffa, P., Wever, D.A.Z., Picchioni, F., Broekhuis, A.A., 2015. Polymeric surfactants: Synthesis, properties, and links to applications. Chemical Reviews 115, 8504–8563.

Ray, S., Moulik, S.P., 1995. Phase behavior, transport properties, and thermodynamics of water/AOT/alkanol microemulsion systems. Journal of Colloid and Interface Science 173(1), 28–33.

Redlich, O.J.D.L., Peterson, D.L., 1959. A useful adsorption isotherm. Journal of Physical Chemistry 63(6), 1024–1024.

Reham, S.S., Masjuki, H.H., Kalam, M.A., Shancita, I., Fattah, I.R., Ruhul, A.M., 2015. Study on stability, fuel properties, engine combustion, performance and emission characteristics of biofuel emulsion. Renewable & Sustainable Energy Reviews 52, 1566–1579.

Rosen, M.J., 1978. Surfactant and Interfacial Phenomena. John Wiley & Sons, pp. 137–190.

Rosen, M.J., 1989. Surfactants and Interfacial Phenomena. Wiley-Interscience.

Rosen, M.J., 2004. Surfactant and Interfacial Phenomena, 3rd ed. John Wiley & Sons.

Rossen, W.R., 2017. Foams in Enhanced Oil Recovery, in: Foams. Routledge, pp. 413–464.

Royer, M., Nollet, M., Catte, M., Collinet, M., Pierlot, C., 2018. Towards a new universal way to describe the required hydrophilic lipophilic balance of oils using the phase inversion temperature of C10E4/n-octane/water emulsions. Colloids and Surfaces A: Physicochemical and Engineering Aspects 536, 165–171.

Salathiel, R.A., 1973. Oil recovery by surface film drainage in mixed wettability rocks. Journal of Petroleum Technology 225, 1216–1224.

Saleh, L.D., Wei, M., Bai, B., 2014. Data analysis and updated screening criteria for polymer flooding based on oilfield data. SPE Reservoir Evaluation & Engineering 17(1), 15–25.

Samanta, A., 2011. Studies on characterization of alkali-surfactant-polymer system and its use in enhanced oil recovery (PhD Thesis, IIT (ISM), Dhanbad).

Samanta, A., Ojha, K., Mandal, A., 2011. Interaction between acidic crude oil and alkali and their effects on enhanced oil recovery. Energy & Fuels 25(4), 1642–1649.

Santanna, V.C., Curbelo, F.D.S., Dantas, T.C., Neto, A.D., Albuquerque, H.S., Garnica, A.I.C., 2009. Microemulsion flooding for enhanced oil recovery. Journal of Petroleum Science and Engineering 66(3–4), 117–120.

Saxena, N., 2020. Synthesis of surfactants from natural resources and their characterization for application in enhanced oil recovery (PhD Thesis, IIT (ISM), Dhanbad, India).

Saxena, N., Kumar, A., Mandal, A., 2019. Adsorption analysis of natural anionic surfactant for enhanced oil recovery: The role of mineralogy, salinity, alkalinity and nanoparticles. Journal of Petroleum Science and Engineering 173, 1264–1283.

Schramm, L.L., Wassmuth, F., 1994. Foams: Basic Principles. Foams: Fundamentals and Applications in the Petroleum Industry, in: Schramm, L.L. (ed.) Advances in Chemistry Series 242. American Chemical Society, pp. 3–45.

Schulman, J.H., Stoeckenius, W., Prince, L.M., 1959. Mechanism of formation and structure of micro emulsions by electron microscopy. The Journal of Physical Chemistry 63(10), 1677–1680.

Scott, A.J., Romero-Zerón, L., Penlidis, A., 2020. Evaluation of polymeric materials for chemical enhanced oil recovery. Processes 8(3), 361.

Sen, R., 2008. Biotechnology in petroleum recovery: The microbial EOR. Progress in Energy and Combustion Science 34(6), 714–724.

Shah, B.N., Lawrence, G., Willhite, P., Green, D.W., 1978, 1–3 October. The Effect of Inaccessible Pore Volume on the Flow of Polymer and Solvent Through Porous Media. In Proceedings of the SPE Annual Fall Technical Conference and Exhibition, Houston, TX.

Shaker Shiran, B., Skauge, A., 2013. Enhanced oil recovery (EOR) by combined low salinity water/polymer flooding. Energy & Fuels 27, 1223–1235.

Sharma, M.M., Gao, B., 2014. Salt-tolerant anionic surfactant compositions for enhanced oil recovery (EOR) applications. US Patent Application 13/963,460, University of Texas System.

Shen, P.P., Yu, J.Y., 2002. Fundamental Research on Enhanced Oil Recovery in Large Scale. Petroleum Industry Publication Company, Beijing.

Sheng, J.J., 2010. Modern Chemical Enhanced Oil Recovery: Theory and Practice. Gulf Professional Publishing.

Sheng, J.J., 2013a, April. A Comprehensive Review of Alkaline-Surfactant-Polymer (ASP) Flooding. In SPE Western Regional & AAPG Pacific Section Meeting 2013 Joint Technical Conference. OnePetro.

Sheng, J.J., 2013b. Surfactant–Polymer Flooding, in: Enhanced Oil Recovery Field Case Studies. Gulf Professional Publishing, pp. 117–142.

Shupe, R.D., 1981. Chemical stability of polyacrylamide polymers. Journal of Petroleum Technology 33, 1513–1529.

Shupe, R.D., Maddox Jr, J., 1978. Surfactant oil recovery process usable in high temperature, high salinity formations. No. US 4077471.

Simoni, L.D., Lin, Y., Brennecke, J.F., Stadtherr, M.A., 2008. Modeling liquid–liquid equilibrium of ionic liquid systems with NRTL, electrolyte-NRTL, and UNIQUAC. Industrial & Engineering Chemistry Research 47, 256–272.

Singh, B.P., Parulkar, S., Kumar, A., 2017, April. Successful Pilot Implementation of ASP Flooding Through CEOR/IOR Application – A Case Study. In SPE Oil and Gas India Conference and Exhibition. OnePetro.

Sips, R., 1950. On the structure of a catalyst surface. II. The Journal of Chemical Physics 18(8), 1024–1026.

Skauge, A., Fotland, P., 1990. Effect of pressure and temperature on the phase behavior of microemulsions. SPE Reservoir Engineering 5(4), 601–608.

Skauge, A., Salmo, I., 2015, April. Relative Permeability Functions for Tertiary Polymer Flooding. In IOR 2015–18th European Symposium on Improved Oil Recovery (pp. cp-445). European Association of Geoscientists & Engineers.

Skauge, A., Zamani, N., Gausdal Jacobsen, J., Shaker Shiran, B., Al-Shakry, B., Skauge, T., 2018. Polymer flow in porous media: Relevance to enhanced oil recovery. Colloids and Interfaces 2(3), 27.

Skelland, A.H.P., 1967. Non-Newtonian Flow and Heat Transfer (Book on quantitative relationships for Non-Newtonian systems, considering classification and fluid behavior of materials with anomalous flow properties). John Wiley & Sons, Inc., 469 pp.

Somasundaran, P., Zhang, L., 2006. Adsorption of surfactants on minerals for wettability control in improved oil recovery processes. Journal of Petroleum Science and Engineering 52, 198–212.

Somerton, W.H., Radke, C.J., 1983. Role of clays in the enhanced recovery of petroleum from some California sands. Journal of Petroleum Technology, 643–654.

Sorbie, K.S., 2013. Polymer-Improved Oil Recovery. Springer Science & Business Media.

Squires, F., 1917. Method of recovering oil and gas. US Patent No. 1238355.

Standnes, D.C., Austad, T., 2000. Wettability alteration in chalk: 2. Mechanism for wettability alteration from oil-wet to water-wet using surfactants. Journal of Petroleum Science and Engineering 28, 123–143.

Stiles, W.E., 1949. Use of permeability distribution in waterflood calculations. Transactions of the AIME 186, 9–13.

Taber, J.J., Martin, F.D., Seright, R.S., 1997. EOR screening criteria revisited—Part 2—Applications and impact of oil prices. SPE Reservoir Engineering 12(3), 199–206. http://dx.doi.org/10.2118/39234-PA.

Taylor, K.C., Nasr-El-Din, H.A., 1998. Water-soluble hydrophobically associating polymers for improved oil recovery: A literature review. Journal of Petroleum Science and Engineering 19(3), 265–280.

Teletzke, G.F., Wattenbarger, R.C., Wilkinson, J.R., 2010. Enhanced oil recovery pilot testing best practices. Journal of Petroleum Technology 61, 1–2.

Temkin, M.J., Pyzhev, V., 1940. Recent modifications to Langmuir isotherms. Acta Physiochim URSS 12, 217–225.

Thomas, A., 2016. Polymer Flooding, in: Chemical Enhanced Oil Recovery (cEOR)–A Practical Overview. IntechOpen.

Thomas, A., Gaillard, N., Favero, C., 2012. Some key features to consider when studying acrylamide-based polymers for chemical enhanced oil recovery. Oil & Gas Science and Technology - Revue de l IFP 67, 887–902.

Tondo, D.W., Leopoldino, E.C., Souza, B.S., Micke, G.A., Costa, A.C.O., Fiedler, H.D., Bunton, C.A., Nome, F., 2010. Synthesis of a new zwitterionic surfactant containing an imidazolium ring. evaluating the chameleon-like behavior of zwitterionic micelles. Langmuir 26, 15754–15760.

Torres, L., Moctezuma, A., Avendano, J.R., Munoz, A., Gracida, J., 2011. Comparison of bio- and synthetic surfactants for EOR. Journal of Petroleum Science and Engineering 76, 6–11.

Turta, A.T., Singhal, A.K., 2002. Field foam applications in enhanced oil recovery projects: Screening and design aspects. Journal of Canadian Petroleum Technology 41(10).

Volokitin, Y., Shuster, M., Karpan, V., Koltsov, I., Mikhaylenko, E., Bondar, M., Podberezhny, M., Rakitin, A., Batenburg, D.W., Parker, A.R., Kruijf, S.D., 2018, March. Results of Alkaline-Surfactant-Polymer Flooding Pilot at West Salym Field. In SPE EOR Conference at Oil and Gas West Asia. OnePetro.

Wadle, A., Förster, T., Von Rybinski, W., 1993. Influence of the microemulsion phase structure on the phase inversion temperature emulsification of polar oils. Colloids and Surfaces A: Physicochemical and Engineering Aspects 76, 51–57.

Wagner, O.R., Leach, R.O., 1959. Improving oil displacement efficiency by wettability adjustment. Transactions of the AIME 216(1), 65–72.

Wang, D., Cheng, J., Wu, J.G., Wang, F., Li, H., Gong, X., 1998, 27–30 September. An Alkaline/Surfactant/Polymer Field Test in a Reservoir with a Long-Term 100% Water Cut. In SPE 49018 Presented at the 1998 SPE Annual Technical Conference and Exhibition, New Orleans, LA.

Wang, D., Cheng, J., Yang, Q-Y., 2000. Viscous-elastic polymer can increase microscale displacement efficiency in cores. Acta Petrologica Sinica 21(5), 45–51.

Wang, D., Xia, H., Liu, Z., Yang, Q., 2001, 17–19 April. Study of the Mechanism of Polymer Solution With Visco-Elastic Behavior Increasing Microscopic Oil Displacement Efficiency and the Forming of Steady "Oil Thread" Flow Channels. In SPE 68723, SPE Asia Pacific Oil and Gas Conference and Exhibition, Jakarta, pp. 1–9.

Wang, D., Zhang, Z., Cheng, J., Yang, J., Gao, S., Li, L., 1997. Pilot Tests of Alakali/Surfactant/Polymer Flooding in Daqing Oil Field. In SPERE, 229.

Wang, J., Han, M., Fuseni, A.B., Cao, D., 2015, March. Surfactant Adsorption in Surfactant-Polymer Flooding for Carbonate Reservoirs. In SPE Middle East Oil & Gas Show and Conference. OnePetro.

Wang, W., Kwak, J.C.T., 1999. Adsorption at the alumina–Water interface from mixed surfactant solutions. Colloids Surfaces A: Physicochemical and Engineering Aspects 156, 95–110.

Wasan, D.T., Shah, S.M., Aderangi, N., Chan, M.S., McNamara, J.J., 1978. Observations on the coalescence behavior of oil droplets and emulsion stability in enhanced oil-recovery. SPE Journal 18(6), 409–417.

Wever, D.A.Z., Picchioni, F., Broekhuis, A.A., 2011. Polymers for enhanced oil recovery: A paradigm for structure-property relationship in aqueous solution. Progress in Polymer Science 36, 1558–1628.

Wilson Jr., L.A., 1976. Physico-Chemical Environment of Petroleum Reservoirs in Relation to Oil Recovery Systems. In AIChE Symposium on Improved Oil Recovery by Surfactant and Polymer Flooding, Kansas City, KS.

Winsor, P.A., 1954. Solvent Properties of Amphiphilic Compounds. Butterworths Scientific Publications.

Wu, X., Wang, Y., Al Naabi, A., Xu, H., Al Sinani, I., Al Busaidi, K., Al Jabri, S., Dhahab, S., Zhang, J., Xiong, C., Ye, Y., 2019, November. A New Polymer Flooding Technology for Improving Low Permeability Carbonate Reservoir Recovery – From Lab Study to Pilot Test – Case Study from Oman. In Abu Dhabi International Petroleum Exhibition & Conference. OnePetro.

Xia, H., Zang, Y-X., Zang, Y-L., 1999. Rheological behavior of alkali-surfactant-polymer solution in porous media. Journal of Daqing Petroleum Institute 23(4), 18–22.

Xiao, Y., Malhotra, S.V., 2005. Friedel-Crafts Acylation reactions in pyridinium based ionic liquids. Journal of Organometallic Chemistry 690, 3609–3613.

Yan, W., Miller, C.A., Hirasaki, G.J., 2006. Foam sweep in fractures for enhanced oil recovery. Colloids and Surfaces A: Physicochemical and Engineering Aspects 282, 348–359.

Yang, B., Mao, J., Zhao, J., Shao, Y., Zhang, Y., Zhang, Z., Lu, Q., 2019. Improving the thermal stability of hydrophobic associative polymer aqueous solution using a "triple-protection" strategy. Polymers 11, 949.

Yang, C-Z., Han, D-K., Song, W-C., 1995. The Alkaline–Surfactant–Polymer Combination Flooding and Application to Oil Field for EOR. In 8th European Symposium on Improved Oil Recovery, Vienna.

Yang, Y., Pu, W.F., 2020. Low interfacial tension emulsion flooding under harsh reservoir conditions: The effect of phase inversion behavior on enhanced oil recovery. Journal of Dispersion Science and Technology 41(7), 1065–1074.

Ye, Z., Zhang, F., Han, L., Luo, P., Yang, J., Chen, H., 2008. The effect of temperature on the interfacial tension between crude oil and Gemini surfactant solution. Colloids and Surfaces 322, 138–141.

Yoon, R.H., Aksoy, B.S., 1999. Hydrophobic forces in thin water films stabilized by dodecylammonium chloride. Journal of Colloid and Interface Science 211(1), 1–10.

Young, T., 1805. An essay on the cohesion of fluids. Philosophical Transactions of the Royal Society of London 95, 65–87.

Yu, H., Wang, Y., Zhang, Y., Zhang, P., Chen, W., 2011. Effects of displacement efficiency of surfactant flooding in high salinity reservoir: Interfacial tension, emulsification, adsorption. Advances in Petroleum Exploration and Development 1, 32–39.

Zana, R., 2002. Dimeric and oligomeric surfactants. Behavior at interfaces and in aqueous solution: A review. Advances in Colloid and Interface Science 97, 205–253.

Zhang, Q.Q., Cai, B.X., Xu, W.J., Gang, H.Z., Liu, J.F., Yang, S.Z., Mu, B.Z., 2015. The rebirth of waste cooking oil to novel bio-based surfactants. Scientific Reports 5, 9971.

4 Miscible Flooding

4.1 INTRODUCTION

Miscible flooding is, at present, considered one of the most capable methods in enhanced oil recovery (EOR). Miscible flooding is a general term for injection processes that introduce miscible gases into the reservoir. A miscible displacement process maintains reservoir pressure and improves oil displacement by reducing the interfacial tension between the oil and displacing fluids and swelling of the oil as well as by reducing the density of the crude oil. The process leads to the fluid capillary forces being reduced to zero, and the resultant displacement of oil is essentially complete in those pore spaces through which the injected solvents move. Mostly, hydrocarbon and CO_2 are injected as solvents to achieve miscibility with the trapped crude oil at the desired pressure; thus, it is also called the "solvent injection" method. Generally, there are three types of solvent drives: high-pressure gas drives, enriched gas drives, and miscible slug drives. The first two drives are mostly deployed for light oil reservoirs, and the third one is applied for moderate to heavy oil reservoirs. However, the success of these techniques for EOR will depend almost entirely on proper engineering and geological evaluation. The engineer must understand the basic principles involved in miscible-drive techniques and the factors on which the success of the process depends, considering the proper design and conduct of the complicated operations.

4.2 THERMODYNAMICS OF MISCIBILITY AND PHASE BEHAVIORS

The mechanism of miscible flooding is based on the distribution of intermediate hydrocarbons (C_2–C_6) between the injected solvent and the crude oil. Crude oil is composed of n number of different hydrocarbon components starting from C_1 to C_n, where n tends to infinity. The state of the hydrocarbons differs depending on the number of molecules and their bonding, in addition to the prevalent pressure and temperature conditions. The Gibbs phase rule for nonreacting systems provides the most convenient method for determining the variables important for describing the phase equilibria. Being a composition of numerous molecules, we need an infinite number of variables to define a crude oil system as per the Gibbs phase rule. The Gibbs phase rule states the following:

$$F = C - P + 2 \tag{4.1}$$

where:
F = number of intensive variables (e.g., pressure, temperature, single-phase composition) required to define the system (known as the degrees of freedom)
C = number of components
P = number of phases

If we try to handle crude oil with all the individual n components, the number of variables in the system will be very large; we will then have to put in a huge computational effort to characterize the system. Therefore, it is more realistic to club the components having similar properties into a single category, which are frequently called pseudo components. In most reservoir engineering studies, we broadly divide the different hydrocarbons into three pseudo components with similar thermodynamic properties. These are as follows:

(a) Light components: consisting mainly of methane (C_1) and possibly N_2.
(b) Intermediate components: mostly, intermediate hydrocarbons (C_2–C_6), possibly nonhydrocarbon gases like CO_2, H_2S, etc.
(c) Heavier components: C_{7+} (C_{7+} and heavier hydrocarbons).

DOI: 10.1201/9781003098850-4

The phase behavior of crude oil and injection fluids is an important phenomenon to understand in the displacement mechanisms of EOR processes. The phase behavior of a multicomponent hydrocarbon mixture is more complicated than that of a pure hydrocarbon (say heptane). Figure 4.1 shows an idealized P-T diagram for a multicomponent reservoir fluid with a fixed overall composition. The area surrounded by the bubble-point and dew-point curves characterizes the pressure and temperature combinations for which both the gas and liquid phases coexist. The two-phase envelope is divided by an infinite number of quality lines, each of which represents the percentage of the total hydrocarbon volume that is liquid at the given temperature and pressure conditions. The hydrocarbon mixture, at any pressure and temperature above the bubble-point curve will be a liquid phase. Similarly, at any pressure and temperature located above or to the right of the dew-point curve, the hydrocarbon mixture will be a gas phase. The bubble-point and dew-point curves meet at the critical point. Unlike a pure component system, two phases can exist at a pressure greater than critical pressure and at a temperature greater than critical temperature for a multicomponent system. The maximum pressure at which two phases (vapor-liquid) can coexist in equilibrium is called cricondenbar and the maximum temperature at which two phases can coexist in equilibrium is termed cricondentherm.

The hydrocarbon mixture may exist either as a gas or a liquid in the reservoir as well as at the surface condition based on its relative volatilities and vapor pressures. When the reservoir pressure falls below the bubble-point pressure, the lighter components of the crude oil start to vaporize and remain in equilibrium with the oil. Similarly, if the hydrocarbon gas is injected into the reservoir to come in contact with crude oil, an equilibrium condition is established after an exchange of mass between the phases. The distribution of any component between these two phases is governed by the

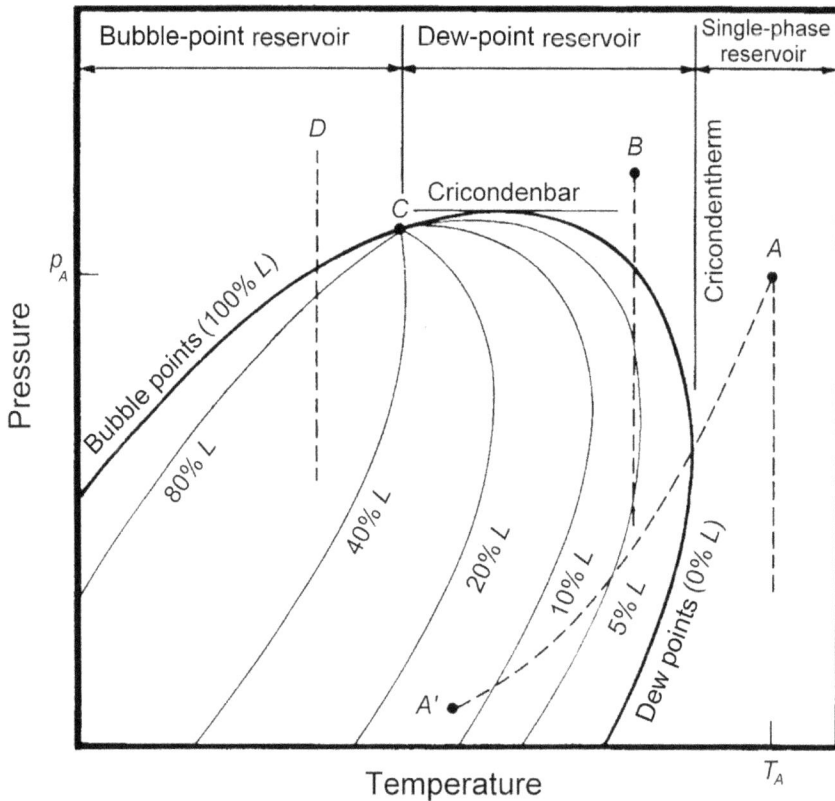

FIGURE 4.1   Typical P-T diagram for a multicomponent hydrocarbon system.

Source: petrowiki.spe.org.

equilibrium constant, K_i, for each component i. K_i is defined as y_i/x_i, where y_i is the mole fraction of component i in the vapor phase and x_i is the mole fraction of the same component in the liquid phase. Generally, K_i is also a function of temperature T and pressure P, besides the composition (x_i and y_i).

When the pressure is less than 100 psia, Raoult's and Dalton's laws for ideal solutions provide a simplified means of predicting equilibrium ratios. According to Raoult's law:

$$p_i = X_i \, p_{vi} \tag{4.2}$$

where p_i = partial pressure of the component i, psia
p_{vi} = vapor pressure of the component i, psia
X_i = mole fraction of the component i in the liquid phase

Again, according to Dalton's law, $p_i = Y_i P$; where P = total system pressure, psia.

$$\text{Therefore, } K_i = \frac{Y_i}{X_i} = \frac{p_{vi}}{P} \tag{4.3}$$

For a real solution, the equilibrium ratios are no longer a function of the pressure and temperature alone but also a function of the composition of the hydrocarbon mixture. Thus, K_i can be expressed as follows:

$$K_i = f(P, T \, z_i) \tag{4.4}$$

Again, when a hydrocarbon mixture of a fixed overall composition (z_i) is held at a constant temperature, with increase in pressure, the equilibrium K values of all components converge toward a common value of unity at a certain pressure. This pressure is designated as the convergence pressure p_k of the hydrocarbon mixture. The concept of the convergence pressure can be better understood for a hydrocarbon mixture, as presented in Figure 4.2. It can be seen that the K_i values of all the compo-

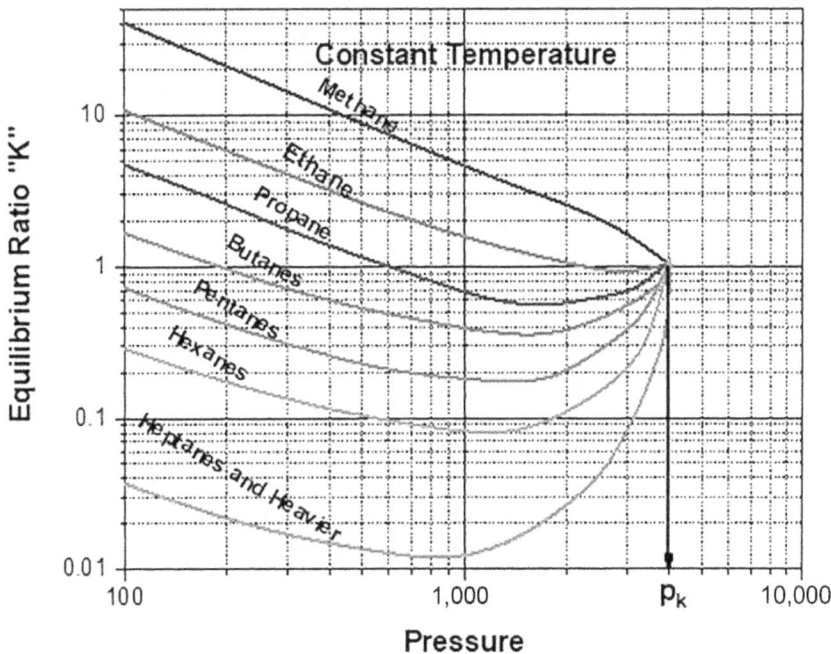

FIGURE 4.2 Equilibrium ratios for a hydrocarbon system.

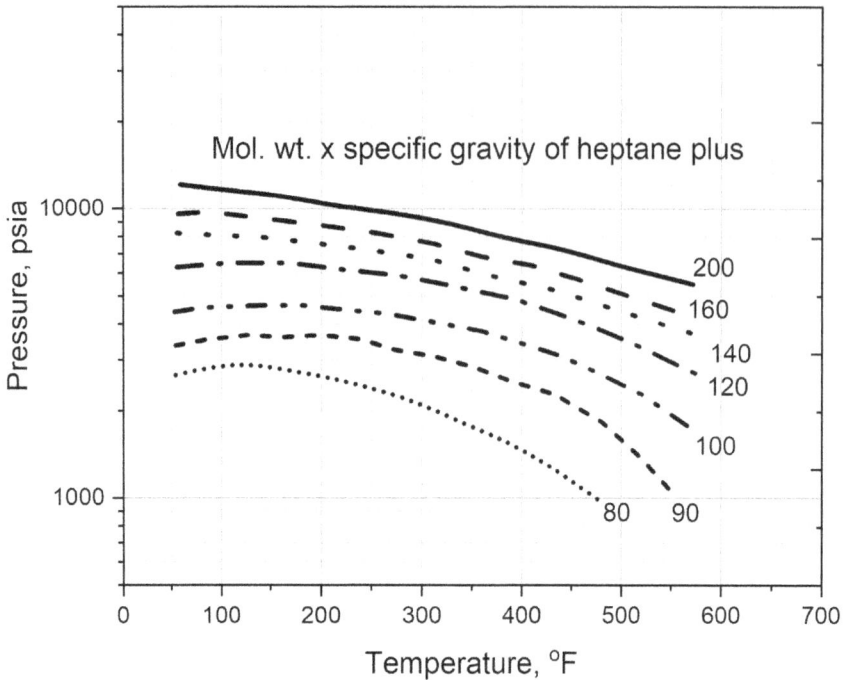

FIGURE 4.3   Rzasa's convergence pressure correlation.

Source: Rzasa et al., 1952.

nents tend to converge isothermally to the value of 1.0 at a specific pressure, known as convergence pressure (p_k). The convergence pressure essentially is used to correlate the effect of the composition on equilibrium ratios.

The hydrocarbon mixture with different compositions may exhibit different convergence pressures. It also depends on temperature. Figure 4.3 shows the Rzasa convergence pressure correlation (Rzasa et al., 1952) of heptanes-plus fraction at different temperatures. They used the temperature and the product of the molecular weight and specific gravity of the heptane-plus fraction as correlating parameters.

Thus, K_i may be expressed as follows:

$$K_i = f\left(P, T, p_k\right) \tag{4.5}$$

4.2.1   FLUID PHASE BEHAVIOR: TERTIARY DIAGRAM

The thermodynamic miscibility can be best explained by a ternary phase diagram, where three pseudo components (C_1, C_2–C_6 and C_{7+}) are conventionally used to describe the overall composition. The advantage of using a ternary plot for depicting compositions is that three variables can be conveniently plotted in a 2D graph, and the mixture of different components can be easily represented. Each side corresponds to 0% of the component represented by the opposite apex. The overall composition of a hydrocarbon mixture with composition of 50% C_1, 20% C_2–C_6, and 30% C_{7+} is shown in Figure 4.4a. The composition of any component in a ternary phase diagram can be obtained by the inverse lever arm rule, as shown in Figure 4.4b. The phase rule can be applied to the combination of any two mixtures represented on the ternary (Fig. 4.4c). The mixture composition of two hydrocarbon systems, M_1 and M_2, would be along the line M_1/M_2 and represented by

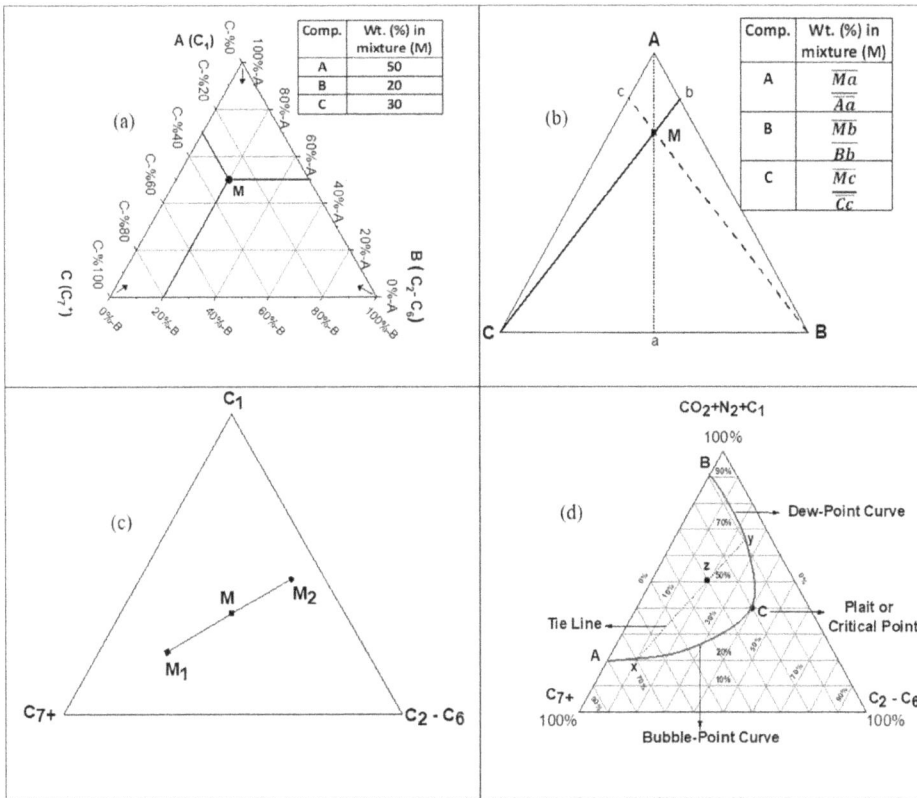

FIGURE 4.4   Ternary phase diagram of a hydrocarbon system.

M. The final composition of M would again be determined by applying the inverse lever arm rule along the line.

The typical phase behavior of multicomponent hydrocarbon mixtures at a particular temperature and pressure is shown as a pseudoternary phase diagram in Figure 4.4d. An area in the graph surrounded by the curve "ACB" is called the phase envelope, which represents the phase behavior of mixtures with varying combinations of the three pseudo components. The section "AC" of the phase envelope passing through the critical point "C" is called the bubble-point curve, which represents the composition of the saturated liquid. On the other hand, the segment "BC" of the phase envelope is called the dew-point curve, representing the composition of the saturated gas. The dew-point curve and the bubble-point curve merge together at a point called the critical (plait) point. At a critical point, the compositions and properties of the gas and liquid are identical. The position of the plait point changes with temperature at a fixed pressure. Any composition (say Z) within the diphasic envelope splits into two equilibrium compositions (X and Y). The line connecting these two equilibrium points through Z is called tie tine. The relative amount of two phases can be calculated using the inverse lever arm rule. The points outside the two-phase envelope are representative of a single-phase composition.

Phase diagram is very important to understand the miscible flooding mechanism. For successful designing of the miscible flooding system, one must identify the miscible zone—i.e., the single-phase zone. As the pressure increases, the two-phase region shrinks, or in other words, light-heavy miscibility increases (Fig. 4.5a). Though there is no direct relationship, the two-phase region generally grows with increasing temperature (Fig. 4.5b), which is the reverse in the case of pressure. The high pressure and low temperature conditions are favorable for miscible displacement because of a larger miscible zone—i.e., higher miscibility.

(a)

C_1

$P < P_m$
P_m
$P > P_m$

P increasing

O

C_{7+}
T = Reservoir Temperature
C_2-C_6

(b)

C_1

80°C
100°C
120°C

C_{7+}
P = Constant
C_2-C_6

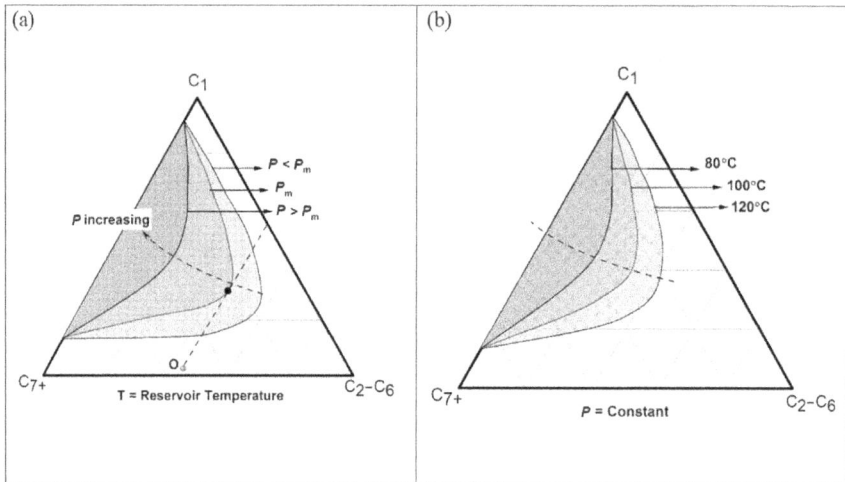

FIGURE 4.5 Effect of (a) pressure and (b) temperature on the diphasic region of a hydrocarbon system.

Example 4.1: In a phase behavior study, 40 g of n-heptane, 20 g of iso-propanol and 40 g of water are mixed together at a particular pressure and temperature to form a binary phase containing an oil-rich phase and a water-rich one (Refer Fig. 4.6). Determine the following:

   (i) The relative volume of the phases and their composition.
  (ii) The degree of freedom at point Z, as shown in the given figure.

Solution:
Total weight of the mixture = 100 g

   (i) So, the composition is
      n-heptane = 40/100 = 0.40
      iso-propanol = 20/100 = 0.20 and
      water = 40/100 = 0.40

A ternary phase diagram is drawn and shown in Figure 4.6.
Weight of oil-rich phase/Weight of water-rich phase = ZY/XZ = 2.2/3.0
Therefore, the weight of oil-rich phase = 100g × 2.2/5.2 = 42.30 g
The weight of water-rich phase = 100g × 3/5.2 = 57.69 g

  (ii) From the ternary phase diagram, the composition of oil-rich phase (X) and water-rich phase (Y) can be measured as given in the following:

The composition of the oil-rich phase (X):
n-heptane = 89%
IPA = 10.2%
Water = 0.8%

The composition of water-rich phase (Y):
Water = 67%
IPA = 27%

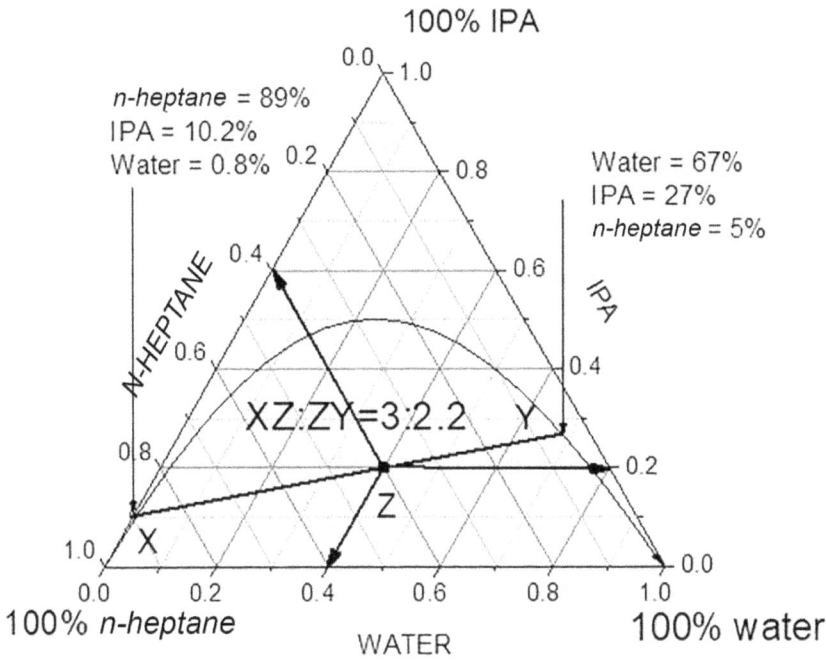

FIGURE 4.6 Ternary phase diagram of Example 4.1.

n-heptane = 5%
Degrees of freedom, $F = C - P + 2$
For the point Z, the number of components, $C = 3$.
The mixture with composition Z will split into two phases with composition Y and Z.
So, number of phases, $P = 2$.
Therefore, $F = 3 - 2 + 2 = 3$; i.e., to define the point Z completely, we have to mention the pressure, temperature, and composition of any one component.

4.3 MECHANISMS OF MISCIBLE DISPLACEMENT

Miscibility can be defined as the physical condition between two fluids that permits the fluids to mix in all proportions without the formation of any interface. The interfacial tension between two fluids is zero when miscibility is achieved. If the two fluids do not mix in all proportions, they are considered immiscible. For example, if methane is mixed with crude oil at atmospheric conditions, it may dissolve into oil up to a certain extent, but they do not mix in all proportions at that condition. So crude oil and methane are thus considered immiscible. Under similar conditions, propane is also immiscible with oil, but upon increasing the pressure, the interface between the two phases becomes indistinguishable at a certain pressure, and they become miscible. At 2,000 psi and 150°F, methane gas and liquid propane are miscible in all proportions. Capillary pressure and interfacial tension play very crucial roles in the miscibility of oil/gas as well as in their displacement in the reservoir. The role of capillary pressure and interfacial tension in either assisting or retarding oil displacement by gas and water drive is demonstrated by many authors (Clark et al., 1958; Srivastava and Huang, 1997; Asgarpour, 1994). Under water-drive displacement, the capillary forces assist oil recovery from low-permeability pore zones and dead-end pore channels when the overall oil saturation in the sand is high. At high saturation, oil, mostly a nonwetting phase, moves through the larger capillaries easily. As oil saturation is reduced to a pendular ring saturation or lower by water invasion, the interfacial forces between oil and water reverse the action and tend to cause oil to break away from the displaced oil flow and become trapped as residual oil (Clark et al., 1958). The miscible drive process reduces the capillary forces to zero, leading to significant recovery by the displacing solvents. In miscible flooding, there is no fluid/fluid interface and capillary force. Thus, theoretically, the residual oil saturation may be reduced to zero in the miscible process with very high pore-scale recovery efficiencies.

The main mechanisms responsible for the oil displacement by gas injection above the minimum miscibility pressure (MMP) of injected gases include oil swelling, reduction of viscosity and IFT, and increasing the injectivity index due to the solubility of gas in water (Orr et al., 1982; Jarrell et al., 2002). The important phenomena that occur during EOR by miscible flood are extraction, dissolution, vaporization, solubilization, condensation, or other phase behavior changes involving crude oil.

4.3.1 Swelling of Crude Oil

For the miscible gas injection process, the injected gas (hydrocarbon, CO_2, etc.) interacts with the residual crude oil before the miscibility is achieved. Based on the pressure and temperature, the oil properties and other thermodynamic conditions, the gases begin to solubilize in the crude oil upon contact (Mullken and Sandler, 1980). This results in an increase in the volume of the oil, conventionally known as oil swelling, and causes the residual oil to mobilize. This is considered to be an important EOR mechanism (Hatzignatiou and Lu, 1994). The relative permeability of the oil is also enhanced by increasing the oil saturation (increased volume) and reducing the viscosity, and hence the mobility of the oil through small capillaries increases (Yang and Gu, 2006). In general, the oil swelling factor and the respective saturation pressure are functions of the moles of injected gas, and there exists a positive relationship as shown in Figure 4.7. The saturation pressure and oil swelling factor increase with the mole fraction and then cease upon attaining equilibrium. The swelling factor of the oil-injected CO_2 system is higher than that of the oil-injected methane

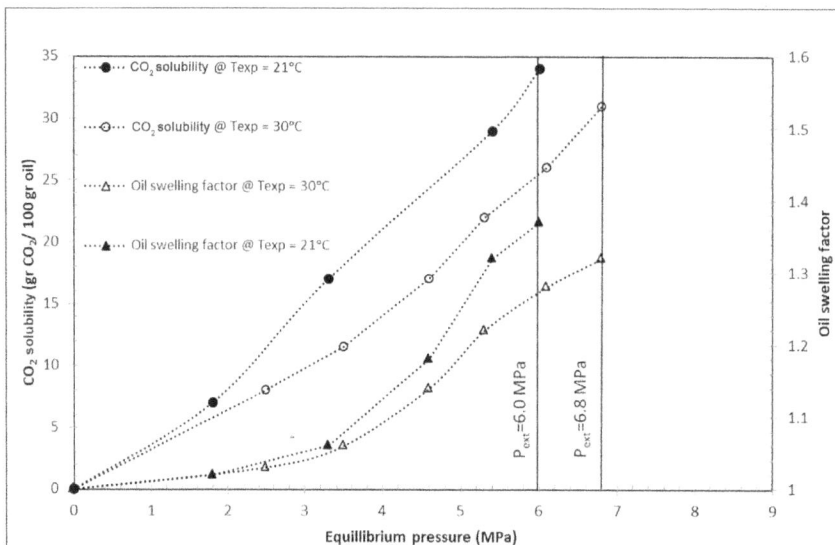

FIGURE 4.7 CO_2 solubility in crude oil and oil swelling factor at various operating pressures and constant temperatures.

Source: Mosavat et al., 2014.

gas system at the same saturation pressure (Choudhary et al., 2019), mainly owing to the higher solubility of CO_2 in oil.

4.3.2 Reduction of Interfacial Tension

Interfacial tension and viscosity of oil are the key controlling parameters in miscible flooding in addition to the viscous force. IFT is considered to be the most sensitive and easily modifiable variable in controlling the efficacy of miscible flooding. The integrated effect of viscous force and interfacial force is expressed in terms of the capillary number, which is defined by Equation 4.6:

$$N_C = \frac{Viscous\ forces}{Capillary\ forces} = \frac{v\mu}{\gamma} \tag{4.6}$$

where v and μ depict the velocity and viscosity of the displacing fluid, respectively, and γ represents the IFT between the oil and displacing fluid. To increase the capillary number even by one order of magnitude by increasing the velocity or viscosity is much more difficult and costly. When gas is injected, an in-situ mass transfer occurs between the crude oil and the injected gas at the reservoir pressure and temperature, which leads to a significant reduction of IFT between the two phases. The IFT value may be reduced many folds by changing the composition of the gas phase and oil phase. The reduction in capillary pressure due to the decrease in IFT results in the gas being able to access pore throats that were essentially isolated from the flowing gas phase at a higher interfacial tension level. The capillary pressure, P_c, is defined by Equation 4.7:

$$P_c = \frac{2\gamma\cos\theta}{r} \tag{4.7}$$

where θ is contact angle and r is the pore throat radius. This reduction in capillary forces between the oil and the injection gas results in a more efficient sweep and reduced residual oil saturation (S_{or}). Before implementation in field, the variation of IFT by dissolution of gas in oil has to be tested in a laboratory.

The variation of IFT at different pressures is illustrated in Figure 4.8, where CO_2 is injected for miscible flooding of representative light crude oil. The experiment was carried out by Cao and Gu, 2013 in a laboratory. As observed in the experiment, the equilibrium IFT reduced almost linearly with equilibrium pressure. Change in slope was clearly distinguished at two pressures pressure ranges: Range I (P_{eq} = 1.7–8.3 MPa) and Range II (P_{eq} = 8.3–19.1 MPa) (Fig. 4.8).

Two different mechanisms are involved in this process. In Range I, the equilibrium IFT reduction is exclusively attributed to the increasing solubility of CO_2 in crude oil at equilibrium pressure. In Range II, the reduction of IFT is also linear, but the mechanism is reversed. The reduction is due to the extraction of light hydrocarbons from the crude oil to the gas phase (CO_2) (Cao and Gu, 2013; Hemmati-Sarapardeh et al., 2014). The first line is the indication of lighter components located at the interface with CO_2, while the second line demonstrates the presence of heavier components after extraction of the light components by CO_2 (Zolghadr et al., 2013). On the basis of the measured data, the equilibrium IFT γ_{eq} (mJ/m^2) is correlated to the equilibrium pressure P_{eq} (MPa) by applying the linear regression in the aforementioned two equilibrium pressure ranges, respectively (Cao and Gu, 2013):

$$\gamma_{eq} = -2.04P_{eq} + 21.63\left(1.7\ \text{MPa} \le P_{eq} \le 8.3\ \text{MPa}\right) \tag{4.8}$$

$$\gamma_{eq} = -0.31P_{eq} + 7.15\left(8.3\ \text{MPa} \le P_{eq} \le 19.1\ \text{MPa}\right) \tag{4.9}$$

The linear regression equation depicts that the IFT is zero at P_{eq} = 10.6 MPa, which is the minimum miscibility pressure (MMP) of the light crude oil–CO_2 system at temperature of 53.0°C. On the other hand, the miscibility pressure between the intermediate to heavy hydrocarbons of this light crude oil and CO_2 is found to be P_{max} = 23.1 MPa from the linear regression equation for Range II.

4.3.3 Viscosity Reduction

The typical oil and gas viscosity ratio may vary widely, ranging from a minimum of 5 up to 10,000, which is a major drawback of gas injection. As the fluid velocity is inversely proportional to the viscosity, an applied differential pressure results in preferential gas flow. In miscible flooding, the exchange of mass between the injected gas/solvent and the crude oil causes a significant reduction in oil viscosity, particularly, when miscibility is achieved. On the other hand, the injected gas is enriched gradually with intermediate hydrocarbons, and its viscosity increases. Thus, the viscosity contrast is significantly reduced during the miscible injection process. Ultimately, the end point relative permeabilities to oil and gas are usually of the same order of magnitude. Figure 4.9 demonstrates the viscosity reduction correlation (viscosity ratio) with the saturation pressure of carbon dioxide.

4.3.4 Miscible Displacement Processes

A schematic of the miscible flooding process is shown in Figure 4.10. In practice, only a certain pore volume of solvent is injected to obtain the miscible zone and subsequent oil bank, depending on the reservoir pressure, temperature, and composition of the trapped crude oil. However, the pressure

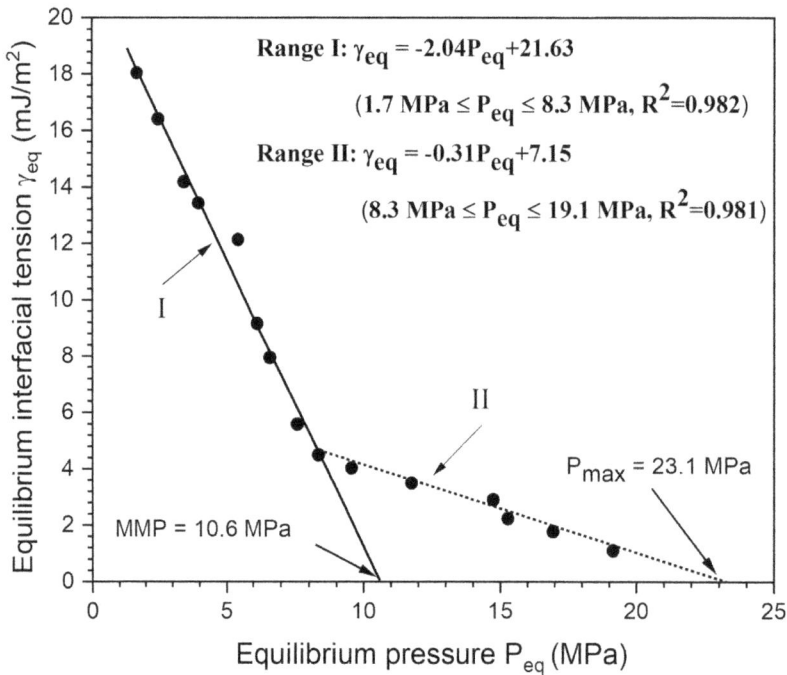

Range I: $\gamma_{eq} = -2.04P_{eq} + 21.63$

$(1.7 \text{ MPa} \le P_{eq} \le 8.3 \text{ MPa}, R^2 = 0.982)$

Range II: $\gamma_{eq} = -0.31P_{eq} + 7.15$

$(8.3 \text{ MPa} \le P_{eq} \le 19.1 \text{ MPa}, R^2 = 0.981)$

P_{max} = 23.1 MPa

MMP = 10.6 MPa

FIGURE 4.8 Measured equilibrium interfacial tension of the Pembina Cardium light crude dead oil–pure CO_2 system at different equilibrium pressures and T_{res} = 53.0°C.

Source: Cao and Gu, 2013.

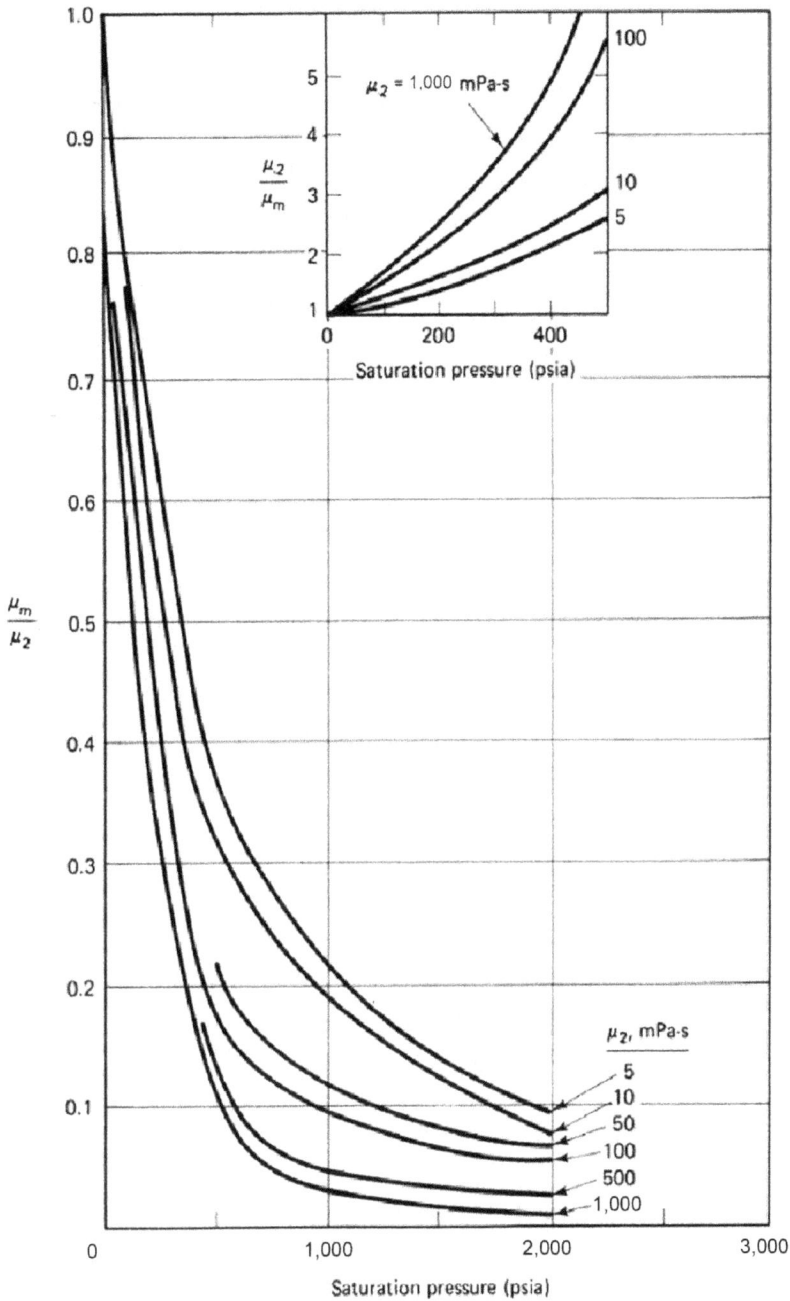

FIGURE 4.9   Viscosity-pressure correlation for carbon dioxide-oil mixture.

Source: Simon and Graue, 1965.

must be maintained by a chase fluid, which should be miscible with the injected solvent but immiscible with the oil. The typical processes include the following combinations:

a) Oil, LGP, gas
b) Oil, alcohol, water

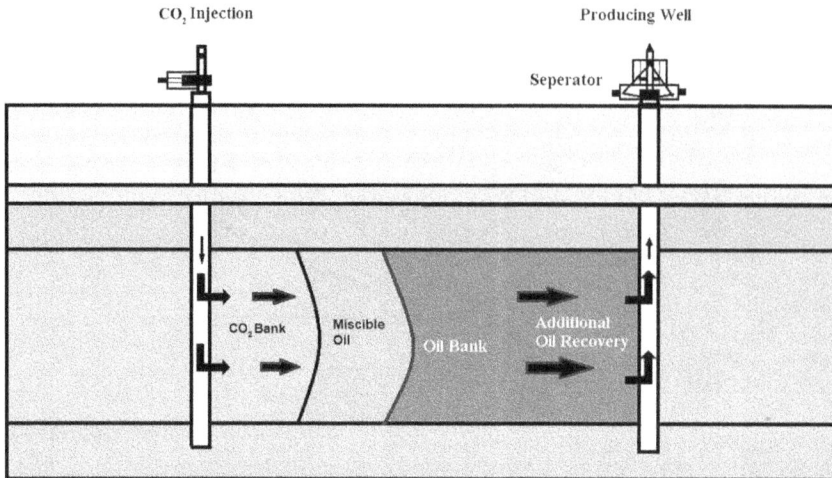

FIGURE 4.10 Schematic of carbon dioxide miscible flooding process.

During the flooding process, because of the miscibility of the solvent with both the chase fluid and the oil, the injected solvent may disappear before its breakthrough because of improper design. Under this condition, the chase fluid, which is immiscible with the oil will come in contact with it, thus converting the miscible flooding into an immiscible flooding mechanism. This condition is called *miscibility rupture*. So, for a successful miscible flooding oil recovery, the proper design of the injected miscible slug volume is very much important.

Depending on the achievement of miscibility, the miscible displacement processes are classified as follows:

a) First-contact miscible (FCM) flooding
b) Multi-contact miscible (MCM) flooding

In FCM, the injected solvent becomes miscible with the oil in all proportions when they come into contact. Injection of LPG is an example of this category. In MCM, the miscibility is achieved only after a continuous exchange of different components between the injected solvent and the trapped crude oil with the subsequent formation of a miscible oil bank. Injection of natural gas or CO_2 falls into the category where miscibility is achieved through the alteration of the composition of injected fluid and crude oil under MMP conditions.

4.4 FIRST-CONTACT MISCIBILITY PROCESS

In the FCM process, two fluids when mixed together in any proportion form a single phase at a given condition. In oil field practice, relatively small slugs of expensive LPG, accounting approximately for 1%–12% hydrocarbon pore volume (HCPV) are used, followed by the injection of less expensive natural gas. FCM can be best explained by a ternary phase diagram as shown in Figure 4.11. If a straight line between the solvent slug and crude oil passes through the single-phase zone in a ternary phase diagram, then the solvent at that composition will be first-contact miscible with oil. The continuous exchange of mass between the reservoir oil and solvent occurring through the solvent-oil mixing zone is responsible for the change in the composition of the oil as well as solvent. There is a range of solvent compositions that will be first-contact miscible with the crude oil at a specific temperature and pressure. In field practice, natural gas is mainly composed of methane mixed with propane or butane, so the mixture hydrocarbon mixture becomes miscible with crude oil at the reservoir condition with first-contact miscibility.

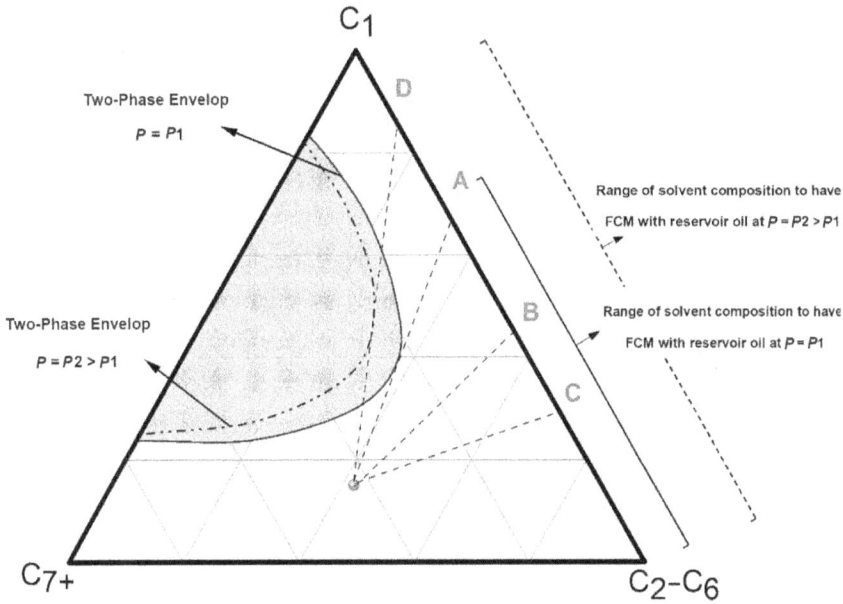

FIGURE 4.11   First-contact miscibility.

Laboratory research has shown that mixing of oil/solvent/drive gas by dispersion (Perkins and Johnston, 1963), aggravated by viscous fingering, gravity tonguing, and channeling caused by stratification, can rapidly dilute small solvent slugs to concentrations that are no longer miscible (Koonce and Blackwell, 1965). CO_2 and crude oil are not generally miscible at first contact, owing to differences in their properties and compositions at a given pressure and temperature condition.

Generally, first-contact miscibility pressure is significantly greater than MMP, which is greater than or equal to BPP (bubble-point pressure).

Most of the CO_2 floods in the United States operate under the premise of MMP (and not the first-contact miscibility pressure, as it would be hard to achieve it without exceeding the original reservoir pressure).

4.4.1   LPG Slug Miscible Injection

Higher molecular weight hydrocarbons, such as propane or LPG, are miscible at first contact with crude oil, at most reservoir conditions. If we inject an LPG slug having composition L to displace an oil (composition O) at the reservoir pressure-temperature condition as shown in Figure 4.12, they will mix in any proportion and form a single phase. This mixture will form an oil bank. However, during the displacement process, the composition of the LPG slug will change continuously along the line LO as it mixes with the heavier component of the crude oil. As soon as the composition of LGP crosses the limiting concentration S, the chase gas G will be immiscible with the LPG slug with modified composition. Thus, the process will be changed from miscible to immiscible displacement. The line joining S" (LPG slug with modified composition) and G passes through the two-phase zone which indicate S" and G are immiscible. The SG line, a tangent on the two-phase boundary indicates the changes between the miscible and immiscible zone.

Hence, the proper design of the LPG slug volume is very much important for successful recovery. Generally, the minimum pore volume of the LPG slug is proportional to the square root of the distance to be covered (Kieschnick, 1960). However, larger than the required minimum volume of LPG slugs is needed to account for the loss of fluid caused by viscous fingering, stratification, and gravity override.

The LPG slug process may be either gas driven or water driven. Field application of the gas-driven LPG flooding process is mostly limited to the deeper reservoirs, because of the high-pressure requirement to attain a miscibility between gas and LPG. The water-driven LPG slug process normally exhibits good sweep efficiency. However, displacement of LPG by water is poor. An improvement in this process appears possible by injecting a slug of carbon dioxide between the LPG slug and the water drive.

The schematic representation of miscible displacement by propane (component of LPG) is given in Figure 4.13. Near to the producing well, a zone containing only oil is followed by a single-phase mixture of propane and oil. The concentration of propane increases in the direction of the injection well and ultimately remain as a pure propane slug. This is followed by a zone containing a mixture of propane and carbon dioxide in continuously varying proportions until 100% CO_2 is reached. In

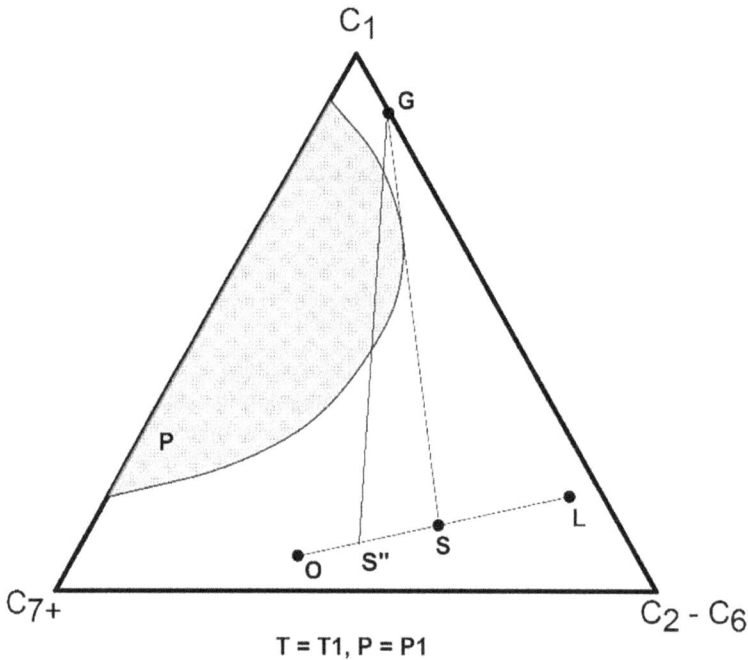

FIGURE 4.12   Ternary phase diagram for miscible displacement by LPG slug.

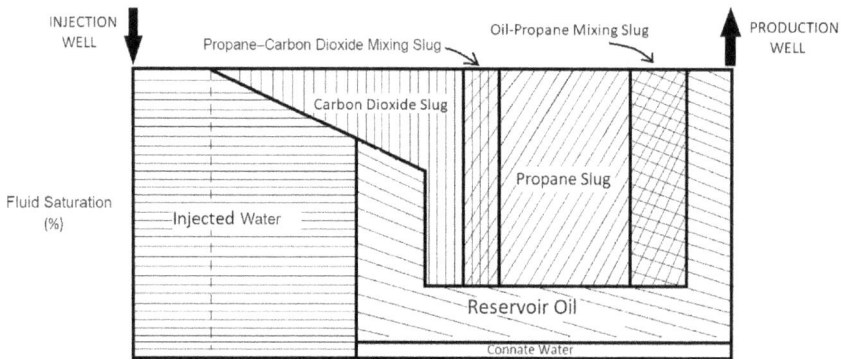

FIGURE 4.13   Schematic of a miscible displacement by the improved water-driven LPG slug process.

Source: Modified after Thompson, 1967.

this zone, the concentration of CO_2 in the immobile water phase also varies from zero to full saturation (Thompson, 1967).

4.5 MULTIPLE-CONTACT MISCIBILITY PROCESS

In the multi-contact miscibility process, the injected solvent is not miscible with crude oil at first contact at the reservoir condition, but miscibility may be achieved through multiple contacts and mass transfer of different components between the reservoir oil and injected fluid. After a continuous exchange of masses, the injected solvent is enriched with intermediate hydrocarbons or crude oil may convert to light oil leading to complete miscibility of the injected solvent and trapped oil, provided the MMP is maintained. Multiple-contact miscible injection fluid normally contains natural gas at high pressure, enriched natural gas, flue gas, nitrogen, or CO_2.

The multiple-contact miscibility processes are categorized as follows:

a) Vaporizing lean gas drive, also called "high pressure lean gas injection"
b) Condensing rich gas drive
c) Combined vaporizing-condensing drive
d) CO_2 displacement

4.5.1 VAPORIZING LEAN GAS DRIVE

In vaporizing lean gas drive, a relatively lean gas containing methane as the main component along with some other lighter hydrocarbons are injected. The injected gas occasionally contains inert gas such as nitrogen. The MMP of the injected gas is generally high, requiring a high injection pressure, and hence, sometimes, the process is called "high pressure lean gas injection." When the injected gas comes in contact with the trapped crude oil in the reservoir, the gas is continuously enriched with the intermediate hydrocarbons transferred from the crude oil and becomes gradually miscible with oil at a certain point in the reservoir. The miscibility in the process can be achieved with natural gas, flue gas, CO_2, or nitrogen, only if the reservoir pressure is maintained above the MMP.

The process is best explained by a ternary phase diagram and thermodynamic equilibrium at every stage of contact. For example, let us consider a representative injection gas (point "G" in Figure 4.14) composed of mainly a light component C_1 and a light reservoir oil "O"; to make the injected gas miscible with oil, the injection pressure should be higher than MMP. Here, the virgin reservoir oil "O" is not miscible with the injected gas at first contact, which is indicated by the passing of the dilution straight line (OG) through the two-phase envelope.

The development of miscibility may be explained conceptually as follows:

- When the injected gas "G" comes in contact with the oil "O" near the injection well, an exchange of mass takes place and the gas is enriched with the intermediate hydrocarbon. The mixture with composition "M_1" splits into two equilibrated phases of liquid O_1 and gas G_1, determined by the equilibrium tie-line (Fig. 4.14).
- As the pressure gradient toward the production well is maintained higher, the gas with altered composition G_1 moves forward and makes further contact with the virgin oil, O (Fig. 4.15) to form mixture M_2. Again, after an exchange of mass, a thermodynamic equilibrium condition is achieved, and the mixture with composition M_2 gets separated into gas G_2 and liquid O_2. The gas G_2 is further enriched with the intermediates, and the process continues.
- The line joining G_2 and O passes through the two-phase zone, which means that G_2 is still immiscible with the oil. However, G_2 again moves forward and comes in contact with the virgin oil and forms G_3. The exchange of mass transfer continues, and finally when enriched gas with composition G_c is formed, it does not pass through the two-phase region—i.e., it enters the single-phase zone, and the gas become miscible with oil at point "G_c." Under

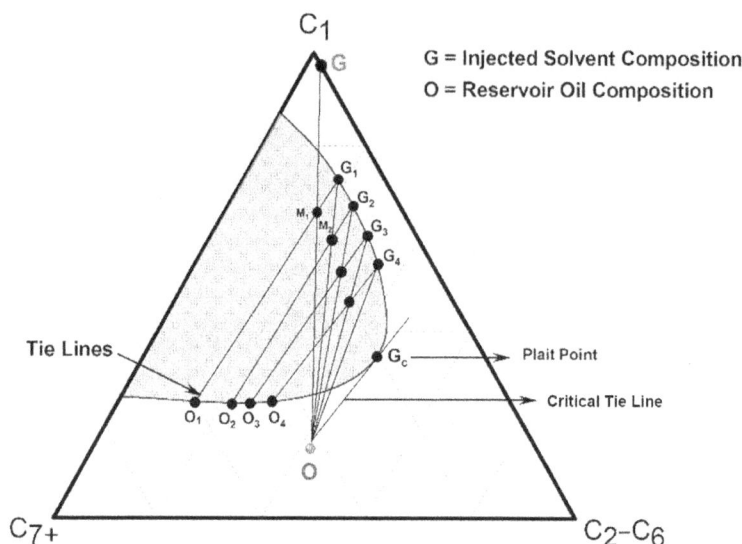

FIGURE 4.14   Representative phase diagram of vaporizing gas drive under multiple-contact miscibility.

this condition, the interfacial tension between the enriched gas phase and oil becomes zero, and after subsequent mixing of the two phases, a miscible oil bank is formed as shown in Figure 4.15. On the other hand, every time the oil comes into contact with the fresh gas, it loses its intermediate and higher hydrocarbon components and becomes progressively heavier with no further exchange of mass with the fresh gas.

4.5.2   CONDENSING GAS DRIVE

In the condensing gas drive, generally gas, enriched with intermediate hydrocarbons—predominantly, ethane through butane—is injected. The dynamic miscibility results from the transfer of intermediate hydrocarbons from the injected fluid into the reservoir oil at the prevailing reservoir pressure-temperature condition. The oil enriched with the intermediate hydrocarbon becomes lighter and swells. On continuous mass transfer, the modified reservoir oil starts to be miscible with the injected gas and forms a single phase. The mechanism of the process may be better understood from the ternary phase diagram shown in Figure 4.16 and the explanation given in the following. As the injected gas is richer in intermediate hydrocarbons, the process is also known as "enriched gas injection." The process may be applicable to heavy oil reservoirs also.

The miscibility development operates conceptually as follows:

> The line connecting between the crude oil "O" and injection solvent/gas "G" (Fig. 4.16) passes through the two-phase zone, which indicates that the injected solvent and oil are not miscible at first contact. However, when the injected gas "G" comes in contact with the oil "O" near the injection well, an exchange of mass takes place, and the intermediate gas is transferred from the gas phase to the crude oil (Fig. 4.17). After the exchange of mass, a thermodynamic equilibrium condition is achieved, and the composition of each phase may be represented at the opposite side of the tie-line: liquid O_1 and gas G_1. The composition of the mixture is represented by M_1. The oil with composition O_1 is enriched with the intermediate hydrocarbon and is lighter than the virgin oil, O. The gas phase G_1 with intermediate hydrocarbon less than that of the originally injected gas, G, moves faster because of its higher mobility and leaves the oil phase enriched with the intermediate hydrocarbon,

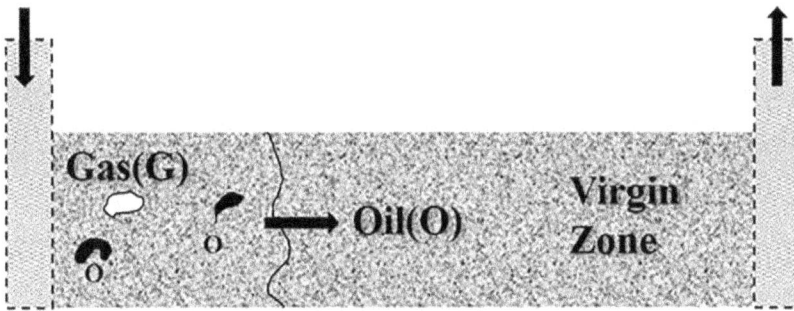

State 1:- Start of HP injection

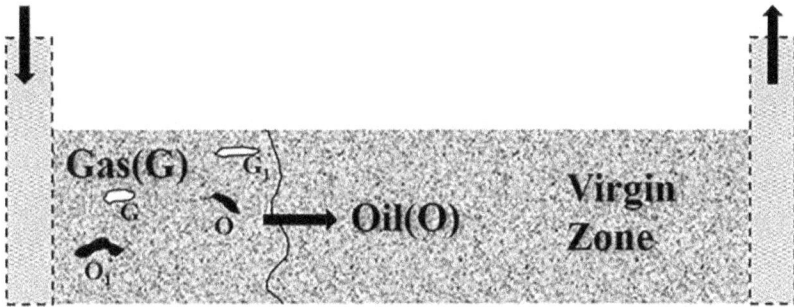

State 2:- Start of the formation of the
miscible bank

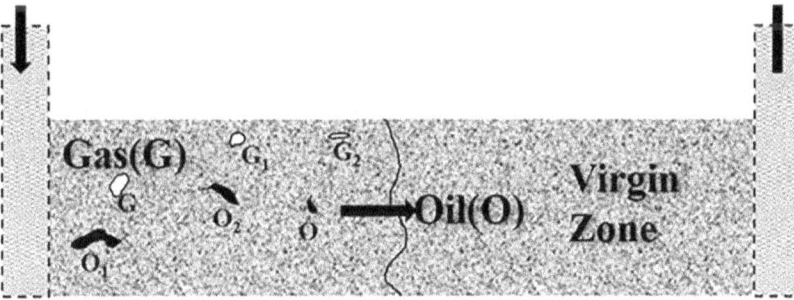

State 3:- Start of the formation of the
miscible bank

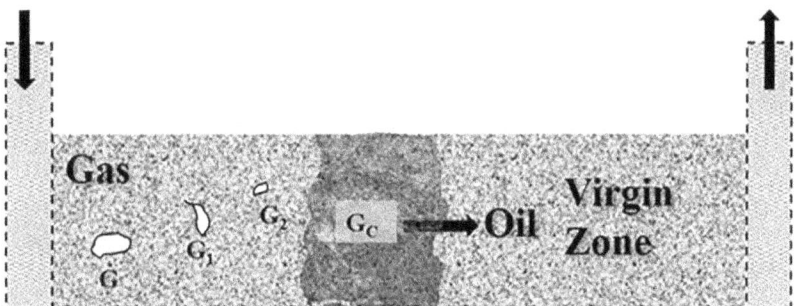

State 4 :- Miscible bank formed

FIGURE 4.15   Schematic of the vaporizing gas drive process.

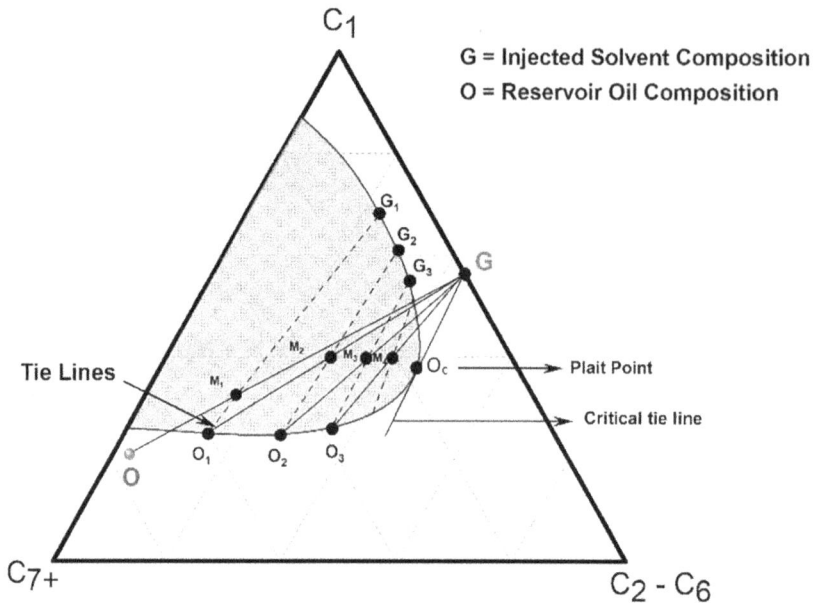

FIGURE 4.16   Multiple-contact miscibility: Condensing gas drive.

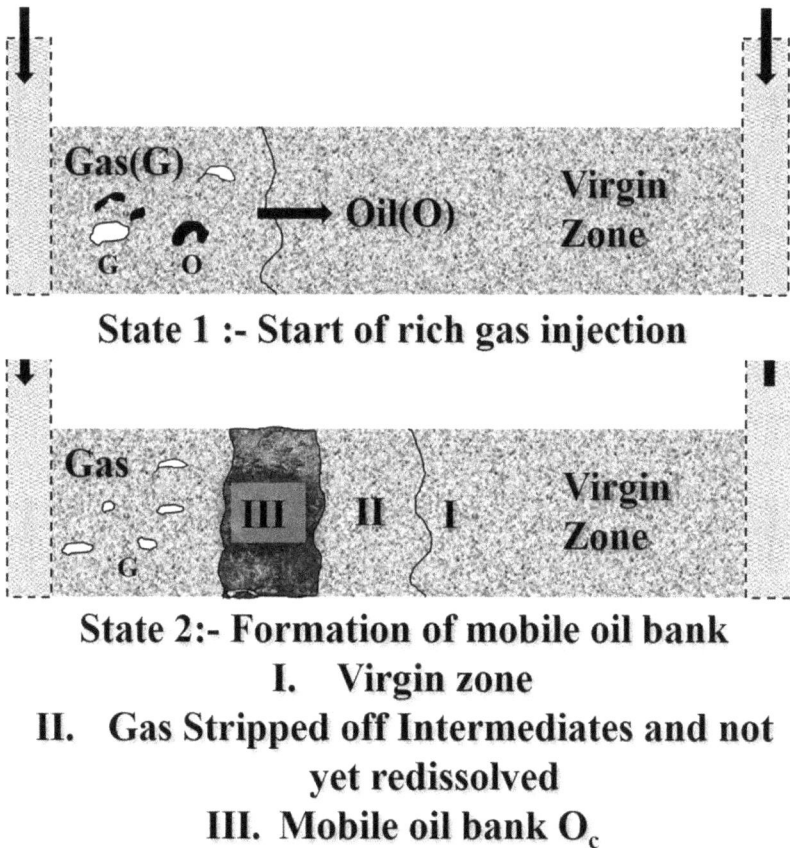

State 1 :- Start of rich gas injection

State 2:- Formation of mobile oil bank
I.   Virgin zone
II.   Gas Stripped off Intermediates and not yet redissolved
III. Mobile oil bank O_c

FIGURE 4.17   Schematic of a condensing gas drive.

O_1, to mix with the fresh injected gas, G, and form mixture M_2. The newly formed mixture now splits into liquid and gas with the composition at O_2 and G_2, respectively. The composition of the liquid at O_2 is now closer to the critical point than that at O_1 and is richer in intermediate components. The gas moves forward, and the oil phase, O_2, contacts with the fresh solvent to form M_3, and so forth. In this way, the oil phase becomes progressively richer until it attains the composition O_c, and the line connecting O_c and the solvent passes through the single-phase zone indicating the complete miscibility of the two phases. On the other hand, every time the gas is in contact with the virgin oil, it loses its intermediate components and becomes progressively drier than G (Latil, 1980).

When the composition of oil changes from O to O_c, it becomes miscible with the injected gas. Subsequently, the saturation of the mobile oil increases and forms a mobile oil bank III as shown in Figure 4.17. Since gas is more mobile than oil phase, the leaner gas stripped off its intermediate components moves ahead of the mobile oil bank (III) and makes contact with the virgin oil and remain immiscible shown as zone II in Figure 4.17. As the pressure is maintained, the oil bank moves forward and captures the trapped oil ahead of the front. If the process is designed and implemented appropriately, a very high recovery factor can be obtained compared to other EOR processes.

4.5.3 Condensing-Vaporizing Mechanism

The condensing/vaporizing mechanism is best illustrated if we consider an oil/gas system that is composed of principally four groups of components (Zick, 1986):

a) The first group is composed of the lean components like methane, nitrogen, and CO_2 having the equilibrium K values greater than one.

b) The second group consists of the light intermediate hydrocarbons such as ethane, propane, and butane. These gases are considered to be the enriching components and can be condensed to oil at high pressure.

c) The middle intermediate components, ranging from C_4 through C_{10} on the low molecular weight side up to C_{30} on the high side are the main constituents of the third group. Generally, these components are not present in the injected gas but in the reservoir oil and may be vaporized from the oil to the gas phase.

d) The fourth group consists of everything else—i.e., those heavy components in the oil that are very difficult to vaporize.

When the enriched gas of the second group comes into contact with the oil, the light intermediates condense from gas to oil, making the oil lighter. The injected gas after exchanging mass with the oil becomes lighter and moves forward. The fresh gas rich in intermediate components again comes into contact with the oil, making the oil even lighter. The process continues till the oil becomes too light to be miscible with the injection gas and establishes the condensing gas drive mechanism.

The process may differ from ideal behavior for a real reservoir oil, and generally, a countereffect exists. The middle intermediates that were not originally present in the injected gas are stripped from the oil to the gas phase. As the injection gas contains none of these middle intermediates, they cannot be replenished in the oil. These middle intermediates are carried forward by the mobile gas phase. After the first few contacts, the oil becomes lighter, as the intermediates are transferred to the oil phase by a net condensation. In subsequent contacts, the oil gets heavier, by net vaporization of the middle intermediates. Once this begins to occur, the oil no longer has a chance of becoming miscible with the gas. Under this condition, the process of oil recovery will not be very efficient.

However, at this stage, there is another mechanism, if we see the picture slightly downstream from the injection point. In the downstream, the oil will be in contact with the equilibrium gas, which is relatively lean due to the loss of its light intermediates. As this gas moves forward and comes into contact with the virgin oil that is saturated in the light intermediates, it will be again enriched with the light intermediates by the vaporizing gas drive mechanism. Therefore, this gas will have about the same amount of light intermediates as the injection gas. It also contains a small amount of middle intermediates, picked up from the oil at an earlier stage of interaction. Thus, it will actually be a little richer than the original injection gas and will have slightly more condensable intermediates than the one upstream. As this process continues and moves downstream, at a favorable condition, the combined vaporizing/condensing process occurs within the transition zone where the compositional path goes through the critical point. When the oil downstream is in contact with this gas, a miscible oil bank will be formed, leading to oil recovery by the vaporizing gas drive mechanism. This process can be imagined as a combination of the condensing process upstream and the vaporizing process downstream. This process is called the "condensing/vaporizing gas drive."

4.5.4 CO_2 MISCIBLE FLOODING

Field experience has shown that CO_2 miscible flooding is an effective method to improve oil recovery from low-permeability, light oil reservoirs. EOR with CO_2 injection was first reported by Whorton and Brownscombe in 1952. And continuous development in this field has been reported through numerous published literary works (Huang and Dyer, 1993; Obi and Blunt, 2006; Zhang et al., 2019). Over a long period of development, CO_2 is now considered a potential injection fluid both for miscible displacement and immiscible displacement. According to IEA (2019), more than 200 CO_2 EOR projects have been successfully implemented in various oil fields across the world. Coupled with enhancing the recovery, it serves the carbon capture and storage requirements that most countries are pursuing to earn carbon credit and reduce CO_2 emissions. Consequently, the current industry interest in CO_2 miscible flooding is high, as evidenced by the level of activity in field testing and CO_2 source development.

The injected CO_2 completely dissolves through the crude oil at MMP, which is determined experimentally through slim tube tests or by mathematical correlations. Hawthorne et al. (2017) reported that compared to CO_2, methane requires much higher pressures to achieve MMP with a typical Bakken crude oil, while ethane achieves MMP at much lower pressures than CO_2.

4.5.4.1   Mechanism of CO_2 Miscible Recovery

EOR mechanisms by CO_2 miscible flooding may be explained as follows:

(i) First-contact miscibility: If the injection pressure is higher than the MMP, the injected CO_2 may be mixed completely with the reservoir oil, leading to oil recovery by miscible displacement.
(ii) Vaporized gas drive process: The reservoir pressure is above the MMP, and miscibility between CO_2 and reservoir oil is achieved through multiple-contact or dynamic miscibility unlike at the first contact; the intermediate and higher molecular weight hydrocarbons from the reservoir oil vaporize into the CO_2.
(iii) Condensed gas drive process: As CO_2 has good miscibility with crude oil, a part of the injected CO_2 dissolves into the oil, which leads to swelling of crude oil and improvement in its mobility.

According to multi-contact miscibility with the exchange of mass transfer between the oil and CO_2, the two phases become completely miscible to each other in the absence of any interface. During the progress, there is a transition zone between the oil in the front and CO_2 in the back (Fig. 4.18).

This is to mention here that in the real field, many a time, the MMP may not be maintained to minimize the expenditure incurred, especially for shallow reservoirs where the reservoir pressure is less. The pressure may be maintained near MMP, and partial miscibility may be achieved by multiple contacts.

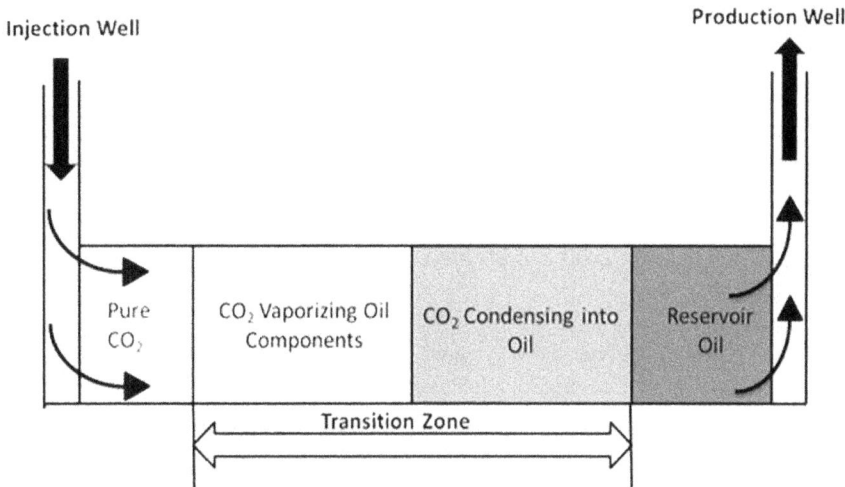

FIGURE 4.18 Schematic of the CO_2 miscible process, showing the transition zone between the injection and production wells.

4.5.4.2 CO_2 Flood/Injection Designs and Classification

Based on reservoir geology, rock and fluid properties, pressure, temperature, and the well-pattern configuration, CO_2-based EOR may be of differentiated as follows (Verma, 2015):

I. ***Continuous CO_2 injection:*** According to this process, a predetermined volume of CO_2 is injected continuously. Sometimes a lighter gas, such as nitrogen, follows the CO_2 slug to maximize gravity segregation. This approach is implemented after primary recovery and is generally suitable for reservoirs containing medium to light oil as well as reservoirs that are strongly water wet or sensitive to waterflooding. The efficiency of the process is increased for the reservoir where gravity drainage is favorable.

II. ***Continuous CO_2 injection followed with water:*** In this process, continuous CO_2 gas is injected followed by chase water. This process works well in reservoirs with low permeability or moderate homogeneity.

III. ***Conventional water alternating gas:*** The water alternating gas (CO_2) (WAG) injection technique is nothing but alternating the CO_2 gas injection with water injection. The process may be repeated for several cycles depending on its efficacy and reduction in residual oil. Water alternating with CO_2 injection helps overcome the gas override and reduces CO_2 channeling, thereby improving the overall sweep efficiency. This technique is mostly suitable for the reservoirs with permeability contrasts among various layers.

IV. ***Tapered WAG:*** This technique uses a varied duration of gas injection. In the beginning, gas is injected for a longer time, which shortens with a progressive WAG cycle. Tapered WAG is found to be more effective than the uniform WAG application with an equal duration of water and gas injection cycle. The technique reduces the response time—i.e., the oil bank arrives earlier. It also uses gas injection more efficiently—i.e., produces more incremental oil per unit of injected gas (Khan et al., 2016). The process is equally applicable to homogeneous and heterogeneous reservoirs.

V. ***WAG followed with gas:*** This process is a conventional WAG process followed by a chase of less expensive gas (e.g., air or nitrogen) after the full CO_2 slug volume has been injected.

4.6   FACTORS INFLUENCING EFFICIENCY OF MISCIBLE DRIVE

In miscible flooding, the microscopic displacement efficiency (E_D) is a measure of the effectiveness of miscible slugs to mobilize/displace. It indicates the fraction of oil that is displaced by the slugs from pores to which the slug comes in contact with. There are a large number of factors that influence the efficiency of displacement of hydrocarbons from a reservoir by miscible drive. The most important factors are as follows (Clark et al., 1958):

a) Pressure, temperature, and oil composition
b) Phase behavior
c) Fluid properties
d) Saturation history of the rock-fluid system
e) Diffusion, solvent flow rate, and residence time
f) Slug size
g) Dispersion
h) Dead-end pore volume
i) Irregularity of porosity and permeability of the reservoir rock
j) Reservoir size and shape
k) Structural dip of the reservoir, etc.

The oil composition affects CO_2 miscible flooding in the following manner (Holm and Josendal, 1974):

a) As the concentration of extractable hydrocarbons ($C_5–C_{30}$) in the reservoir oil sample increases, the MMP decreases.
b) Oil API gravity is also inversely related to the MMP.
c) MMP will be higher if the heavy molecular weight component of the oil increases.
d) If the asphaltene content increases, the MMP will also rise.
e) As the reservoir temperature increases, the MMP increases.

The presence of methane, nitrogen, and oxygen in CO_2 increases the pressure requirement to develop miscibility. On the other hand, the presence of LPG components (C_3 and C_4) and H_2S reduces the MMP.

Factors like phase behavior, diffusion, and dispersion play important roles in developing and sustaining the miscible displacement conditions. Similarly, factors like slug size, dead-end pore volume, pore geometry, and structure are also crucial in determining the microscopic displacement efficiency for field applications of the miscible flood process.

4.7   DETERMINATION OF MISCIBILITY CONDITION

To design a miscible displacement process, it is very important to determine the miscibility condition, which indicates the state at which the miscibility between the injected solvent and the reservoir oil is achieved during the FCM or MCM process. The three major parameters that determine the miscibility condition are temperature, pressure, and injected solvent composition. In general, the average reservoir temperature is set as the process temperature, and the pressure is changed gradually for a particular oil composition. The displacement efficiency for a crude oil system by gas injection is very sensitive to pressure, and miscible displacement can be achieved only when the reservoir pressures exceed MMP. At MMP, the interfacial tension across the interface between the injected gas and reservoir fluid approaches zero and results in potential transfer of molecules across the interface. The phenomenon finally leads to mutual miscibility and the formation of a homogeneous fluid (Green and Willhite, 1998).

The miscibility conditions (MMP or, minimum miscibility enrichment, MME) may be determined by the following techniques:

1. Experimental measurements.
2. Empirical correlations based on experimental results.
3. Phase-behavior calculations based on equation of state and computer modeling.

While the experimental methods directly measure the MMP for a given system, the others predict these values from various empirical correlations.

4.7.1 Experimental Measurements

Slim tube, vanishing interfacial tension (VIT), and rising bubble apparatus (RBA) are the primary industry-recommended laboratory methods to measure the MMP.

4.7.1.1 Determination of Miscibility Condition: Slim Tube Test

The slim tube test performs dynamic miscibility studies at simulated reservoir conditions of temperature and pressure. The MMP, or minimum miscibility concentration, of a given solvent and reservoir oil can be estimated from this study. A representative slim tube apparatus, along with its accessories, is demonstrated in Figure 4.19. The slim tube is made of a long-coiled tube packed with specific mesh-size sand or similar porous material. While flowing through the packed porous media, the injected solvent comes in contact with the saturated oil, similar to that in the reservoir. Thus, it provides multiple equilibrium contacts between simultaneously flowing fluids.

At the beginning of each test, the tube is fully saturated with the reservoir oil at reservoir temperature, followed by the solvent injection at several test pressures (constant downstream/upstream pressure). The exit-end pressure is maintained by a back pressure regulator. The pressure drop across the coiled tube is generally a small fraction of the applied pressure on the system, so the entire displacement pressure is considered to be at a single constant pressure. The coiled tube is placed inside a constant temperature bath. The fluid production rate, density, and composition are noted as the functions of the injected volume downstream.

The tube is coiled in a manner that the flow is basically horizontal and the gravity effects are insignificant. Oil recovery after the injection of a specific number of pore volumes (PV) such as 1.2 PV of solvent is the test criterion for miscibility. The miscibility criterion may be observed visually from the color change of the effluent stream and confirmed accurately from the recovery versus displacement pressure data. Below MMP, the recovery is boosted by the increased displacement pressure, which flattens once MMP is reached. Thus, two distinct trend lines are observed if we plot

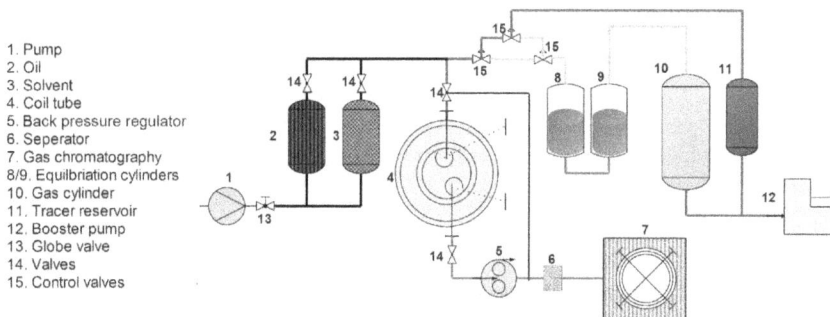

FIGURE 4.19 Schematic diagram of the slim tube apparatus.

Source: Kantzas et al., 2012.

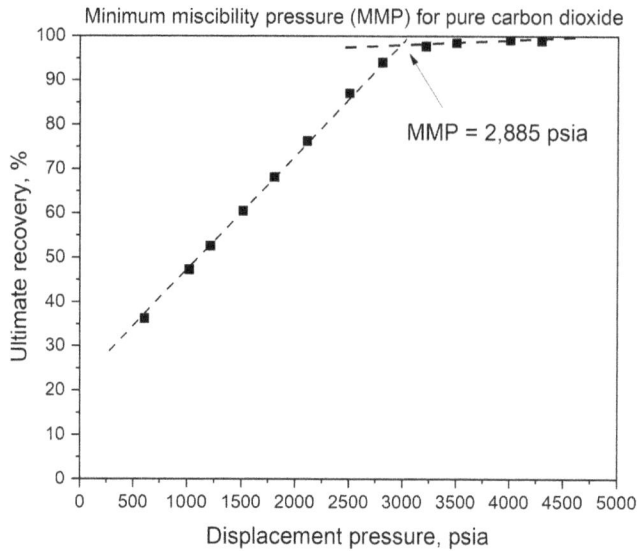

FIGURE 4.20 Determination of MMP for pure CO$_2$ displacement.

Source: Choubineh et al., 2019.

oil recovery against pore pressure for several slim tube tests. The point of intersection of those trend lines is the estimated MMP for the given oil-solvent system (Fig. 4.20).

In general, MMP is achieved when the oil recovery attains a value between 90% and 95%. The achievement of miscibility is expected to accompany a gradual change of color of the flowing fluids from that of the oil. On the other hand, the observation of a two-phase flow is an indication of an immiscible displacement. The effluent fluid is cooled, and the separated gas is evaluated by a gas chromatograph. The liquid properties like density, composition, and viscosity are measured, and the data are used as inputs to fine-tune a fluid equation of state for the reservoir simulation.

Example 4.2: A slim tube experiment was conducted at 66°C to determine the minimum miscibility pressure of CO$_2$. During the injection process, oil recovery was measured at different injection pressures as follows:

Pressure, psi	1,064	1,211	1,368	1,429	1,697	1,832	1,929	2,067
Recovery factor (%)	58.3	66.4	71.9	78.6	92.4	94.6	95.4	96.7

Determine the MMP for the system.

Solution:

By plotting the recovery factor against the injection pressure, one can get the plot as shown in Figure 4.21. The plot has two different slopes at the lower and higher pressure ranges. The pressure corresponding to the intersection of the two lines gives the MMP. From the figure, the MMP may be measured as 1,730 psi. It can be observed that increase in recovery is minimal, beyond the MMP.

4.7.1.2 Vanishing Interfacial Tension

According to the definition of MMP, no interface exists between the crude oil and injection fluid—i.e., the interfacial tension approaches zero at MMP. The vanishing interfacial tension (VIT) method offers a direct method to measure the MMP from the interfacial tension (IFT) values. Oil recovery can be improved with gas injection near MMP, where IFT approaches zero. VIT is deployed to determine MMP for a given gas-oil system from the measured IFT values at various pressures. The IFT values are generally measured by pendant drop shape method in combination with an optical tensiometer in a high-pressure view cell. Crude oil is introduced as a drop phase into the chamber filled with the injection fluid. The pressure is gradually increased by pumping more injection fluid into the chamber with simultaneous measurement of IFT. The IFT data are plotted against the pressure and extrapolated to zero IFT value to get the MMP. A representative plot is illustrated in Figure 4.22.

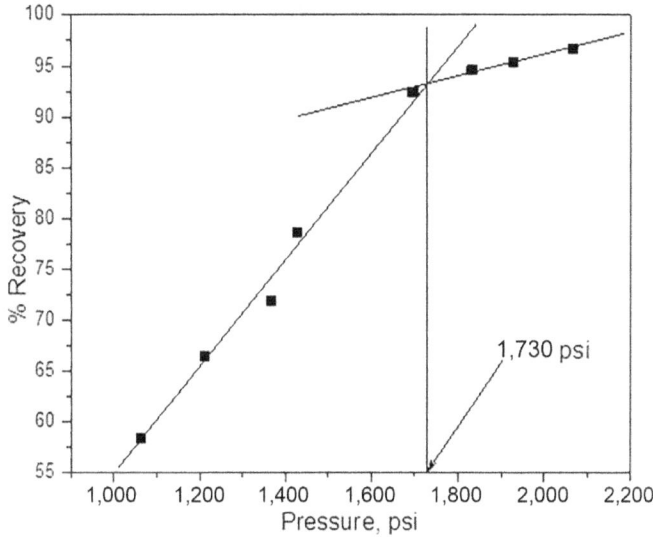

FIGURE 4.21 Oil recovery versus pressure plot with reference to Example 4.2.

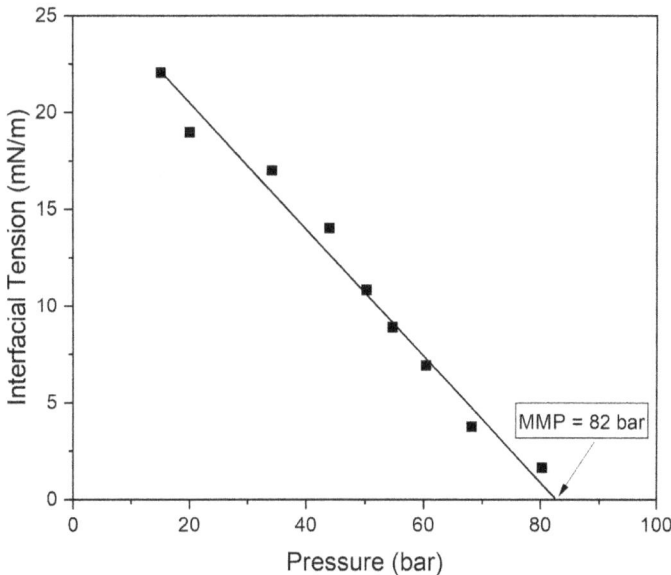

FIGURE 4.22 Example of a VIT measurement graph.

4.7.1.3 Rising Bubble Apparatus Method (Forward Method)

A typical RBA apparatus for measuring MMP is shown in Figure 4.23. In a rising bubble test, a gas bubble having the desired composition is released into the live oil column from the bottom of the apparatus at a particular pressure. Throughout the experiment, the bubble's shape and motion are continuously monitored and recorded with high-precision cameras. The MMP is determined from direct observations of the bubble behavior during the rising process in the test. The change in shape of the rising bubble indicates its miscibility with the oil at those conditions. Below the MMP, the bubble holds its shape as it rises. Above the MMP, the bubble shape changes as it rises; it may disintegrate, dissolve, or disappear into the oil. Testing at several pressures helps determine the MMP between the gas and oil. In this process, two pressure data are very important: first is the pressure at which the bubble under surveillance just rises to the oil column, and the second is the point at which it disappears. The average of these two pressures is estimated to be the MMP.

The rising bubble test represents a forward-contacting miscibility process and, therefore, may not accurately estimate the MMP for a backward or combined contact mechanism.

4.7.2 MINIMUM MISCIBLE PRESSURE CORRELATIONS

Empirical correlations for determining the MMP for a set of reservoir fluids and injection solvent at a given reservoir condition may be developed by regression analyses of experimental data. Empirical correlations are easy to use and always serve as a tool for quick MMP prediction (Chen et al., 2017). Numerous studies reveal that the MMP has a weak relationship with the molecular weight, moderate relationship with mole fractions, and a strong relationship with system temperature. Mostly, regression approaches are applied to develop the empirical correlations. Examples of such common correlations include the Glaso correlation, Sebastian et al. correlation, and Khazam et al. correlation (Sebastian et al., 1985; Glaso, 1985; Khazam et al., 2016). Generally, the accuracy of the empirical correlations increases as the mathematical complexity of the equation increases. Most of the empirical correlations are used mainly for fast-screening applications (Yuan et al., 2005). In addition, analytical methods have been coupled with the equation of state (EOS) to estimate the MMP.

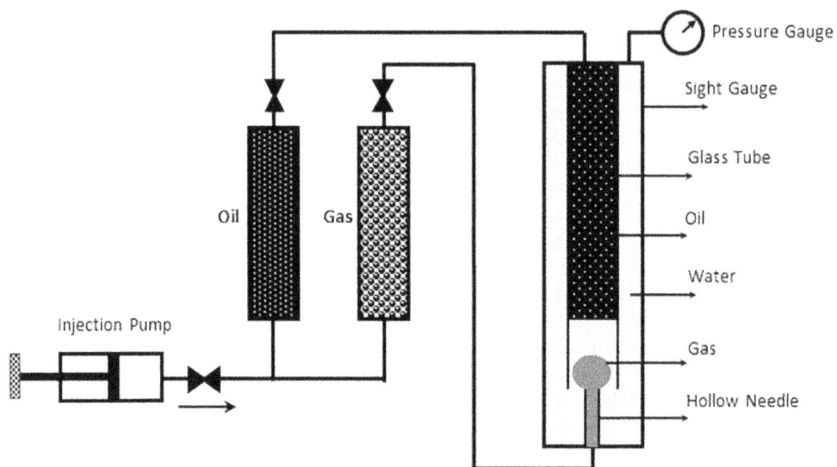

FIGURE 4.23 Rising bubble apparatus.

Source: Kantzas et al., 2012.

Though the correlations are less accurate than other methods, they are frequently used because of their ability to quickly predict the MMP and screen a potential reservoir for miscible gas flooding, particularly, when detailed fluid characterizations are not available. The main advantage of analytical methods is that they can determine the MMP without introducing uncertainties associated with the condensing or vaporizing displacement (CV) process. As the displacement processes are associated with complex miscibility mechanisms and reduce the reliability of MMP prediction, the analytical gas theory is applied to estimate the CO_2 MMP. Some such correlations are discussed in the following.

Firoozabadi and Aziz (1986) performed a series of experimental studies on the MMP for N_2 and methane systems. Based on the experimental data and some simulation runs, they developed a simple and reliable correlation for the prediction of MMP. They developed a correlation for MMP as a function of the mole fraction of the intermediates, molecular weight of the C_{7+} fraction, and temperature, as given in Equation 4.10. The effect of temperature is less pronounced, and MMP is inversely related to temperature in degrees Fahrenheit.

$$P_{mmp} = 9433 - 188 \times 10^3 \left(\frac{C_{C_2} - C_{C_5}}{M_{C_{7+}} T^{0.25}} \right) + 1430 \times 10^3 \left(\frac{C_{C_2} - C_{C_5}}{M_{C_{7+}} T^{0.25}} \right)^2 \tag{4.10}$$

where P_{mmp} = MMP (psia), T = temperature (°F), $C_{C_2} - C_{C_5}$ = concentration of intermediates (mol%), and $M_{C_{7+}}$ = molecular weight of heptane-plus.

Johnson and Pollin (1981) proposed a correlation for pure CO_2-displacement in which MMP is expressed as a function of molecular weight of gas, critical temperature of injection gas, reservoir temperature, and molecular weight of oil as given here:

$$P_{mmp} - P_{ci} = 0.006939 \alpha_i \left(T - T_{ci} \right) + I \left(\beta M - M_i \right)^2 \tag{4.11}$$

where P_{mmp} = minimum miscibility pressure (MPa); P_{ci} = critical pressure of injection gas (MPa); T_{ci} = critical temperature of injection gas (K); T = reservoir temperature (K); M = molecular weight of oil; M_i molecular weight of injection gas; and β = constant, $\beta = 0.235$. α_i is a constant, when the content of impurities in CO_2 (Y2) is less than 10 mol%, and 300 K < T < 410 K. I is the characterization index of oil, which is determined as a function of API gravity and the molecular weight of oil.

Sebastian et al. (1985) proposed a correlation for impure CO_2-displacement in the following form:

$$P_{mmp}^{imp} = P_{mmp}^{pure} \times F_{imp} \tag{4.12}$$

where P_{mmp}^{pure} is the miscibility pressure for impure CO_2 displacement, P_{mmp}^{pure} is miscibility pressure for pure CO_2 displacement, P_{imp}^{imp} is a correction factor to the MMP calculated for pure CO_2-displacement and determined from Equation 4.13:

$$F_{imp} = 1.0 - 0.0213 \left(T_{cm} - 304.2 \right) + 2.51 \times 10^{-4} \left(T_{cm} - 304.2 \right)^2 - 2.35 \times 10^{-7} \times \left(T_{cm} - 304.2 \right)^3 \tag{4.13}$$

where T_{cm} is the molar average critical temperature of the driving gas:

$$T_{cm} = \sum_i^m X_i T_{ci} \tag{4.14}$$

X, and T_c denote the mole fraction and critical temperature (K) of component i in the injection gas, respectively.

Depending on the characteristics of reservoir fluids, injection fluids, and pressure-temperature condition, many other correlations developed by various researchers and systems are reported in the literature (Zuo et al., 1993).

The analytical equation for estimating CO_2 MMP by Yuan et al. (2004):

$$MMP = a_1 + a_2 M_{C7+} + a_3 P_{C2-6} + \left(a_4 + a_5 M_{C7} + \frac{a_6 P_{C2-6}}{M_{C7+}} \right) T$$
$$+ \left(a_7 + a_8 M_{C7+} + a_9 M_{C7+}^2 + a_{10} P_{C2-6} \right) T^2 \tag{4.15}$$

where MMP is the predicted minimum miscibility pressure for CO_2 injection, M_{C7+} is the molecular weight of C_{7+}, P_{C2-6} is the $\%C_2$ to C_6, and a_1–a_{10} are the fitting coefficients: $a_1 = -1,463.4$, $a_2 = 6.612$, $a_3 = -44.979$, $a_4 = 2.139$, $a_5 = 0.11667$, $a_6 = 8,166.1$, $a_7 = -0.12258$, $a_8 = -0.0012283$, $a_9 = -4.0152E-6$, and $a_{10} = -9.2577E-4$.

Glaso correlation (1985):
The empirical correlation for pure CO_2 injection is given by:

For $C_{2-6} > 18\%$,

$$MMP_{pure} = 810 - 3.404 M_{C7+} + 1.700 \times 10^{-9} M_{C7+}^{3.730} e^{786.8 M_{C7+}^{-1.058}} T \tag{4.16}$$

For $C_{2-6} < 18\%$,

$$MMP_{pure} = 2947.9 - 3.404 M_{C7+} + 1.700 \times 10^{-9} M_{C7+}^{3.730} e^{786.8 M_{C7+}^{-1.058}} T - 121.2 C_{2-6} \tag{4.17}$$

where MMP_{pure} is the estimated minimum miscibility pressure in psia for pure CO_2 injection, C_{2-6} is the mole fraction of C_2–C_6, and MC_{7+} is the molecular weight of the heptane-plus fraction.

4.7.3 MMP FROM PHASE-BEHAVIOR CALCULATIONS

Another method of MMP calculation with good accuracy is by computational approaches that use fluid flow and thermodynamic phase-equilibrium principles. Phase-equilibrium data and accurate MMP may be obtained from multi-contact mixing-cell experiments (Benmekki and Mansoori, 1988; Ahmadi and Johns, 2011) for vaporizing or condensing gas floods. The principle of the process relies on P/T flash calculations using any cubic EOS for each contiguous step, where the equilibrium phase of the present step (downstream) comes in contact with the injected gas and moves forward for the second stage. According to the method, the first two cells are maintained at a fixed temperature and pressure; the injection gas is located in the upstream cell, and the reservoir fluid is in the downstream cell (Fig. 4.24).

If a reservoir oil (x^O) and injection gas (y^G) are mixed at the desired ratio (α), then the resultant mole fraction will be obtained using the mass-balance equation $z = x^O + \alpha(y^G - x^O)$. As long as the pressure is below the MMP, the resulting overall composition z will be either in the two-phase region or in the region of tie-line extensions. Thus, a flash or negative flash (Whitson and Michelsen, 1989) with a cubic EOS can be performed at this overall composition, resulting in two equilibrium compositions: one for liquid (x) and one for vapor (y). The equilibrium vapor moves ahead of the equilibrium liquid because gas is injected. This is the first contact.

The second series of contacts contain both upstream and downstream contacts (Fig. 4.25). The pressure values in the subsequent steps are increased systematically. There are varieties of ways to determine the next applicable pressure. For the second step, the pressure may be simply increased by a small amount, ~200 psia. To determine the third pressure, one can use a linear extrapolation of the tie-line lengths (from the previous two pressures) to zero tie-line length to give the first MMP estimate. The downstream contact mixes the equilibrium vapor (y) with fresh oil, and the upstream contact mixes the equilibrium liquid (x) with fresh injection gas. For mixing vapor and liquid, the

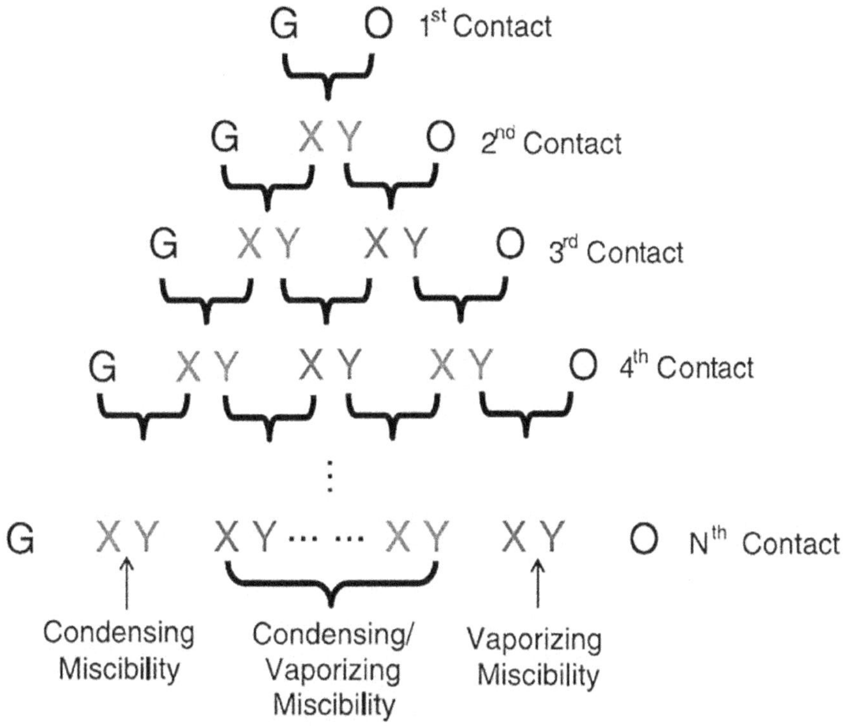

FIGURE 4.24 Illustration of repeated contacts in the multiple-mixing-cell method. G: injecting gas composition, O: oil composition, Y: equilibrium gas composition, X: equilibrium liquid composition.

same material-balance equation, $z = x + \alpha(y - x)$ is used. Two new sets of equilibrium liquid and vapor phases result from these flash calculations so that there are six cells, including the reservoir oil and injection gas. This completes the second contact.

The mixing is continued to make additional contacts until all key tie-lines develop and converge to within a specified tolerance. Thus, after N contacts, there will be a total of $2N + 2$ cells. A key tie-line is developed when three successive cells have a slope of zero, to within a specified tolerance. A zero slope means that two neighboring cells have the same tie-line length within a specified tolerance.

A multiple-parameter regression of the minimum tie-line lengths is performed to determine the exponent n in $TL^n = aP + b$, where TL is the tie-line length, a is the slope, and b is a constant. The three parameters are determined when the correlation coefficient exceeds 0.999. Then, a first estimate of the MMP as the pressure at which the power-law extrapolation gives zero length can be calculated.

The following steps may be used to calculate the MMP.

1. Specify the reservoir temperature and an initial pressure that is substantially below the MMP.
2. Start with two cells filled with injection gas and reservoir oil. Mix the gas and oil, and flash the resulting overall composition (using any cubic EOS) to get two new equilibrium compositions, liquid x and vapor y. If a reservoir oil (x^O) and injection gas (y^G) are mixed at the desired ratio (α), then the resultant mole fraction will be obtained using the mass-balance equation $z = x^O + \alpha(y^G - x^O)$.
3. Mix the resulting equilibrium liquid(s) (x) with equilibrium vapor(s) (y), assuming gas moves ahead of the oil phase. Each of the contacts results in new compositions for the next set of contacts. The composition is again determined from the mass-balance equation.

4. Continue with additional contacts by mixing neighboring cells, as shown in Figure 4.25, until all NC-1 key tie-lines develop and converge to a specified tolerance. The slopes of the tie-lines are determined as the function of cell number. The convergence will be achieved only when there is a key tie like where three successive cells have a slope of zero, to within a specified tolerance. A zero slope means two neighboring cells have the same tie-line length to within a specified tolerance.

5. Calculate the tie-line length of each key tie-line (found in step 4), and store the minimum tie-line length (TL).

6. Increase the pressure, and repeat steps 2–5. There are a variety of ways to determine the next pressure to use. For the second pressure, you may increase the pressure by a small amount, say 200 psia. To determine the third pressure, a linear extrapolation of the tie-line lengths (from the previous two pressures) may be used to zero tie-line length to obtain the first MMP estimate. The increment in pressure, ΔP, is determined by dividing the difference in the estimated MMP and the current pressure by three or by however many additional pressure points are desired. Smaller increases are possible but at the expense of time and with little improved accuracy in the final MMP estimate. For subsequent pressures, ΔP is based on a power-law extrapolation of the minimum tie-line lengths found at the previous three pressures; that is, a multiple parameter regression of the minimum tie-line lengths is done to determine the exponent n in $TLn = aP + b$, where a is the slope and b is a constant. The three parameters are determined when the correlation coefficient exceeds 0.999. The first estimate of the MMP as the pressure is determined at which the power-law extrapolation gives zero length. Typical 45 values for n are between 1.5 and 10, although the extrapolated MMP near the actual MMP changes little over this range.

7. Repeat step 6 for two or three additional pressures, fitting the last three pressure points, until the MMP is estimated to within the desired accuracy, say 20 psia. If the accuracy is not satisfactory, then use smaller pressure increments in step 5. This approach not only yields the final MMP but also gives an estimate of the uncertainty in MMP based on the MMP estimate from the previous regression. We take the absolute value of the difference in the previous MMP estimate and the final MMP estimate as the uncertainty of the MMP. It is likely that this MMP range is larger than the true error in the final MMP.

4.8   NEAR-MISCIBLE SOLVENT FLOODING

Near-miscible gas flood refers to the injection of gases that do not develop complete miscibility with the oil but come closer (Sohrabi et al., 2008). For instance, condensing-vaporizing gas drives at enrichment slightly below minimum miscibility enrichment or at pressures slightly below MMP are near-miscible processes. From the economic and operational standpoints, the technique is very attractive to operators. Cost of the leaner injectant (lower LPG or NGL enrichment in dry gas) is less compared to a richer injectant. In addition, a lower-pressure process also makes the process cheaper because of the lower injectant density and reduced cost of compression at near-miscibility. Specially for the reservoirs at pressures lower than MMP, and the miscible-flood projects that are experiencing pressure decline, a near-miscible operation may be more feasible than a miscible one.

The main mechanisms of oil recovery by near-miscible solvent flooding are (1) oil phase swelling, as the oil becomes saturated with CO_2; (2) viscosity reduction of the swollen oil and CO_2 mixture; (3) extraction of the lighter hydrocarbon in the CO_2 phase and reduction of the gas-oil IFT; and (4) fluid drive plus pressure (Pande, 1992; Thomas et al., 1994). Burger et al. (1994) showed that economically optimum enrichment in high viscosity ratio secondary gas floods could be operated at pressures below the MMP. Shyeh-Yung (1991) demonstrated that tertiary gas flood recoveries below MMP do not decrease as severely as predicted by slim tube tests for CO_2. In near-miscible gas

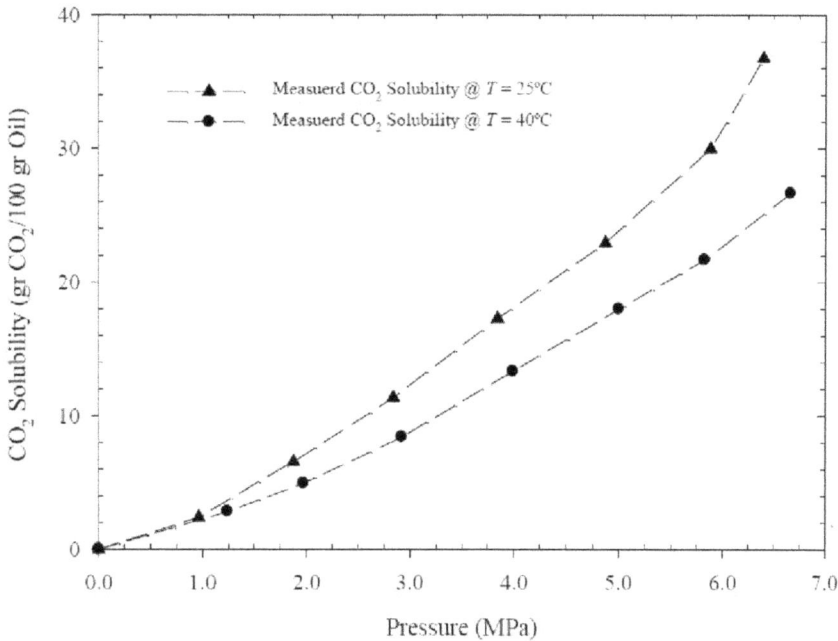

FIGURE 4.25 Solubility of CO_2 in crude oil sample at different equilibrium pressures and two constant temperatures of $T = 25°C$ and $40°C$.

Source: Mosavat and Torabi, 2014.

injection, at the pore scale, significant cross flow of oil into the main flow stream takes place behind the gas front. This can lead to total recovery of the contacted oil.

The solubility of CO_2 and other gases (light hydrocarbon, nitrogen) plays an important role in achieving the desirable properties—i.e., reduction of IFT, crude oil swelling, and improvement of mobility ratio. A little increase in the viscosity of the carbonated water improves the motility ratio and, hence, sweep efficiency (Islam and Carlson, 2012).

Figure 4.25 presents the solubility of CO_2 in crude oil at different equilibrium pressures at $T = 25°C$ and $40°C$. In general, the solubility of gas in increases with pressure but varies with temperature.

Carbon dioxide also has good solubility in water, though it is much less in contrast to that in oil. The solubility of CO_2 in water depends widely on water salinity, in addition to the pressure and temperature conditions (Holm, 1963), as shown in Figure 4.26. The solubility change at a lower pressure is generally high, but the increase is only marginal after a pressure of 70 bar. Solubility decreases with an increase in temperature and salinity.

Generally, first-contact miscibility pressure is significantly greater than MMP, which is greater than or equal to BPP.

Most CO_2 floods in the United States operate under the premise of MMP (and not the first-contact miscibility pressure, as that would be hard to achieve, without exceeding the original reservoir pressure).

4.8.1 Effect of CO_2 on Carbonate Rock

Being an acidic gas, CO_2 reacts with water to form carbonic acid as per Equation 4.18. The pH of the system generally varies in the range of 5–6. In carbonate reservoirs, the injection of CO_2 has a significant impact in changing the petrophysical properties of rocks. It reacts with carbonate rocks to form soluble bicarbonates according to Equations 4.19 and 4.20 and thus solubilizes the rocks

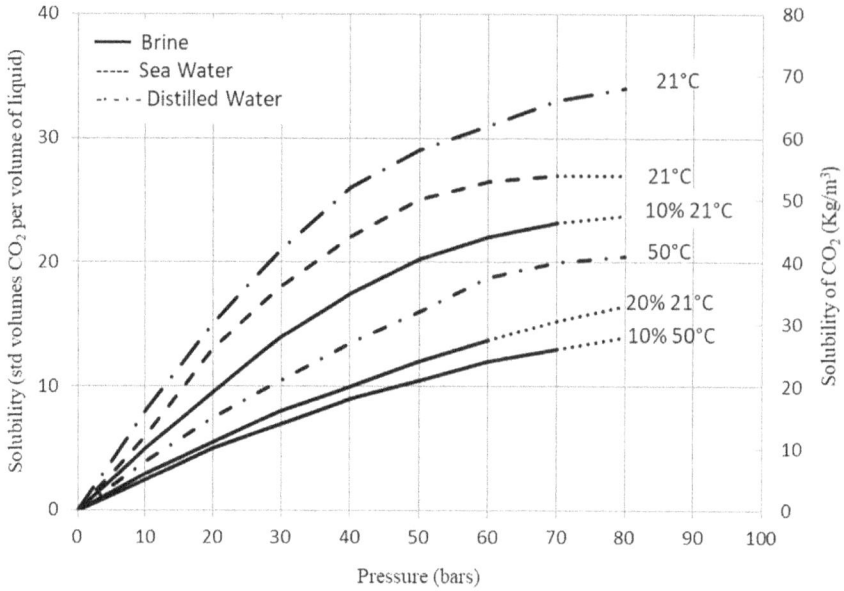

FIGURE 4.26 Solubility of CO_2 in brine, distilled water, and seawater with different pressures and temperatures.

Source: Holm, 1963.

partially. This effect leads to an increase in the permeability of the carbonate reservoir, especially around the wellbore.

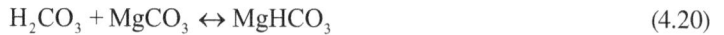

$$CO_2 + H_2O \leftrightarrow H_2CO_3 \tag{4.18}$$

$$H_2CO_3 + CaCO_3 \leftrightarrow CaHCO_3 \tag{4.19}$$

$$H_2CO_3 + MgCO_3 \leftrightarrow MgHCO_3 \tag{4.20}$$

4.9   ALCOHOL FLOODING

Miscibility flooding oil recovery by gas injection is generally dependent on pressure. A high reservoir pressure, at least at the order of 1,500 psi is required for successful miscible flooding (Latil, 1980). These pressure limitations may prohibit the application of miscible flooding to shallow reservoirs. There are also many areas where large quantities of LPG and natural gas are not readily available. These factors emphasize the need for improved techniques, particularly for shallow reservoirs. Although these techniques hold great promise, certain drawbacks do exist. Poor areal sweep efficiency, inherent in any displacement having a highly unfavorable mobility ratio, is one disadvantage.

In the alcohol miscible flooding method, a relatively small volume (slug) of an alcohol (such as isopropyl) is injected into the system. Water is then used to drive the slug through the medium. Thus, in a three component—alcohol, oil, and connate water—system, the alcohol slug is miscible with both oil and chase water. The miscibility is obtained at a certain alcohol concentration, this being dependent on the solubility of the particular system. The volume of the alcohol slug must be designed properly based on the reservoir fluid properties to avoid miscibility rupture. When miscibility is lost, the process reverts to a waterflood, making it an immiscible displacement process. Isopropyl alcohol (IPA) and tertiary butyl alcohol (TBA) are the conventional alcohols used in this process. The main disadvantage of the method is the relatively high cost of alcohols.

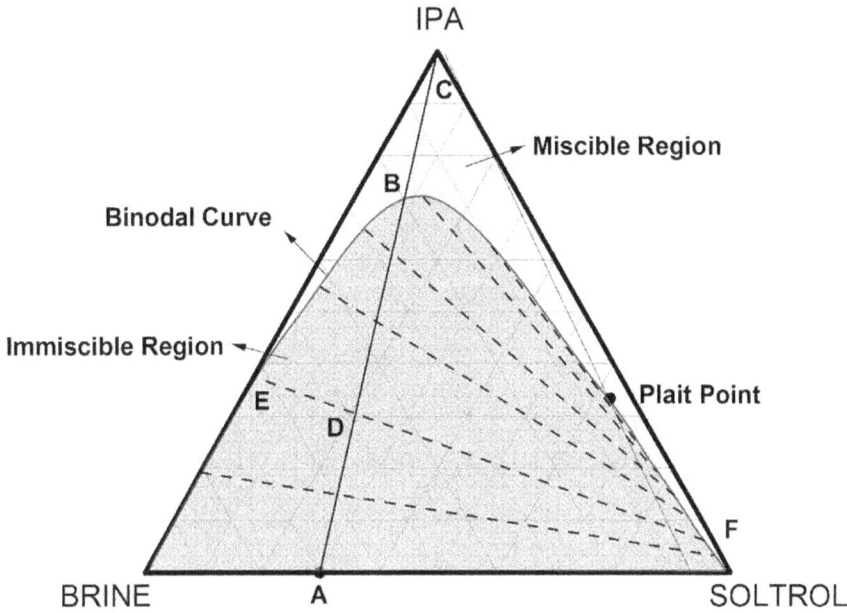

FIGURE 4.27 Equilibrium phase diagram, IPA-Soltrol-2% CaCl₂ brine.

If an adequate volume of alcohol is added to an oil-water system, the alcohol may partition into both the oil and water phases and cause significant changes to oil viscosity, density, solubility, and interfacial tension (Lunn and Kueper, 1996). Figure 4.27 represents the ternary phase diagram of an IPA-oil-water system. If we keep on adding alcohol to the immiscible mixture of water and oil having initial composition at A, the equilibrium composition of the oil will continually change along the path ABC and pass from the immiscible region to the miscible region by crossing the binodal curve at B. The quantity and composition of the mixture at an equilibrium may be determined from the intersection point of the tie-lines (e.g., tie-line EF and intersection point D) with the line ABC and points E and F, respectively. Compositions of the two phases are specified by points E and F. The quantity of the oil phase is proportional to the segment ED, whereas that of the aqueous phase is proportional to DF.

Ternary phase diagrams with tie-lines sloping down toward the oil end point signify systems where the alcohol is preferentially soluble in water. If alcohol concentrations are kept below the binodal curve, the primary recovery mechanism in such systems will be oil dissolution. Oil mobilization may also occur if the interfacial tension is sufficiently reduced. If alcohol concentrations are increased to above the binodal curve, the interfacial tension will be reduced to zero, resulting in complete miscibility and, therefore, complete mobilization of the oil.

It is reported that the IPA slug requirement necessary for complete oil recovery is a linear function of M where M = mobility ratio, as defined by Equation 4.21 (Gatlin and Slobod, 1960).

$$M = \frac{\mu_o S_{o2}}{\mu_s k_{ro}} \qquad (4.21)$$

where μ_o = oil viscosity;
μ_s = viscosity of alcohol slug;
S_{o2} = oil saturation in the zone where the oil and water saturations are the same; and
k_{ro} = relative permeability to oil in the zone where the oil and water saturations are the same.

From the ternary phase diagram, it may be observed that ~70% IPA is required in the water-rich layer to achieve 10% IPA in the conjugate oil-rich layer, assuming equilibrium between the phases. This means that the front portion of the IPA slug is being used largely to displace water. It, therefore, would appear desirable to inject a small slug of a cheaper, more preferentially water-soluble alcohol ahead of the IPA slug to displace water and, at the same time, reduce the IPA requirement. Generally, an alcohol slug of 13% displaceable pore volume is injected to ensure the maximum recovery of oil. Methyl alcohol, which is completely miscible with water, but shows only limited solubility with oil in the presence of water, is generally chosen as the buffer-zone material. To make the process economically viable, the slug is made up of three equal parts, each being 4% of displaceable pore volume, with the central part being isopropyl alcohol and the outer parts being methyl alcohol.

However, this process has received only limited study, presumably, due to the lack of industrial interest, caused by the seemingly prohibitive cost of alcohols.

4.10 WATER ALTERNATING GAS INJECTION

Despite the many successes of the gas injection technique during secondary oil recovery, it often results in poor sweep efficiency. The main reason is the significant difference in viscosity as well as density of the injected gaseous phase (displacing fluid) and the oleic phase (displaced fluid) present in the reservoir (Kumar and Mandal, 2017). Continuous injection of light- and low-viscosity gas into the reservoir cannot push the oil front in a pistonlike manner as a consequence of high mobility ratio and causes viscous fingering. This fluid front (oil-gas) instability is commonly known as gas overriding, in which gas moves faster, bypassing the oil left behind, which further leads to an early breakthrough of the gaseous phase. Overall, a considerable fraction of crude oil remains unswept during the continuous gas injection (CGI) process and requires some strategic changes to overcome such issues.

The alternate water and gas injection technique can be used to reduce the problem and enhance the sweep efficiency of the flood. The technique is called water alternating gas (WAG) injection. Owing to the higher density, the injected water can sweep the crude oil present at the bottom part of the reservoir in a pistonlike manner as a result of the favorable mobility ratio at the oil-water interface. At the same time, the injected gas tends to displace the crude oil present in the upper portion of the reservoir. In addition, the wetting-phase water displaces oil from the smaller pores, whereas gas being nonwetting favors oil movement through the larger pores, irrespective of its depth-wise location. Thereby, the process sweeps oil throughout the pay zone and results in enhancement of the macroscopic sweep efficiency.

The lower IFT of the gas-oil system (as compared to water-oil systems) also helps in improvement of the oil recovery unlike the case of only waterflooding (Alkhazmi et al., 2018; Alzayer and Sohrabi, 2018; Shahverdi et al., 2018). Therefore, the WAG technique has the combined benefits of both water injection and gas injection at the same time, which enhances both microscopic and macroscopic efficiencies—i.e., volumetric sweep efficiency of the flood (Afzali et al., 2018; Christensen et al., 2001; Kumar and Mandal, 2017). A schematic representation of the WAG injection, including the flow pattern of the injected water and gas, is shown Figure 4.28. In addition, this technique also reduces the net amount of gas required and, hence, the cost of the compressor also, compared to that in CGI. The WAG technique has been found effective in most cases with an average incremental oil recovery of 10% (Christensen et al., 2001; Sanchez, 1999).

It is now obvious to ask which should be injected first, water or gas? In another way, why not gas alternating water (GAW) instead of water alternating gas (WAG)? Injection of gas prior to the waterflooding (GAW) results in severe viscous fingering and channeling due to the ultralow viscosity of the gases (supercritical), resulting in a relatively low volumetric sweep efficiency (Kumar and Mandal, 2017). Sometimes, CGI followed by water injection works well, specifically, in less permeable reservoirs and moderately homogenous reservoirs.

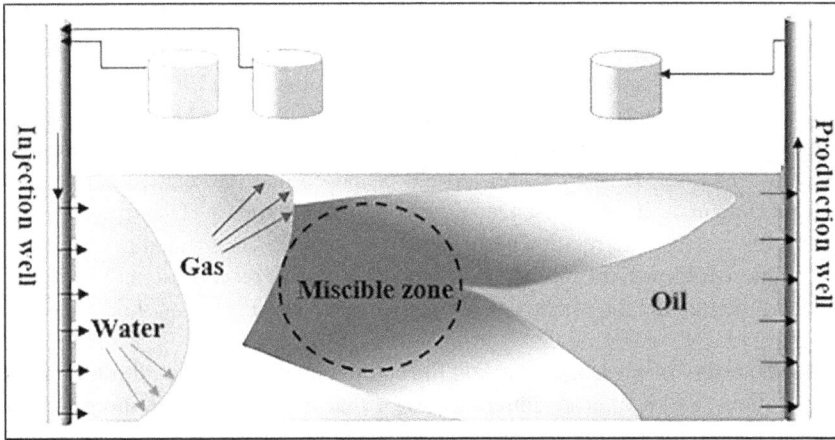

FIGURE 4.28 Schematic of the WAG injection process.

The performance of the WAG injection technique is significantly affected by numerous factors such as fluid and rock properties (water salinity, wettability, capillary pressure, relative permeability, and reservoir heterogeneity) and availability and composition of the gas, as well as the compatibility of fluid phases (Zahoor et al., 2011). However, proper injection strategies may enhance the overall oil recovery efficiency of the WAG injection process. The success of the WAG process depends on various parameters as described in the following:

(i) The injection strategies include optimization of the injection well pattern, WAG cycles, WAG ratio, dipping angle, injection rate and pressure, and the mode of injection (secondary or tertiary) (Christensen et al., 2001; Kumar and Mandal, 2017).

(ii) Moreover, alternate injection of water and gas results in successive imbibition and drainage in each cycle, and therefore, a clear understanding of relative permeability and hysteresis might be helpful for an optimal design, by considering each critical parameter (Afzali et al., 2020; Alkhazmi et al., 2018; Takeuchi et al., 2016).

(iii) It is also important to regulate the volume of injected water and gas throughout the flood process to achieve an optimum volume fraction for both the phases in the injection stream. Water in an excess amount leads to poor microscopic sweep efficiency, while gas in excess results in poor volumetric sweep efficiency.

(iv) The effectiveness of the WAG injection may be improved further, according to the injection approach, which mainly includes: miscible WAG (MWAG), immiscible WAG (IWAG), simultaneous WAG (SWAG), selective simultaneous WAG (SSWAG), and hybrid WAG (HWAG) (Christensen et al., 2001; Darvishnezhad et al., 2010; Kumar and Mandal, 2017).

Although WAG techniques have many advantages such as low mobility ratios, improved volumetric sweep, and increased ultimate oil recovery, there are some common disadvantages that make this technique challenging, such as early gas breakthrough, injectivity loss, corrosion, scale formation, asphaltene precipitation, and hydration formation (Afzali et al., 2020; Christensen et al., 2001; Kumar and Mandal, 2017).

4.10.1 WAG CLASSIFICATION

WAG processes can be classified into four major categories based on the displacement behavior of both the injected phases.

4.10.1.1   Miscible WAG Injection

Miscible WAG (MWAG) injection is defined as the process where the reservoir pressure is maintained higher than MMP by manipulating the injection pressure. At miscible condition, no interface exists between the phases because the IFT becomes zero at this stage. This reduction in IFT toward ultralow values (that finally becomes zero at complete miscible state) affects the relative permeability curvatures (Jamaloei, 2015). Leverett (1939) proposed that the relative permeability curves become straight diagonals in the presence of zero interfacial forces. Therefore, at this instance (completely miscible state), the relative permeability becomes a linear function of saturation only, with a slope of one. Moreover, miscibility is a strong function of reservoir conditions such as pressure, temperature, depth, and the properties of the displaced and displacing fluids (Lyons and Gary Plisga, 2004).

Maintaining the reservoir pressure to sustain the completely miscible condition is very crucial in real field applications, and it often results in behavior that is between single-point miscibility and complete miscibility like multi-contact miscibility (Christensen et al., 2001). Enhanced microscopic sweep, as a consequence of reduced oil viscosity as a result of the gaseous phase dissolving into the oleic phase, is one of the major advantages of MWAG injection techniques.

4.10.1.2   Immiscible WAG Injection

The WAG process where the injected gas cannot create miscibility with the oleic phase in the reservoir, because the reservoir pressure is lower than MMP, is called immiscible WAG (IWAG). Increased displacement efficiency and sweep efficiency as a consequence of improved trapped gas saturation makes the IWAG process a preferred one (Khanifar et al., 2015). Theoretically also, the IWAG process has been proposed as an improved technique to enhance the volumetric sweep efficiency (specifically the sweep conformance of gas) along with displacement efficiency compared to the case of the continuous water and continuous gas injection processes. During water injection, oil displacement occurs by the gas that had been bypassed by water while relative permeability alteration, viscosity reduction, and swelling come into action during the gas injection period. The cyclic nature of the IWAG injection process often leads to gas trapping, which further results in reduced gas mobility and, hence, favors oil mobilization or recovery (Holtz, 2016). However, the extent of gas trapping significantly depends on reservoir wettability. It has been found that the maximum residual gas saturation or minimum oil saturation occurs in the case of a water-wet reservoir as compared to an oil-wet reservoir. One of the major objectives of IWAG is to improve the frontal stability (three-phase zone). Another objective is to create oil film flow, which further performs as a pathway for oil movement once the gas sweeps the oil out, specifically from the larger pores. However, when IWAG processes tend to achieve miscibility criteria (MWAG), the transformation (disappearance) of the interface and the relative permeability estimation (model) are crucial to further estimating the fluid distribution (three-phase) in the reservoir (Christensen et al., 2001). Consideration of hysteresis during the cyclic injection of water and gas during IWAG may correctly forecast the performance of the injection process. Oil swelling, vaporization, viscosity reduction, and three-phase relative permeability—including hysteresis, capillary pressure, wettability effect, and oil film flow—are the major governing mechanisms that control the performance of the IWAG injection technique (Holtz, 2016).

4.10.1.3   Simultaneous WAG Injection

In the simultaneous WAG (SWAG) injection technique, water and gas are mixed at the surface or injected together through a single wellbore into a portion or the entire thickness of the formation. This technique is considered economically viable for mobility control by combining water and gas injection lines, which reduces both capital and operating costs (Shetty et al., 2014). This method has also shown better oil recovery efficiency as well as better gas utilization than CGI and WAG injection. The higher oil recovery efficiency is attributed to the increase in effective viscosities, which helps to contact maximum oil, even in the smaller pores. However, the extent of oil recovery efficiency and gas utilization factor have been found to be a function of the gas fraction.

Another strategy of this technique includes the injection of the two phases separately (without mixing) using a dual completion injector called selective simultaneous water alternating gas (SSWAG) (Kumar and Mandal, 2017). Usually, this method takes advantage of the gravity segregation and comprises injection of gas into the bottom portion and water into the upper portion of the formation.

4.10.1.4 Hybrid WAG Injection

In light of the early favorable single-slug injection and the overall higher recovery by the WAG technique, hybrid WAG was introduced. In this process, a large fraction of the pore volume of the CO_2 is continuously injected to about 20%–40% HCPV, and the remaining fraction is then injected using the WAG technique at a specific WAG ratio (Hadlow, 1992). Improved injectivity, better gas utilization, and reduced chance of water blocking are the main advantages of the HWAG process (Bagrezaie et al., 2014). However, evaluation of the optimum values of the initial slug volume, time, and the WAG ratio in the latter stage are crucial and might help in improving the overall oil recovery efficiency with maximum gas utilization factor. The hybrid WAG process produces about the same amount of oil as the single-slug case in the first ten years of injection. However, much less CO_2 is injected in the hybrid case. After ten years of injection, the hybrid case consistently outperforms the single-slug case. Considering the advantages of the accelerated oil response in the early years, the hybrid process is preferred against the WAG process on the basis of economic conditions (Lin and Poole, 1991).

4.11 DESIGN ASPECTS OF MISCIBLE FLOODING

Miscible flooding is an EOR technology, in which EOR is achieved through improvement in both macroscopic and microscopic sweep efficiencies. However, miscible flooding faces some application issues at the field level due to an adverse impact by gravity override, mobility control, viscous fingering, and volumetric and displacement sweep efficiencies. Generally, fluid front instability (viscous fingering), early breakthrough, and bypassing of a part of the oil are commonly encountered due to the high mobility of the injected gases (low viscosity). To overcome these disadvantages, the WAG method is used in the field, which reduces the consumption of the injected gas (economic benefits) and improves the volumetric sweep efficiency due to the stabilization of the displacing front by creating a more favorable mobility ratio. However, the optimization of the process parameters is essential to maximize the net present values in miscible flood. The parameters that affect the miscible flood design include the reservoir characteristics (rock type, fluid properties, heterogeneity, wettability, etc.) and well or operational control parameters (miscibility conditions, trapped gas, injection pattern, injection rates, injection techniques, bottom-hole pressures, WAG or cycle ratio, cycle time, and the composition of the injection gas). However, the reservoir characteristics are usually either uncontrollable or too costly to modify. Therefore, operational/well control parameters need to be optimized to attain the maximum profit from the miscible flooding process. Nonoptimal well control parameters result in low oil recovery, less profit, high GOR, and early breakthrough.

A brief description of the factors that are considered in the design of miscible flooding are given in the following:

- **Availability and composition of injected gas:** In miscible flood design, it is important to know the identified miscible solvent in terms of the available quantity and composition. These parameters play vital roles in designing an effective flooding scenario and injection strategy. The MMP or MME values are also directly dependent on composition. Also, it is important to make a plan to purchase the required quantity of solvent and consider the option for reinjection as a recycled gas, produced after breakthrough. The common miscible solvents are enriched methane, CO_2, N_2, and, less often, exhaust or flue gases.

- **Injection pattern and well spacing:** The pattern and spacing of the injection wells play very crucial role in controlling the sweep efficiency of a WAG process. Well spacing is a strong indicator of the average reservoir pressure. The injection/production well pairs are often considered to be critical in the WAG process design. It is obvious that the average reservoir pressure will be maintained higher for the greater the ratio of injectors to producers. Usually, a five-spot injection pattern is very popular, as it provides better control on frontal displacement. Owing to the higher drilling cost, the injections wells are generally placed according to the geological factors in case of offshore operations, unlike in the fixed injection pattern.
- **Injection pressure:** The injection pressure (below fracture pressure) is important to maintain a reservoir pressure higher or nearer to the MMP to achieve miscibility, to ensure a smooth and stable displacement, and thereby, to attain a higher sweep efficiency. A higher average reservoir pressure also helps in maintaining the high drawdown, resulting in a higher production rate with greater oil recovery.
- **Injection rate:** The injection rate of a solvent, especially of gas, is a strong function of the fracture pressure, injectivity, formation flow capacity, and the necessity of voidage balance, in addition to the capacity of the pump/compressor. Assigning an optimized injection rate to each injector has a strong effect on the recovery and success of the WAG process. However, the limitations must be considered in terms of equipment capability, stable displacement, and reservoir fracture pressure. The oil production will be increased with the increase of injection rate below the critical point; a further increase in rate beyond this may lead to phase segregation and early breakthrough, especially in the case of dipping reservoirs with high vertical permeability. Therefore, the optimum injection rate safeguards the stable production-injection and economics in view of the system capacity, processing costs, and supply.
- **WAG ratio:** The WAG ratio can be defined as the ratio of the injected water slug to the injected gas slug in a single cycle. The WAG ratio has a significant effect on the optimum recovery in a miscible flood design due to its important role in mobility control, gas/oil production performance, and gas utilization. The uniform WAG ratio can vary from zero to five for different fields based on reservoir rock type, fluid type, viscosity, and wettability. However, the WAG ratio of 1:1 is generally used in miscible flood design for most fields. A miscible flood with high WAG ratio decreases the peak oil production rate, delays the peak time, and decreases gas utilization, as it behaves like a waterflood and causes a hindrance in the interactions of injected gases with residual oil. The production performance behaves as a gas flood with a very small WAG ratio and can lead to an early breakthrough and a rapid decline of reservoir pressure as well as an increase in operating cost. Therefore, an optimum WAG ratio needs to be determined in miscible flooding to control the water cut and gas breakthrough and improve sweep efficiency. However, the industry recognizes the benefit of a tapered WAG ratio over the one due to efficient utilization of the gas, faster oil recovery, faster response, and reduction of the overall production GOR. In tapered WAG, the WAG ratio changes over time. WAG cycle and WAG ratio are best defined by Equations 4.22 and 4.23, respectively. The details of WAG cycle and WAG ratio are shown in Figures 4.29a and 4.29b.

$$WAG\,cycle\,length = Duration_{water} + Duration_{gas} \qquad (4.22)$$

$$WAG\,ratio = \frac{Volume\,of\,water\,injected}{Volume\,of\,gas\,injected} = \frac{\left(Inj.rate \times Duration\right)_{water}}{\left(Inj.rate \times Duration\right)_{gas}} = \frac{R_w \times T_w}{R_g \times T_g} \qquad (4.23)$$

A comparison of the recovery efficiency of the miscible CO_2 WAG process to that of the miscible CO_2 gas flood and waterflood is shown in Figure 4.30. The scenario of WAG of CO_2 flooding has

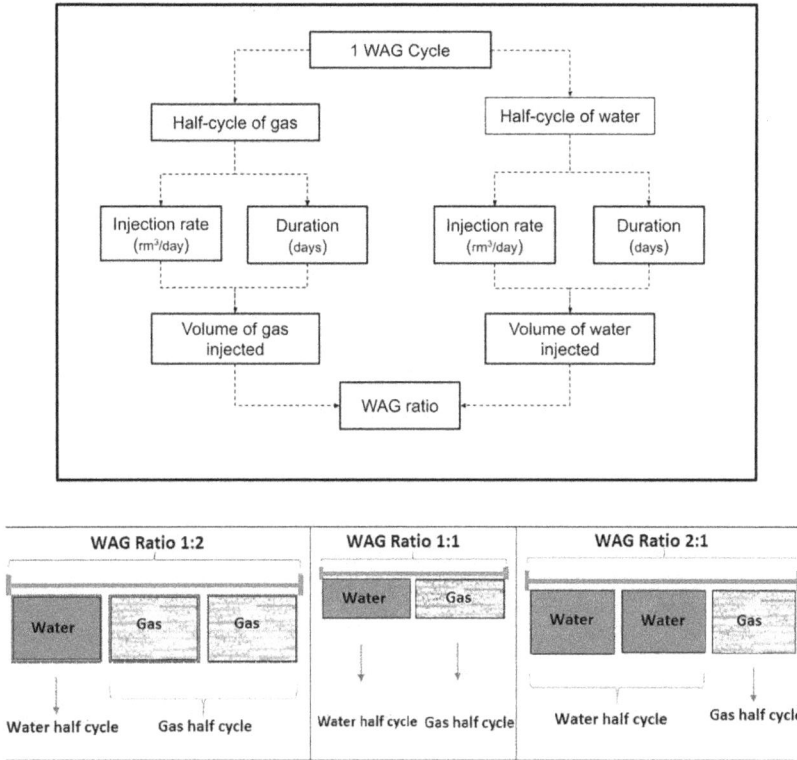

FIGURE 4.29   Details of (a) WAG cycles and (b) WAG ratio.

the highest oil recovery, while simple waterflooding results in the lowest recovery. The WAG of CO_2 flooding can offset the shortcomings of the low oil displacement efficiency of waterflooding, and it can also improve the low sweep efficiency of successive CO_2 flooding. To slow down the speed of the CO_2 breakthrough and maintain the formation pressure, it should be appropriate to increase the water slug. The MMP for CO_2 and crude oil is a key factor for the WAG of CO_2 flooding, and it is better to maintain the formation pressure 1–2 MPa higher than MMP. CO_2 injection volume is proportional to the oil recovery, but there is an optimal value. Considering the same recovery from both the processes, WAG requires a lesser amount of CO_2 compared to successive CO_2 flooding, which reduces the cost of gas flooding.

Example 4.3: In a WAG process, the injection rate of gas is 1,125 m³/d with duration 60 days, and water injection rate is 2,250 m³/d with duration 30 days. Calculate the WAG ratio and WAG cycle length.

Solution:
From Equation 4.23,

$$WAG\,ratio = \frac{Volume\,of\,water\,injected}{Volume\,of\,gas\,injected} = \frac{\left(Inj.rate \times Duration\right)_{water}}{\left(Inj.rate \times Duration\right)_{gas}}$$

$$= \frac{R_w \times T_w}{R_g \times T_g} = \frac{2250\ \text{m}^3/\text{d} \times 30\ \text{d}}{1125\ \text{m}^3/\text{d} \times 60\ \text{d}} = 1:1$$

From Equation 4.22, the $WAG\,cycle\,length = Duration_{water} + Duration_{gas} = 30\,\text{days} + 60\,\text{days} = 90\,\text{days}$

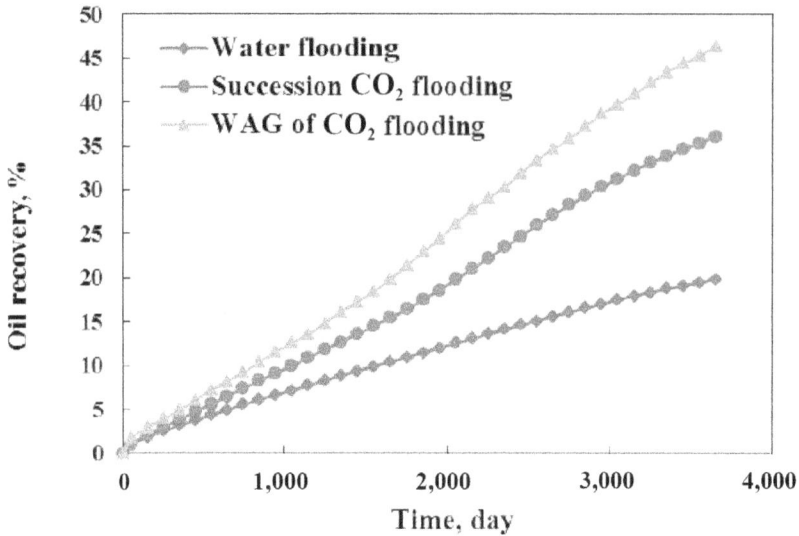

FIGURE 4.30 Comparison of oil recoveries by miscible CO_2, miscible CO_2 WAG, and waterflooding.
Source: Zhao et al., 2015.

- **Gas slug size:** The slug size is defined as the cumulative gas injected in a single cycle. The selection of the optimum gas slug size is crucial, as it can improve the oil recovery on a microscopic level and prevent an early breakthrough in addition to reducing the cost of the injected solvent. The larger slug size escalates the project expenditure due to early breakthrough and viscous fingering, which aggravates in high vertical permeability reservoirs. On the other hand, a smaller slug size makes on-site operations complex. Therefore, an optimum slug size is required to design a miscible flood, which depends on various economic factors such as injected gas cost, incremental oil recovery, crude oil price, etc. A typical range of slug size is considered as 0.1–3 pore volumes.
- **WAG cycle:** The WAG cycle is defined as the length of time of an injection cycle—i.e., the sum of the time for water and gas injection in a single injection cycle. The long cycle time provides higher gravity segregation compared to that of a shorter one. Therefore, the long cycle might reduce the incremental oil recovery due to potential viscous fingering and gravity segregation.
- The design of a miscible flood also considers the start of infill drilling and start of the WAG injection process to overcome shortfalls in the required target production. The miscible flood design should also consider surveillance and operational activities like monthly adjustments of WAG ratios in individual patterns based on the observed performance of offset producers, monthly to quarterly studies of WOR/GOR trends, and required WAG ratios to offsetting injectors to achieve the desired performance.
- **Surface facilities:** In the miscible flooding process, the surface facilities required to be designed are huge. The common operations include gas extraction through separator, gas processing to purify the extract at the required specification level after its extraction from the separator gas, dehydration before compression, and gas compression to raise its pressure for injection.

4.12 FEASIBILITY AND SCREENING CRITERIA

The feasibility analysis of miscible flooding is generally based on detailed geological and simulation studies, composition design, and solvent/chase gas size determination. The geological study consists of the evaluation of the structural and stratigraphic cross sections to understand the effects of stratification for the horizontal miscible flood or the impact of shale barriers for the vertical miscible flood. The porosity and permeability distribution data are also essential for this evaluation. These data are input to the simulation studies.

The simulation studies are conducted by the black oil model and compositional model. Black oil simulation is carried out to reconcile the geological and reservoir data by history-matching the pool performance under primary and secondary drive mechanisms. The compositional simulation is essential for the evaluation and prediction of changes in the pressure and composition of the hydrocarbon phases. The model considers the cross flow between different layers, gravity segregation, channeling, and the variable mobility effects. The degree of dispersion between solvent and oil, viscous fingering, gravity segregation, etc. needs to be modeled in miscible flooding (Asgarpour, 1994).

Based on industry experiences and practices, the screening criteria for hydrocarbon miscible flooding and CO_2 miscible flooding are given in Tables 4.1 and 4.2, respectively (Taber et al., 1997).

TABLE 4.1
Screening of Miscible Hydrocarbon Flooding

Parameters	Recommended Screening Criteria
Crude Oil	
a) Gravity, °API	> 23
b) Viscosity, cp	< 3
c) Composition	High percentage of light hydrocarbons
Reservoir	
d) Oil saturation, % PV	> 30
e) Type of formation	Sandstone or carbonate with minimum fracture and high-permeability streaks
f) Net thickness	Relatively thin, unless formation is dipping
g) Average permeability	Not critical if uniform
h) Depth, ft	> 4,000
i) Temperature, °F	The effect of temperature on MMP should be considered

Limitation: The minimum depth is set by the pressure needed to maintain the generated miscibility
Problem: Viscous fingering results in poor vertical and horizontal sweep efficiency

TABLE 4.2
Screening of Miscible CO_2 Flooding

Parameters	Recommended Screening Criteria
Crude Oil	
a) Gravity, °API	> 22
b) Viscosity, cp	< 10
c) Composition	High percentage of light hydrocarbons (especially C_5–C_{12})
Reservoir	
d) Oil saturation, % PV	> 20
e) Type of formation	Sandstone or carbonate and relatively thin unless dipping
f) Average permeability	Not critical if sufficient injection rates can be maintained
g) Depth and temperature	For miscible displacement, depth must be high enough to allow injection pressure greater than the MMP, which increases with temperature and heaviness of the oil.

Limitation: A good source of low-cost CO_2 is required
Problem: Corrosion can cause problems, especially if there is early breakthrough of CO_2 in the producing wells

4.13 CASE STUDIES

Case Study 1: The San Andres Formation at Wasson Field is Permian (Guadalupian) in age and was deposited near the rim of the Northwest Shelf of the Midland Basin, North America. The reservoir and fluid properties are given in Table 4.3 (Hindi et al., 1992). Initially, production was obtained by waterflooding above bubble-point pressure. Composition gradually changed, and production below bubble-point pressure was evidenced by the declining GOR. To maintain the production after waterflooding, optimization of the project by tertiary production was aimed at using three different methods of CO_2 injection: 1) continuous injection; 2) water alternating gas injection; and 3) hybrid injection. Economic analyses on the compositional simulation results indicated that the optimum mode of CO_2 injection in a hybrid mode consisting of a 10% HCPV initial CO_2 slug with an ensuing 1:1 WAG for a total CO_2 injection of 40% HCPV was the best recovery mechanism. The simulation studies predicted the ultimate tertiary oil recovery of ~14.1% of OOIP with CO_2 injection.

Case Study 2: Arbuckle reservoirs are a significant resource in Kansas for improved oil recovery (IOR). Several oil fields producing from Arbuckle reservoirs were considered in the selection of a candidate well for a pilot test with CO_2 injection near to miscible pressure. Prior to CO_2 injection, these reservoirs produced an estimated 2.2 billion barrels of oil, representing 35% of the 6.1 billion barrels of total Kansas oil production (Franseen et al., 2004). The parameters used in prescreening include the depth of well, reservoir temperature, current reservoir operating pressure, and MMP at reservoir conditions. Among all the wells prescreened, three wells were selected for further evaluation of their potential for near-miscible CO_2 injection application. Table 4.4 summarizes the selected parameters of these three wells. The pilot test was conducted on well #3 (Dreiling field). Approximately 17 tons of CO_2 was injected at pressures below MMP, with a follow-up water displacement, followed by a second tracer test performed to determine the remaining oil saturation. The pilot test led to 13% improvement of oil displacement, which resulted from CO_2 injection at near-miscible conditions (Tsau and Ballard, 2014).

Case Study 3: Miscible gas flooding and the WAG process have been applied successfully in the Prudhoe Bay Oil Field on the north slope of Alaska. The reservoir properties are shown in Table 4.5 (Wu et al., 2004). Three drive mechanisms—waterflooding, miscible gas displacement, and WAG processes—were tested. For miscible and WAG process, CO_2 (85%)/NGL (15%) mixture was injected with an injection rate of 1,200 mole/day. The parameters investigated include injection solvent type, optimal producer bottom-hole pressure, WAG ratio, cycle length, and WAG timing.

TABLE 4.3
Reservoir and Fluid Properties in Target Area
(Hindi et al., 1992)

Reservoir Temperature, °F	114.0
Average absolute permeability, md	4.0
Average porosity, %	10.0
Connate water saturation, %	17.5
Residual oil saturation to water, %	38.0
Average oil saturation after fill up, %	66.0
Current bubble-point pressure, psia	1120
Oil formation volume factor, at 2,000 psia	1.213
Oil viscosity, at 2,000 psia, cp	1.13
Primary recovery, % OOIP	14
Ultimate secondary recovery, % OOIP	19
Ultimate tertiary recovery (estimated), % OOIP	14.1

TABLE 4.4
Candidate Well List

Well #	1	2	3
Field	Ogallah	Joe Lang	Dreiling
Well depth (ft)	4,000	3,902	3,450
Reservoir temperature (°F)	110	125	106
Reservoir pressure (psia)	1,200	1,127	1,145
MMP (psia)	1,350	1,650	1,500

TABLE 4.5
Reservoir Properties of Prudhoe Bay Oil Field (Wu et al., 2004)

Length, ft	1,000	Reservoir pore volume, bbl	136,600
Width, ft	50	Viscosity of oil in place, cp	42
Thickness, ft	55	Oil gravity, °API	14–21
Depth at the top of formation, ft	4,100	Residual water saturation	0.25
Initial reservoir pressure, psia	1,750	Horizontal to vertical permeability ratio	0.1
Initial water saturation	40%	Residual oil saturation	0.2
Initial oil saturation	60%	Residual gas saturation	0.05
Reservoir temperature, °F	86	Water end point relative permeability	0.21
Average porosity	0.279	Oil end point relative permeability	0.71
Average permeability, md	170.9	Gas end point relative permeability	1
Horizontal to vertical permeability ratio	0.1	Total injection, pore volumes	2

The oil recovery factor from the WAG process is about 70% for CO_2 (85%)/NGL (15%) alternating with water injection. Gas flooding resulted in 48% cumulative oil recovery for CO_2 (85%)/NGL (15%) mixture and cumulative recovery of about 20% for lean gas injection. Waterflooding yielded the lowest cumulative oil recovery, only about 20%. So the WAG process was recommended as the optimal recovery mechanism for this specific reservoir, and the ultimate oil recovery factor is expected to be about 70% or higher.

Case Study 4: An investigation on immiscible and miscible CO_2 injection southwest of the Iranian oil field is reported by Fath and Pouranfard (2014). The field has two reservoirs: Gurpi and a shallower Asmari reservoir. The main reservoir in this field is the Asmari formation with Oligocene and Miocene ages, which is divided into seven zones. Only the Asmari formation has been producing oil at commercial levels. It consists of fractured carbonates with a low-permeability matrix. The matrix has a porosity and permeability of about 0.088% and 3.4 mD, respectively. The oil has the API gravity of 20.93°, and the reservoir temperature is 250°F. A reservoir modeling approach was used to evaluate immiscible and miscible CO_2 flooding in the fractured oil field. The MMP for CO_2 injection obtained by slim tube simulation was about 4,630 psia. The simulation study for immiscible CO_2 flooding resulted in an ultimate recovery factor of ~34.45% for the injection rate of 17,000 Mscf/day. For the miscible CO_2 injection scenario with an injection rate of 30,000 Mscf/day at the end of 20 years, the total oil production, average field pressure, and oil recovery factor were estimated as 1.041×10^8 stb, 5,095 psia, and 36.59%, respectively. They concluded that for such a heavy oil reservoir, miscible displacement was very hard; therefore, they recommended for an immiscible CO_2 injection for these reservoirs.

Case Study 5: The Gaoqing 89 production block is located in Shandong Province, Bohai Basin, Eastern China. The depth of the Gao 89–1 reservoir is medium to deep at 2,950 meters on average. The rocks in this reservoir are mainly characterized by gray mudstone and a thin layer of gray matter with shale as well as dolomitic and silty sandstone of different thicknesses. The reservoir compaction is strong with low porosity and extra low permeability. The average porosity for this reservoir is 13.8%; the permeability range is 0.43–7.1×10^{-3} μm^2, and the average is 3.9×10^{-3} μm^2.

According to the result of slim tube test (Qin et al., 2011), MMP is 28.94 MPa for the injected CO_2 multiple-contact miscible with the formation oil from Gao 89–4. The original formation oil pressure of this reservoir is 42 MPa, which facilities miscible flooding with CO_2.

CO_2 miscible flooding in low-permeability and thin-interbedded reservoir Gao 89–1 obtains a significant result. CO_2 can be injected readily compared with water, and the production wells produced much oil. The study indicates that CO_2 miscible flooding is an effective technique in low-permeability and thin-interbedded reservoir. Production for a single well rises from 9.4 tons to 10.2 ton per day; cumulatively, the total increasing oil amounts to 9,870 tons compared with the reservoir development with original formation energy.

REFERENCES

Afzali, S., Ghamartale, A., Rezaei, N., Zendehboudi, S., 2020. Mathematical modeling and simulation of water-alternating-gas (WAG) process by incorporating capillary pressure and hysteresis effects. Fuel 263, 116362. https://doi.org/10.1016/j.fuel.2019.116362

Afzali, S., Rezaei, N., Zendehboudi, S., 2018. A comprehensive review on enhanced oil recovery by Water Alternating Gas (WAG) injection. Fuel 227, 218–246. https://doi.org/10.1016/j.fuel.2018.04.015

Ahmadi, K., Johns, R.T., 2011. Multiple-mixing-cell method for MMP calculations. SPE Journal, 16(4), 733–742.

Alkhazmi, B., Farzaneh, S.A., Sohrabi, M., Buckman, J., 2018, September. A Comprehensive and Comparative Experimental Study of the Effect of Wettability on the Performance of Near Miscible Wag Injection in Sandstone Rock. In Proceedings of the SPE Annual Technical Conference and Exhibition.

Alzayer, H., Sohrabi, M., 2018, March. Water-Alternating-Gas Injection Simulation-Best Practices. In SPE EOR Conference at Oil and Gas West Asia. Society of Petroleum Engineers.

Asgarpour, S., 1994. An overview of miscible flooding. Journal of Canadian Petroleum Technology 33(2).

Bagrezaie, M.A., Pourafshary, P., Gerami, S., 2014. Study of different water alternating carbon dioxide injection methods in various injection patterns in an Iranian non fractured carbonate reservoir. Proceedings of the Annual Offshore Technology Conference 2, 1061–1070.

Benmekki, E.H., Mansoori, G.A., 1988. Minimum miscibility pressure prediction with equations of state. SPE Reservoir Engineering 3(2), 559–564.

Burger, J.E., Bhogeswara, R., Mohanty, K.K., 1994. Effect of phase behavior on bypassing in enriched gas-floods. SPE Reservoir Engineering, 112.

Cao, M., Gu, Y., 2013. Oil recovery mechanisms and asphaltene precipitation phenomenon in immiscible and miscible CO_2 flooding processes. Fuel 109, 157–166.

Chen, H., Zhang, X., Yuan, C., Tang, H., Shen, X., 2017. Study on pressure interval of near-miscible flooding by production gas re-injection in QHD offshore oilfield. Journal of Petroleum Science and Engineering 157, 340–348.

Choubineh, A., Helalizadeh, A., Wood, D.A., 2019. The impacts of gas impurities on the minimum miscibility pressure of injected CO_2-rich gas–Crude oil systems and enhanced oil recovery potential. Petroleum Science 16(1), 117–126.

Choudhary, N., Nair, A.K.N., Ruslan, M.F.A.C., Sun, S., 2019. Bulk and interfacial properties of decane in the presence of carbon dioxide, methane, and their mixture. Scientific Reports 9(1), 1–10.

Christensen, J.R., Stenby, E.H., Skauge, A., 2001. Review of WAG field experience. SPE Reservoir Evaluation & Engineering 4, 97–106. https://doi.org/10.2118/71203-pa

Clark, N.J., Shearin, H.M., Schultz, W.P., Garms, K., Moore, J.L., 1958. Miscible drive-its theory and application. Journal of Petroleum Technology 10(6), 11–20.

Darvishnezhad, M.J., Moradi, B., Zargar, G., Jannatrostami, A., Montazeri, G.H., 2010. Study of Various Water Alternating Gas Injection Methods in 4- and 5-Spot Injection Patterns in an Iranian Fractured Reservoir. In Trinidad and Tobago Energy Resources Conference, SPE TT 2, 588–595. https://doi.org/10.2523/132847-ms

Fath, A.H., Pouranfard, A-R., 2014. Evaluation of miscible and immiscible CO_2 injection in one of the Iranian oil fields. Egyptian Journal of Petroleum 23(3), 255–270.

Firoozabadi, S., Aziz, K., 1986. Analysis and correlation of nitrogen and lean-gas miscibility pressure. SPE Reservoir Engineering, 575–582.

Franseen, E.K., Byrnes, A.P., Cansler, J.R., Steinhauff, D.M., Carr, T.R., 2004. The geology of Kansas— Arbuckle Group. *Midcontinent Geoscience*, 1–43.

Gatlin, C., Slobod, R.L., 1960. The alcohol slug process for increasing oil recovery. Transactions of the AIME 219(1), 46–53.

Glaso, O., 1985. Generalized minimum miscibility pressure correlation. Society of Petroleum Engineers Journal 25, 927–934.

Green, D.W., Willhite, G.P., 1998. Enhanced Oil Recovery. SPE Textbook Series. Society of Petroleum Engineers.

Hadlow, R.E., 1992. Update of industry experience with CO_2 injection. Society of Petroleum Engineers 2492.

Hatzignatiou, D., Lu, Y., 1994. Feasibility study of CO_2 immiscible displacement process in heavy oil reservoirs. In PETSOC Annual Technical Meeting. https://doi.org/10.2118/94-90

Hawthorne, S.B., Miller, D.J., Grabanski, C.B., Sorensen, J.A., Pekot, L.J., Kurz, B.A., Gorecki, C.D., Steadman, E.N., Harju, J.A., Melzer, S., 2017, February. Measured Crude Oil MMPs with Pure and Mixed CO_2, Methane, and Ethane, and Their Relevance to Enhanced Oil Recovery From Middle Bakken and Bakken Shales. In SPE Unconventional Resources Conference. Society of Petroleum Engineers.

Hemmati-Sarapardeh, A., Ayatollahi, S., Ghazanfari, M.H, Masihi, M., 2014. Experimental determination of interfacial tension and miscibility of the CO_2–Crude oil system; temperature, pressure, and composition effects. Journal of Chemical & Engineering Data 59(1), 61–69.

Hindi, R., Cheng, C.T., Wang, B., 1992, January. CO_2 Miscible Flood Simulation Study, Roberts Unit, Wasson Field, Yoakum County, Texas. In SPE/DOE Enhanced Oil Recovery Symposium. Society of Petroleum Engineers.

Holm, L.W., 1963. CO_2 Slug and carbonated water oil recovery processes. Producers Monthly 27(9), 6–8.

Holm, L.W., Josendal, V.A., 1974. Mechanisms of oil displacement by carbon dioxide. Journal of Petroleum Technology, 1427–1438.

Holtz, M.H., 2016, 11–13 January. Immiscible Water Alternating Gas (IWAG) EOR: Current State of the Art. In SPE—DOE Improved Oil Recovery Symposium. https://doi.org/10.2118/179604-m

Huang, S.S., Dyer, S.B., 1993. Miscible displacement in the Weyburn reservoir: A laboratory study. Journal of Canadian Petroleum Technology 32(7).

IEA. (2019). World Energy Outlook 2019, IEA, Paris. https://www.iea.org/reports/world-energy-outlook-2019, License: CC BY 4.0.

Islam, A.W., Carlson, E.S., 2012. Viscosity models and effects of dissolved CO_2. Energy & Fuels 26(8), 5330–5336.

Jamaloei, B.Y., 2015. The effect of interfacial tension on two-phase relative permeability: A review. Energy Sources, Part A: Recovery, Utilization, and Environmental Effects 37, 245–253. https://doi.org/10.1080 /15567036.2011.557708

Jarrell, P.M., Fox, C.E., Stein, M.H., Webb, S.L., 2002. Practical Aspects of CO_2 Flooding (Vol. 22). Society of Petroleum Engineers.

Johnson, J.P., Pollin, J.S., 1981. Measurement and Correlation of CO_2 Miscibility Pressures. In SPE 9790, Presented at SPE/DOE Enhanced Oil Recovery Symposium, Tulsa, OK.

Kantzas, A., Bryan, J., Taheri, S., 2012. Fundamentals of Fluid Flow in Porous Media (Pore size distribution). PERM Inc.

Khan, M.Y., Kohata, A., Patel, H., Syed, F.I., Al Sowaidi, A.K., 2016. Water Alternating Gas WAG Optimization using Tapered WAG Technique for a Giant Offshore Middle East Oil Field. p. SPE-183181-MS.

Khanifar, A., Raub, M.R.A., Tewari, R.D., Zain, Z.M., Sedaralit, M.F., 2015. Designing of Successful Immiscible Water Alternating Gas (IWAG) Coreflood Experiment. https://doi.org/10.2523/iptc-18555-ms

Khazam, M., Arebi, T., Mahmoudi, T., Froja, M., 2016. A new simple CO_2 minimum miscibility pressure correlation. Oil & Gas Research 2.

Kieschnick Jr, W.F., 1960. What Is Miscible Displacement. The University of Texas at Austin, pp. 63–69.

Koonce, K.T., Blackwell, R.J., 1965. Idealized behavior of solvent banks in stratified reservoirs. Society of Petroleum Engineers Journal 5(4), 318–328.

Kumar, S., Mandal, A., 2017. A comprehensive review on chemically enhanced water alternating gas/CO_2 (CEWAG) injection for enhanced oil recovery. Journal of Petroleum Science and Engineering 157, 696– 715. https://doi.org/10.1016/j.petrol.2017.07.066

Latil, M., 1980. Enhanced Oil Recovery. Editions Technip.

Leverett, M.C., 1939. Flow of oil–Water mixtures through unconsolidated sands. Transactions of the AIME 132, 149–171

Lin, E.C., Poole, E.S., 1991. Numerical evaluation of single-slug, WAG, and hybrid CO_2 injection processes, Dollarhide Devonian Unit, Andrews County, Texas. SPE Reservoir Engineering 6(4), 415–420.

Lunn, S., Kueper, B., 1996. Removal of DNAPL Pools Using Upward Gradient Ethanol Floods, in: Non-Aqueous Phase Liquids (NAPLs) in Subsurface Environment: Assessment and Remediation. ASCE, pp. 345–356.

Lyons, W., Gary Plisga, B., 2004. Standard Handbook of Petroleum and Natural Gas Engineering, 2nd ed. Gulf Professional Publishing.

Mosavat, N., Abedini, A., Torabi, F., 2014. Phase behaviour of CO_2–Brine and CO_2–Oil systems for CO_2 storage and enhanced oil recovery: Experimental studies. Energy Procedia 63, 5631–5645.

Mosavat, N., Torabi, F., 2014. Application of CO_2-saturated water flooding as a prospective safe CO_2 storage strategy. Energy Procedia 63, 5408–5419.

Mullken, C., Sandler, S., 1980. The prediction of CO_2 solubility and swelling factors for enhanced oil recovery development, American Chemical Society. Industrial & Engineering Chemistry Process Design and Development 19(4),709–711.

Obi, E.O.I., Blunt, M.J., 2006. Streamline-based simulation of carbon dioxide storage in a north sea aquifer. Water Resources Research 42(3).

Orr Jr, F.M., Silva, M.K., Lien, C.L., Pelletier, M.T., 1982. Laboratory experiments to evaluate field prospects for CO_2 flooding. Journal of Petroleum Technology 34(4), 888–898.

Pande, K.K., 1992, 4–7 October. Effects of Gravity and Viscous Crossflow on Hydrocarbon Miscible Flood Performance in Heterogeneous Reservoirs. In SPE 24935, Annual Technical Conference and Exhibition, Washington, DC.

Perkins, T.K., Johnston, O.C., 1963. A review of diffusion and dispersion in porous media. Society of Petroleum Engineers Journal 3(1), 70–84.

Qin, L., Hailong, L., Yin, X.Q., 2011, January. Application of CO_2 Miscible Flooding on Gao 89–1 Low Permeability Reservoir. In SPE Asia Pacific Oil and Gas Conference and Exhibition. Society of Petroleum Engineers.

Rzasa, M.J., Glass, E.D., Opfell, J.B., 1952, January. Prediction of Critical Properties and Equilibrium Vaporization Constants for Complex Hydrocarbon Systems. In Chemical Engineering Progress Symposium Series (Vol. 48, No. 2). American Institute of Chemical Engineers, pp. 28–37.

Sanchez, N.L., 1999. Management of Water Alternating Gas (WAG) Injection Projects. In Latin American and Caribbean Petroleum Engineering Conference. Society of Petroleum Engineers. https://doi.org/10.2523/53714-ms

Sebastian, H.M., Wenger, R.S., Renner, T.A., 1985. Correlation of minimum miscibility pressure for impure CO_2 streams. Journal of Petroleum Technology, 2076–2082.

Shahverdi, H., Fatemi, S., Sohrabi, M., 2018. Gas/oil IFT, three-phase relative permeability and performance of Water-Alternating-Gas (WAG) injections at laboratory scale. Journal of Oil Gas and Petrochemical Sciences 1, 10–16. https://doi.org/10.30881/jogps.00005

Shetty, S., Hughes, R.G., Afonja, G., 2014. Experimental Evaluation of Simultaneous Water and Gas Injection Using Carbon Dioxide. In SPE EOR Conference at Oil and Gas West Asia.

Shyeh-Yung, J.J., 1991, 6–9 October. Mechanisms of Miscible Oil Recovery: Effects of Pressure on Miscible and Near-Miscible Displacements of Oil by Carbon Dioxide. In SPE 22651, Annual Technical Conference and Exhibition, Dallas, TX.

Simon, R., Graue, D.J., 1965. Generalized correlations for predicting solubility, swelling and viscosity behavior of CO_2-crude oil systems. Journal of Petroleum Technology, 102–106.

Sohrabi, M., Danesh, A., Tehrani, D.H., Jamiolahmady, M., 2008. Microscopic mechanisms of oil recovery by near-miscible gas injection. Transport in Porous Media 72(3), 351–367.

Srivastava, R.K., Huang, S.S., 1997. Laboratory Investigation of Weyburn CO_2 Miscible Flooding. In Technical Meeting/Petroleum Conference of the South Saskatchewan Section. Petroleum Society of Canada.

Taber, J.J., Martin, F.D., Seright, R.S., 1997. EOR screening criteria revisited—Part 2: Applications and impact of oil prices. SPE Reservoir Engineering 12(3), 199–206.

Takeuchi, J., Sumii, W., Fujihara, M., 2016. Modeling of fluid intrusion into porous media with mixed wettabilities using pore-network. International Journal of GEOMATE 10, 1971–1977. https://doi.org/10.21660/2016.22.5190

Thomas, F.B., Holowach, N., Zhou, X., Bennion, D.B., Benion, D.W., 1994, 17–20 April. Miscible or Near-Miscible Gas Injection, Which Is Better? In SPE 27811, SPE/DOE Symposium on Improved Oil Recovery, Tulsa, OK.

Thompson, J.L., 1967. A laboratory study of an improved water-driven LPG slug process. Society of Petroleum Engineers Journal 7(3), 319–324.

Tsau, J.S., Ballard, M., 2014. Single Well Pilot Test of Near Miscible CO_2 Injection in a Kansas Arbuckle Reservoir. In SPE Improved Oil Recovery Symposium. Society of Petroleum Engineers.

Verma, M.K., 2015. Fundamentals of Carbon Dioxide-Enhanced Oil Recovery (CO_2-EOR): A Supporting Document of the Assessment Methodology for Hydrocarbon Recovery Using CO_2-EOR Associated with Carbon Sequestration. US Department of the Interior, US Geological Survey, p. 19.

Whitson, C.H., Michelsen, M.L., 1989. The negative flash. Fluid Phase Equilibria 53, 51–71.

Whorton, L.P., Brownscombe, E.R., Dyes, A.B., 1952. Method for producing oil by means of carbon dioxide. US Patent 2,623,596.

Wu, X., Ogbe, D.O., Zhu, T., Khataniar, S., 2004, January. Critical Design Factors and Evaluation of Recovery Performance of Miscible Displacement and WAG Process. In Canadian International Petroleum Conference. Petroleum Society of Canada.

Yang, C., Gu, Y., 2006. Diffusion coefficients and oil swelling factors of carbon dioxide, methane, ethane, propane, and their mixtures in heavy oil. Fluid Phase Equilibria 243, 64–73

Yuan, H., Johns, R.T., Egwuenu, A.M. Dindoruk, B., 2004, April. Improved MMP Correlations for CO2 Floods Using Analytical Gas Flooding Theory. In SPE Improved Oil Recovery Conference? SPE-89359. SPE.

Yuan, H., Johns, R.T., Egwuenu, A.M., Dindoruk, B., 2005. Improved MMP correlation for CO_2 floods using analytical theory. SPE Reservoir Evaluation & Engineering 8, 418–425.

Zahoor, M.K., Derahman, M.N., Yunan, M.H., 2011. Wag process design – An updated review. Brazilian Journal of Petroleum and Gas 5, 109–121. https://doi.org/10.5419/bjpg2011-0012

Zhang, N., Yin, M., Wei, M., Bai, B., 2019. Identification of CO_2 sequestration opportunities: CO_2 miscible flooding guidelines. Fuel 241, 459–467.

Zhao, X., Liao, X., Wang, H., 2015. The research of gas flooding of horizontal well with SRV in tight oil reservoir. Journal of Petroleum & Environmental Biotechnology 6(1), 1.

Zick, A.A., 1986. A Combined Condensing/Vaporizing Mechanism in the Displacement of Oil by Enriched Gases. In SPE Annual Technical Conference and Exhibition. Society of Petroleum Engineers.

Zolghadr, A., Escrochi, M., Ayatollahi, S., 2013. Temperature and composition effect on CO_2 miscibility by interfacial tension measurement. Journal of Chemical & Engineering Data 58(5), 1168–1175.

Zuo, Y-X., Chu, J-Z., Ke, S-L., Guo, T-M., 1993. A study on the minimum miscibility pressure for miscible flooding systems. Journal of Petroleum Science and Engineering 8(4), 315–328.

5 Thermal Recovery Processes

5.1 INTRODUCTION

Considerable interest has been observed recently in the implementation of the thermal EOR technique, especially in heavy oil reservoirs to fulfill the growing energy demand. According to Mokheimer et al. (2019), the thermal EOR process accounts for around 67% of the total EOR projects worldwide. Thermal recovery techniques are preferably used for highly viscous, heavy oil with a °API ≤ 20. A thermal recovery process involves the injection of heat into the reservoir formation or generation of heat within the reservoir by combustion of oil in situ to increase the temperature of the formation rock and fluids. The rise in temperature helps to reduce the oil viscosity, thereby increasing the mobility of the oil significantly. Besides oil viscosity, the increased temperature also causes various physical and chemical changes in the in situ crude oil, including swelling of the oil, reduction of interfacial tension (IFT), and specific gravity of oil. The chemical changes that a crude oil undergoes include C-C bond scission, dehydrogenation, and polymerization through the combustion and cracking process. This chapter describes the various thermal recovery techniques applied to oil reservoirs, with their detailed mechanisms of oil recovery by the application of heat.

5.2 THERMAL RECOVERY TECHNIQUES AND THE CANDIDATE RESERVOIR

According to the EIA's International Energy Outlook 2017, the total reserves of crude oil are estimated at 9–11 trillion barrels in the world, with more than two-third of the volume accounting for heavy oil and bitumen. The global distribution of heavy oil and bitumen is shown in Figure 5.1, which shows that Canada and Venezuela have the most heavy oil reserves. Compared to conventional oil resources, these specific types of hydrocarbons are highly viscous under reservoir conditions of temperature and pressure. This necessitates the use of thermal recovery methods to reduce the viscosity of the oil and improve mobility control significantly. The temperature of such reservoirs is increased by injecting a hot fluid or by generating thermal energy in situ by combustion of a part of oil. Depending on the viscosity range of crude oil and reservoir properties, different types of thermal EOR methods are applied—i.e., hot water flooding, steam flooding, cyclic steam simulation (CSS), steam-assisted gravity drainage (SAGD), in situ combustion, toe-to-heel air injection (THAI), etc.

Hot water flooding is relatively simple and less expensive. The process is preferred in shallow reservoirs containing oil with the viscosity range between 100-1000cP. Steam flooding is effective for producing heavy oils; CSS is applicable for extra heavy oils, and SAGD is mainly employed for bitumen recovery, in addition to other heavy oils.

Another method of heavy oil extraction is in situ combustion (ISC), in which heat is generated by controlled burning of a fraction of the heavy hydrocarbons in place. The process induces several phenomena like cracking of heavier hydrocarbons and vaporization of lighter fractions and reservoir water, in addition to the deposition of coke. As the fire front moves ahead, it pushes the mixture of hot combustion gases, steam, and hot water forward. This results in a reduction in the oil viscosity and displacement of the oil bank toward the production wells. Alvarado and Manrique (2010) reported that most thermal EOR projects are so far implemented in sandstone reservoirs. The main technical problems associated with thermal recovery processes include the poor sweep efficiencies, heat energy losses to nearby formations, and poor steam/air injectivity issues. In addition, environmental concerns related to stack gas formations are also considered prior to its application. The mechanism of the movement of the generated heat front lies in an in-depth understanding of various heat, mass, and momentum transfer mechanisms, including the different chemical reactions, along with the variations

DOI: 10.1201/9781003098850-5

Heavy Oil and Bitumen
5.6 trillion barrels

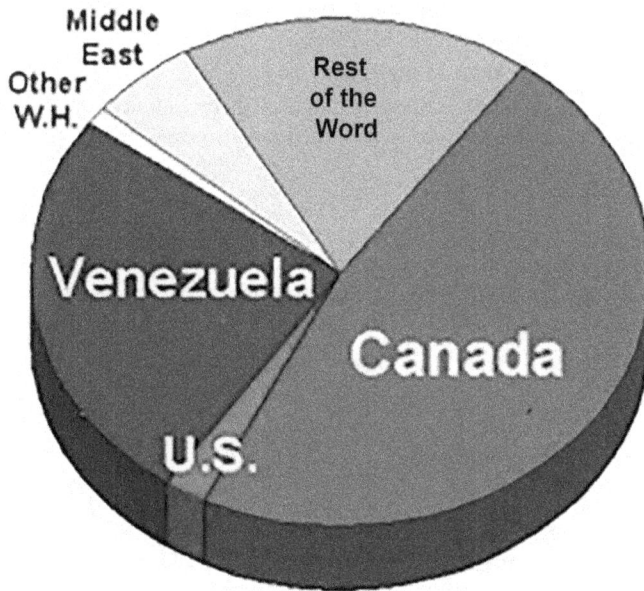

FIGURE 5.1 Reserves of heavy oil and bitumen.

W.H., Western Hemisphere.

Source: IEA, 2017.

in IFT and phase behaviors. The types of chemical reactions and intensity of the heat and mass transfer largely depend on the properties of the crude oil and rock type besides other process conditions. It is, therefore, necessary to conduct a detailed review of the success of earlier thermal recovery techniques, the gradual evolution of the process, and practical implications in the petroleum industry today.

5.3   GENERAL PRINCIPLE AND THERMODYNAMICS

In thermal processes, the heat may be applied from the surface by specially designed hot water heaters or steam generators. In in situ combustion, a portion of the oil is burned by the injection of compressed air to generate heat within the reservoir. It is more efficient than other thermal methods, as chance of heat loss is minimal. The knowledge of the physical and thermal properties of steam, reservoir fluids, and reservoir rock are important for evaluating the efficacy of the thermal recovery processes by exact determination of heat losses to reservoir rock and computation of the heating capacity of the equipment.

5.3.1   THERMAL PROPERTIES OF STEAM

5.3.1.1 Quality of Steam

Steam is produced by increasing the temperature of water above its saturation temperature, T_s, at a constant pressure, P_s. For the constant pressure of 1 atm, the saturation temperature of water is

100°C. The amount of heat required to increase the temperature from its initial temperature, T_i, to saturation temperature, T_s, is known as "sensible heat" and is determined for unit mass as follows:

$$h_w = C_w (T_s - T_i) \tag{5.1}$$

where C_w represents the specific heat of water, and h_w is the heat content of the saturated water. When the saturated water is provided more heat, the heat is utilized in the change of phase of water from liquid to steam without change in the temperature. This heat required to convert the liquid water to steam is known as enthalpy of vaporization, L_v, or the latent heat of steam. The total heat content, h_s, of a unit mass of steam can be determined as follows:

$$h_s = h_w + L_v \tag{5.2}$$

Heating of steam beyond its saturation temperature, T_s, at constant pressure, P_s, leads to the generation of superheated steam. The heat content of the unit mass of superheated steam, h_{sup}, at a temperature T_{sup} is given as follows:

$$h_{sup} = h_s + C_s \left(T_{sup} - T_s \right) \tag{5.3}$$

where C_s signifies the specific heat of steam.

In thermodynamics, steam quality is defined as the mass fraction of steam in a saturated mixture. The saturated steam has a "quality" of 100%, and saturated water has a "quality" of 0%.

The steam quality, χ, can be calculated by dividing the mass of the steam by the total mass of the mixture:

$$\chi = \frac{m_{steam}}{m_{total\ mixture}} \tag{5.4}$$

where m indicates mass.

Alternatively, the quality of steam can be defined as follows:

$$\chi = \frac{h - h_f}{h_{fg}} \tag{5.5}$$

where h is the mixture specific enthalpy, defined as follows:

$$h = \frac{m_f h_f - m_g h_g}{m_f - m_g} \tag{5.6}$$

The subscripts f and g refer to saturated liquid and saturated gas, respectively, and f_g refers to vaporization. The quality of steam may be best understood from Figure 5.2. The term "dry steam" is used for steam that remains at its saturation temperature with no or negligible water particles in suspension. Therefore, in heating applications, dry steam is preferable because of its better energy exchange capacity but negligible corrosivity. Wet steam contains more than 5% liquid water in addition to steam and exists at the saturation temperature. Superheated steam is generated by further heating of the wet or saturated steam beyond the saturated point. Superheated steam has a lower density than saturated steam at the same pressure.

Example 5.1: In a steam injection project, it is decided to inject superheated steam at 3,000 psia and 450°C. How much more enthalpy does this superheated steam have compared to saturated steam at 100°C and 1 atm pressure?

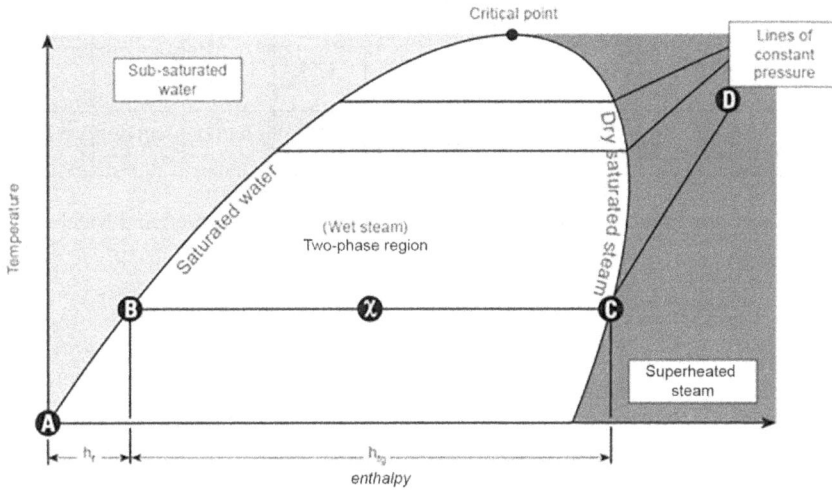

FIGURE 5.2 Temperature enthalpy phase diagram.

Source: *Courtesy*: www.spiraxsarco.com.

Solution:

Specific enthalpy of saturated steam at 100°C (14.7 psia) = 2,676 kJ/kg (from saturated steam tables).

Specific enthalpy of superheated steam at 450°C and 3,000 psia = 3047 kJ/kg (from superheated steam tables).

Extra enthalpy that the superheated steam contains in comparison to the saturated one is:

= (3047 − 2676) kJ/kg = 371 kJ/kg

5.3.2 THERMAL PROPERTIES OF RESERVOIR FLUIDS

The increase in the reservoir temperature has a significant effect on the viscosity of reservoir fluids. The viscosity of crude oil and water decrease with an increase in the temperature, whereas the viscosity of gases increases. However, the reduction of the viscosity of crude oil, particularly, heavy crude oil is much more effective. For example, crude oil with a viscosity of about 50,000 cSt at 40°C has a viscosity of less than 20,000 cSt at 50°C. The oil viscosity at various temperatures during thermal recovery can be determined experimentally in a laboratory or by using different established correlations. For applying these correlations to predict viscosity, it must be ensured that the increase in the temperature does not lead to the cracking or distillation of crude oil. The Andrade (1930) viscosity correlation given in Equation 5.7 is frequently used to calculate the change in viscosity at different reservoir temperatures.

$$\mu_i = Ae^{B/T} \tag{5.7}$$

where μ_i is the viscosity of the oil component in cP at a temperature T (K), A and B are the constants obtained from the regression analysis of the data.

Example 5.2: Draw the plot of viscosity of pentane versus temperature between 20°C and 90°C. The values of the constants A and B for pentane are 0.0191 cP and 722.2/K, respectively.

Solution: The Andrade equation may be used to solve the problem.

Temperature (°C)	20	30	40	50	60	70	80	90	100
$e^{B/T}$	11.76	10.84	10.05	9.36	8.75	8.21	7.74	7.31	6.93
$\mu_{Pentane} = Ae^{B/T}$ (cp)	0.225	0.207	0.192	0.179	0.167	0.157	0.149	0.140	0.132

Another viscosity correlation with °API of heavy oil is given as (Ahmed and Meehan, 2012):

$$\mu_o = 220.15 \times 10^9 \left(\frac{5T}{9}\right)^{-3.556} \left[\log(API)\right]^z \tag{5.8}$$

where

$$z = \left[12.5428 \times \log(5T/9)\right] - 45.7874 \tag{5.9}$$

and T is the temperature in °R. The viscosity of water at the temperature, T, can be determined by:

$$\mu_w = \left(\frac{2.185}{0.04012(T-460) + 0.0000051535(T-460)^2 - 1}\right) \tag{5.10}$$

In addition to the heat content of steam, the heat contents of water and crude oil are equally important to evaluate the thermal recovery processes. The specific heat of water, C_w, in BTU/lb.-°F, is determined by (Ahmed and Meehan, 2012):

$$C_w = 1.3287 - 0.000605T + 1.79 \times 10^{-6}(T-460)^2 \tag{5.11}$$

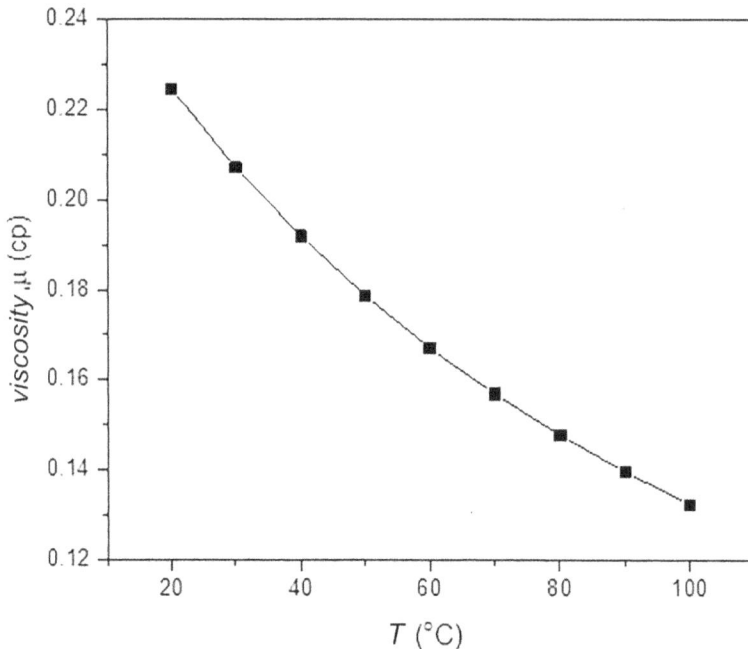

FIGURE 5.3 Variation of viscosity of pentane with temperature.

Similarly, the specific heat of crude oil, C_o, in BTU/lb.-°F, is calculated as follows:

$$C_o = \left(\frac{0.022913 + 56.9666 \times 10^{-6} (T - 460)}{\sqrt{\rho_o}} \right) \qquad (5.12)$$

where T is the temperature in °R, and ρ is the density of crude oil in lb./ft.^{3}.

The general relationships between the viscosity of heavy crude oil with temperature and °API are shown in Figure 5.4a and 5.4b.

5.3.3 THERMAL PROPERTIES OF FORMATION

The heat capacity of formation containing water and oil in the pores of the rock is the sum of their individual heat capacities. Thus, the total heat capacity of the formation, including rock and fluid, M, in BTU/ft.^{3}-°F, can be given as follows:

$$M = \phi \left(S_o \rho_o C_o + S_w \rho_w C_w \right) + \left(1 - \phi \right) \rho_r C_r \qquad (5.13)$$

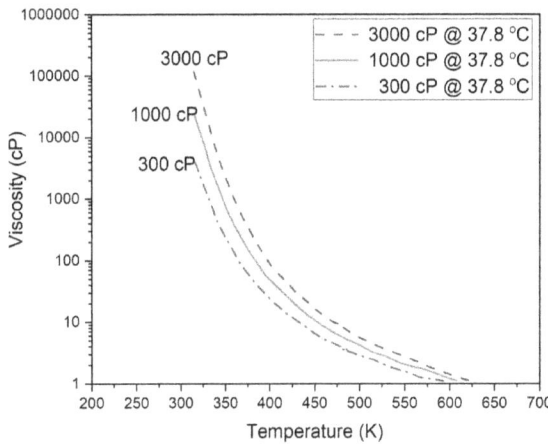

FIGURE 5.4 (a) Variation of viscosity with increase in temperature for three crude oil samples with different viscosities at 37.8°C.

Source: Barillas et al., 2008.

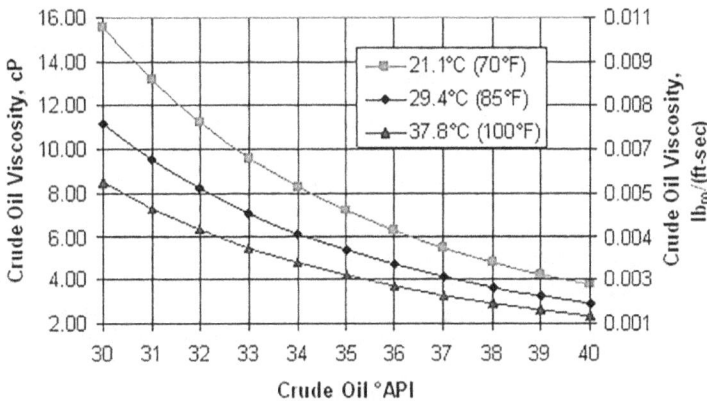

FIGURE 5.4 (b) Typical variation of crude oil viscosity with API and temperature.

where ϕ is the porosity of the rock, S signifies the saturation of respective phase, ρ is the density at the reservoir temperature in lb./ft.^{3}, and C is the heat capacities in BTU/lb.-°F. The subscripts o, w, and r, respectively, represent oil, water, and rock. The heat capacity of the formation rock is around 75% of the total heat capacity of the formation; thus, the maximum injected/generated heat is consumed by the rock for increase in an equal degree of temperature.

Example 5.3: The rock-fluid properties of a reservoir are given in the following. Calculate the percentage of total heat capacity within the formation rock.
Porosity of the reservoir = 25%
Oil saturation, So = 70%
Volumetric heat capacity of the reservoir rock = 33.2 BTU/ft.^{2}-°F
Density of crude oil = 54.3 lb./ft.^{3}
Specific heat capacity of oil = 0.5 BTU/lb.-°F

Solution:
The total heat capacity can be calculated from Equation 5.13.
The density and heat capacity of water are 62 lb./ft.^{3} and 1.0 BTU/lb.-°F, respectively.
Therefore,

$$M = 0.25\left(0.70\times54.3\frac{\text{lb}}{ft^3}\times0.5\frac{BTU}{\text{lb}°F}+0.30\times62\frac{\text{lb}}{ft^3}\times1\frac{BTU}{\text{lb}°F}\right)+\left(1-0.25\right)\times33.2\frac{BTU}{ft^3°F}$$

$$= 0.25(19.00 + 18.6)\frac{BTU}{ft^3°F}+24.9\frac{BTU}{ft^3°F}$$

$$= (9.4+24.9)\frac{BTU}{ft^3°F}$$

$$= 34.3\frac{BTU}{ft^3°F}$$

Therefore, the percentage of total heat capacity within the formation rock = $(24.9/34.3)\times100=72.6\%$.

5.3.3.1 Thermal Expansion
Thermal expansion of reservoir rock due to an increase in temperature causes a reduction in porosity, which in turn, expels oil from the matrix block to the fracture media. The thermal expansion of liquid in a confined permeable porous space increases the pressure of nongaseous liquid significantly, and a part of the liquid is ejected under the effect of thermal expansion.

According to Cosma and Mehaysen (1990), the expansion of the solid medium volume of the reservoir, which causes the ejection of an equal volume of oil, is defined as follows:

$$\Delta Nd_s = V_r\left(1-\varnothing\right)S_{oi}\alpha_s\Delta t \tag{5.14}$$

where V_r is the reservoir volume, $\varnothing$ is porosity, S_{oi} is initial oil saturation, α_s is the coefficient of thermal expansion of solid, and Δt is the increase in temperature. However, it is to be remembered that the entire reservoir volume is not heated at a time to the final temperature t; only a fraction (f_t) of the entire reservoir volume is heated. Inclusion of f_t, which is termed as the thermal flooding factor, results in Equation 5.15.

$$\Delta Nd_s = V_r\left(1-\varnothing\right)S_{oi}\alpha_s f_t\Delta t \tag{5.15}$$

So the thermal flooding factor is defined as the ratio of reservoir volume heated at temperature t to the total reservoir volume.

Similarly, the oil volume ejected from the porous permeable space as a result of thermal expansion of oil may be written as follows:

$$\Delta Nd_o = V_r \emptyset S_{oi} \alpha_o f_t \Delta t \tag{5.16}$$

Thus, the total amount of oil ejected from the porous and permeable space as an overall effect of thermal expansion is given as follows:

$$\Delta Nd = \Delta Nd_s + \Delta Nd_o = V_r S_{oi} \left[(1 - \emptyset) \alpha_s + \emptyset \alpha_o \right] f_t \Delta t \tag{5.17}$$

The typical values of α_o and α_s are in the order of 10^{-4} / °C and 10^{-5} / °C, respectively. The thermal expansion of the fluids is believed to contribute up to 10%–18% of the pore volume for an increase in matrix block temperature of 200°C (Mollaei and Maini, 2010).

Example 5.4

Calculate the percentage of oil ejected from the porous permeable space as an overall effect of thermal expansion by increase in reservoir temperature of 50°C. The following reservoir data are given:
Well spacing = 40 acres
Pay thickness = 15 ft.
Porosity, $\emptyset$ = 20%
S_{oi} = 70%
The values of α_o and α_s are in the order of 3×10^{-4} / °C and 5×10^{-5} / °C, respectively. Assume f_t = 80%.

Solution:
Initial oil volume of the reservoir, $V_r \times \emptyset \times S_{oi}$ = 7758 × 40 × 15 × 0.20 × 0.70 RB
= 651,672 RB
The total amount of oil ejected from the porous permeable space as an overall effect of thermal expansion is (using Equation 5.17):

$$\Delta Nd = \Delta Nd_s + \Delta Nd_o = V_r S_{oi} \left[(1 - \emptyset) \alpha_s + \emptyset \alpha_o \right] f_t \Delta t$$

Therefore,

$$\Delta Nd = 775840150.70 \times \left[(1 - 0.20) \times 5 \times 10^{-5} + 0.20 \times 3 \times 10^{-4} \right] \times 0.80 \times 50$$

$$= 13033.44 \text{ RB}$$

Therefore,
Percentage of oil recovery by thermal expansion = (13033.44 RB/651,672 RB) × 100 = 2.0%.

5.3.4 INTERFACIAL TENSION AND WETTABILITY

Temperature has a significant influence on IFT and wettability of rock. The IFT between oil and water inversely varies with temperature. Wei and Babadagli (1917) reported that an increase in oil temperature from 25°C to 85°C may reduce ~90% of its IFT (Fig. 5.5). Thus, not only thermal expansion but a reduction in IFT during the thermal EOR process also helps in improving the oil recovery.

Contact angle, which indicates the preferential adherence of a liquid on rock also undergoes changes with increasing temperature. In general, the contact angle is changed at higher temperatures, improving the water-wet characteristics of the rock (Flock et al., 1986). Desorption of the

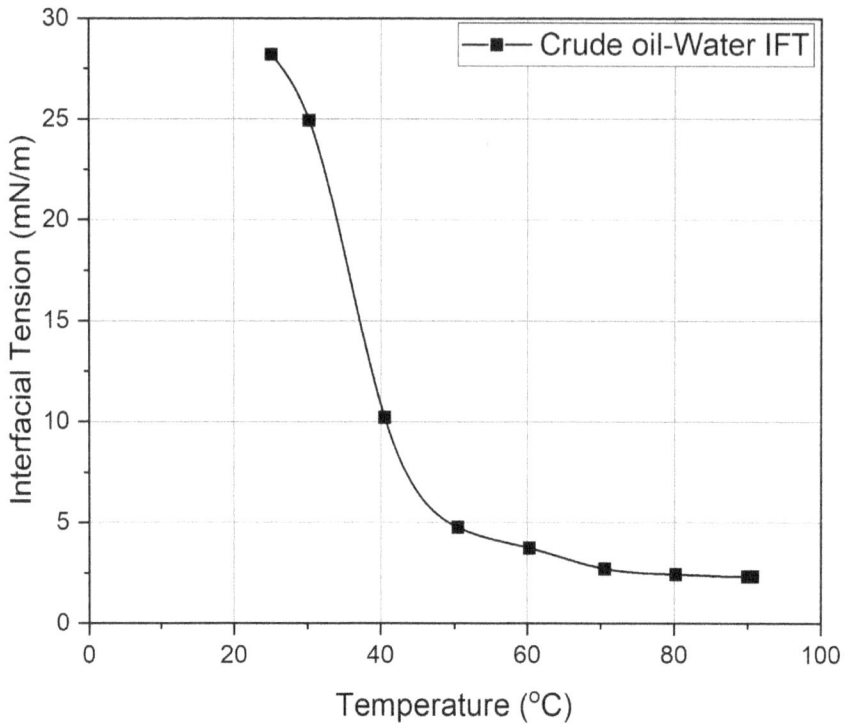

FIGURE 5.5 Typical decrease of IFT with temperature for typical crude oil–water system.

polar compounds from the rock surface at higher temperatures is the primary reason for the wettability change toward water wet. However, some rock and fluid combination may show more variations. An investigation by Petke et al. (1969) showed that the contact angle of an organic liquid droplet in equilibrium with saturated vapor on polymeric solid surfaces reaches the maximum value of 100° when the temperature varied between 5°C and 160°C.

5.3.5 Thermal Conductivity

The thermal conductivity of the rock, which defines its ability to conduct heat, is very important in determining the amount of heat transfer by conduction. The heat flow rate by conduction is positively related to the thermal conductivity (k) and temperature gradients. Consider a block of reservoir rock with a temperature difference ΔT across two opposing faces separated by a distance Δz (Fig. 5.6). The heat flow across the rock is proportional to the temperature difference divided by the distance (Fig. 5.6, top).

Generally, reservoir rocks display anisotropic thermal conductivity owing to their heterogeneity—i.e., variation in lithology, porosity, and fluid saturation. The magnitude of heat in such cases is determined by summing up the heat flow across each layer as shown in the bottom part of Figure 5.6. Calculations are similar to those used for determining the fluid flow for a reservoir with varying bed permeability in series/parallel or determination of the voltage differential in a circuit having wires arranged in series/parallel.

The success of the thermal recovery process in a reservoir depends largely on the thermal conductivity and heat capacity of the formation. Somerton et al. (1973) presented a correlation of thermal conductivity of a rock as a function of porosity, oil and water saturation, and temperature as given here:

$$K = a + b\varnothing + c\sqrt{S_w} + d\sqrt{S_o} + eT \tag{5.18}$$

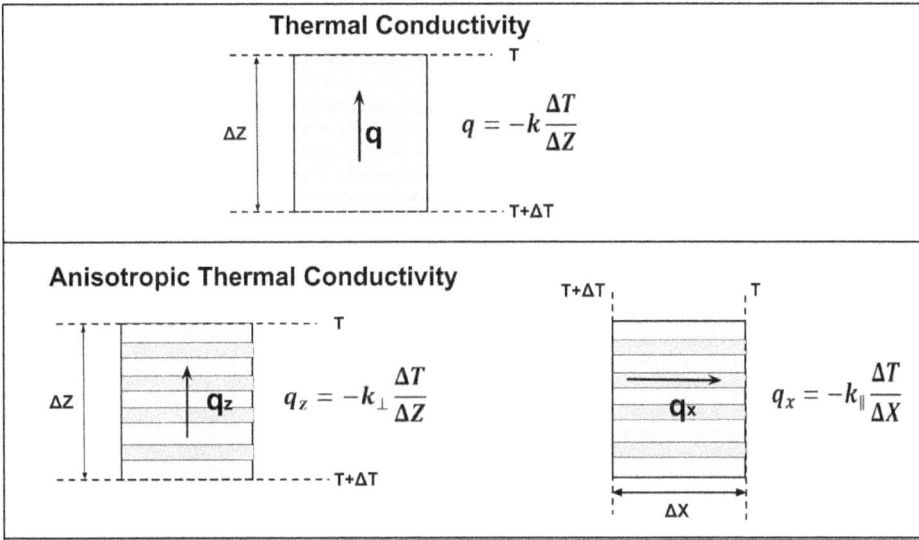

FIGURE 5.6 Thermal conductivity and heat flow through isotropic and anisotropic reservoir geometry.

where
K = thermal conductivity (W/(m°C))
$\emptyset$ = porosity (fraction)
S_w = water saturation (fraction)
S_o = oil saturation (fraction)
T = Temperature (°C)

Here, a, b, c, d, e are constant coefficients whose values are 4.318, 4.883, 0.474, 0.987, and 0.0024, respectively, as obtained by regression analysis. The values of the coefficients depend on the rock properties.

The thermal conductivities of water and crude oil are ~0.6 and 0.13 W/mK, respectively. The thermal conductivity of rocks is generally higher, in a range from about 0.5 to 6.5 W/mK.

5.3.6 Residual Oil Saturation and Relative Permeability

The residual oil saturation and relative permeability are two deciding parameters in predicting the cumulative oil production and the production rate of an oil field. As mentioned already, a rise in temperature will result in a decrease in residual oil saturation (S_{or}) and an increase in irreducible water saturation. Various studies reported that the relative permeability of oil at a particular saturation increases with temperature and that of water decreases with temperature, which in turn improves the mobility of oil. The changes to residual oil saturation and relative permeabilities with temperature are because of very complex phenomena, and some of these are the following (Mollaei and Maini, 2010; Shabani et al., 2014):

a) thermal expansion of reservoir rock and fluids;
b) interactions between the injected fluids and the matrix;

c) chemical changes in the constituents of heated rocks; and
d) change of wettability and IFT.

Figure 5.7 shows the typical variation of residual oil saturation and relative permeability of oil and water with respect to temperature.

Example 5.5: Compare the oil recovery at breakthrough for waterflood (60°C) and steam injection (120°C) from a five-spot well pattern; the following data are given:

Well spacing	= 50 acres	μ_o at 60°C = 8.0 cp
Pay thickness	= 15 ft.	μ_w at 60°C = 0.8 cp
Porosity	= 25%	μ_o at 120°C = 3.0 cp
S_{oi}	= 70%	μ_w at 120°C = 0.65 cp
S_{or}	= 30%	

The relative permeabilities to oil at S_{oi} at 60°C and 120°C are 0.30 and 0.75, respectively.
The relative permeabilities to water at S_{or} at 60°C and 120°C are 0.175 and 0.475, respectively.

Solution:
Displaceable oil = $7758 \times A \times h \times \emptyset \times (S_{oi} - S_{or})$
= $7758 \times 50 \times 15 \times 0.25 \times (0.7 - 0.3)$ rb
= 581,850 rb

Mobility ratio, $M = \dfrac{k_w/\mu_w}{k_o/\mu_o}$

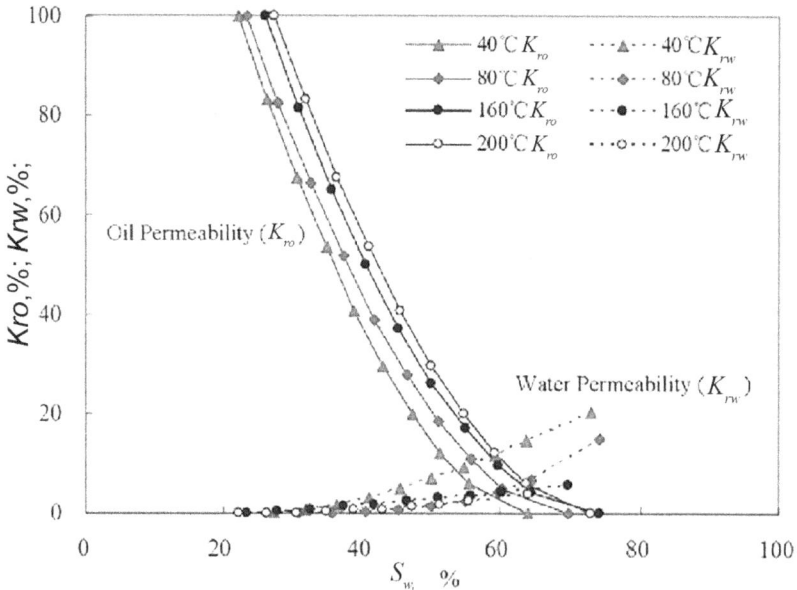

FIGURE 5.7   Water-oil relative permeability curves.

Source: Xie et al., 2020.

From the given data, the end point relative permeabilities at 60°C are:

$k_w = 0.175$ (at $S_{or} = 0.30$)

$k_o = 0.75$ (at $S_{oi} = 0.70$)

M for waterflooding = 5.833

From Figure 5.8, $E_A = 0.516$

Oil recovery for waterflooding at breakthrough

= Displaceable oil (movable oil) × E_A

= 581850 rb × 0.516

= 300,234.6 rb

From the given data, the end point relative permeabilities at 120°C are:

$k_w = 0.475$ (at $S_{or} = 0.30$)

$k_o = 0.30$ (at $S_{oi} = 0.70$)

M for steam flooding = 1.077

From Figure 5.8, $E_A = 0.705$

Oil recovery for steam flooding at breakthrough

= 581850 rb ×0.705

= 410,204.25 rb

5.3.7 THERMOCHEMICAL REACTIONS

Reservoir oil generally undergoes two kinds of chemical reactions: (1) pyrolysis and (2) oxidation. A pyrolysis reaction occurs in the absence of oxygen, while as the name signifies, oxidation occurs only in the presence of oxygen. Pyrolysis is common for steam injection and in situ combustion, while oxidation takes place only during in situ combustion. The reactions that occur in thermal EOR under specified conditions can be classified into three types: (1) cracking, (2) dehydrogenation, and (3) condensation. Temperature has a strong effect on the type of reaction and the amount of the

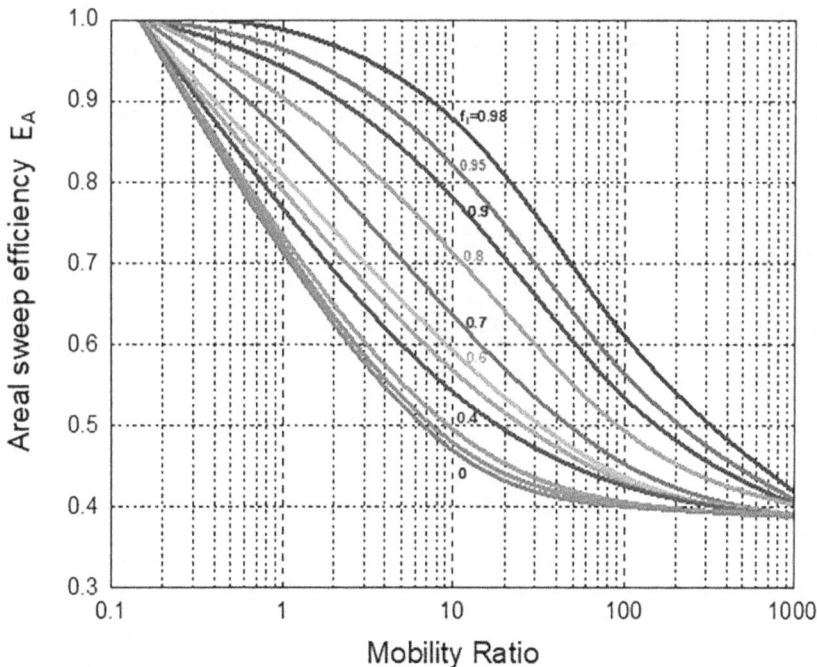

FIGURE 5.8   Areal sweep efficiency versus mobility ratio for various values of f.

products. Reactions that occur at low temperature (< 350°C) are referred to as visbreaking, while high-temperature reactions are called cracking.

5.4 HOT WATER INJECTION

From an operational point of view, hot water injection is the most attractive thermal recovery process. Relatively simple and inexpensive equipment and facilities are required for generating and handling hot water (Farouq Ali, 1974b), as it only involves sensible heat. It resembles conventional waterflooding, with the difference that the injection of hot water offers better oil recovery. Improved mobility ratio due to a drop of the oil phase viscosity at higher temperature is the main reason for enhanced recovery by steam injection compared to that by simple water injection. The other mechanisms of hot waterflooding include the reduction of IFT and residual oil saturation, which lead to a potentially higher recovery factor. Obviously, it is less effective in reducing oil viscosity compared with steam due to the absence of latent heat. However, for thin heavy oil reservoirs, hot waterflooding has advantages over steam flooding. The parameters that are important for designing a hot waterflood project include the temperature of the injected water, slug volume, injection rate, starting time, geological and physical characteristics of the oil field, etc. A representative temperature and oil saturation distribution for a hot waterflooding project is presented in Figure 5.9. The temperature distribution in the heated zone depends on the heat losses to over and underlying formations, but the velocity of its leading edge is independent of such losses.

The sweep efficiency of hot waterflooding is better compared to steam, as water viscosity is much larger than that of steam (Diaz-Munoz and Farouq Ali, 1975). It permits the use of a much higher injecting pressure than steam flooding at a given temperature. The heat losses to the overburden and understrata will be substantially lesser than that encountered in steam flooding (Zao et al., 2013).

5.4.1 CASE STUDIES

A hot waterflood is especially applicable for the shallow depth, viscous, and crude-like conditions that prevail in the North Sea. Some shallow reservoirs in the North Sea region contain oil with viscosities of several hundred centipoise. The reservoirs have very high horizontal and vertical permeabilities in the range 1–10 Darcy. Because of the adverse mobility ratio, recovery

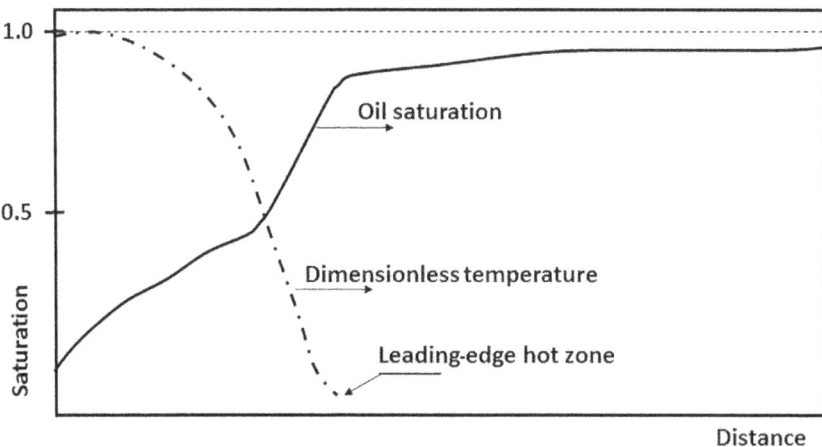

FIGURE 5.9 Typical temperature and saturation distribution in a hot water drive.

TABLE 5.1
Base Case Model Parameters, North Sea
Reservoirs (Goodyear et al., 1996)

Property	Values
Water/oil density different (kg/m³)	98*
Temperature (°C)	30.6
Oil viscosity (cp)	400*
Water viscosity (cp)	0.79*
Permeability (Darcy)	5.0
Porosity (fraction)	0.33
k_v/k_h	0.5
Residual oil saturation	35%
Connate water saturation	10%
Water end point relative permeability	0.16
Well position (depth from top, ft)	256

*at initial reservoir temperature

efficiencies from conventional waterflooding are likely to be low. Increasing the temperature of the injected water reduces the viscosity contrast between oil and water in the heated region with an improvement of sweep efficiency and subsequent oil recovery (Goodyear et al., 1996). The simulation studies predict an incremental oil recovery, up to 18% of STOOIP. Hot waterflooding may also increase the economic life of individual wells by as much as a factor of two. The thermal expansion of water plays an important role in the incremental oil recovery mechanism, reducing the density of the injected water relative to the aquifer water. Table 5.1 summarizes the reservoir properties.

Hot water injection in the Mangla field (Rajasthan, India) also showed a similar result. Hot water injection above the wax appearance temperature (WAT) of crude oil improves the well injectivity and recovery (Kumar et al., 2008). The WAT of crude is very high ~60°C, whereas the reservoir temperature is ~65°C. The average permeability is 5 Darcy, crude oil gravity is 23°API, and the oil viscosity at in situ condition varies between 9.3cP and 50cP.

The Pelican Lake Field is located ~250 km north of the city of Edmonton, Alberta, Canada, where the studies were carried out for prospective hot waterflooding (Duval et al., 2015). The Wabiskaw "A" sand in the Pelican Lake area comprises both lower shoreface and middle shoreface sediments with an average porosity of 29%, average water saturation of 27%, and a net pay of 5 meters. The oil viscosity in the Pelican Lake field ranges from 600 centipoise at 15°C to over 200,000 cp at 15°C. Hot water injection on a pilot scale shows that hot water circulation can be an effective way to develop the heavier oil accumulations of the Pelican Lake field.

5.5 STEAM INJECTION

Steam injection is a thermal recovery process similar to hot injection with the difference that steam is injected into the reservoir in place of hot water to improve fluid mobility within a reservoir. It has been proved that heavy oil reservoirs benefit significantly from steam injection compared to water injection, as extra heat is injected into the reservoir. The heat transferred from the injected steam causes a reduction in the viscosity of heavy oil. Again, compared to saturated steam, oil recovery can be further improved using superheated steam, which contains more latent heat than the saturated one, under equivalent conditions.

5.5.1 Mechanisms

Oil recovery by steam injection involves a number of mechanisms. These are as follows:

(1) reduction of the oil viscosity and increase in relative permeability to oil result in improvement of the mobility ratio and, subsequently, the sweep efficiency;
(2) reduction of the residual oil saturation;
(3) increasing the formation volume factor;
(4) vaporizing and distilling condensable hydrocarbons from the crude;
(5) providing a gas drive mechanism; and
(6) solution gas drive.

The first three mechanisms are discussed in Section 5.3 along with the other mechanisms of thermal recovery. The last three mechanisms are discussed in the following.

5.5.1.1 Vaporizing and Distilling Condensable Hydrocarbons from the Crude

One of the important mechanisms of steam flooding is vaporization and steam distillation of condensable hydrocarbons from the crude. Steam distillation occurs especially in reservoirs containing light oils. Injected steam heats the formation and eventually forms a steam zone, which grows with continued steam injection. A fraction of the crude oil in the steam zone vaporizes into the steam phase when the total vapor pressure of steam and the hydrocarbon component of crude oil equals the reservoir pressure. The hydrocarbon vapor is transported through the steam zone by the flowing steam. Both the steam and hydrocarbon vapor condense at the steam front to form a hot water zone and a hydrocarbon distillate bank. The dynamic process involving vaporization, transport, and condensation of the hydrocarbon fractions displaces the lighter hydrocarbon fractions and generates a distillate bank that miscibly drives reservoir oil to producing wells (Duerksen and Hsueh, 1983). A schematic of the steam distillation mechanism in steam flooding is shown in Figure 5.10.

In the steam distillation of crude oils, the distillate yield Y, is expressed by a correlation parameter, V_w / V_{oi}, which is defined as the ratio of the cumulative water distilled to the initial oil volume. Y depends on the distillation temperature and steam throughput volume. The temperature determines the vapor pressures of the hydrocarbon species present in the crude oil. The steam throughput volume determines how much of the vaporized hydrocarbon is swept from the distillation zone, assuming vapor/liquid equilibrium. The steam throughput volume, V_s, can be calculated by the following equations:

$$V_s = V_w \left(V_s\right)_{sat} \left(\frac{T_s}{T_{sat}}\right) \tag{5.19}$$

where
V_w = volume of ambient-temperature water put through the distillation zone;

$\left(V_s\right)_{sat}$ = volume of saturated steam per unit volume of ambient-temperature water (from steam tables); and

$\left(\dfrac{T_s}{T_{sat}}\right)$ = ratio of steam temperature, T_s, to saturated steam temperature, T_{sat}, in °R (to account for the effect of superheat on steam volume).

The effects of pressure and temperature on the steam volume throughput and hydrocarbon vapor pressure were combined to give a distillation factor, F, given as follows:

$$F_{T,p} = \frac{\left(V_s\right)_{T,p}}{V_{oi}} \frac{p_T}{p_{T_r}} \tag{5.20}$$

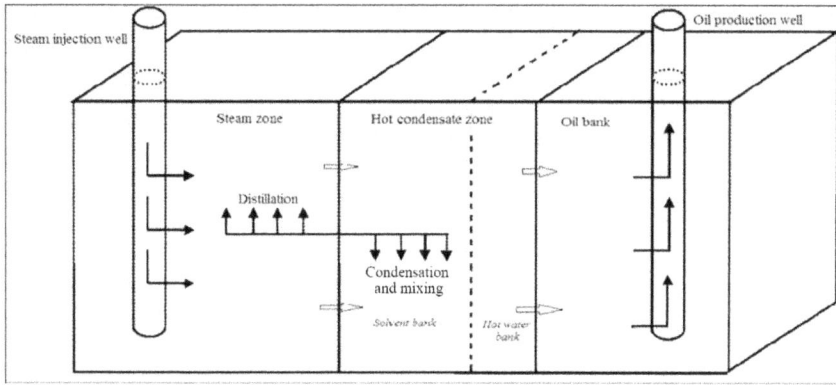

FIGURE 5.10   Steam distillation mechanism in steam flooding.

Source: Vafaei et al., 2009.

where

$F_{T,p}$ = distillation factor for the distillate fraction produced at temperature T and total pressure P;

$(V_s)_{T,p}$ = steam volume at T and P;

V_{oi} = initial oil volume;

p_T = vapor pressure of distillate fraction at T; and

p_{T_r} = vapor pressure of distillate fraction at reference temperature, T_r.

5.5.1.2   Gas Drive Mechanism

During flooding, steam undergoes various water/oil or water/matrix chemical reactions (Holladay, 1966), generating a significant amount of gas. The generated gas volume is high and even may exceed the pore volume of the affected reservoir. This gas can displace oil from the matrix as its volume increases. Besides lighter hydrocarbon, these gases include CO_2 and H_2S. The temperature for gas generation may be as low as 450°F (232°C).

5.5.1.3   Solution Gas Drive

Reactivation of existing solution gas drive may be considered as a mechanism for increasing oil recovery during steam injection. Exsolution and expansion of the dissolved gases in the oil and water provide the maximum driving force in steam flooding because of increased temperature. However, contribution of a solution gas drive is considerable only when the reservoir pressure is at or below the bubble point and there is a sufficient amount of gas dissolved in the oil. The free gas present so in the reservoir will expand upon heating and behave similar to an active solution gas drive reservoir.

5.5.2   Application of Steam Injection

Not only for heavy oil, steam injection is applied as an efficient enhanced recovery technique for light oil reservoirs also. It helps in increasing oil production by reducing the residual oil saturation remaining after other recovery methods such as waterflooding. For heavy oil reservoirs, reduction of oil viscosity by steam injection increases oil recovery; whereas vaporization of light oil is the primary recovery method in light oil reservoirs (LORs). Even though steam injection processes in LOR promise a significant potential, studies have shown that field development and technical analysis are very sparse (Dehghani and Ehrlich, 1998). As per the mechanisms of injection, steam flooding is broadly divided into two main categories: CSS and steam flooding. Both are most commonly applied to the oil reservoirs that are relatively shallow and contain highly viscous crude oils at the in situ temperature.

Steam injection is widely applied to the oil fields located in the northern bank of the Orinoco River and that extend from east to west along the Orinoco petroleum belt (Venezuela), the San Joaquin Valley of California (USA), and the Lake Maracaibo area. The oil sands of northern Alberta (Canada) are also considered as ideal candidates for steam injection. As mentioned earlier, steam injections were also applied to the Barmer field in India, which contains waxy crude at a very shallow depth.

5.5.3 Cyclic Steam Injection

The cyclic steam injection process is a very effective thermal recovery process. It was first developed by Imperial Oil in the late 1950s at Cold Lake (Speight, 2012). Cyclic steam injection is also known as cyclic steam stimulation, steam soak, or huff-n-puff. This technique is quite effective even for extra heavy crude oil like bitumen or pitch. The steam's role is to dissolve that solid and to allow it to flow through the reservoir.

A typical cyclic steam injection process consists of three stages: injection period, soak period, and production period.

Stage 1, Steam Injection: During this period, the steam is continuously injected for up to a month into the reservoir. Very high injection rates are maintained to reduce heat losses. The temperature of the steam is maintained at ~572°F–644°F (300°C–340°C). However, care must be taken to avoid fracture development. The formation injectivity and its variation with time (which depends on reservoir conditions) are closely monitored. Heat from the injected steam raises the reservoir temperature, resulting in a pronounced increase in the mobility of heavy oils and a corresponding improvement in production rates. Care must be taken to prevent the channeling of steam due to gravity segregation, presence of high-permeability strata, and adverse viscosity ratios.

Stage 2, Soak Period: Prior to putting a steam-stimulated well on production, soak time is required to allow the injected steam to heat the oil around the wellbore. This period is aimed at attaining partial steam condensation to heat the rock and fluids and bring about a more uniform injected heat distribution. In actual operations, the soak time may vary from days to weeks or even months, depending on the reservoir rock and fluid properties. Excessive soak time may cause the well productivity to decline because of the continued heat loss to the formation. On the other hand, an insufficient soak period will result in incomplete heat transfer to the formation and a large amount of steam will still exist. In such a case, heat will be removed from formation, with the production stream containing the remaining steam and condensate water once the well is switched to production. Thus, it is desirable to determine the optimal soak time that would maximize oil production or net profit.

Stage 3, Production Stage: After the soak stage, the well is put on production. The initial production rates are typically very high for a short period of time and then declines gradually over several months. In case the reservoir pressure depletes fast, resulting in very low production rates with further production no longer being economic, the well shall be put on steam injection stage again, and the whole process repeats for another injection-soak-production cycle (Fig. 5.11).

Steam stimulation is also used to stimulate the producers and to clean up the formation around the wellbore, specifically for the formation damage caused by the deposition of solids, paraffin, or asphaltene near the production well. In such cases, steam makes the precipitates soluble in oil and reduces the viscosity of the material that ties paraffins and asphaltenes to the rock surfaces; thus, the pores are cleaned and the precipitates are removed from the formation, improving the permeability of near-wellbore zones. Other positive benefits that may contribute to production stimulation

Stage 1
Steam Injection

Stage 2
Soak Phase

Stage 3
Production

Water table

Casing
Concrete

Diatomite

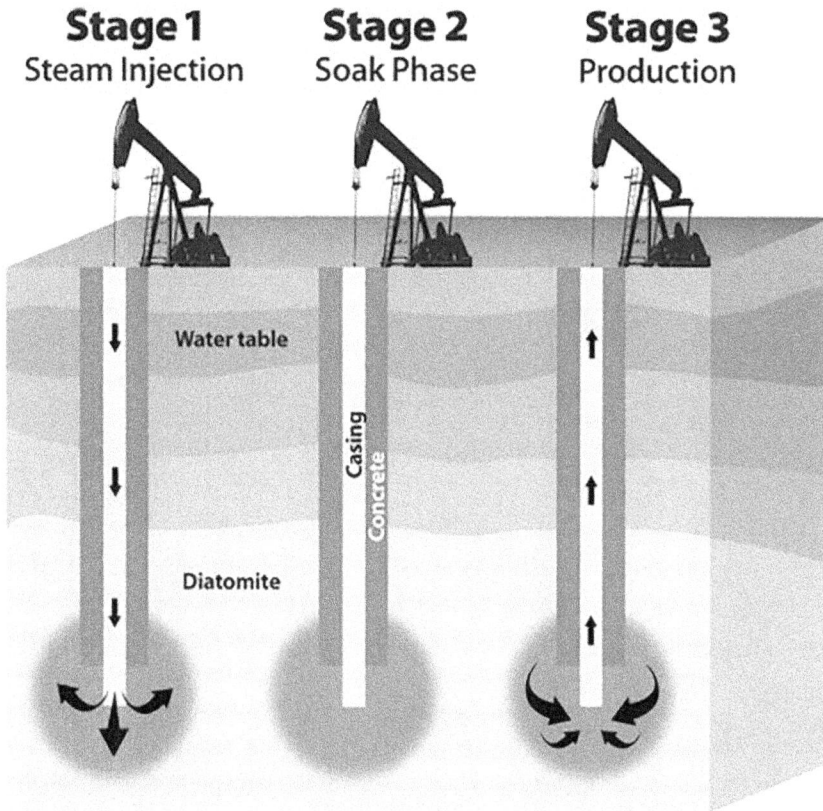

FIGURE 5.11   Schematic of cyclic steam injection.

Source: www.independent.com/2014/06/26/anatomy-cyclic-steaming.

include (i) thermal expansion of fluids; (ii) compression of solution gas; and (iii) reduction in resid-ual oil saturation. The technique has gained wide acceptance because of the quick payout result from successful applications. However, many project failures have been reported globally, principally due to improper design and implementation.

The production patterns for successive cycles are shown in Figure 5.12. It may be seen from the figure that the oil rate decreases in subsequent cycles. Beyond a certain period, the project will fall below the economic limit, as the cost of steam injection will be more than the incremental oil recov-ery. So it is desirable to determine the number of cycles that will maximize the oil recovery from a cyclic steam injection project. From various case studies, it is observed that eight or nine cycles can give the maximum cumulative oil.

A significant characteristic of steam stimulation is that the injected heat is concentrated close to the well, where the flow lines converge and pressure gradients are the highest. Unlike a continuous steam injection process where the oil passes through colder areas of the reservoir, the displaced oil becomes and remains warm, as it flows to the producing well in cyclical stimulation.

5.5.4   STEAM FLOODING

Steam flooding is the process of injection of steam into the injection wells and production of oil from the producing wells. The zones around the injection wells are heated by the steam, and the high-temperature zones spread toward the production wells, leading to production of oil and con-densed water from the heated zones. The much lower viscosity of steam with respect to the viscosity

Oil rate (m³ std/day)

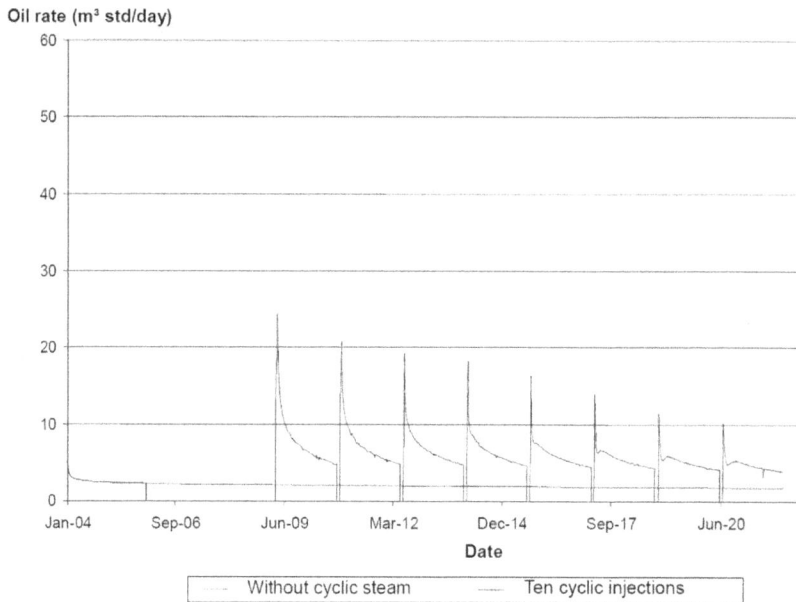

FIGURE 5.12 Oil rate in 20 years of production in a numerical model with and without cyclic steam injection.
Source: Barillas et al., 2008.

of crude oil limits the effective conduction of heat as well as the sweeping of oil and leads to a lower recovery by steam injection (Lee and Lee, 2019). Steam is injected at higher rates in comparison to steam stimulation to offset the lower thermal efficiency. It is usually recommended to initially recover the oil by steam stimulation followed by conversion of the wells to specific injection and production wells for the pattern drive of steam flooding (Speight, 2012). The fields with efficient steam stimulation recovery are often considered good candidates for steam flooding. This leads to an initial high economic production of oil, leading to more than 50% recovery of oil by the steam flooding process. In a dipping reservoir, the steam is injected from the wells into the upper zones to prevent the steam from bypassing and driving the crude oil downward. A typical injected fluid has a steam quality of 80%—i.e., 80% steam and 20% water (Satter et al., 2008).

The mechanism of oil recovery by steam injection can be well understood by considering steam in a long, porous medium initially containing oil and connate water. The following processes occur in sequence between the injection and production wells:

(i)   The oil in the immediate vicinity of the injection end is vaporized and pushed ahead.
(ii) A fraction of the oil is not vaporized and is left behind but at a higher temperature due to the heat provided by the steam.
(iii) The advancing steam eventually condenses into hot water due to heat losses into the over-burden and attains the steam temperature, thereby generating a hot condensate bank.
(iv) This hot water bank drives oil ahead as it moves and eventually cools down to reservoir temperature.
(v)   Ahead of this point, the displacement process continues as it would in a conventional waterflood.

As the injected steam moves forward from the injection well to the production well, it has undergone several physicochemical interactions with the reservoir rock and fluids. Based on the physical state, temperature, and oil saturation, the reservoir is divided into five zones as shown in Figure 5.13.

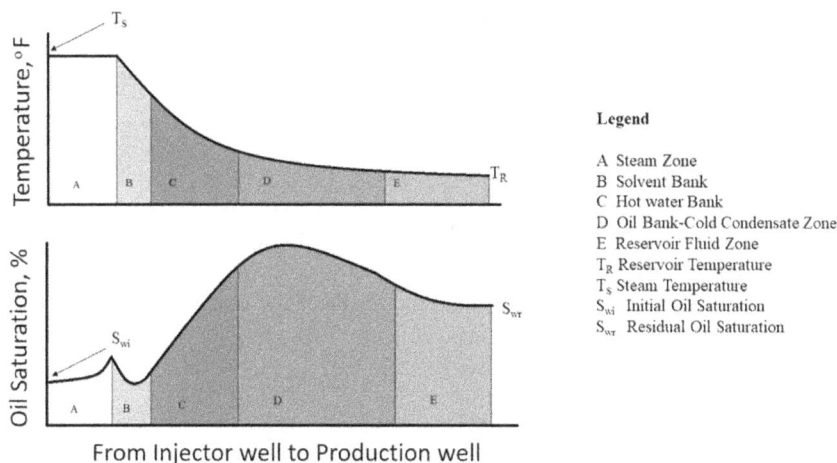

FIGURE 5.13   Typical temperature and saturation profile for steam flooding.

5.5.4.1   Steam Zone

In the steam zone (A), the predominant effect is steam distillation. High temperature and the presence of a gas phase lead to the vaporization of the light ends, which are carried forward by the advancing steam until they condense in the cooler portion of the reservoir. The relatively heavier components of the oil, characterized by a high vapor pressure, are left behind. The actual oil recovery by steam distillation is determined by the composition of the oil involved. The additional recovery by steam distillation may account for as much as 20% of the oil in place for a 50% distillable crude at 340°F (Willman et al., 1961). The oil ahead of the steam zone gets richer in lighter ends, serving to extract and displace the original oil, thus enhancing oil recovery. In the steam zone (A), oil saturation reaches its lowest value because the oil is subject to the highest temperature.

5.5.4.2   Solvent Bank

As mentioned earlier, the lighter components of the crude oil are vaporized by steam distillation in the steam zone (A), ahead of which, the steam condenses into water as it loses heat to the formation and forms a hot condensate zone. Along with the steam, the lighter hydrocarbon vapors also condense and thus form a hydrocarbon solvent bank (B) and extract additional oil from the formation to form an oil-phase miscible drive. The high temperature in this zone reduces the oil viscosity and expands the oil to produce saturations lower than those found in a conventional waterflood.

5.5.4.3   Hot Water Zone

The oil recovery for the hot water zone (C) is largely governed by the thermal characteristics of the oil involved. If the viscosity of the oil exhibits a sharp decrease with an increase in temperature, a considerable amount of oil will be recovered by the hot waterflood. The thermal expansion of oil accounts for the recovery of 3%–5% of the oil in place. The displacement efficiency is related to the reduction in residual oil saturation with temperature. The change in residual oil saturation, wettability, and thermal expansion finally lead to an extra recovery, which may account for as much as 10%–20% of the oil in the underlying zones of the reservoir not swept by the steam (Goyal and Kumar, 1989). The injection of chemicals in the unswept zone may improve the recovery further, and this, thus, draws the interest of researchers.

5.5.4.4   Cold Condensate Zone

As the mobilized oil is pushed ahead by the advancing steam (A) and hot water (C) fronts, the temperature of the condensate further decreases. This zone with an oil bank is referred to as the cold

condensate zone (D). The temperature of this zone lies between the temperature of the hot condensate zone and the initial reservoir temperature. The oil saturation in this zone is actually higher than
initial oil saturation. The oil recovery from the cold water zone is approximately equal to that for an
equivalent waterflood.

5.5.4.5   Reservoir Fluid Zone

The temperature of the reservoir ahead of the cold condensate zone has been unaltered by the steam
injection process and approaches the initial reservoir temperature. The oil saturation in this reservoir
fluid zone (E) is the initial oil saturation, S_{oi}.

The oil production by steam flooding is thus the result of the combination of various recovery
mechanisms (Wu, 1977) as outlined here:

- Viscosity reduction
- Thermal expansion and swelling of oil
- Steam distillation
- Solution gas drive
- Miscible displacement

In a macroscopic sense, the process of steam flooding depends upon the following parameters:

(1) Change of in situ fluid properties: densities, viscosities, composition, compressibility, and
    phase behavior.
(2) Change of the petrophysical properties of reservoir rocks in the presence of steam.
(3) Variations in wettability, capillary pressures, relative permeabilities, IFT, etc.
(4) Thermal properties of the reservoir rock and fluids.
(5) Reservoir heterogeneity, faults, initial oil saturation, temperature, pressure.
(6) Well spacing, injection pattern, and producing-injecting interval location and thickness.
(7) Operating parameters such as steam injection rate, steam quality, injection pressure (tem-
    perature), cumulative amount of injection, etc.

The challenges associated with the steam flooding are the adverse mobility ratio and channeling the
steam through high-permeability zones, as the mobility of steam is higher than that of oil. Further,
the lower density of steam leads to gravity override, which in turn, leads to early breakthrough and
reduces the amount of contacted oil in the reservoir. Thus, the portion of the reservoir that is swept
by the steam has low residual oil saturation, whereas the other portions have higher residual oil
saturation (Green and Willhite, 1998).

5.5.4.5.1   Heat Loss in Steam Flood

The heat loss in the wellbore during steam injection has a significant impact on final recovery and
project economics as well as the flooding conformance of a steam flooding project. Maintaining
steam quality is a big challenge in steam flooding and depends on many factors such as completion
design, injection rate, etc. The heat loss in the wellbore can be measured by steam quality change
from wellhead to bottom hole. The heat loss incurred during steam injection may be determined
using the following equation developed by Wang et al. (2010):

The basic equation to calculate heat loss per length of pipe, Q_{ls}, is:

$$Q_{ls} = \left.\left(T_b - T_a\right)\middle/ R_h\right. \tag{5.21}$$

where Q_{ls} is the rate of heat loss per unit length in Btu/ft-hr
R_h is the specific heat resistance in 1/Btu/ft-hr-°F

T_b is the steam or hot fluid temperature flowing in the tubing
T_a is the temperature of the surrounding rock away from T_b
The total heat flow rate, Q_{to} can be written as follows:

$$Q_{to} = U_{to} A_{to} (T_b - T_a)$$ (5.22)

where
U_{to} is the overall heat transfer coefficient Btu/Hr-ft.^{2}-°F;
A_{to} = is surface area; and
$A_{to} = 2\pi r L_{to}$, where L_{to} is the length of the pipe and r is the radial distance from T_b to T_a.
Considering the condition of loss of the total flowing heat, i.e. $Q_{is} = Q_{to}$, the R_h may be expressed as $1/(2\pi r U_{to})$.
Based on Equation 5.22, once U_{to}, T_b, T_a, and the radius r are known, the total heat loss can then be analytically computed per unit length.

5.5.4.5.2 Heat Distribution
For the purpose of determining the heat distribution, the injected heat is distributed among the three zones of regions: the steam zone, the hot water zone, and the overburden and substratum (Baker, 1969). The volumetric heat capacity (ρc) of the hot water zone, overburden, and substratum—i.e., the effective heat capacity of the formation is defined as follows:

$$\rho c = \varphi \rho_w c_w + (1-\varphi) \rho_R c_R$$ (5.23)

In the steam zone the temperature is constant, and the heat content is:

$$H_s = \left\{ \left[(1-\varphi) \rho_R c_R + \varphi S_w \rho_w c_w + \varphi (1-S_w) \rho_s c_s \right] (T_s - T_o) + (1-S_w) \rho_s \lambda \right\} V_s$$ (5.24)

where T_s is the saturation temperature, T_o is initial temperature, φ is the porosity, V_s is the volume of steam zone and λ is the latent heat of vaporization.
The total accounted heat $H_T = H_S + H_W + H_L$, where H_S is the heat in the steam, H_W is the heat in the hot water zone, and H_L is heat in the overburden and substratum. The amount of heat injected during time t is:

$$H_I = W_s \left[c_w (T_s - T_O) + \lambda \right] t$$ (5.25)

where W_s is the mass injection rate. Figure 5.14 shows the typical percentage of heat distribution among the three zones.

5.5.4.5.3 Criteria of Steam Flooding Project
For implementing a steam flooding project, the following criteria are generally used (Lyons and Plisga, 2005):

1. Thickness of the pay zone should be a minimum of 20 ft. to minimize heat losses to the adjacent formations. The integrity of caprock must be checked.
2. Steam flooding is primarily recommended to recover viscous oil from large fields with high-permeability sandstone or unconsolidated sands. Carbonate reservoirs are not the right choice for this purpose.
3. A high percentage of water-sensitive clay content in the formation may cause problems in a steam flood. So the candidate reservoir should contain a low percentage of water-sensitive clay to ensure good injectivity.

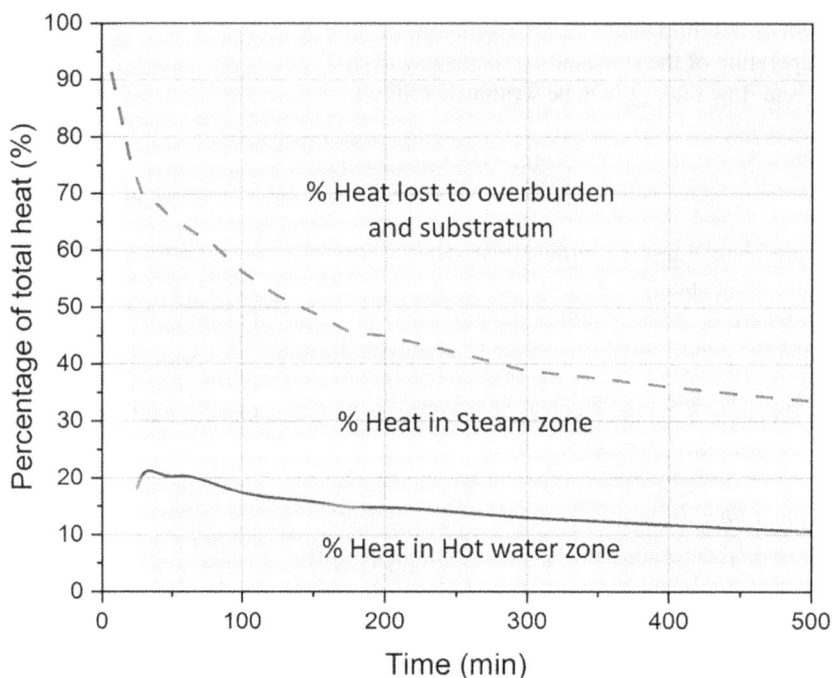

FIGURE 5.14 Typical heat distribution among steam zone, hot water zone, and overburden and substratum as percent of total heat accounted for a certain injection rate.

4. In addition to the heavy oil, steam flood may be utilized for less viscous crude oils in case of an unsuccessful waterflood.
5. In a steam flood, maintaining the injected steam quality is a critical issue. In deeper wells, huge heat loss and the requirement of a high injection pressure are the main hurdles in using steam flood. So shallow reservoirs, with viscous and/or heavy oil, are the main target candidates for steam flood.
6. Steam generation is a costly affair. Approximately one-third of the revenues obtained from the additional oil recovered is consumed to generate the required steam. So it must be ensured that the oil saturation is high enough to fulfill the economic criteria.

5.5.5 STEAM-ASSISTED GRAVITY DRAINAGE

The steam-assisted gravity drainage (SAGD) process is one of the most promising approaches for the development of large heavy oil and bitumen accumulations. The mechanisms working in this method are similar to the steam flood, with an additional recovery caused by gravity. Besides, the SAGD process requires the combination of an injection and production well drilled in the same pay zone. Butler et al. (1981) were the first ones to derive the classical mathematical model of SAGD by coupling Darcy's law and heat transfer theory.

SAGD can be carried out by using different well patterns as follows:

(i)   a group of vertical wells for joint production;
(ii) clustered wells/slant horizontal wells for joint production;
(iii) u-shaped wells, vertical wells for injection and horizontal wells for production; and
(iv) parallel horizontal wells, where steam is injected into the upper well, while hot fluids are produced from the lower well.

Out of these four, the parallel horizontal well pattern contributes most to the recovery rate, with the unit production rate of horizontal footage at 0.1–0.2 m^3/d-m, according to a simulation analysis (Rui, 2009). The process has several features (Butler, 1994):

a) High recovery rate.
b) Higher steam to oil ratio compared to conventional steam flooding.
c) The process is useful even for heavy bitumen reservoirs without expensive preheating.

This process creates a chamber caused by condensing steam at the chamber boundary. Heated oil and water are drained by gravity along the chamber walls toward the lower production well.
   The mechanisms of SAGD can be better understood from Figures 5.15a and 5.15b.

- Steam is injected from a well, either a horizontal well or, sometimes, one or more vertical wells, above the horizontal producer.
- A steam-saturated zone is formed in which the temperature is essentially that of the injected steam.
- The steam flows from the perimeter of the steam chamber and condenses.
- The heat from the steam is transferred by thermal conduction into the surrounding reservoir.
- The water condensate from the steam and the heated oil flow, driven by gravity, to the production well below.
- As the oil flows away and is produced, the steam chamber expands.

The steam/oil ratio represents the volume of steam necessary to produce one volume unit of oil. SAGD is very promising economically because it combines high flow rates with a favorable energy balance. Expected oil recovery is ~50%–75%. To be efficient, SAGD requires careful production monitoring to adjust well parameters to optimize the development of the steam chamber development.
   A theoretical calculation of the oil flow rate at the producing well in terms of reservoir rock, fluid, and steam properties, as given by Equation 5.26 (Butler, 1994):

$$q = 2L\sqrt{\frac{1.5Kg\alpha\Delta S_o h}{m\gamma_s}} \tag{5.26}$$

and

$$\frac{1}{m\gamma_s} = \int_{T_R}^{T_S}\left(\frac{1}{\gamma} - \frac{1}{\gamma_R}\right)\frac{dT}{(T - T_R)} \tag{5.27}$$

where
g = acceleration due to gravity, m s^{-2}
h = chamber height, m
k = effective permeability for oil flow, m^2
L = length of horizontal well, m
m = a dimensionless parameter
q = oil flow rate to production well, m^3 s^{-1}
α = thermal diffusivity of oil reservoir, m^2 s^{-1}
T_R = original reservoir temperature
T_S = temperature of steam zone
ΔS_o = initial fractional oil saturation less residual oil saturation in the steam chamber
γ = kinematic viscosity of oil at temperature T, °C m^2 s^{-1}
γ_R = kinematic viscosity of oil at temperature T_R, °C m^2 s^{-1}
γ_S = kinematic viscosity of oil at temperature T_S, °C m^2 s^{-1}

Normal SAGD Process

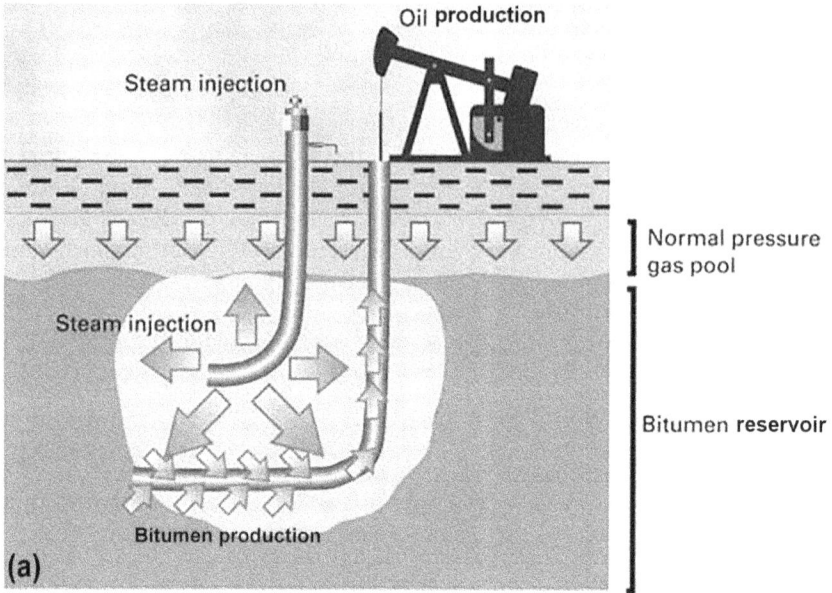

(a)

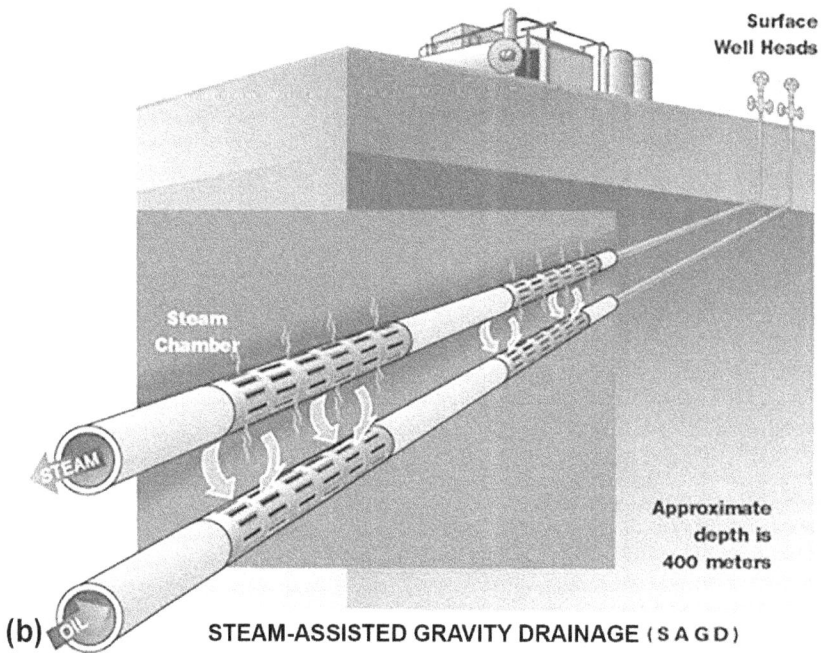

(b)   STEAM-ASSISTED GRAVITY DRAINAGE (S A G D)

FIGURE 5.15   (a) and (b) Conceptual diagram of the steam-assisted gravity drainage process.

Source: www.hcl.ca/services/steam-assisted-gravity-drainage-sagd.

If we consider that there is no aquifer support, the pressure maintenance is assured when steam injection rate compensates for water production. As oil is replaced by steam, it requires the injection of a small additional volume of water. Then the injection rate at surface conditions can be approximately expressed by Equation 5.28 (Egermann et al., 2001):

$$q_{injection}^{surface} \approx \frac{q_o}{B_w}(WOR) \tag{5.28}$$

The WOR for most SAGD projects lies between 1.5 and 2, and the corresponding injecting rate varies between 130 and 173 m^3/d at surface conditions.

5.5.6 STEAM AND GAS PUSH

Steam and gas push (SAGP) is very much similar to SAGD, with the difference that a small amount of non-condensable gas is injected along with the steam. The SAGP concept was first defined and modeled by Butler (1999). For gravity to be an effective force and displace the oil downward, it is necessary that a lighter fluid (e.g., gas) be allowed to move upward. The non-condensate gas gradually rises and gathers at the top of the reservoir, which can effectively decrease heat transfer between the steam chamber and the caprock. Thus, the expansion of steam chamber toward the upper side gradually decreases, and steam is mainly utilized to heat oil and sands on either side of the steam chamber. A comparative picture of SAGD and SAGP is illustrated in Figure 5.16. The non-condensate gas plays the vital role in the maintenance of the steam chamber pressure and, hence, to reduce the steam consumption. The fingering of non-condensate gas is also beneficial to increase the mobility of the steam front.

5.5.7 GAS-ASSISTED STEAM INJECTION

The major challenges of the steam injection process involve thermal losses to the surrounding rocks, huge consumption of steam, high water cut, and concomitant poor production performance. Numerous studies have suggested that these problems can be mitigated to some extent by injecting flue gas or CO_2 gas (Wan et al., 2020; Wang et al., 2017). The main components of flue gas are 80%–85% of N_2 and 15%–20% of CO_2. The heat loss of CO_2 is much lower than that of steam, and it can penetrate deep into the formation and expand the steam chamber. CO_2 also has good solubility with the oil and, thus, reduce the viscosity and IFT, resulting in an improvement of displacement efficiency. N_2 has the effect of increasing reservoir energy and heat insulation and volume expansion. To inject a flue gas with steam into the reservoir not only improves the oil recovery of steam flooding but also reduces greenhouse gas emissions.

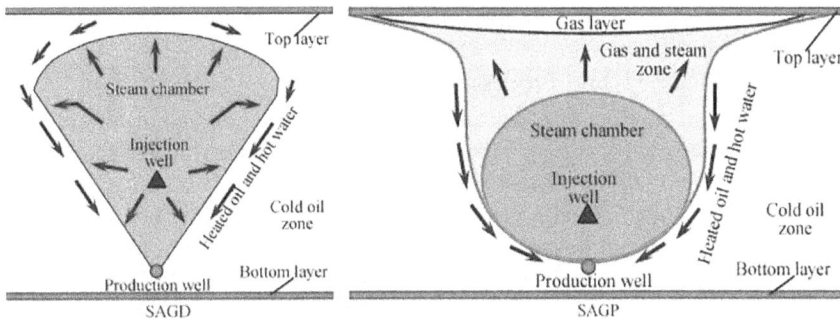

FIGURE 5.16   Mechanisms of steam and gas push.

Source: Pang et al., 2017.

5.5.8 Hybrid Steam Injection

The process that includes the injection of hot water, solvents, and other chemicals like surfactants/alkalis, etc. after steam flooding to improve efficiency of steam flooding and reduce water consumption and heat loss is called hybrid steam injection.

5.5.8.1 Steam Flooding Followed by Hot Water Injection

In the hybrid steam/hot water injection process, steam injection is followed by hot water injection to displace the bypassed residue from the reservoir. During the late stages of steam injection, the formation of steam channels leads to stream override, resulting in a decrease in the oil production and a subsequent increase in the water cut (Wu and Liu, 2019). Thus, the injection of hot water after steam flooding is an effective EOR technique for the production of heavy oil, to sweep the oil bypassed by the steam.

The injection of hot water maintains the increased temperature of the reservoir and utilizes the residual heat of the steam. The reduction of the viscosity of heavy crude oil with increase in temperature is the main mechanism for EOR in a hybrid steam/hot water injection. The reduction of the viscosity of crude oil also improves the mobility ratio and reduces the chances of viscous fingering in the reservoir. Thus, the areal sweep efficiency of the EOR process is significantly improved in comparison to steam flooding. The heat added to the reservoir by a hybrid steam/hot water injection also improves the relative permeability, leading to a reduction of residual oil saturation (Wang et al., 2018). The heat loss to the surrounding region is also lower in comparison to steam flooding due to the lower injection temperature of hot waterflooding. In addition, hot water also provides the displacement energy, leading to maintenance of the reservoir pressure.

The hybrid steam/hot water injection can also incorporate cyclic injection of hot water and steam into the reservoir, also known as water alternating steam process (WASP) (Lee and Lee, 2019). WASP prevents the early breakthrough of steam, as the injected water reduces the channeling tendencies of steam (Hong and Stevens, 1992). The coupling of WASP method to other methods such as low-salinity waterflooding (LSWF) has also shown encouraging results, as it combines their respective EOR mechanisms (Al-Saedi et al., 2019).

5.5.8.2 Solvent-Steam Hybrid Process

In the hybrid steam injection process, different solvents are used in addition to steam. The energy requirement for the conventional steam injection processes is quite high, and requirement of fresh water is also high. Combining technologies in the form of hybrid steam-solvent processes offers dual benefits: the potential of higher oil rates and recoveries but at a lower energy and water consumption level than conventional steam injection processes such as SAGD. Steam and solvent are generally injected in continuous or alternating schemes. The solvent choice for hybrid steam/solvent injection is not solely dependent on the mobility improvement capability of the solvents but also the reservoir properties and operational conditions such as operating pressure and injection strategy. Intermediate hydrocarbons are generally used as solvents for such hybrid processes.

5.5.8.3 Chemical Steam Hybrid Process

The efficiency of steam flooding also can be improved by injecting chemical additives such as alkalis, surfactants, and polymers along with the steam. The additional mechanisms of the hybrid steam alkaline process (HASP) over conventional steam flooding involve emulsification, wettability alteration, IFT reduction, and rigid film breaking (Okoye and Tiab, 1982).

In the hybrid steam surfactant process (HSSP), a small amount of surfactant is co-injected with the steam, and the mechanisms involve IFT reduction, wettability alteration, oil relative permeability enhancement, and in situ emulsification (Wu et al., 2018).

Some experimental studies reported that the injection polymer can improve the SAGD performance in oil sand reservoirs with top water, where a stable high viscosity layer can be developed at the bottom of the top water (Zhou and Zeng, 2014).

Wu et al. (2015) reported a case study of hot water foam flooding in the Wutonggou formation in Block A of the Xinjiang oil field with the reservoir properties given in the following. Based on detailed geological modeling, they performed the pilot test and forecast that the oil recovery could reach 34.8%, which is 14.49% higher than in conventional waterflooding.

Reservoir characteristics:

Heterogeneous under fluvial environment
Buried depth 1,635–1,676 m
Average pay thickness = 26.13 m
Average permeability = 34 mD
Average porosity = 18.96%
In situ oil viscosity = 424.87 cP (mPa s)

CO_2-assisted steam flooding in J6 block in Xinjiang, China (Xi et al., 2019), was conducted, and it was found that the process is significantly better than pure steam injection, with the steam oil ratio decreased from 8.4 m³ (CWE)/t to 6.25 m³ (CWE)/t, which reduced by 25.60%, with the recovery degree increasing by 5.4% from 16.1% to 21.5%, and the ultimate recovery rate of CO_2-assisted steam flooding being 66.5%.

5.5.9 SCREENING OF STEAM INJECTION

The specific screening criteria for steam flood varies widely based on the combination of the deciding factors—i.e., depth, porosity, permeability of the reservoir, API gravity, viscosity, density of crude oil, etc. Several authors have reported different screening criteria for steam flood projects as summarized in Table 5.2.

5.5.10 CASE STUDIES

(a) The feasibility study of steam flooding at the "Llanos Orientales" and "Magdalena valley medium" basin of Colombia is reported by Trigos et al. (2010). The field was developed using cyclic steam injection, showing recovery factors of ~15%, indicating huge residual oil saturation. There are two producing zones: zone A (depth: 1,400–1,800 ft., permeability: greater than 1,080 mD, porosity up to 29%) and zone B (depth: 1,920–2,050 ft., permeability: greater than 780 mD, porosity up to 28%) with maximum value of viscosity 1,500 cp. The maximum recovery is reported as 39.26% with injection rate 600 bbl/day. Zone B with low permeability was suggested for selective steam injection.

(b) A simulation study of the steam flooding process in the shallow, thin, extra heavy oil reservoir of Junggar basin is reported by Zan et al. (2010). The depth of the reservoir is less than 600 m, and average thickness is around 25 m, with the maximum oil viscosity around 100,000 cP at 20°C. By steam flooding with vertical injection–vertical production well scheme and vertical injection–horizontal production well scheme, the recovery reaches ~70% at the end of eighth year.

(c) The prospect of steam flooding in a low-permeability sandstone formation of the Daqing oil field was reported by Wu et al. (2019). The reservoir has the following properties: depth is 800 m; average permeability is around 8 mD, typical porosity of 16%, and oil viscosity roughly of 40 cP (m-Pa-s). Initially, waterflooding was carried out, but the recovery declined rapidly. Thereafter, promising results were obtained for steam flooding in a pilot test with predicted incremental recovery over 10%.

(d) The steam flooding development for an oil field discovered by the former Soviet Union in the Buzachi uplift of the North Ustyurt basin in the 1970s is reported by Tao et al. (2019). From top to bottom, stratigraphic sequence mainly develops: quaternary (Q), neogene

TABLE 5.2

General Screening Guide for Steam Flooding (Hama et al., 2014)

Author	Year	°API	μ_o cP	Ø %	S_o saturation, %	K md	T °F	Depth ft.	Height ft.	kh/μ_o (md-ft./cP)	$\varnothing S_o$
Farouq Ali (1974a)	1974	12–25	<1,000	≥30	1,200–1.700 bbl/ac-ft	~1,000		<3,000	≥30	-	0.15–0.22
Lewin & Assocs (1976)	1976	>10	NC	-	>50	NC	NC	<5,000	>20	>100	>0.065
Iyoho (1978)	1978	10–20	200–1,000	≥30	>50	>1,000		2,500–5,000	30–400	>50	-
Chu (1985)	1985	<36		>20	>40			>400	>10		>0.08
Brashear and Kuuskraa (1978)	1978	>10	NC	-	42	NC	NC	<5,000	>20	>100	>0.065
Taber et al. (1997a)	1997	8–25	<100,000	-	>40	>200	NC	<5,000	>20	>50	-
Dickson et al. (2010)	2010	8–20	1,000–10,000	-	>40	>250		400–4,500	15–150	-	-
Aladasani and Bai (2010)	2010	8–30	5×10^{-6}–10^3	12–65	35–90	1–15,000	10–350	200–9,000	>20	-	-

Note: NC, Not critical

(N2), and early paleogene (E3) of the cenozoic, lower cretaceous (K1), middle jurassic (J2) and lower triassic (T1) of the mesozoic. The reservoir properties are as follows: porosity is between 29% and 39%, with an average of 34.5%; permeability is between 330 and 1,936 mD, with an average of 950 mD; oil saturation ranges from 45% to 85%, with an average of 65%; crude oil viscosity is 200 cP (m-Pa-s) at 50°C. The analysis and evaluation of various measures to enhance recovery efficiency at the medium and late period of steam flooding development are summarized in Table 5.3.

Figure 5.17 shows one example of cumulative oil and oil rate versus time with and without steam injection for an inverted five-spot pattern. It can be seen that cumulative oil after 16 years of

TABLE 5.3

Simulation Results of Different Development Modes (Tao et al., 2019)

Development Mode	Production Time, d	Cumulative Steam Injection, 10³ ton	Cumulative Hot Water Injection, 10³ ton	Cumulative Oil Production, 10³ ton	Oil Recovery Ratio, %
Depleting development mode	3,653	–	–	272.5	21.91
Continue steam flooding	3,653	882	–	553.6	44.51
Intermittent steam injection of steam flooding	3,653	648.7	–	597.1	48.01
Hot waterflooding	3,653	526.4	119.4	587.4	47.23
Water steam alternating slug flooding	3,653	2,114.7	1,057	635.7	51.11

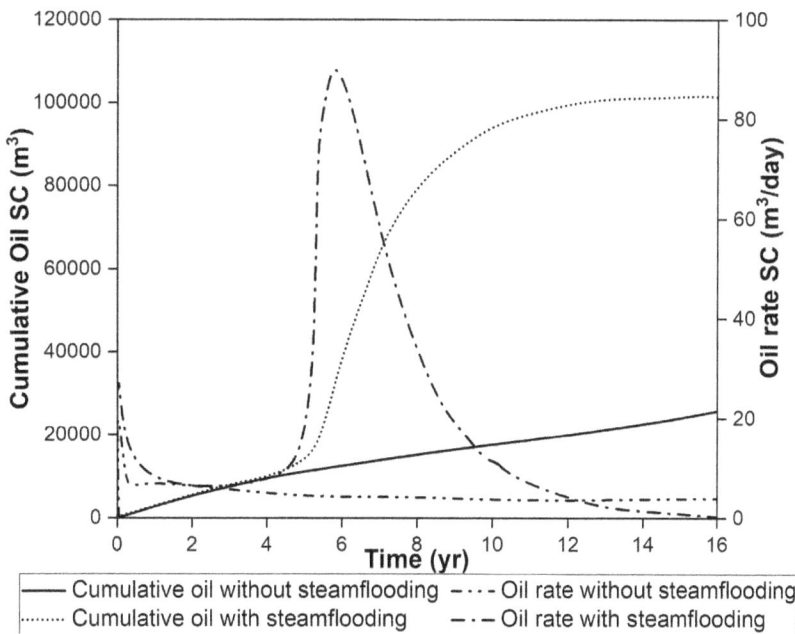

FIGURE 5.17   Cumulative oil and oil rate versus time with and without steam injection for an inverted five-spot pattern.

Source: Barillas et al., 2008.

production is much higher (100,000 SCM) in the case of steam injection as compared to that without steam injection (20,000 SCM). Oil saturations behind the steam zone can be as low as 5%.

Safety measures: It is very important that the oil field staff is familiarized with the operation of the steam generator to maintain efficiency. High-temperature operations are always risky from a safety point of view. Thermal operations require higher attention from engineering and operation staff to keep them efficient.

Economics: High-viscosity oils, usually considered for thermal projects, are also the ones sold at a lower price. Successful projects generally use centralized facilities to reduce the cost of production and steam generation (Abdulkadir et al., 2017).

5.6 IN SITU COMBUSTION

In situ combustion, also known as fireflooding, is a thermal recovery process, where the heat is generated within the reservoir, leading to heat production at a cheaper rate in comparison to the surface steam generators (Moore et al., 1984). It has been used for more than nine decades with many economically successful projects. In situ combustion is regarded as a high-risk process by many, primarily because of the many failures of early field tests. Most of those failures came from the application of a good process to the wrong reservoirs or the poorest prospects.

The performance of in situ combustion process depends on the (1) proper understanding of the constitutive chemical and physical processes that determine oil displacement through the swept area of the reservoir; and (2) understanding how fluid flow and reservoir characteristics determine swept volume.

Heat is produced by the ignition of crude oil and the injection of air or air enriched with oxygen to causes the fire to sustain as a result of the chemical reaction between oxygen and the oil. In situ combustion combines several driving mechanisms, which include the following (Ahmed and Meehan, 2012):

a) steam drive;
b) hot and cold waterflood; and
c) miscible and immiscible flood.

The oil that has not been contacted by the actual combustion front is subjected to the following:

a) condensing steam drive;
b) gas drive;
c) miscible drive; and
d) thermal drive.

There are two types of in situ combustion: forward combustion and reverse combustion. Also considering the presence and absence of water, the in situ combustion may also be wet combustion or dry combustion, respectively.

A schematic with the several distinct zones formed in an oil reservoir during the forward combustion process is shown in Figure 5.18.

5.6.1 FORWARD COMBUSTION

In forward combustion, the direction of the injection of air is the same as that of the advancement of the fire (Jamaluddin et al., 2018). This method involves the generation of heat within the reservoir in the form of a hot front moving away from the injection well. Initially, a heater or igniter is operated until ignition starts. The heater/igniter is subsequently withdrawn, and air injection is continued to maintain the advancement of a combustion front. The exothermic combustion reactions can provide enough heat to mobilize heavy crude oil.

During the process of forward combustion, several regions of different temperatures are formed between the injection and production well. Figure 5.19 demonstrates the oil displacement

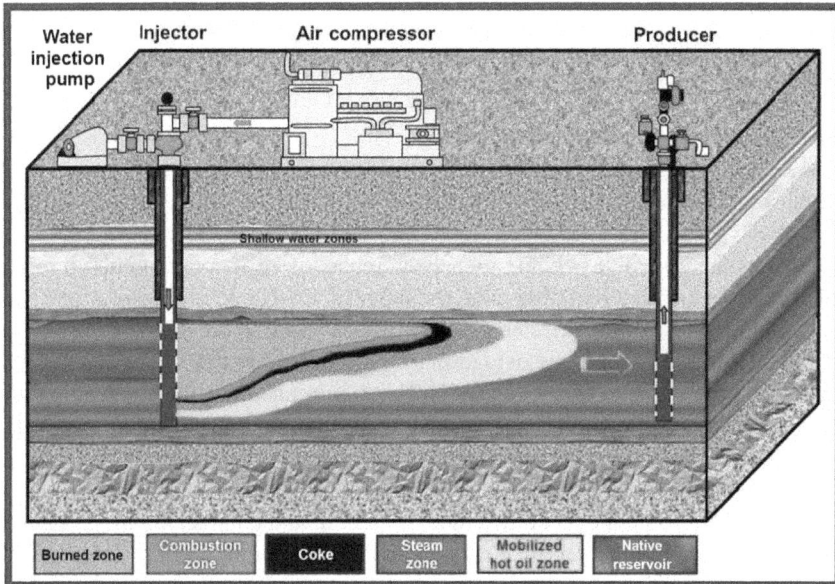

FIGURE 5.18 Schematic of forward in situ combustion.

Source: https://slideplayer.com/slide/13124249/.

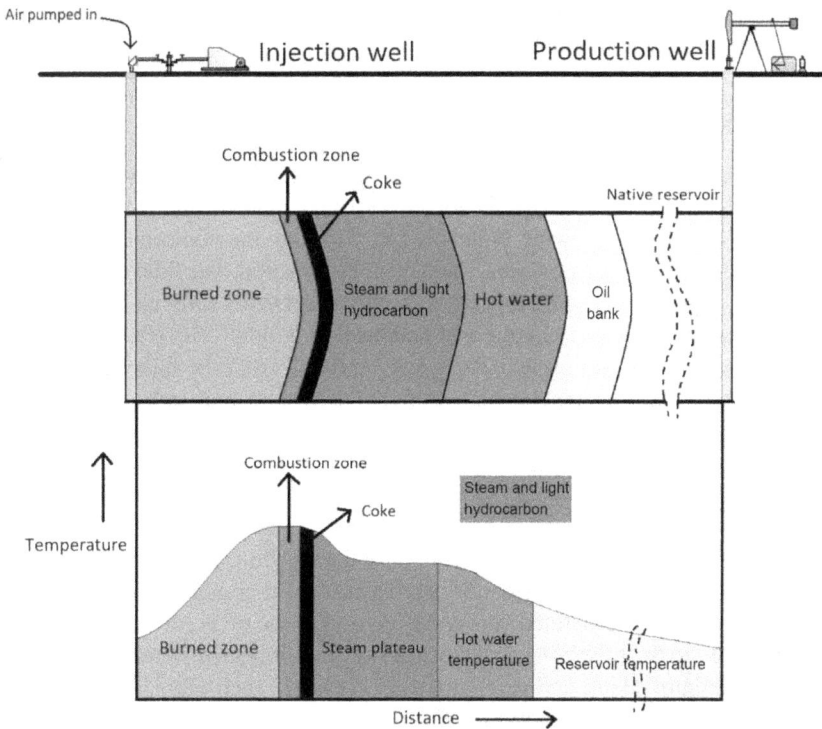

FIGURE 5.19 Oil displacement mechanism and temperature profile of in situ combustion.

mechanism and temperature profile associated with each of these zones. The zones are distinguished as follows:

A. Burned zone
B. Combustion front zone
C. Coke zone
D. Vaporizing zone
E. Condensing zone
F. Oil bank zone
G. Original reservoir zone

The burned zone is the zone left behind, after the combustion of its oil contents. It might contain residual unburned organic material along with the injected air. The temperature of the burned zone varies from the temperature of the injected air at the injector well to the temperature of the combustion zone. In the combustion zone, the oxidation of the crude oil with oxygen takes place, causing the temperature to reach around $600°F–1,200°F$. The region ahead of the combustion zone represents the coke zone, where coke has been deposited by the thermal cracking of crude oil. The coke is usually the residual crude oil components that have a high boiling temperature and consists of high molecular weight hydrocarbons. These represent around 20% of the crude oil.

The vaporized lighter components of the crude oil, steam, and combustion by-products are accumulated ahead of the coke in the vaporizing zone, and the temperature of this zone is high enough for the reservoir water to be in the vapor phase. Proceeding further, a reduction in the temperature causes the steam to condense and form a hot waterflood. The lighter hydrocarbons are also condensed in this zone and form a miscible drive with the crude oil (Moore et al., 1995). The combustion gases also provide additional driving force by a gas drive mechanism. These drive mechanisms lead to the formation of an oil bank, where the temperature is slightly above initial reservoir temperature. The initial reservoir temperature is maintained in the foremost section of the reservoir, which is unaffected by the combustion process.

5.6.2 DRY COMBUSTION

In the dry forward combustion process, much of the heat generated during burning is stored in the burned sand behind the burning front and is not used for oil displacement. No water is injected into the reservoir in dry combustion. Owing to its low heat capacity, the injected dry air cannot transfer heat from the sand matrix as fast as it is generated from combustion. The fluid around the production well remains at the original reservoir temperature and, hence, is very difficult to displace because of its high viscosity. So the efficiency of the dry combustion is poor. The temperature distributions of dry combustion are shown in Figure 5.20. It may be found from the figure that the temperature of the reservoir ahead of the combustion zone drops suddenly and approaches the original reservoir temperature.

5.6.3 WET COMBUSTION

To overcome the limitation of poor heat transfer in dry combustion, water is injected along with air in the wet combustion process. Contrary to dry combustion, wet combustion offers better recovery, as the maximum amount heat generated during the burning of the oil is carried by the injected water and utilized efficiently to heat the zone ahead of the fire front. Injected water absorbs heat from the burn zone, vaporizes into steam, and transports heat forward much more efficiently than dry air. Much of the heat stored in the burned sand can thus be recovered and transported forward together by water and air. Thus, the process combines the advantages of dry combustion and steam injection. The injection of water during the forward in situ combustion significantly increases the amount of

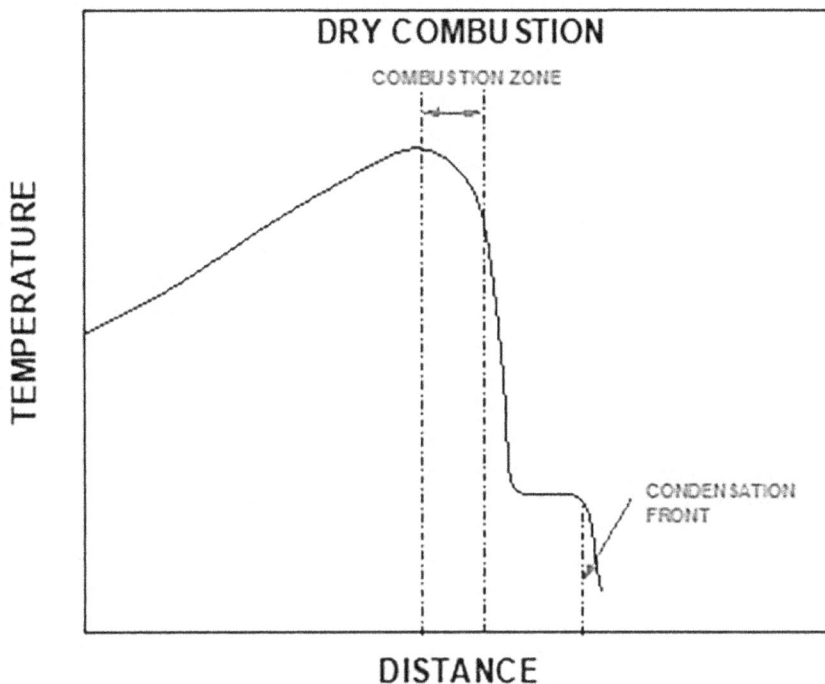

FIGURE 5.20 Schematic of the temperature profile for dry combustion.

heat transported by the fluids from the burned-out zone to the region downward from the combustion front. As a result of this mode of operation, the viscosity of the crude oil is substantially reduced, thereby permitting the transport of the viscous oil at low pressures. However, the process should not be used in formations where flow resistance is marginally acceptable for dry combustion, because the addition of water will increase the flow resistance further.

Depending on the variations of the water/air ratio in increasing order in the injection fluids, the process is categorized as normal wet, incomplete wet, and super wet combustion. In the case of wet combustion (normal or incomplete), the water that flows through the combustion zone is in the gas phase; that is, the peak temperature is higher than the vaporization temperature of water.

(a) Incomplete Wet Combustion Process: At a low water to air ratio, the injected water is converted to superheated steam as it moves toward the combustion front. In this case, the injected water fails to recover all the heat from the burned zone. As can be observed from Figure 5.21, the vaporization of water is delayed, and the temperature difference between the steam and hot water coming out of the combustion zone and the condensation front is less compared to that in the case of normal wet combustion.

(b) Normal Wet Combustion: In the case of normal wet combustion, the water flowing through the combustion zone absorbs the heat fully from this zone. The steam zone ahead of the combustion front is larger, and thus, the reservoir is swept more efficiently than with air alone. The improved displacement from the steam zone results in lower fuel availability and consumption in the combustion zone, so a greater volume of the reservoir is burned for a given volume of air injected. All the heat generated during burning is recovered and quenched by the injected water in this case. Water is generally injected at a higher rate in this process. The temperature profile of normal wet combustion is shown in Figure 5.22.

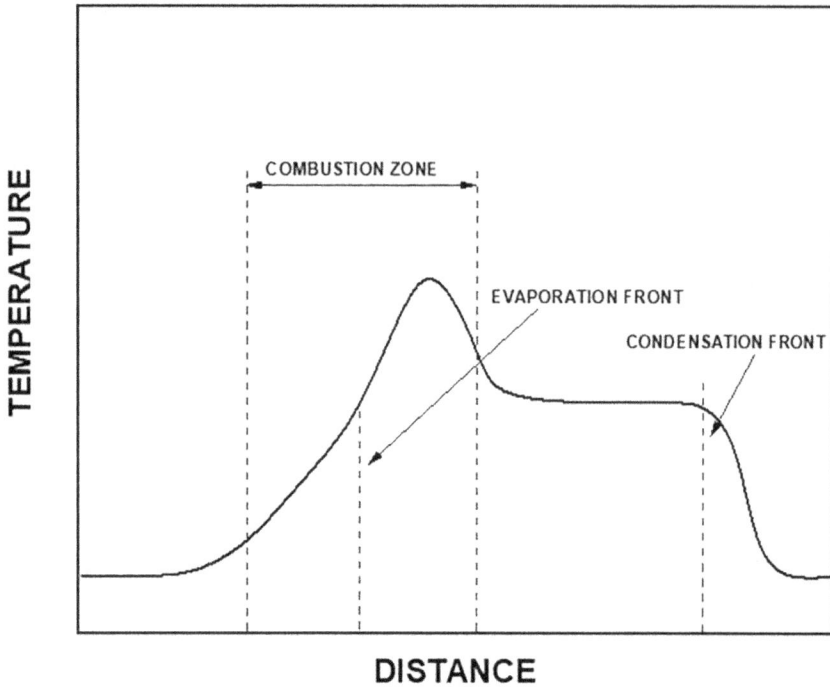

FIGURE 5.21   Schematic of the temperature profile for an incomplete (partially quenched) wet combustion process.

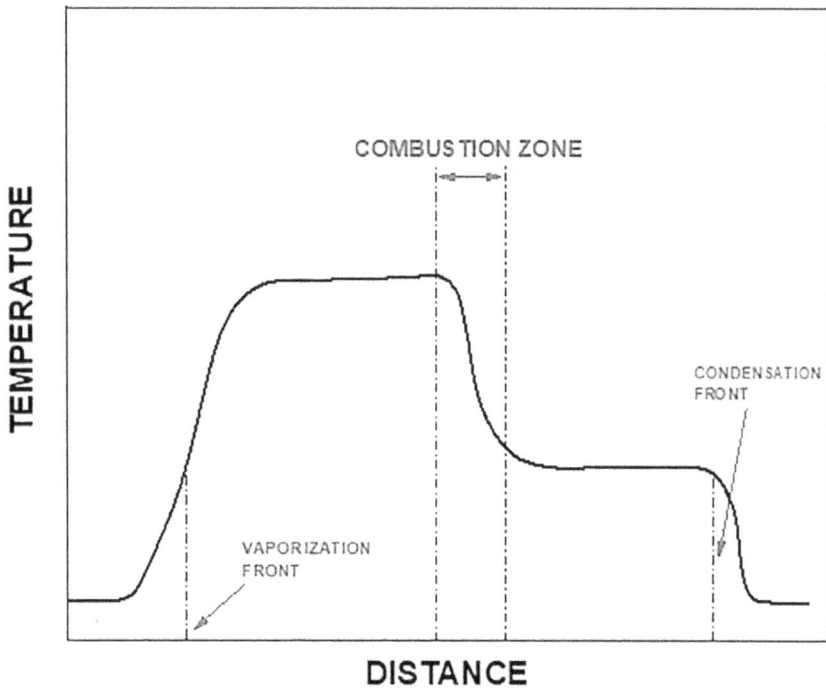

FIGURE 5.22   Schematic of the temperature profile for a normal wet combustion process without convective heat front.

(c) Super wet combustion: Super wet combustion results when the amount of heat available in the burned-out rock is too low to vaporize the water reaching the combustion
front; then the temperature peak disappears, and a vaporization-condensation front moves
through the porous media (Fig. 5.23). The process significantly decreases the air requirement but does not increase the oil recovery. The increased heat, transported by the steam
as it passes through to burn front, causes combustion temperatures to decrease. Super wet
combustion is more applicable in heavy oil reservoirs and less feasible for high gravity
oils with low fuel deposits.

Thus, it must be remembered that the influence of the water/air ratio on the amount of air required
for the combustion front to sweep a defined volume of the porous medium is of great importance for
estimating the benefit obtained by water injection (Burger and Sahuquet, 1973).

Basically, in a forward combustion process, air is injected to initiate the combustion and is then
followed by the injection of water in to the burning zone to create an in situ hot waterflood and
steam drive. This way, the residual heat behind the combustion front is utilized, thereby improving
the displacement efficiency. The combustion zones are established near the injection well and are
propagated through the formation by the continued injection of air, thus forcing the oil in the formation toward the producing well. The temperatures reached are often in the range of 1,000°F–1,100°F.
Partially quenched combustion is defined as the process where part of the available fluid is left
behind unburned. Usually, a super wet condition indicates that more water flows through the hot
zone than can be evaporated.

In wet combustion, the water injection results in decreased fuel content as a result of efficient
sweep in the steam plateau, which in turn, facilitates in maintaining the combustion front velocity
and increased oil recovery. Complete burning of coke is possible as long as a high-temperature zone
exists. However, it is to be kept in mind that too much water injection may absorb the heat and extin-
guish the burning front. The air volume requirement in wet combustion may be reduced up to 63%

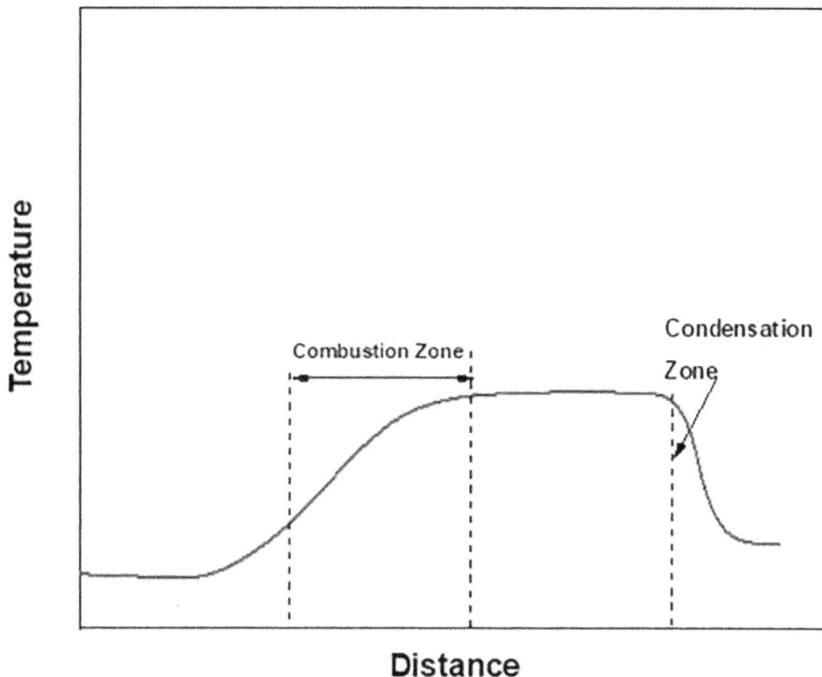

FIGURE 5.23   Schematic of the temperature profile for a super wet combustion process.

compared to dry combustion. As the major portion of the cost for a combustion project is to operate and maintain the air compressor, wet combustion offers significant economic incentives.

In situ combustion is recommended for reservoirs with high oil saturation, good porosity and permeability, and oils of moderate viscosity. Reservoirs with low ultimate primary production, with the crude occurring at a reasonable depth in a homogeneously consolidated sand, are particularly suitable to this method of recovery. Bottom water zone, lack of gas saturation, and low reservoir temperature are some of the detrimental aspects to the application of in situ combustion.

A properly designed and engineered in situ combustion can recover 50%–60% of the original oil. The gas produced in the process usually contains nitrogen, carbon dioxide, carbon monoxide, and unreacted oxygen. The injection gas is usually air. However, in a few cases, fuel gas and recycled gas, low in oxygen content, have been used to control the fireflood velocity.

Failure to stimulate production using in situ combustion may be due to a number of reasons such as 1) inability of the oil to deposit enough fuel to support combustion, 2) low air infectivity, 3) gross channeling and leaking of the injected air from the formation, 4) excessive air requirements, 5) low air saturation, and 6) plugging of porous rock, leading to the flow of a limited supply of air.

5.6.4 REVERSE COMBUSTION

The reverse combustion process is achieved by converting the injection well of a forward combustion process to a producer well. After initial ignition and burning of oil near the well, the injector well is put on production, and the air injection is continued form another adjacent well. The air injected from the well drives the oil through the combustion zone, whereas the combustion zone moves toward the injection well, opposite to the direction of air injection. The oil flows toward the production well through the combustion zone. A schematic of the temperature distribution for reverse in situ combustion is shown in Figure 5.24. Zone-1 represents the porous medium that has its

---

TABLE 5.4
Dry Combustion versus Wet Combustion

Dry Combustion	Wet Combustion
• In the dry forward combustion process, much of the heat generated during burning is stored in the burned sand behind the burning front and is not used for oil displacement.	• In wet combustion, water, on the other hand, can absorb and transport heat many times more efficiently than air.
• The heat capacity of dry air is low, and consequently, the injected air cannot transfer heat from the sand matrix as fast as it is generated.	• As water is injected together with air, much of the heat stored in the burned sand can be recovered and transported forward.
• The temperature of the reservoir after the combustion zone drops suddenly and approaches the original reservoir temperature.	• A higher temperature of the reservoir is retained after the combustion zone in contrast to dry combustion.
• Air volume requirement in dry combustion is much higher, and hence, the operational cost is high.	• Air volume requirement in wet combustion may be reduced up to 63% compared to dry combustion.

• Injection of water simultaneously or intermittently with air is commonly known as wet, partially quenched combustion.

• The ratio of the injected water rate to the air rate influences the rate of advance of the burning front and the oil displacement behavior.

• The injected water absorbs heat from the burned zone, vaporizes into steam, passes through the combustion front, and releases the heat as it condenses in the cooler sections of the reservoir.

• The growth of the steam and water banks ahead of the burning front are accelerated, resulting in faster heat movement and oil displacement.

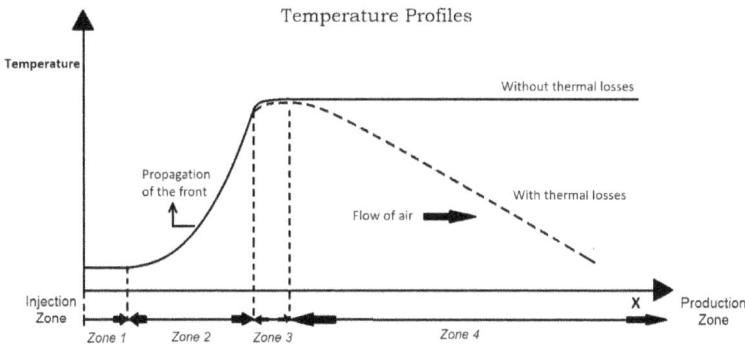

FIGURE 5.24 Schematic of the temperature distribution for reverse in situ combustion.

initial characteristics. As this area is swept by the oxygen containing gas, certain oxidation reactions can take place, if the desirable temperature is achieved. In zone-2, the temperature begins to increase due to conduction of heat from the hot zone. As the temperature in this zone increases water vaporization, distillation of the light oil fractions and oxidizing and cracking of some of the hydrocarbons may take place. Zone-3 is the combustion zone where the temperature reaches its maximum. The preheated oxygen from the previous zones is used up in the oxidation and combustion reactions of the more reactive hydrocarbon molecules. The unburned coke remains in zone-4 behind in the porous medium. The fluids in their gaseous and condensed states flow toward the producer. As no oil bank is formed, the total flow resistance decreases with time, and thus, this method is particularly suitable for reservoirs containing very viscous crude oils (Land et al., 1977). This method has the advantage that oil flows through a rock preheated from 500°F to 700°F during the initial combustion stage, thus reducing the viscosity of oil by a factor of 1,000 or more. Reverse combustion requires about twice the amount of air required in forward combustion but recovers about half the residual oil. However, reverse combustion has been difficult to apply and is economically unattractive in general. This method is usually resorted to when all other methods prove impractical.

5.6.5 CHEMICAL REACTIONS ASSOCIATED WITH IN SITU COMBUSTION

In in situ combustion, the composition of oil and rock mineralogy play a very important role in its success. In situ combustion is associated with a series of chemical reactions between the crude oil and the injected air within the reservoir at different temperature ranges. The reservoir rock minerals and clay content of the reservoirs influence the combustion reactions. The reactions that take place in in situ combustion can be categorized as follows:

1. Low-temperature oxidation (LTO);
2. Intermediate temperature, fuel formation reactions; and
3. High-temperature oxidation (HTO).

5.6.5.1 Low-Temperature Oxidation

LTO reactions generally occur at temperatures below 400°F. The reactions are heterogeneous (gas/liquid), and yield water and partially oxygenated hydrocarbons such as carboxylic acids, aldehydes, ketones, alcohols, and hydroperoxides (Burger and Sahuquet, 1972). Thus, LTO can be considered as an oxygen addition reaction. LTO occurs by dissolution of oxygen into crude oil. The rate of dissolution of oxygen depends on the degree of diffusion of the oxygen molecule into the crude oil at reservoir temperature. Light oils are more susceptible to LTO than heavy oils.

In LTO reactions, the lightweight molecules undergo condensation to form higher molecular weight components. Thus, the viscosity, boiling range, and densities of the original oil increase significantly in this process (Alexander et al., 1962). LTO reactions mostly yield asphaltene, with a reduction in the aromatic and resin content of the crude oil. The process increases the amount of fuel available for combustion and causes a substantial decline in the recoverable oil from the distillation and cracking zone.

5.6.5.2 Intermediate Temperature, Fuel Formation Reactions

It involves mostly cracking and pyrolysis of hydrocarbons, which leads to the formation of coke. The pyrolysis of crude oil in porous media often occurs in three overlapping stages: distillation, visbreaking, and coking. During distillation, most of the lighter and part of the medium gravity fractions of crude oil are distilled. At higher temperatures (400°F–540°F), mild cracking of the oil (visbreaking) occurs, in which the hydrocarbon loses small side groups and hydrogen atoms to form less-branched compounds that are more stable and less viscous. At still higher temperatures (>550°F), the oil remaining in the porous medium cracks into coke.

5.6.5.3 High-Temperature Oxidation

High-temperature oxidation (HTO) reactions occur between the oxygen of injected air and deposited coke at temperatures above 650°F. The principal products of these reactions are carbon dioxide (CO_2), carbon monoxide (CO), and water (H_2O). HTO reactions are heterogeneous (gas-solid and gas-liquid) reactions and are characterized by the consumption of all the oxygen in the gas phase. The combustion reaction is represented by the following equation:

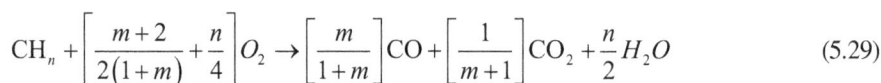

$$CH_n + \left[\frac{m+2}{2(1+m)} + \frac{n}{4} \right] O_2 \rightarrow \left[\frac{m}{1+m} \right] CO + \left[\frac{1}{m+1} \right] CO_2 + \frac{n}{2} H_2O \qquad (5.29)$$

where n denotes the atomic ratio of hydrogen to carbon, and m is the molar ratio of produced CO to CO_2.

The value of m is zero for complete combustion. The heat generated from these reactions provides the thermal energy to sustain and propagate the combustion front.

5.6.6 In Situ Combustion in Light Oil Reservoir

Though in situ combustion is most common for a heavy oil reservoir, it is increasingly getting attention in recovering light oil from deep reservoirs. Due to the differences of oil properties between light oil and heavy oil, the mechanisms of oil recovery are also different. For example, a reduction in oil viscosity by thermal effects is a critical factor to enhance the recovery of heavy oil, whereas the vaporization of light oil is the primary recovery method in LORs. The application of in situ combustion for LORs has significant practical merit, as oil mobility is not a limiting factor. The main concern with light oils is whether the residual oil saturation available to the combustion zone and the oxidation kinetics are such as to provide sufficient fuel to make the process self-sustaining. For light oils, the peak temperatures are expected to decline rapidly, as the burn progresses away from the injection well. Peak temperatures are expected to be much lower than those seen with heavier oils.

5.6.7 Air Requirement in In Situ Combustion

Several authors reported the performance of in situ combustion in terms of oxygen/air requirement (Nelson and McNeil, 1961; Gates and Ramey, 1980; Brigham et al., 1980). Nelson and McNeil (1961) provided a relatively simple method for air requirement in the combustion process based on experimental combustion tube data, which are discussed here. The different parameters used for calculations are as follows:

D = Internal diameter of the combustion tube, ft.
L = Length of the pack burned

$\varnothing$ = Porosity, fraction
V_g = Volume of produced gas, scf
N_{2a} = Volume fraction of nitrogen in injected air
O_{2a} = Volume fraction of oxygen in injected air
N_{2g} = Volume fraction of nitrogen in produced gas
O_{2g} = Volume fraction of oxygen in produced gas
CO_{2g} = Volume fraction of carbon dioxide in produced gas
CO_g = Volume fraction of carbon monoxide in produced gas

In the calculation it is assumed that nitrogen is inert and does not take part in the combustion reaction; hence, all the injected nitrogen gas is produced.

The general combustion reaction may be written as follows:

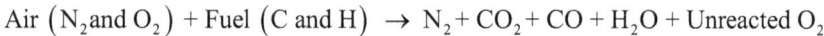

$$\text{Air } (N_2 \text{ and } O_2) + \text{Fuel } (C \text{ and } H) \rightarrow N_2 + CO_2 + CO + H_2O + \text{Unreacted } O_2$$

Weight of the carbon in the fuel burned, W_c = (CO_2 produced + CO produced) × (12/379*)
(*One mole of any gas occupies the same volume at standard condition: 379 cu. ft./lb.)

$$= \left[\left(V_g \times CO_{2g} \right) + \left(V_g \times CO_g \right) \right] \times \left(\frac{12}{379} \right) \text{ lb} \qquad (5.30)$$

Water formed by combustion, W_w

$$= 2[(\text{volume of oxygen injected} - \text{volume of unreacted oxygen produced})$$
$$- (\text{volume of } CO_2 \text{ produced}) - 0.5(\text{volume of CO produced})] \times 18/379$$

$$= 2 \left[\left(V_g \times N_{2g} \times O_{2a} / N_{2a} \right) - \left(V_g \times O_{2g} \right) - \left(V_g \times CO_{2g} \right) - 0.5 \left(V_g \times CO_g \right) \right] \times \left(\frac{18}{379} \right) \text{ lb} \qquad (5.31)$$

Hydrogen in the fuel burned, W_H

$$= 2 \left[\left(Oxygen\ injected - Unreacted\ oxygen\ produced \right) - \left(CO_2\ produced \right) \right.$$
$$\left. - 0.5 \left(CO\ produced \right) \right] \times \left(\frac{2}{379} \right)$$

$$= 2 \left[\left(V_g \times N_{2g} \times \frac{O_{2a}}{N_{2a}} \right) - \left(V_g \times O_{2g} \right) - \left(V_g \times CO_{2g} \right) - 0.5 \left(V_g \times CO_g \right) \right] \times \left(\frac{2}{379} \right) \text{ lb} \qquad (5.32)$$

$$\text{Total fuel consumed, } W_F = (W_c + W_H) \text{ lb} \qquad (5.33)$$

$$\text{Volume of sand burned, } V_b = \left(\pi \times \frac{D^2}{4} \right) \times L \qquad (5.34)$$

$$\text{Pounds of fuel consumed per cu. ft. of sand burned, } W = \left(\frac{W_F}{V_b} \right) \qquad (5.35)$$

Pounds of fuel consumed per acre-ft. of reservoir burned (W_R),

$$W_R = (43560 W) \times (1 - \varnothing_R) / (1 - \varnothing_P) \qquad (5.36)$$

where $\varnothing_R$ = porosity of the reservoir and $\varnothing_P$ = porosity of the sand pack

$$\text{Total air injection } V_a = \left(N_2 \text{ injected} + O_2 \text{injected} \right) \text{ scf}$$

$$= \left[\left(V_g \times N_{2g} \right) + \left(V_g \times N_{2g} \right) \left(O_{2a} \big/ N_{2a} \right) \right] \text{scf} \qquad (5.37)$$

The air injected per pound of fuel consumed $= V_a \big/ W_F \left(\dfrac{scf}{lb} \right) \qquad (5.38)$

Air injected per cubic feet of reservoir sand burned (A),

$$A = \left(V_a \big/ W_F \right) \times (W) \times \left[\left(1 - \varnothing_R \right) / \left(1 - \varnothing_P \right) \right]$$

$$= \left(4 V_a F \right) \big/ \left(\pi D^2 L \right) \left(\dfrac{scf}{cu.ft} \right) \qquad (5.39)$$

where $F = \left(1 - \varnothing_R \right) / \left(1 - \varnothing_P \right)$

Assuming an areal sweep efficiency of 62.6%, the air required in MMscf per acre-ft. in the five-spot pattern is computed as follows:

$$\text{Air injected per acre-ft of pattern} = \left(0.626 \times 43560 \times A \times 10^{-6} \right) \left(\dfrac{MMscf}{acre\text{-}ft} \right) \qquad (5.40)$$

Total air requirement for a given pattern will be as follows:

$$\left(0.626 \times 43560 \times A \times 10^{-6} \right) \times Acres \times h \left(MMscf \right)$$

where h is the thickness of the reservoir.

Example 5.6: For lab studies on in situ combustion in combustion tube, the following data are available:
Combustion tube internal diameter = 1 ft.
Length of burned pack = 10 ft.
Porosity of the pack = 30%
Produced gas volume (dry basis) = 200 scf
Composition of injected air (volume percent): Nitrogen = 79; Oxygen = 21
Composition of produced gas (volume percent): Nitrogen = 83.9; Oxygen = 1.5; Carbon dioxide = 12.1 and Carbon Monoxide = 2.5.
Reservoir data:
Reservoir area = 7 acres
Formation thickness = 70 ft.
Porosity of reservoir = 22%
Volumetric sweep of burned zone = 60%
Calculate the volume of air required for the pattern in in situ combustion.
From Equation 5.37, the total air injection $V_a = \left(N_2 \, injected + O_2 \, injected \right) scf$

$$= \left[\left(V_g \times N_{2g} \right) + \left(V_g \times N_{2g} \right) \left(O_{2a} \big/ N_{2a} \right) \right] \text{scf}$$

$$= \left[\left(200 \, scf \times 0.839 \right) + \left(200 \, scf \times 0.839 \right) \left(21 \big/ 79 \right) \right]$$

$$= 167.8 + 44.605$$

$$= 212.405 \text{ scf}$$

From Equation 5.31, the water formed by combustion

$$= W_w = 2\left[\left(Oxygen\,injected - Unreacted\,oxygen\,produced\right) - \left(CO_2\,produced\right)\right.$$
$$\left. - 0.5\left(CO\,produced\right)\right] \times \left(18/379\right)$$

$$= 2\left[\left(V_g \times N_{2g} \times O_{2a} / N_{2a}\right) - \left(V_g \times O_{2g}\right) - \left(V_g \times CO_{2g}\right) - 0.5\left(V_g \times CO_g\right)\right] \times \left(18/379\right) \text{ lb}$$

$$= 2\left[\left(200 \times 0.839 \times 21/79\right) - \left(200 \times 0.015\right) - \left(200 \times 0.121\right) - 0.5\left(200 \times 0.025\right)\right] \times \left(18/379\right) \text{ lb}$$

$$= 2\left[44.605 - 3.00 - 24.20 - 2.50\right)\right] \times \left(18/379\right) \text{ lb}$$

$$= 1.4157 \text{ lb.}$$

From Equation 5.32, the hydrogen in the fuel burned $= W_H =$

$$= W_H = 2\left[\left(\left(Oxygen\,injected - Unreacted\,oxygen\,produced\right) - \left(CO_2\,produced\right)\right.\right.$$
$$\left.\left. - 0.5\left(CO\,produced\right)\right] \times \left(2/379\right)\right.$$

$$W_H = 2\left(V_g \times N_{2g} \times O_{2a}/N_{2a}\right) - \left(V_g \times O_{2g}\right) - \left(V_g \times CO_{2g}\right) - 0.5\left(V_g \times CO_g\right) \times \left(2/379\right) \text{ lb}$$

$$= 2\left[\left(200 \times 0.839 \times 21/79\right) - \left(200 \times 0.015\right) - \left(200 \times 0.121\right) - 0.5\left(200 \times 0.025\right)\right] \times \left(2/379\right) \text{ lb}$$

$$= 0.1573 \text{ lb}$$

From Equation 5.30, the carbon in the fuel burned $= W_c = (CO_2$ produced $+ CO$ produced$) \times (12/379)$

$$= \left[\left(V_g \times CO_{2g}\right) + \left(V_g \times CO_g\right)\right] \times \left(12/379\right) \text{ lb.}$$

$$= \left[\left(200 \times 0.121\right) + \left(200 \times 0.025\right)\right] \times \left(12/379\right) \text{ lb.}$$

$$= \left[24.20 + 5.0)\right] \times \left(12/379\right) \text{ lb.}$$

$$= 0.9245 \text{ lb.}$$

Therefore, the total fuel consumed, $W_F = (W_c + W_H)$ lb. $= (0.9245 + 0.1573)$ lb. $= 1.0818$ lb.

Volume of sand burned, $V_b = \left(\pi \times D^2/4\right) \times L = \left(3.14 \times 1/4\right) \times 10$ ft.^{3} = 7.85 ft.^{3}

$F = \left(1 - \varnothing_R\right) / \left(1 - \varnothing_P\right) = \left(1 - 0.22\right) / \left(1 - 0.30\right) = 1.1142$

From Equation 5.39, the air injected per cubic feet of reservoir sand burned

$$A = \left(V_a/W_F\right) \times (W) \times \left[\left(1 - \varnothing_R\right) / \left(1 - \varnothing_P\right)\right]$$

$$= \left(V_a F\right)/\left(V_b\right) \left(\frac{scf}{cu.ft}\right)$$

$$= \frac{(212.405 \times 1.1142)}{(7.85)} \left(\frac{scf}{cu.ft} \right)$$

$= 30.1479$ scf/cu. ft.

Therefore, air injected per acre-ft of pattern $= \left(0.60 \times 43560 \times A \times 10^{-6}\right) \left(\frac{MMscf}{acre-ft} \right) \times 7 \ acres \times 70 \ ft$

$= 386.09$ MMscf

5.6.8 TOE-TO-HEEL AIR INJECTION

Toe-to-heel air injection (THAI) is a new method within the category of in situ combustion for extracting oil from heavy oil deposits, which may have significant advantages over existing methods (Xia et al., 2003). The main limitations of conventional in situ combustion process are as follows:

a) gravity segregation, or gas overriding;
b) oil banking downstream in the cold region; and
c) permeability heterogeneity.

Also, in common in situ combustion processes, the producer and injectors are vertical, which leads to a limited sweep efficiency caused by overrunning or channeling. These problems can be resolved by drilling a horizontal well in combination with a vertical well. The THAI process consists of at least one vertical air or air/water injector, perforated in the upper part of the oil layer, and a horizontal producer placed at the lowest possible part of the oil layer, with its toe facing the injector.

A typical schematic of the THAI process is shown in Figure 5.25. The basic operations of THAI are similar to in situ combustion—i.e., a combustion front is created, where part of the oil in the

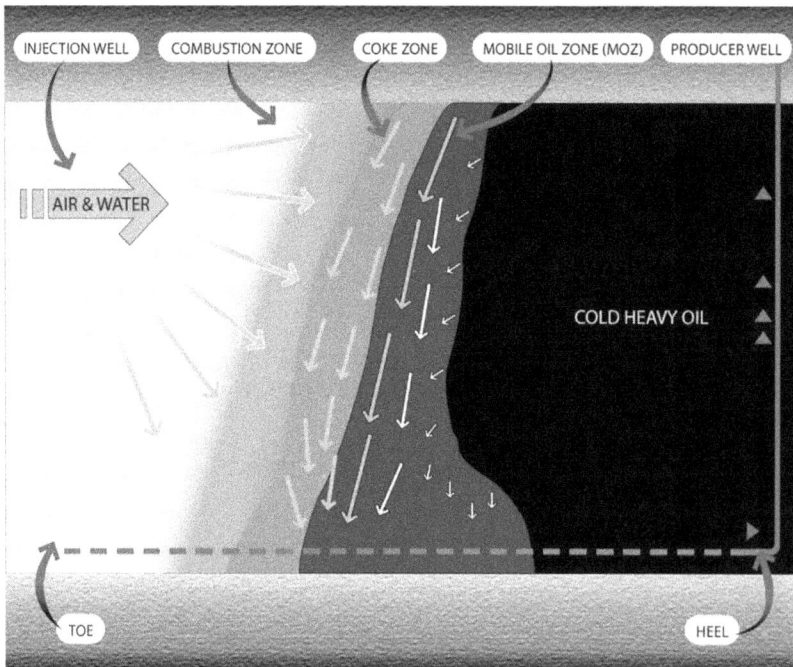

FIGURE 5.25 Mechanism for stability of toe-to-heel air injection.

Source: www.vistaprojects.com/projects/toe-to-heel-air-injection-thai/.

reservoir is burned, generating heat, which reduces the viscosity of the oil, allowing it to flow by gravity to the horizontal production well. In THAI, the temperature of the mobile oil zone is much higher than the initial reservoir temperature. Thus, the relative mobility ratio is much more favorable, because the viscosity of the displaced oil is hundreds of times lesser than that of the corresponding cold downstream region. In this process, the front propagation is more under control. The front sweeps the oil from the toe to the heel of the horizontal producing well, recovering an estimated 80% of the original oil in place while partially upgrading the crude oil in situ.

The benefits of THAI include the following:

- Increased control over the ISC front propagation, compared to the conventional ISC process; ISC front breakthrough is always at the toe, and there is progressive advancement toward the heel.
- Relatively less sensitive to rock heterogeneity (mainly to stratification).
- Easier to implement when heavy oil has some mobility at reservoir conditions; otherwise initial hot communication is needed.
- More resilient to thermal production operations (if damage is occurring, it is progressive from toe to heel).
- Easy to implement using existing facilities and horizontal wells (HW), especially if HWs are placed near the bottom of the oil layer.

5.6.8.1   Criteria for THAI Application

The selection of reservoirs for the application of the THAI process is based on the following conditions:

- The reservoir rock should be preferably sandstone, like in conventional in situ combustion.
- Existence of natural or hydraulic fracturing should be avoided.
- Pay zone thickness must be more than 6 m.
- The oil viscosity and density should be more than 200 m-Pa-s and 900 kg/m³, respectively.
- The horizontal and vertical permeability of the reservoir should be more than 200 and 50 mD, respectively, with the value of K_V/K_H greater than 0.25 to facilitate gravity segregation.
- Presence of bottom water is to be avoided. If there exists a bottom water zone, its thickness should not be more than 30% of the oil zone thickness.
- Water cut should be less than 70%.

5.6.9   Screening Criteria of In Situ Combustion

Proper selection of reservoirs for an in situ combustion project is dependent on the properties of formation rock and fluids, geology, and previous performance of the reservoir. A large number of successful in situ combustion projects has been reported for a wide variety of reservoir characteristics. Thus, it is very difficult to mention a selective guideline to screen reservoirs for the in situ combustion process. Still a general guideline is given in Table 5.5. The management should take the decision on implementation of the project after a successful result at the laboratory and pilot scales.

5.6.10   Case Studies

Case 1: The Suplacu de Barcau field in northwestern Romania is constituted of slightly shaly unconsolidated sands with an average to coarse grain size. The reservoir contains a heavy and viscous oil. Its properties are listed in Table 5.6 (Cadelle et al., 1981). Initially, the pilot test was conducted in dry combustion mode. The pilot test was successful, and the production rate of some wells increased from 3 to 100 m³/d (19 to 630 BID). Then the process was extended for commercial production, and the reported incremental recovery was more than 20% of the oil in place above the conventional method for the whole reservoir.

TABLE 5.5
Screening Criteria of In Situ Combustion (Sarathi, 1999)

Screening Criteria

Oil:

Viscosity: Preferably less than 5,000 cp at reservoir condition.

Gravity: 10°–40°API

Composition: low asphaltic, low heavy metal content crude. Heavy metal (Va, Ni, etc.) should be preferably less than 50 ppm

Water:

Connate water properties are not critical

Lithology:

Heavy oil: Low clay content; low in minerals that promote increased fuel formation such as pyrite, calcite, and siderite; low in heavy metals

Light oil: Lithology that tends to promote fuel deposition is preferred

Reservoir:

Depth: 300–12,500 ft.

Thickness: 5–50 ft.

Permeability: Not critical, but higher permeability is preferred; all successful projects have high permeability

Porosity: > 0.18

Oil concentration: 700 bbl/ac-ft, $S_o > 0.45$

Transmissibility: $\dfrac{kh}{\mu_o} \geq 20\,mD$ -ft./cp

Favorable Factors

1. High reservoir temperature
2. Low vertical permeability, except THAI
3. Good lateral continuity
4. Multiple thin sand layers
5. Good overburden competence
6. High dip
7. Uniform permeability profile

Risk Factors

- Extensive fractures
- Large gas caps
- Strong water drive
- Highly heterogeneous reservoir
- Early breakthrough of the combustion front

Operational problems, such as severe corrosion caused by low pH hot water, serious oil-water emulsions, increased sand production, deposition of carbon or wax, and pipe failures in the producing wells as a result of the very high temperatures

Limitations (Taber et al., 1997a, 1997b)

- If sufficient coke is not deposited from the oil being burned, the combustion process will not be sustained; this prevents the application for high-gravity paraffinic oils.
- If excessive coke is deposited, the rate of advance of the combustion zone will be slow and the quantity of air required to sustain combustion will be high.
- Oil saturation and porosity must be high to minimize heat loss to the rock.
- Process may tend to sweep through the upper part of a thick reservoir, resulting in poor sweep.

TABLE 5.6
Suplacu De Barcau Reservoir Properties (Cadelle et al., 1981)

Sand character	Unconsolidated
Depth, m (ft)	50–200 (164–656)
Net pay thickness, m (ft)	10 (32.8)
Reservoir temperature, °C (°F)	18 (64.4)
Porosity	0.32
Absolute permeability, md	1,700
Oil saturation	0.85
Oil specific gravity	0.96
Oil viscosity at 18°C (64.4°F) Pa·s (cp)	2 (2,000)
Dip angle (°)	2–4

TABLE 5.7
Reservoir Properties of Suplacu de Barcau Field
(Panait-Patica et al., 2006)

Type of trap	Fault-dependent closure
Depth, m	35–220
Dip, degree	5–8
Net pay thickness, m	4–24
Initial reservoir pressure, bar	4–22
Initial reservoir temperature, °C	18
Porosity, %	32
Initial oil saturation, %	85
Absolute permeability, mD	2,000
Oil dynamic viscosity, cP	2,000
Oil density, kg/m^3	960
Oil type	asphalt-base-oil

Case 2: The Suplacu de Barcau structure is situated in the northeast margin of the Panonian Depression. The structure forms a monocline, trending east-west, with a fault-dependent closure to the south (Panait-Patica et al., 2006). The properties of the reservoir are shown in Table. 5.7. The initial production by conventional solution gas drive mechanism declined rapidly. After successful pilot test results, commercial production was started with a predicted ultimate oil recovery of 55%.

Case 3: In situ combustion is implemented in Balol field, India (Dayal et al., 2012). The field is located in the heavy oil belt of the North Cambay basin in North Gujarat, India. It was discovered in 1978 and put on production in 1985. The peak oil production of 1,050 m^3/d was achieved by mid-1991. Subsequently, with increase in water cut to 75%, oil production from the Balol field gradually decreased to 400 m^3/d. The in situ combustion process was successfully applied in the southern part of the Balol field for many years. The pilot was initiated in 1990 followed by semicommercialization in 1992 and commercialization in 1997. It has been instrumental in enhancing the oil recovery from 13% to 50%.

The pay sands of the field has the following properties: loose formation; permeability in the range of 8–10 Darcy; porosity between 20% and 30%; the gravity of the oil is 15° API, while the viscosity varies from 150 cp in the south to 1,000 cp in the north.

5.7 ELECTROMAGNETIC HEATING: A METHOD OF THERMAL EOR

The significant presence of unconventional hydrocarbons plays a role in mitigating the issue of increasing energy demand on a global scale. Heavy crude oils with < 20° API gravity may be effectively produced by a promising method of thermal EOR, which employs the use of electromagnetic (EM) heating. This process involves the conversion of electrical energy at the surface to heat energy, generated either within the wellbore or formation. This energy is transmitted via metallic structures/cables into the reservoir, and consequently, the viscosity of heavy and extra heavy crude is reduced for beneficial displacement (Sahni et al., 2000; Sivakumar et al., 2020). EM heating is more eco-friendly and fast-acting compared to other conventional methods, which make it a potential route toward extraction of shale oil, heavy oil, bitumen, etc. This energy may be constricted to the pay zone of interest, while controlling the uniformity of the heating process from the surface with ease. A schematic heavy oil recovery by electrical heating is shown in Figure 5.26. All factors—including electricity supply from fossil fuel–based power plants, temperature control, heating distribution, and multidisciplinary issues—must be considered for effective research and development (Sivakumar et al., 2020; Chilingar et al., 1997; Ji et al., 2020). The EM heating method relies on the preferential adsorption of materials on the rock surfaces. With the application of EM waves, the mobile molecular dipole moments of polar groups within the porous media orient themselves with oscillating electric fields due to the torque effect. This causes association of the oscillating polar molecules with neighboring constituents to generate frictional heat and increase temperature (Bera et al., 2015). This behavior is gradually distributed throughout other materials present in the reservoir rock system and aids in the favorable extraction of heavy oil/bitumen from highly clayey reservoirs or shale or other unconventional reservoirs.

EM heating–assisted oil recovery technology is applicable in a wide range of heterogeneous reservoirs with varying permeability levels. In addition, it is characterized by a decreased usage of water/aqueous fluids, cost-profitability, and minimal environmental footprint. Unlike conventional thermal EOR techniques, this method can be employed in thin pay zones and is not controlled by the lithology and depth of the reservoir (Vinsome et al., 1994; Wilson, 2012). As a viable alternative to steam injection, EM heating commands a higher depth of penetration, with a strong crude oil attracting and displacing capacity. The field-scale performance of this method cannot be properly assessed with a limited number of laboratory tests alone. Further advances in the field of research are required to implement the enhanced metal-nanoparticle incorporating electromagnetic heating (EMNIEH) at pilot- and field-scale tests.

The characteristics of EM heating are different in a tight/shale environment containing light, medium, or heavy oils. This leads to a more homogeneous temperature distribution during the lifetime of a reservoir (Chakma and Jha, 1992; Mata and Mata, 2001). Studies by Bridges et al. (1979) and Kasevich et al. (1994) showed favorable recoveries of the order of 20%–35% in shale and tar sands under temperature conditions of 473–673 K in heavy oil fields of the United States. Sayakhov and coworkers (1980, 1992) investigated the radio-frequency mode of heating in the high-porosity, low-permeability conditions of Ishimbayskoye oil at the Bashkortostan and Yultimirovskoye bitumen fields, Russia. In more successful applications, the Wildmere field in Alberta successfully

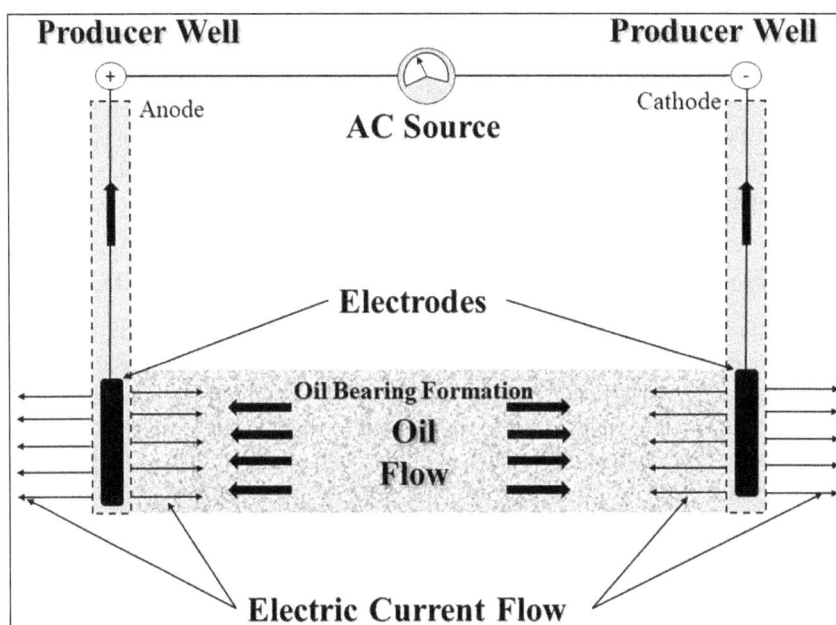

FIGURE 5.26 Schematic heavy oil recovery by electrical heating.

produced at a rate of 3.18 tons/day with EM heating by Spencer (1987). Another work by Davidson (1995) discussed the field recovery of sparky heavy oil formation to achieve 20 m³/day production rates in Northminster and Lashburn using 30 kW power EM-heating technique. The EM heating process is efficient in the respect that ~92% of the energy applied can be transferred to the reservoir (Sahni et al., 2000; Bera et al., 2015; Sivakumar et al., 2020). Thus, EM heating–based EOR results in a higher recovery factor and better production rate and pressure control as compared to other types of thermal recovery methods.

REFERENCES

Abdulkadir, U., Hashim, J., Alkali, M., Kumar, A., 2017. Application of thermal methods for heavy oil recovery: Phase one. International Journal for Advance Research and Development 2(5), 102–120.

Ahmed, T., Meehan, D.N., 2012. Advanced Reservoir Management and Engineering. Gulf Professional Pub.

Aladasani, A., Bai, B., 2010. Recent developments and updated screening criteria of enhanced oil recovery techniques. Society of Petroleum Engineers. DOI: 10.2118/130726-MS.

Alexander, J.D., Martin, W.L., Dew, J.N., 1962. Factors affecting fuel availability and composition during in-situ combustion. Journal of Petroleum Technology 14(10), 1156–1164.

Al-Saedi, H.N., Flori, R.E., Alkhamis, M., Brady, P.V., 2019. Coupling of low-salinity water flooding and steam flooding for sandstone unconventional oil reservoirs. Natural Resources Research 28, 213–221. https://doi.org/10.1007/s11053-018-9407-2

Alvarado, V., Manrique, E., 2010. Enhanced oil recovery: An update review. Energies 3(9), 1529–1575.

Andrade, E.N.C., 1930. The viscosity of liquids. Nature 125(3148), 309–310.

Baker, P.E., 1969. An experimental study of heat flow in steam flooding. Society of Petroleum Engineers Journal 9(1), 89–99.

Barillas, J.L.M., Dutra Jr, T.V., Mata, W., 2008. Improved oil recovery process for heavy oil: A review. Brazilian Journal of Petroleum and Gas 2(1).

Bera, A., Babadagli, T., 2015. Status of electromagnetic heating for enhanced heavy oil/bitumen recovery and future prospects: A review. Applied Energy 151, 206–226. https://doi.org/10.1016/j.apenergy.2015.04.031.

Brashear, J.P., Kuuskraa, V.A., 1978. The potential and economics of enhanced oil recovery. Journal of Petroleum Technology 30(9), 1231–1239.

Bridges, J., Taflove, A., Snow, R., 1979. Method for in situ heat processing of hydrocarbonaceous formation. US Patent Application, Serial No. 4,140,180 filed on 20 February.

Brigham, W.E., Satman, A., Soliman, M.Y., 1980. Recovery correlation for in-situ combustion field projects and application to combustion pilot. Journal of Petroleum Technology, 2132–2138.

Burger, J.G., Sahuquet, B.C., 1972. Chemical aspects of in-situ combustion—Heat of combustion and kinetics. Society of Petroleum Engineers Journal 12(5), 410–422.

Burger, J.G., Sahuquet, B.C., 1973. Laboratory research on wet combustion. Journal of Petroleum Technology 25(10), 1–137.

Butler, R., 1999. The steam and gas push (SAGP). Journal of Canadian Petroleum Technology 38(03).

Butler, R.M., 1994. Steam-assisted gravity drainage: Concept, development, performance and future. Journal of Canadian Petroleum Technology 33(2), 44–50.

Butler, R.M., Mcnab, G.S., Lo, H.Y., 1981. Theoretical studies on the gravity drainage of heavy oil during in-situ steam heating. Canadian Journal of Chemical Engineering 59(4), 455–460.

Cadelle, C.P., Burger, J.G., Bardon, C.P., Machedon, V., Carcoana, A., Petcovici, V., 1981. Heavy-oil recovery by in-situ combustion – Two field cases in Rumania. Journal of Petroleum Technology 33(11), 2057.

Chakma, A., Jha, K.N., 1992, 1 January. Heavy-oil recovery from thin pay zones by electromagnetic heating. Society of Petroleum Engineers. DOI: 10.2118/24817-MS.

Chilingar, G.V., Loo, W.W., Khilyuk, L.F., Katz, S.A., 1997. Electrobioremediation of soils contaminated with hydrocarbons and metals: Progress report. Energy Sources 19, 129–146. http://dx.doi.org/10.1080/00908319708908838.

Chu, C., 1985. State-of-the-art review of steamflood field projects. Journal of Petroleum Technology, 1887–1902.

Cosma, V., Mehaysen, A.M., 1990. The contribution of solid and fluid media thermal expansion to oil recovery by thermal methods. University School of Management Studies (USMS) 020398.

Davidson, R.J., 1995. Electromagnetic stimulation of Lloydminster heavy oil reservoirs: Field test results. Journal of Canadian Petroleum Technology 34(4), 15–24.

Dayal, H.S., Bhushan, B.V., Mitra, S., Pandey, V., Bhandari, A.C., Dwivedi, M.M., 2012, January. Simulation of In-Situ Combustion Process in Balol Pilot. In SPE Oil and Gas India Conference and Exhibition. Society of Petroleum Engineers.

Dehghani, K., Ehrlich, R., 1998, January. Evaluation of Steam Injection Process in Light Oil Reservoirs. In SPE Annual Technical Conference and Exhibition. Society of Petroleum Engineers.

Diaz-Munoz, J., Farouq Ali, S.M., 1975. Simulation of Cyclic Hot Water Stimulation of Heavy Oil Wells. p. SPE 5668-MS.

Dickson, J.L., Leahy-Dios, A., Wylie, P.L., 2010, April. Development of improved hydrocarbon recovery screening methodologies. In SPE Improved Oil Recovery Conference? SPE-129768. SPE.

Duerksen, J.H., Hsueh, L., 1983. Steam distillation of crude oils. Society of Petroleum Engineers Journal 23(2), 265–271.

Duval, K., Gutiérrez, D., Petrakos, D., Ollier, P., Johannson, D., 2015, June. Successful Application of Hot Water Circulation in the Pelican Lake Field-Results and Analyses of the E29 Hot Water Injection Pilot. In SPE Canada Heavy Oil Technical Conference. Society of Petroleum Engineers.

Egermann, P., Renard, G., Delamaide, E., 2001. SAGD Performance Optimization Through Numerical Simulations: Methodology and Field Case Example. In SPE International Thermal Operations and Heavy Oil Symposium. Society of Petroleum Engineers.

Farouq Ali, S.M., 1974a. Current status of steam injection as a heavy oil recovery methods. Journal of Canadian Petroleum Technology, 1–15.

Farouq Ali, S.M., 1974b. Heavy Oil Recovery – Principles, Practicality, Potential, and Problems. In Paper SPE 4935 Presented at the Rocky Mountain Regional Meeting, Billings, MT.

Flock, D.L., Le, T.H., Gibeau, J.P., 1986. The effect of temperature on the interfacial tension of heavy crude oils using the pendent drop apparatus. Journal of Canadian Petroleum Technology 25(2).

Gates, C.G., Ramey Jr., H.J., 1980. A method for engineering in-situ combustion oil recovery projects. Journal of Petroleum Technology, 285–294.

Goodyear, S.G., Reynolds, C.B., Townsley, P.H., Woods, C.L., 1996, 1 January. Hot Water Flooding for High Permeability Viscous Oil Fields. In SPE/DOE Improved Oil Recovery Symposium. Society of Petroleum Engineers.

Goyal, K.L., Kumar, S., 1989. Steamflooding for Enhanced Oil Recovery, in: Developments in Petroleum Science (Vol. 17). Elsevier, pp. 317–349.

Green, D.W., Willhite, G.P., 1998. Enhanced Oil Recovery (Vol. 6). Henry L. Doherty Memorial Fund of AIME, Society of Petroleum Engineers.

Hama, M.Q., Wei, M., Saleh, L.D., Bai, B., 2014. Updated Screening Criteria for Steam Flooding Based on Oil Field Projects Data. In SPE-170031-MS.

Holladay Jr., C.H., 1966, 10–11 November. The Basic Effects of Steam on a Reservoir. In Paper SPE 1666 Presented at the SPE Eastern Regional Meeting, Columbus, OH. DOI: 10.2118/1666-MS.

Hong, K.C., Stevens, C.E., 1992. Water-alternating-steam process improves project economics at west Coalinga field. SPE Reservoir Evaluation & Engineering 7, 407–413. https://doi.org/10.2118/21579-PA

IEA, 2017. World Energy Outlook 2017. IEA, Paris. https://www.iea.org/reports/world-energy-outlook-2017, License: CC BY 4.0

Iyoho, A.W., 1978. Selecting enhanced oil recovery processes. World Oil, 61–74.

Jamaluddin, A., Law, D.H.-S., Taylor, S.D., Andersen, S.I., 2018. Heavy Oil Exploitation. PennWell Corporation.

Ji, D., Harding, T.G., Chen, Z., Dong, M., Liu, H., Li, Z., Lai, F., 2020, 24 September. Feasibility of electromagnetic heating for oil sand reservoirs. Society of Petroleum Engineers. DOI: 10.2118/199910-MS.

Kasevich, R.S., Price, S.L., Faust, D.L., Fontaine, M.F., 1994, 25–28 September. Pilot Testing of a Radiofrequency Heating System for Enhanced Oil Recovery From Diatomaceous Earth. In Paper SPE 28619 Presented at the SPE 69th Annual Technical Conference and Exhibition, New Orleans, LA.

Kumar, S., Kumar, P., Tandon, R., Beliveau, D., 2008, January. Hot Water Injection Pilot: A Key to the Waterflood Design for the Waxy Crude of the Mangala Field. In International Petroleum Technology Conference. International Petroleum Technology Conference.

Land, C.S., Cupps, C.Q., Marchant, L.C., Carlson, F.M., 1977. Field test of reverse-combustion oil recovery from a Utah Tar Sand. Journal of Canadian Petroleum Technology 16(2).

Lee, K.S., Lee, J.H., 2019. Hybrid Thermal Recovery Using Low-Salinity and Smart Waterflood, in: Hybrid Enhanced Oil Recovery Using Smart Waterflooding. Elsevier, pp. 129–135. https://doi.org/10.1016/B978-0-12-816776-2.00006-4

Lewin & Associates, Inc., 1976, April. The Potentials and Economics of Enhanced Oil Recovery. Report B76/221. Federal Energy Administration.

Lyons, W.C., Plisga, G.J., 2005. Standard Handbook of Petroleum and Natural Gas Engineering. Gulf Professional Publishing, pp. 2–20, 22.

Mata, W., Mata, A.L., 2001, 1 January. Electromagnetic heating process combined with water displacement for recovering petroleum reservoirs-a new concept. Petroleum Society of Canada. DOI: 10.2118/2001-020-EA.

Mokheimer, E., Hamdy, M., Abubakar, Z., Shakeel, M.R., Habib, M.A., Mahmoud, M., 2019. A comprehensive review of thermal enhanced oil recovery: Techniques evaluation. Journal of Energy Resources Technology 141(3).

Mollaei, A., Maini, B., 2010. Steam flooding of naturally fractured reservoirs: Basic concepts and recovery mechanisms. Journal of Canadian Petroleum Technology 49(1), 65–70.

Moore, R.G., Bennion, D.W., Millour, J.P., 1984 Comparison of Enriched Air and Normal Air In-Situ Combustion, in: Curtis, F.A. (ed.) Energy Development: New Forms, Renewable, Conservation: Proceedings of ENERGEX 84, Regina, Saskatchewan, Canada, May 14–19. Pergamon Press, pp. 65–70.

Moore, R.G., Laureshen, C.J., Belgrave, J.D.M., 1995. A Comparison of the Laboratory In Situ Combustion Behavior of Canadian Oils. No. CONF-9502114-Vol. 1. UNITAR, New York.

Nelson, T.W., McNeil, J.S., 1961. How to engineer an in-situ combustion project. Producer Monthly, 2–11.

Okoye, C.U., Tiab, D., 1982, 26–29 September. Enhanced Recovery of Oil by Alkaline Steam Flooding. In SPE 11076 Presented at the SPE Annual Technical Conference and Exhibition, New Orleans, LA.

Panait-Patica, A., Serban, D., Ilie, N., Pavel, L., Barsan, N., 2006, January. Suplacu de Barcau Field–A Case History of a Successful In-Situ Combustion Exploitation. In SPE Europec/EAGE annual conference and exhibition. Society of Petroleum Engineers.

Pang, Z.-X., Wu, Z-B., Zhao, M., 2017. A novel method to calculate consumption of non-condensate gas during steam assistant gravity drainage in heavy oil reservoirs. Energy 130, 76–85.

Petke, F.D., Ray, B.R., 1969. Temperature dependence of contact angles of liquids on polymeric solids. Journal of Colloid and Interface Science 31, 216–227.

Rui, Y., 2009. Drilling technology of SAGD parallel horizontal wells in Xinjiang's Fengcheng oilfield. Petroleum Mechanics 37(8), 79–82.

Sahni, A., Kumar, M., Knapp, R.B., 2000, 1 January. Electromagnetic heating methods for heavy oil reservoirs. Society of Petroleum Engineers. DOI: 10.2118/62550-MS.

Sarathi, P.S., 1999. In-Situ Combustion Handbook – Principles and Practices. No. DOE/PC/91008-0374. National Petroleum Technology Office.

Satter, A., Iqbal, G., Buchwalter, J., 2008. Practical Enhanced Reservoir Engineering. PennWell Books.

Sayakhov, F.L., Bulgakov, R., Dyblenko, V., Deshura, B., Bykov, M., 1980. About HF heating of bitumen reservoirs. Petroleum Engineering 1, 5–8 [Russian].

Sayakhov, F.L., Kovaleva, L.A., Fatikhov, M.A., Khalikov, G.A., 1992. Method of thermal effect on oil-bearing formation. SU Patent Application, Serial No. 1723314.

Shabani, B., Kazemzadeh, E., Entezari, A., Aladaghloo, J., Mohammadi, S., 2014. The calculation of oil-water relative permeability from capillary pressure data in an oil-wet porous media: Case study in a dolomite reservoir. Petroleum Science and Technology 32(1), 38–50.

Sivakumar, P., Krishna, S., Hari, S., Vij, R.K., 2020. Electromagnetic heating, an eco-friendly method to enhance heavy oil production: A review of recent advancements. Environmental Technology & Innovation 20, 101100. https://doi.org/10.1016/j.eti.2020.101100.

Somerton, W.H., Keese, J.A., Chu, S.L., 1973. Thermal Behavior of Unconsolidated Oil Sands. In 48th Annual Fall Meeting of the Society of Petroleum Engineers, Las Vegas, NV, paper SPE-4506.

Speight, J.G., 2012. Oil Sand Production Processes. Gulf Professional Publishing.

Spencer, H.L., 1987. Electromagnetic Oil Recovery Ltd.

Taber, J.J., Martin, F.D., Seright, R.S., 1997a. EOR screening criteria revisited—Part 1: Introduction to screening criteria and enhanced recovery field projects. SPE Reservoir Engineering 12(03), 189–198.

Taber, J.J., Martin, F.D., Seright, R.S., 1997b. EOR screening criteria revisited—Part 2: Applications and impact of oil prices. SPE Reservoir Engineering 12(3), 199–206.

Tao, Y., Zhao, L., He, K., Duan, L., Zheng, Q., Zhao, L., 2019, October. The Numerical Simulation Study of Steam Flooding in Shallow Medium Heavy Oil Reservoirs in Kazakhstan. In SPE Russian Petroleum Technology Conference. Society of Petroleum Engineers.

Trigos, E.M., Gonzales, A., Pinilla Torres, J.M., Munoz, S., Mercado Sierra, D.P., 2010, January. Feasibility Study of Applying Steam Flooding in A Reservoir with High Shale/Sand: Teca Field. In Trinidad and Tobago Energy Resources Conference. Society of Petroleum Engineers.

Vafaei, M.T., Eslamloueyan, R., Enfeali, L., Ayatollahi, S., 2009. Analysis and simulation of steam distillation mechanism during the steam injection process. Energy & Fuels 23(1), 327–333.

Vinsome, K., McGee, B.C.W., Vermeulen, F.E., Chute, F.S., 1994, 1 April. Electrical heating. Petroleum Society of Canada. DOI: 10.2118/94-04-04.

Wang, J.J., Ross, M., Zhang, Y.M., 2010, January. Well bore heat loss-options and challenges for steam injector of thermal EOR project in Oman. SPE EOR Conference at Oil & Gas West Asia. Society of Petroleum Engineers.

Wan, T., Wang, X.J., Jing, Z., Gao, Y., 2020. Gas injection assisted steam huff-n-puff process for oil recovery from deep heavy oil reservoirs with low-permeability. Journal of Petroleum Science and Engineering 185, 106613.

Wang, C., Liu, P., Wang, Y., Yuan, Z., Xu, Z., 2018. Experimental study of key effect factors and simulation on oil displacement efficiency for a novel modified polymer BD-HMHEC. Scientific Reports 8, 3860. https://doi.org/10.1038/s41598-018-22259-z

Wang, J.J., Ross, M., Zhang, Y.M., 2010, January. Well Bore Heat Loss-Options and Challenges for Steam Injector of Thermal EOR Project in Oman. In SPE EOR Conference at Oil & Gas West Asia. Society of Petroleum Engineers.

Wang, Z., Li, Z., Lu, T., Yuan, Q., Yang, J., Wang, H., Wang, S., 2017. Research on Enhancing Heavy Oil Recovery Mechanism of Flue Gas Assisted Steam Flooding. In Carbon Management Technology Conference. Carbon Management Technology Conference.

Wei, Y., Babadagli, T., 2017. Alteration of interfacial properties by chemicals and nanomaterials to improve heavy oil recovery at elevated temperatures. Energy & Fuels 31(11), 11866–11883.

Willman, B.T., Valleroy, V.V., Runberg, G.W., Cornelius, A.J., Powers, L.W., 1961. Laboratory studies of oil recovery by steam injection. Journal of Petroleum Technology 13(7), 681–690.

Wilson, A., 2012, 1 June. Comparative analysis of electromagnetic heating methods for heavy-oil recovery. Society of Petroleum Engineers. DOI: 10.2118/0612-0126-JPT.

Wu, C.H., 1977, January. A Critical Review of Steamflood Mechanisms. In SPE California Regional Meeting. Society of Petroleum Engineers.

Wu, Y., Liu, X., Xing, D., 2015. Case Study of Hot Water Foam Flooding in Deep Heavy Oil Reservoirs. In SPE Asia Pacific Enhanced Oil Recovery Conference. Society of Petroleum Engineers.

Wu, Z., Liu, H., 2019. Investigation of hot-water flooding after steam injection to improve oil recovery in thin heavy-oil reservoir. Journal of Petroleum Exploration and Production Technology 9, 1547–1554. https://doi.org/10.1007/s13202-018-0568-7

Wu, Z., Liu, H., Wang, X., Zhang, Z., 2018. Emulsification and improved oil recovery with viscosity reducer during steam injection process for heavy oil. Journal of Industrial and Engineering Chemistry 61, 348–355.

Xi, C., Qi, Z., Liu, T., Zhang, Y., Zhao, F., Yu, Q., Shen, D., Li, X., 2019. CO_2 Assisted Steam Flooding Technology after Steam Flooding–A Case Study in Block J6 of Xinjiang Oilfield. In SPE Russian Petroleum Technology Conference. Society of Petroleum Engineers.

Xia, T.X., Greaves, M., Turta, A., 2003. Main Mechanism for Stability of THAI-Toe-to-Heel Air Injection. In Canadian International Petroleum Conference. Petroleum Society of Canada.

Xie, J., Hu, X., Liang, H.Z., Wang, M.Q., Guo, F.J., Zhang, S.J., Cai, W.C., Wang, R., 2020. Cold damage from wax deposition in a shallow, low-temperature, and high-wax reservoir in Changchunling Oilfield. Scientific Reports 10(1), 1–14.

Zan, C., Ma, D., Wang, H., Li, X.X., Guo, J., Li, M., Jiang, H., Luo, J., 2010, January. Experimental and Simulation Studies of Steam Flooding Process in Shallow, Thin Extra-Heavy Oil Reservoirs. In International Oil and Gas Conference and Exhibition in China. Society of Petroleum Engineers.

Zhao, D.W., Gates, I.D., 2013, 11 June. Stochastic Optimization of Hot Water Flooding Strategy in Thin Heavy Oil Reservoirs. In SPE Heavy Oil Conference-Canada. Society of Petroleum Engineers.

Zhou, X., Zeng, F., 2014, 10–12 June. Feasibility Study of Using Polymer to Improve SAGD Performance in Oil Sands with Top Water. In SPE 170164 Presented at the SPE Heavy Oil Conference Canada, Calgary, AB, Canada.

6 Application of Nanotechnology in EOR

6.1 INTRODUCTION

Nanotechnology is defined as a new technology for making applicable matters, systems, and devices using nanosized materials as well as new phenomena and properties at the nanoscale (1–100 nm) (Cheraghian et al., 2020). The application of nanotechnology in the oil and gas industry is just emerging. Recent research projects have shown that nanotechnology has the potential to solve or manage several problems in the petroleum industry. One of the speculated areas of application is in enhanced oil recovery (EOR). The different mechanisms of EOR are significantly improved by the application of nanotechnology. Nanoparticles and nanofluids have great influence in the field of EOR, with advantages such as wettability alteration, changes in fluid properties, improving the trapped oil mobility, enhancing the consolidation of sands, and decreasing interfacial tension (IFT). Nanoparticles are highly effective in the harsh reservoir conditions of high temperature and high salinity. Another application of nanotechnology in the oil and gas industry is nanoemulsion-based EOR. Nanoemulsions are kinetically stable homogeneous dispersions of immiscible liquids with droplet sizes in the order of 500 nm. The nanosized droplets are advantageous in enhancing nanoemulsion properties such as sturdy stability; large surface area per unit volume; improved interfacial, wetting behavior; and configurable rheology. The oil displacement is improved by nanoemulsion flooding through basic mechanisms of reduction of interfacial tension, wettability alteration of rocks, emulsification of trapped petroleum, and mobility ratio improvement. In this chapter, the application of nanotechnology in the oil and gas industry is discussed with special emphasis on nanoparticles and nanoemulsion-based EOR.

6.2 NANOPARTICLES IN OIL AND GAS INDUSTRY

The use of nanoparticles in EOR is still in the early stages of development, and much of the work is limited to laboratory studies. This section will give a brief introduction to nanoparticles, their types, and their different potential applications in oil industries, especially in EOR application. The discussion on the possible mechanism of EOR with nanoparticles has also been presented. Also, a review of the recent work in nanoparticle EOR is summarized in the subsequent sections.

6.2.1 NANOPARTICLE, NANOTECHNOLOGY, AND OIL FIELD APPLICATION

Nanoparticles are the purest forms of nanomaterials categorized as metallic, metal oxides, or magnetic nanoparticles, etc. Colloidal particles that have a dimension in the range of 1–100 nm are termed as nanoparticles (Bawa et al., 2005). The nanosized ranges increase the effective surface area, which leads to many marked improvements. The term "nanotechnology" refers to technical advancement that uses nanoparticles or nanosized sensors/materials, and it has been widely used in many industries from consumer electronics to health care to telecommunications and as well as in the oil and gas industry. At the nanoscale, it has been found that the properties of many materials tend to change, as the quantum effect dominates. Different properties, such as melting point, chemical activities, etc., change with the changing particle size. These nanoparticles have proven to be useful in different areas of the oil and gas industry, such as exploration, drilling, production, enhanced oil recovery, and refining (Kapusta et al., 2011). Although the potential of nanotechnology has not been thoroughly studied and realized in oil and gas exploration and production, the application of nanotechnology is not entirely new. Nanoparticles have been successfully used in drilling muds and cementing operations for the past 50 years (Cocuzza et al., 2012). Due to better optical, magnetic, and electrical properties and stress tolerance, nanoparticles can be used to develop extremely pressure- and temperature-sensitive nanosensors (Ponnapati et al., 2011; Alaskar et al., 2012). These nanosensors can deliver improved temperature and pressure evaluations in deep wells and hostile environments and could permit in situ measurements within the reservoir.

Complex heterogeneous reservoir and geological changes within the reservoir can be efficiently evaluated using sensors developed with nanoparticle materials. Nanoparticle-based catalysts are being used in the downstream operations of oil refining and petrochemical formation. The high costs and limitations associated with conventional chemical EOR may be overcome by taking advantage of the unique characteristic of nanoparticles (Kong and Ohadi, 2010). The properties of nanoparticles such as their unique surface chemistry, corrosion resistance, and mechanical strength make them a promising prospect for thrust research area in EOR applications. Nanoparticles can be tailored to alter reservoir properties such as wettability, improve mobility ratio, or control the formation of fine migration. The aqueous dispersion of nanoparticles in EOR application is generally referred as a nanofluid. Nanofluids have successfully been developed in laboratories; they have reduced the interfacial tension between the oil and the injected fluid and increased the viscosity of the injected fluid (Cocuzza et al., 2012). However cost-efficient field production of nanofluids has not been yet possible (Kakaç and Pramuanjaroenkij, 2009). Table 6.1 gives a brief review of the recent progress in the use of nanoparticles and nanotechnology in the oil and gas industry.

TABLE 6.1
Application of Nanoparticles and Nanotechnology in the Oil and Gas Industry

Oil Field Process	Nanoparticle/ Nanotechnology	Application/Outcome	References
Drilling fluids	• Nanoparticle • Nanocomposite	• Control of lost circulation and differential pipe sticking • Shale inhibition during drilling in shale section • Improvement of the rheological properties of the drilling fluid • Optimization of filtration loss in harsh HPHT conditions • Improvement in hole cleaning and increased wellbore stability • Reduction in torque and drag during directional drilling and reduced bit bailing problem	Zakaria et al., 2012; Sharma et al., 2012
Well cementing	• Nanoparticles • Carbon nanotubes	• Improved rheological properties and setting characteristics of cement • Quality control application in cement grout curing • Carbon nanotube reinforcement materials for cement instead of conventional fibers • Improvement in crushing strength of cement and increasing the early compressive strength	El-Diasty and Ragab, 2013; Mendes et al., 2015

TABLE 6.1 (*Continued*)

Oil Field Process	Nanoparticle/ Nanotechnology	Application/Outcome	References
Drilling and completion	• Nanocomposites and fibers • Nano-coating materials • Nano-diamond polycrystalline diamond compact (PDC) bit technology	• Construction of an advanced, efficient tool for casing, tubing, and drill bits • Coating materials for improvement in equipment life and performance • Advanced drill bits with resistance to abrasion and corrosion • Improvement of PDC bit performance in very harsh wellbore conditions	Lau et al., 2017; Ponmani et al., 2013
Well formation evaluation and reservoir characterization	• Nano-optics • Carbon fluorescent nanoparticles • Nanostructured glass ceramic • Magnetic particle–based nanosensors • Nano-optical fibers	• Monitoring of the transport phenomenon within the reservoir • Production logging, petrophysical, and production characterization • Better Li-6 scintillation detector with nanomaterials • Magnetic field–based measurement of fluid saturation inside reservoir • Laser light–based measurement of residual oil and bypassed oil in microbial EOR	Levitt and Pope, 2008; Al-Shehri et al., 2013
Production	• Nanoparticles	• Improvement of the viscosity of fracturing fluids • Enhanced cleanup in postfracturing operation • Prevention of fine migration • NP association with visco-elastic surfactant (VES) micelles for stabilization of fluid viscosity	El-Diasty and Ragab, 2013; Liang et al., 2016
Exploration	• Nano-computerized topography • Nanoscale metals	• High-resolution pore imaging in tight gas sand, tight sales, and tight carbonate reservoirs • Geothermal exploration	(El-Diasty and Ragab, 2013; Bustos et al., 2010
Gas hydrates	NP	• Recovery of gas hydrates	Bhatia and Chacko, 2011; El-Diasty and Ragab, 2013
Enhanced oil recovery	• Nano-catalysts • Nanocomposites • Nanoparticles	• Nano-catalyst for in situ upgradation of heavy oil • Nanocomposite materials for better polymer flood performance in harsh reservoir conditions • Polymer-nanoparticle suspension for improvement in polymer rheology • Nanofluids for IFT reduction and wettability alteration of carbonate reservoirs • Nanoparticles for emulsion-based EOR • Nanoparticles for emulsification of crude oil in reservoir conditions	Maurya and Mandal, 2016; Kim et al., 2017
Refining and processing	• Nano-filters • Nanoparticles • Nano-catalysts	• Removal of highly toxic substances in the refining operation • Upgradation of bitumen and heavy oil	Esmaeili, 2011; El-Diasty and Ragab, 2013

6.2.2 Current Status of Nanoparticle-Based EOR

Although nanoparticles have proven useful in many areas of exploration and production, their potential in EOR is yet to be fully realized. Recently, many researchers elevated the use of nanoparticles by identifying their features such as wettability alteration, foam and emulsion stabilization, conformance control, mobility ratio improvement, IFT reduction, reduction of residual oil saturation, and ability to travel long distances inside formation without logjam, all of which can improve oil recovery (Ayatollahi and Zerafat, 2012; Gharibshahi et al., 2015). Nanoparticles are ideally suited to harsh environments due to their mechanical and thermal stability (Ogolo et al., 2012). In spite of much research work in the last decade, the function of the nanoparticle and their application in EOR is not fully understood and requires further laboratory investigation. Nanoparticles, as a result of their smaller size, can travel without hindrance inside the reservoir without getting plugged at the pore throat in a porous medium. However, at higher loads, they can still plug the pore throat. Different metal oxide nanoparticles have been used in EOR research by several authors. The recovery mechanisms are IFT reduction, wettability alteration, and reduction of oil viscosity. However, magnesium and zinc oxide application resulted in permeability impairment. The use of nanoparticles in a different type of EOR process is summarized in the following.

6.2.2.1 Nanofluids

The nanofluid is defined as a uniform nanoparticle (NP) distribution in solution, which generally produces higher physical and thermal stability compared to any conventional fluid. These nanofluids are prepared by dispersing NPs in water using laboratory-based mixers, stirrers, and ultrasonification. The water-based nanofluid showed favorable results for oil recovery and other oil field applications. The formulated silica nanofluid in water shows significant improvement in oil recovery via change in reservoir rock and fluid properties. NPs after dispersion in base fluid show a surface potential measured as ζ-potential, which generally helped NPs to maintain uniform distribution in the nanofluid phase. Nevertheless, a reduction in this ζ-potential allows NPs to show agglomeration in the nanofluid. However, nanofluid formulation in water shows limitations with time and at reservoir conditions of high salinity, high pressure, and high temperature.

6.2.2.1.1 Wettability Alteration and IFT Reduction with Nanoparticles

As discussed earlier, IFT reduction and wettability alteration can be the main driving force for the mobilization of entrapped oil. Research has been done to investigate the same with different nanoparticles. Recent works advocate that nanoparticles can reduce IFT and cause wettability alteration to enhance oil recovery. Li et al. in their studies demonstrated that the silica nanoparticle could reduce IFT and change the wettability of reservoir rocks to more water wet (Li et al., 2013). NPs create a thin layer on the surfactant that distributes between oil and injected fluids. This process leads to a significant reduction in the interfacial tension parameter. In fact, in this process, the capillary numbers increase, and capillary forces significantly decrease. This means that both surfactants and NPs support each other for injected emulsion stability. In addition, the stability against sedimentation increases in solutions due to NPs that can balance surface forces and the force of gravity (Cheraghian et al., 2020). Karimi et al. demonstrated that zirconium oxide could change the wettability status of carbonate rocks from oil-wet to water-wet conditions and improve oil recovery (Karimi et al., 2012). Moreover, experimental investigation suggests that nanoparticle-laden fluid or nanofluid injection has a promising future in EOR application through IFT reduction and wettability alteration (Karimi et al., 2012; Ayatollahi and Zerafat, 2012). There are several experimental works in the oil recovery field with hydrophilic silica NPs; most use silica NPs to achieve wettability alteration. Silica NPs are one of the most used materials for changing wettability. Silica NPs reduce trapped oil in throats and pores and increase heavy oil recovery from reservoirs. Further, silica NPs are a suitable choice for changing wettability in polymer flooding, as well as to increase sweep efficiency and heavy oil recovery. However, experimental results indicated that other NPs have greater wettability alteration ability and IFT in the case of light oil (Rostami et al., 2019). The change of wettability from oil wet to water wet in presence of nanofluids with time is shown in Figure 6.1. The wettability alteration can be attributed to structural disjoining pressure related to the spreading of nanofluids on solids (Fig. 6.2).

The energy responsible for the governing interactions are (Khilar and Fogler, 1998) (i) London-van der Waals attractive potential energy as described by DLVO theory; (ii) electric double-layer repulsion energy as described by the Poisson-Boltzmann distribution; (iii) born repulsion due to the overlapping of electron clouds around the particles and porous medium; (iv) hydrodynamic energy due to relative motion resulting from drag force on the particles and friction forces in the opposite direction of fluid flow when particles adhere to the surface of the porous medium; and (v) acid-base interaction based on the acid-base theory of adhesion of electron pairing.

The structural disjoining pressure helps in the mobilization of oil droplets attached to rock surfaces. The presence of these nanoparticles in a three-phase contact region has a tendency to create a wedge-film structure. The interfacial forces will cause the aqueous phase (nanofluid) contact angle (θ) to decrease to 1°, and the result is a wedge film. This wedge film will act to separate the formation fluid such as oil, paraffin, water, and gas from the formation surface. This way of spreading the nanofluid in a porous medium also helps in the removal of small oil droplets attached to rock surfaces. Wasan et al. (2011). studied the role of disjoining pressure in the wetting and spreading of nanofluid on a solid surface. Disjoining pressure is a pressure that arises when two surface layers reciprocally overlap and is caused by the total effect of forces that are different by nature.

Reservoir salinity and the presence of electrolytes in the dispersion fluid can influence nanofluid flooding in several ways: (1) The presences of electrolytes can alter the surface characteristics of the NPs and change the way they interact with each other, the rock surface, and the oil/water interface. (2) They can reduce the repulsive forces between the NPs causing increased particle sizes through agglomeration and change in iso-electric point (IEP) (Adil et al., 2020). (3) They can be adsorbed

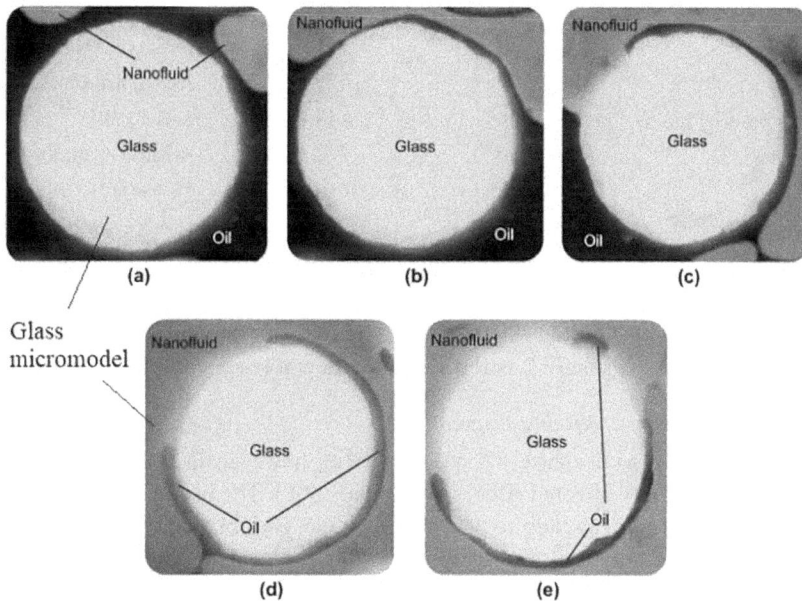

FIGURE 6.1   Surface wettability changes. Pore-scale views of the micromodel during nanofluid flooding for a period: (a) initial time, (b) 1,000 s, (c) 2,000 s, (d) 3,000 s, (e) 5,000 s of nanofluid injection.

Source: Rostami et al., 2019.

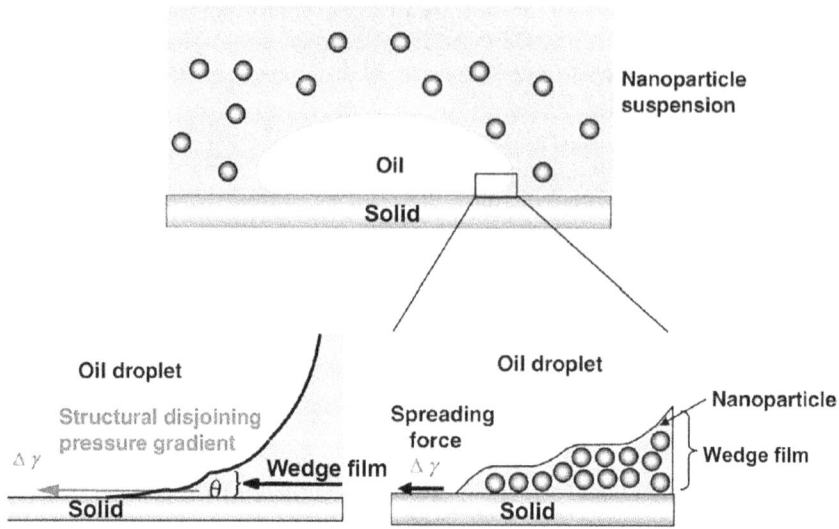

FIGURE 6.2 Spreading of nanofluid on solid surface: structural disjoining pressure.

Source: Wasan et al., 2011.

TABLE 6.2
Effect of Nanoparticles on IFT

Type of NPs	NP Concentration	IFT (mN/m)	References
SiO_2	0.05 wt. %	17.5	Hendraningrat, and Torsæter, 2015
Al_2O_3	0.5–3 g/L	3.5	Joonaki and Ghanaatian, 2014
Fe_2O_3	3 g/L	2.7	Joonaki and Ghanaatian, 2014
Carbon nanoparticles (CNP)	0.1 wt. %	13.4	Li et al., 2017
Iron ore–carbon shell	100 mg/L	2.7	Betancur et al., 2020
ZrO_2	0.05 wt. %	6.6	Moslan et al., 2017
Hyroxylated nanopyroxene	50 ppm	10	Sagala et al., 2020

on the rock surface and at fluid-fluid interface, thereby reducing the total number of NPs that can be adsorbed. The effect of NPs on IFT as reported by different researchers is shown in Table 6.2.

6.2.2.1.2 Nanoparticles as Mobility Improvers

Polymers are widely used to decrease the mobility ratio; however, in harsh reservoir conditions, their performance is severely affected (Levitt and Pope, 2008). The use of nanoparticles and nanocomposites with the polymer can help to address this issue. Recent studies have shown that nanoparticle suspension in the polymer solution improves the rheological performance of the polymer solution. Nanocomposites—where polymeric chains are grafted onto the silica nanoparticle—can provide long-term thermal stability, which is not feasible with a conventional polymer such as polyacrylamide. Hu et al. have shown that the presence of silica nanoparticles in a polyacrylamide solution not only increases its viscosity but also provides stability for long-term aging when hard brines are present (Hu et al., 2017). Moreover, Pu et al. synthesized a nanocomposite that is water soluble and provides excellent performance in reservoir conditions (Pu et al., 2015). Several other researchers have also found nanoparticles to be promising in increasing the macroscopic sweep efficiency (Ponnapati et al., 2011; Ye et al., 2013).

6.2.2.1.3   Nanoparticles as Emulsion and Foam Stabilizers

Foams are extensively used for reducing gas mobility during injection to prevent the adverse effect of low gas viscosity. For example, during CO_2 injection, injected CO_2 tends to bypass oil by floating to the top of the reservoir, especially in thick reservoirs. Also, low viscosity leads to early breakthrough due to viscous fingering of injected gas. In the same way as foam, the emulsion can also be used for conformance control, and enhanced oil recovery (Zhang et al., 2009; Mandal et al., 2010). Although the surfactant can stabilize foams and emulsion, the stability is a concern in harsh reservoir conditions. Nanoparticles can stabilize foam efficiently by irreversibly adsorbing at the interface and providing the stability needed in adverse conditions. Pei et al. reported the stabilization of the emulsion with a cationic surfactant along with a nanoparticle, which provides improved recovery in core flooding studies (Pei et al., 2015). Kim at al. also formed a nanoparticle-based emulsion that provided excellent results in the recovery of heavy oil (Kim et al., 2017). Sun et al. demonstrated that foam stability increased when a silica nanoparticle was added to a surfactant solution, and stability is a function of nanoparticle concentration (Sun et al., 2014). Lau et al. used nanoparticle-stabilized foam in core flooding experiments to demonstrate the incremental recovery of oil in reservoir conditions (Lau et al., 2017). Recently, several other researchers have also reported the use of nanoparticles in foam and emulsion stabilization application in EOR in laboratory core flooding studies (Talebian et al., 2013; Zhang et al., 2011).

6.2.3   Factors Affecting Nanofluid Flooding Oil Recovery Mechanisms

There are a number of parameters such as particle hydrophobicity, particle size, particle morphology, particle concentration, particle surface charge, reservoir temperature, reservoir salinity, reservoir pressure, reservoir pH, and reservoir rock surface charge influencing the fluid-fluid and fluid-rock interactions of nanofluids and, consequently, oil recovery (Shalbafan et al., 2019). Nanofluid concentration is considered one of the major parameters to enhance oil recovery. Thus, when investigating nanofluids for EOR applications, there is a need to pay attention to these parameters to comprehend the variables that can influence the efficiency and feasibility of the nanofluid.

6.3   NANOCOMPOSITES IN EOR

In the context of EOR, nanocomposites (NCs) are defined as the dispersions of nanoparticles in a polymer solution, also known as polymeric NCs. The idea behind NCs is to use building blocks with dimensions in a nanometer range to design and create new materials with unprecedented flexibility and improvement in their physical properties. NCs having higher strength, flexibility, and wear resistance, and other desired characteristics have also been obtained owing to dynamically developing nanotechnology and are widely used by leading oil companies for drilling, oil extraction, transportation, and refining (Singh et al., 2010). The application of NCs to enhance oil production in various EOR methods has been a subject of interest for researchers (Negin et al., 2016). Polymers are mixed with nanoparticles to form nanocomposites (nano-polymeric solution), and different nanoparticles have been reported to have varying effects on polymer viscosity.

Another type of nanocomposite is obtained by grafting the nanoparticle's surface with a polymer. This is a recently developed approach and has been proven to improve the stability in an aqueous solution (Hamidi et al., 2020). NCs are widely recommended for EOR because of their unique properties, which combine the benefits of inorganic nanoparticles and organic polymers (Mishra et al., 2014; Kortam et al., 2017). The mechanisms to enhance and optimize oil recovery by NCs are through alteration of reservoir rock wettability, reduction of oil viscosity, IFT reduction, electrostatic forces, and control of fine migration. Although economic considerations are important for successful field application and commercialization of NCs in EOR processes, improvements are achieved at very small NC concentrations—i.e., effective reduction of IFT and rock surface wettability alteration leads to a notable increase in oil recovery factor from petroleum reservoirs. In addition, there are numerous possibilities to investigate the combination of NCs with other EOR approaches as hybrid EOR methods.

6.4 NANOEMULSION-BASED ENHANCED OIL RECOVERY

Nanoemulsions are isotropic dispersed systems of two immiscible liquids, normally consisting of an oily system dispersed in an aqueous system, or an aqueous system dispersed in an oily system, forming droplets or oily phases of nanometric sizes. Nanoemulsions are typically stabilized by a surface-active agent (surfactant), which forms a protective coating over the oil droplets, preventing their coalescence by imparting an electrostatic repulsion. Nanoemulsions often have a higher loading capacity for lipophilic active ingredients than microemulsions, which can be an advantage in some applications. Nanoemulsion droplet sizes fall in the range of 20–500 nm, depending on the composition and structural arrangement of constituting emulsifiers. Controlling emulsion droplet stability and behavior plays a primary role in structure-property modification and the field of application. Techniques for nanoemulsion synthesis are broadly classified as low-energy and high-energy methods. Low-energy methods employ the IFT reduction ability of nanoemulsions to achieve suitable phase transitions without using excess shear. During high-energy or dispersion methods such as in ultrasonification, a significant energy input (~108 W/kg) is required to cause the droplet to break down under high shear conditions. Ultrasound technology has attracted rapid interest in this field of research due to the formation of homogenous systems in comparison to low-energy mechanical processes and better control during synthesis in comparison to other high-energy methods such as high-pressure homogenization, micro-fluidization, and jet dispersion. For reservoir oil to flow through pore throats, the continuous Brownian motion of nanosized droplets and subsequent slow coalescence rates are desirable attributes for nanoemulsion-assisted oil recovery.

Based on the preparation methods, emulsions are classified as macroemulsions, microemulsions, and nanoemulsions, where the droplet size of the dispersed phase and stability varies. Phase separation occurs in macroemulsions and microemulsions if given sufficient time. They are sensitive to variation in composition and temperature. On the contrary, nanoemulsions are kinetically stable and, thus, remain stable for longer time periods due to their nanosized oil droplets.

The conventional chemicals (surfactants) used for petroleum production are effective; however, they require high turbulent flow and large volumes and are prone to contaminating the formation zone with unsuitable surfactant selection and emulsions due to their poor stability, thereby, leading to overall reduced petroleum production rates with increased operational costs. Therefore, to mitigate the challenges of the conventional chemical slugs (surfactants/emulsions) during EOR operations, nanotechnology offers a solution in the form of nanoemulsions (Dantas et al., 2019; Braccalenti et al., 2017).

- Nanoemulsions once prepared can be flexibly manipulated in a precise manner by incorporating different additives likely polymers and NPs.
- Nanoemulsions are relatively less influenced by physical and chemical alterations in comparison to their counterparts and, therefore, have shown great potential for EOR operations.
- Nanoemulsions are stable to creaming or sedimentation due to the high diffusion rate than sedimentation probably due to the Brownian motion of the nanosize droplets that imparts kinetic stability.
- An oil-water interfacial tension (IFT) value much lower than 0.001 mN/m can be obtained in nanoemulsions favoring small-sized oil droplets formation, leading to an efficient solubilization of a large amount of residual hydrocarbons without needing much turbulence.

- The ability to effectively alter the wetting behavior of rock surfaces, as wettability directly influences the capillarity, determines the affinity of oil to reservoir rock and how easily oil displacement will occur.
- Nanoemulsions, owing to their small droplet size, facilitate their easy passage through the minute capillaries in porous media. Enhanced solubilizing power is shown by nanoemulsions in miscibility tests with hydrocarbons.
- Nanoemulsions demonstrate the existence of dual features of elasticity and flowability, owing to their shear-induced viscoelastic property.
- Nanoemulsion formulation may be costly in comparison to a few chemical slugs; however, low chemical requirement, long-term stability, and beneficial recovery factors, as apparent from findings, make them suitable for petroleum extraction.

6.4.1 Formation and Properties

6.4.1.1 Formation

An insight into the science behind nanoemulsion formulation is vital in controlling the droplet size. Griffin classified surfactants depending on their hydrophilic-lipophilic balance (HLB) value. Surfactants with HLB < 6 are oil soluble and will form water-in-oil (W/O) emulsions, whereas surfactants having HLB > 8 are water soluble and will form oil-in-water (O/W) emulsions (Griffin, 1949). Typically, nanoemulsions are formed via a two-step process. Initially, a macroemulsion is prepared, and then it is transformed into a nanoemulsion. Nanoemulsions are prepared either by low-energy methods or by high-energy methods. The dispersion and stability of nanoemulsions are better in the high-energy method, where dispersion is made by ultrasonification. Figure 6.3 presents schematics, process perception, and the advantages and disadvantages of the most widely employed nanoemulsion formulation methods.

6.4.1.2 Properties

The unique properties of nanoemulsions such as long-term stability (kinetically stable), low IFT, improved wetting behavior, and exceptional rheological performance are not only influenced by their composition but also depend on the formation path and the extent of mixing.

6.4.1.2.1 Stability

The influencing factor of the stability of nanoemulsions is molecular diffusion or Ostwald ripening from the solubility difference of small and large droplets, polydispersity of emulsion, and rates of droplet aggregation and coalescence (Tadros et al., 2004). Nanoemulsion stability can be investigated at three scales: phase separation (macroscopic scale), droplet size by light scattering (mesoscopic level) and scattering/transmitting electron microscopy (microscopic level) (Huang et al., 2016). The stability of nanoemulsions can be tuned to be robust with a long shelf life (few months to years) by controlling the droplet size. The droplet size is influenced by varying the oil to surfactant (O/S) and additives to surfactant (A/S) ratios. A characterization of nanoemulsion systems by Pal et al. (2019) also reported an increase in droplet size by PHPA polymer addition. This is because of the polymer chain entanglement that covers the surfactant-stabilized oil droplets. However, nanoemulsions remain stable owing to the combined effect of the electrostatic repulsion of the surfactant and the steric hindrance of the polymer. Addition of NPs strengthens the electrostatic barrier with improved droplet surface adsorption that contributes to a decrease in droplet size and enhanced nanoemulsion stability.

6.4.2 Mechanisms of Nanoemulsion Action

The physicochemical properties of nanoemulsions suggest that they can be successfully used to recover the residual oil trapped in the fine pore of reservoir rock by capillary forces after primary

Methods of Nanoemulsion Formation	Schematics and Process Perception	Advantages	Disadvantages	References
Phase inversion composition (PIC)	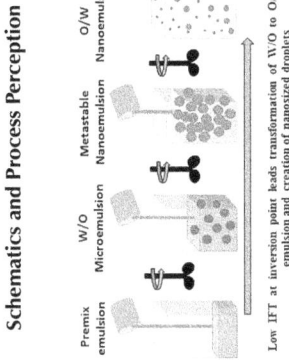 Premix emulsion / W/O Microemulsion / Metastable Nanoemulsion / O/W Nanoemulsion. Low IFT at inversion point leads transformation of W/O to O/W emulsion and creation of nanosized droplets	1. Low-energy process 2. Flexibility in path a. liquid ration variation b. mixed surfactant c. electrolyte change 3. Heat requirement not compulsory 4. Low cost	1. Prone to Ostwald ripening 2. Oleic and aqueous phase solubility must be good 3. Limitation in type of surfactants 4. Limitation in range of compositions compared to high-energy methods	Davies, 1987; Forgiarini et al., 2000, 2001
Phase inversion temperature (PIT)	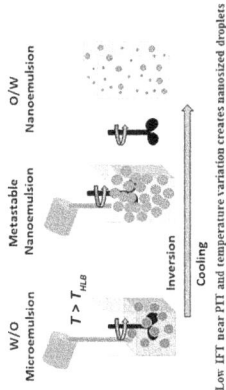 W/O Microemulsion / $T > T_{HLB}$ / Metastable Nanoemulsion / O/W Nanoemulsion. Inversion / Cooling. Low IFT near PIT and temperature variation creates nanosized droplets	1. Low-energy process 2. Low cost	1. Prone to Ostwald ripening 2. Oleic and aqueous phase solubility must be good 3. Limited to temperature-sensitive surfactants—i.e., nonionic surfactants 4. Heat energy input requirement 5. Stable range of T (temperature) is limited	Izquierdo et al., 2005
High energy	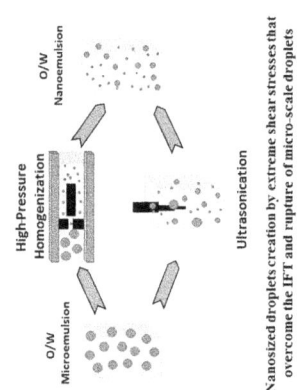 O/W Microemulsion / High-Pressure Homogenization / Ultrasonication / O/W Nanoemulsion. Nanosized droplets creation by extreme shear stresses that overcome the IFT and rupture of micro-scale droplets	1. Fast and easy way of preparation 2. Shear force and composition both control the droplet size distribution 3. Suitable for extensive array of compositions 4. Very large scale production capability	1. In comparison to low-energy methods, it requires a huge energy input 2. Requires an initial significant investment 3. Considerable loss of energy in the form of heat	Davies, 1987; Wooster et al., 2008

FIGURE 6.3 Recapitulates the schematics, process perception and advantages and disadvantages of the most widely employed nanoemulsion formulation methods: PIC, PIT, and high-energy (HPHT and ultrasonification).

Source: Kumar et al., 2021.

and secondary recovery. Some specific surfactants have been successfully used for the formation of oil-in-water microemulsions in pharmaceutical and medical sciences. The same methodology can be adopted to recover residual oil from the reservoir. Nanoemulsions possesses unique properties such as kinetic stability, low IFT, improved wetting behavior, and exceptional rheological performance that significantly influence the petroleum recovery factors. Additionally, nanoemulsion injection reduces surfactant adsorption over the rock surfaces, which improves the chemical carrying ability of the aqueous phase and the blockage of high-permeability zones. Figure 6.4 illustrates the mechanism associated with the nanoemulsion flooding process.

6.4.2.1   Interfacial Activity

Interfacial tension is an important factor for oil displacement. At the interface, the surroundings of molecule are different than in the bulk phase, as the interface is accompanied with surface-free energy. A high IFT indicates the immiscibility of immiscible fluids (oleic and aqueous phases). On the other hand, zero IFT indicates the complete miscibility of the two phases. The wetting behavior or wettability of reservoir rock is also affected by the interfacial activity. Cabaleiro et al. (2019) reported that a decrease in surface tension by surfactant improves the wetting behavior compared to that of water. Pal and Mandal (2020a, 2020b) investigated the IFT and wetting behavior of Gemini surfactant–based nanoemulsions systems to be applied for understanding the oil displacement through porous media through the detachment energy model. The reduction of contact angle measured for nanoemulsions on sandstone rock indicated a rapid spreading and an enhanced wettability alteration ability of nanoemulsion fluids. Another attribute of nanoemulsions not much investigated but directly linked to interfacial activity is miscibility. The interfacial activity of nanoemulsions for crude oil extraction can be discussed through the detachment energy model. The decrease in

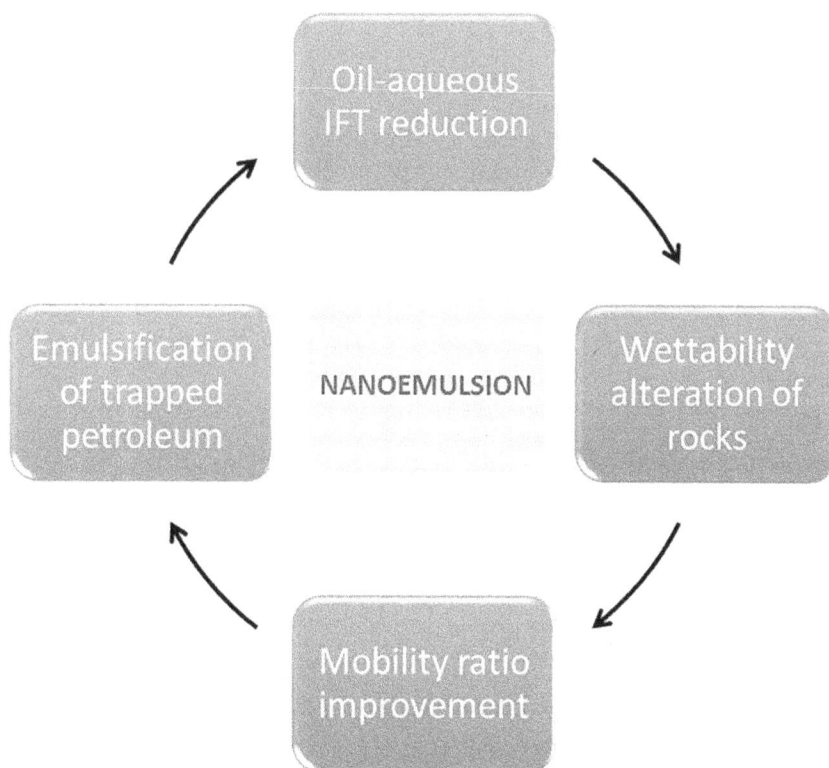

FIGURE 6.4   The mechanisms of nanoemulsion flooding.

detachment energy with time due to surfactant molecules gradual slipping toward the oil phase is caused by the increased hydrophobicity of the adsorbed constituents and is responsible for crude oil detachment from the rock surface. Figure 6.5 shows the different interfacial arrangements of an oil droplet stabilized by a surfactant molecule in an O/W nanoemulsion.

6.4.2.2 Tunable Rheology

From a rheological perspective, nanoemulsions are interesting, as they exhibit notably better rheological behavior than macro- or microemulsions. A major factor that influences the rheological properties of emulsions is the interactions (repulsive or attractive) between the dispersed droplets. The rheology depends on many factors such as packing or volume fraction (ϕ) of the dispersed phase droplets; properties of the continuous phase; and the type, size, and strength of interactions between the dispersed droplets. These rheological characteristics can be controlled and modified as per requirement by governing the volume fraction of the dispersed phase and the droplets' size, by the addition of different additives such as viscosifying agents (polymers) and salts/depletion agents, causing the gelling of emulsion droplets, and by variation in temperature. A higher viscosity of the nanoemulsion compared to that of water improves the mobility ratio and, hence, the sweep efficiency. The characteristic rheology also influences the seepage properties of nanoemulsions such as migration, pore blockage, and displacement efficiency in porous media. The improved recovery factor by injecting more viscous nanoemulsion system would preferentially sweep the low-permeability zone, avoiding the high-permeability formation paths. Figure 6.6 presents the mechanism of crude oil displacement by nanoemulsions induced via mobility control and plugging of high-permeability regions.

6.4.2.3 Wettability Alteration of Rocks

The reservoir rock-wetting behavior directs the position, spreading, and movement of fluids. Wetting behavior affects the petroleum recovery factors such as relative permeability and capillary pressure. Wettability alteration of reservoir rocks from intermediate-wet or oil-wet condition to water-wet condition lessens the adverse effect of the capillary forces within the reservoir pores and increases the permeability of the oleic phase in the reservoir, which is favorable for attaining

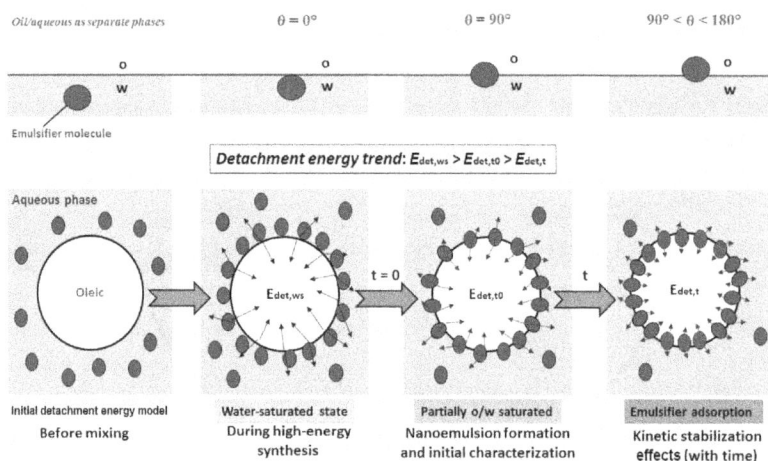

FIGURE 6.5 The different interfacial arrangements for oil/water attained for nanoemulsions according to the detachment energy model.

Source: Adapted from Pal and Mandal, 2020a.

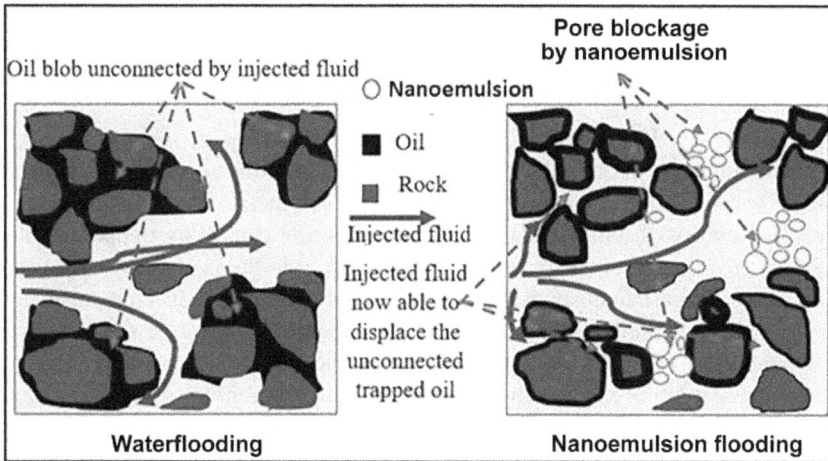

FIGURE 6.6 Representation of conformance improvement in a reservoir induced by nanoemulsions. Nanoemulsion plugs the high-permeability pore spaces and forces oil to flow through low-permeability pore spaces.

Source: Modified from Foroozesh, and Kumar, 2020.

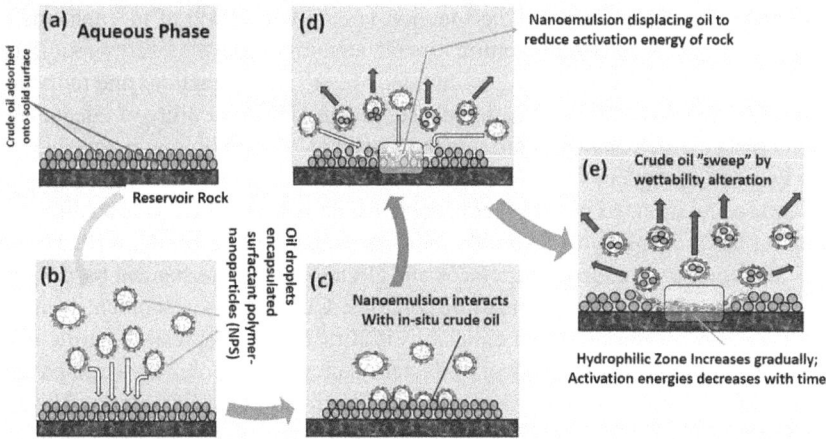

FIGURE 6.7 Schematic presentation of the stages of wettability alteration by nanoemulsions through the Arrhenius surface energy model.

Source: Adapted from Pal and Mandal, 2020a.

desired displacement efficiencies. During nanoemulsion flooding, interactions of ion-pair (cationic/anionic) and electron polarization (nonionic) occurs between the adsorbed oleic phase on the rock surface and surfactant monomers. This is followed by an impulsive imbibition of water into the rock matrix in low-permeability zones, with eventual production of petroleum. The petroleum extraction through wettability alteration of rock surface by nanoemulsions can be illustrated through the Arrhenius surface energy approach. Nanoemulsions interact with the crude oil adsorbed on rock surfaces, and the surfactant molecules detach them into the aqueous phase. Thereby, nanoemulsions favorably sweep the crude oil. Figure 6.7 shows the different wetting stages of the rock surface and crude oil detachment by nanoemulsions, exemplified through the Arrhenius solid-liquid surface energy model.

6.4.2.4 Improvement of Permeability

Injection of nanoemulsions through the porous media may alter the permeability, which results in the plugging of water channels, leading to an improved volumetric sweep of the reservoir and a reduction in permeability of the media due to flow restriction caused by trapped emulsion droplets. Emulsification property of nanoemulsions with reservoir fluids improves recovery factors. The selection of an appropriate nanoemulsion is important to attain enhanced emulsification of the hydrocarbons along with rheological property improvements for effective mobility control. The oleic phase is snapped off from the reservoir surfaces by the present surfactant in nanoemulsions to form a homogeneous solution and thereby improve the petroleum mobilization by viscosity reduction. The oil droplet coalescence in the homogeneous solution thus formed is prevented by the electrostatic repulsion or steric hindrance effect provided by the protective coating of surfactant molecules. The emulsification process is contingent to the surfactant type and its concentration, oleic phase category, salinity, and adsorption behavior of rock.

6.4.3 PICKERING EMULSIONS

Pickering emulsions stabilized by solid particles have gained much attention, which afford high stability, low toxicity, controllable rheological properties, and stimuli-responsive behavior compared to the traditional emulsions formed by surfactants (Liu et al., 2021). Typical emulsifiers consisting of surfactant, polymer, and/or nanoparticle function via a combination of mechanisms to obtain a common effect—i.e., oil droplet stabilization. Surfactants minimize the energy required for emulsion droplet formation, thereby increasing electrostatic repulsive interactions and reducing IFT. Polymers improve the viscosity of the displacing phase, causing lower mobility of injectant than that of the oil phase in place, which aids in improving the sweep of zones with low permeability. Nanoparticles adsorb onto the oil-aqueous interface and further strengthen the dispersed oil droplets against flocculation and coagulation. As oil-displacing vehicles, these systems provide an effective platform for designing biphasic agents with favorable interfacial and wetting behavior. Pickering particles "accumulate" onto crude oil-aqueous interfaces and strengthen the mechanical barrier, thereby preventing flocculation and coalescence of oil dispersions. Currently, development of new stabilization techniques in oil-in-water pickering emulsions is attracting widespread scientific interest in the petroleum sector. The structural stability of surfactant-nanoparticle-polymer-based pickering emulsions of micron-sized droplets is highly beneficial for application in oil recovery. The stabilization and propagation of surfactant-polymer-nanoparticle-stabilized emulsions depend on a myriad number of factors such as crude oil properties, emulsifier(s) type and dosage, rock mineralogy, droplet size and behavior, and wettability. Therefore, the need for incorporation of nanoemulsion route is important from a stability/functional viewpoint in oil recovery operations.

6.4.4 SCREENING CRITERIA FOR NANOEMULSION EOR

The screening criteria are a convenient and quick way to examine suitable candidate reservoirs prior to performing costly reservoir elaborations and economic assessments. Setting of a criterion is helpful in lowering the expenses incurred in operations for reservoir descriptions. The development in the field of EOR and instigation of new EOR projects will commence only if these appear to be profitable. Therefore, screening criteria acts as a cursory guide for the valuation of economic aspects because the upcoming EOR ventures are highly dependent upon the discernment of the future crude oil prices.

  In present days of reservoir management exercises, oil and gas industry engineers opt for different EOR techniques much earlier in the production life of the wells. In such practices, for the majority

TABLE 6.3
Screening Criteria for Nanoemulsion EOR

Crude Oil	Recommended
Gravity (°API)	35°–22.3°
Viscosity (cP)	< 35
Composition	Light/medium oil is desirable
Reservoir	
Type of formation	Sandstones preferred
Porosity (%)	10–35
Oil saturation (PV)	28–35
Permeability (mD)	10–350
Net thickness	Not critical
Depth, ft	625–5300 (see temperature)
Temperature, °F	< 200

of fields, the decision is not whether or not to inject but when. Palpably, the decision to "inject or not to inject" is greatly influenced by the economics (profits); however, having some screening criteria on hand helps in ruling out the less likely reservoir candidates. Moreover, they provide useful data collected from the examinations of a large number of reservoirs helpful in selecting the most likely injection fluid.

As the introduction of nanoemulsions in the oil and gas sector is new and not much work has been performed on nanoemulsion EOR, therefore, limited data is available in literature regarding their screening guidelines. Table 6.3 presents a quick screening criterion of nanoemulsion EOR based on published works (Taber et al., 1996; Adasani and Bai, 2011; Sakthipriya et al., 2015).

REFERENCES

Adasani, A.A., Bai, B., 2011. Analysis of EOR projects and updated screening criteria. Journal of Petroleum Science and Engineering 79, 10–24.

Adil, M., Zaid, H.M., Chuan, L.K., 2020. Electromagnetically-induced change in interfacial tension and contact angle of oil droplet using dielectric nanofluids. Fuel 259, 116274.

Alaskar, M., Ames, M., Connor, S., Liu, C., Cui, Y., Li, K., Horne, R., 2012. Nanoparticle and microparticle flow in porous and fractured media—An experimental study. SPE Journal 17(4), 1160–1171.

Al-Shehri, A.A., Ellis, E.S., Felix Servin, J.M., Kosynkin, D.V., Kanj, M.Y., Schmidt, H.K., 2013. Illuminating the Reservoir: Magnetic NanoMappers. In SPE Middle East Oil and Gas Show and Conference, SPE-164461. SPE.

Ayatollahi, S., Zerafat, M.M., 2012. Nanotechnology-Assisted EOR Techniques: New Solutions to Old Challenges. In SPE International Oilfield Nanotechnology Conference and Exhibition. OnePetro.

Bawa, R., Bawa, S.R., Maebius, S.B., Flynn, T., Wei, C., 2005. Protecting new ideas and inventions in nanomedicine with patents. Nanomedicine: Nanotechnology, Biology and Medicine 1(2), 150–158.

Betancur, S., Olmos, C.M., Pérez, M., Lerner, B., Franco, C.A., Riazi, M., Gallego, J., Carrasco-Marín, F., Cortés, F.B., 2020. A microfluidic study to investigate the effect of magnetic iron core-carbon shell nanoparticles on displacement mechanisms of crude oil for chemical enhanced oil recovery. Journal of Petroleum Science and Engineering 184, 106589.

Bhatia, K., Chacko, L., 2011. Ni-Fe Nanoparticles: An Innovative Approach for Recovery of Hydrates. In SPE EUROPEC/EAGE Annual Conference and Exhibition. pp. 23–26.

Braccalenti, E., Del Gaudio, L., Belloni, A., Albonico, P., Radaelli, E., Bartosek, M., 2017. Enhancing Oil Recovery with Nanoemulsion Flooding. In Offshore Mediterranean Conference. OMC-2017–819.

Bustos, A.V., Eduardo, A., Febres, A., Villavicencio, A., Shen, F., 2010, January. Integrated Fractured Reservoir Characterization and Connectivity Study in the Cantarell Field. In International Oil and Gas Conference and Exhibition in China. Society of Petroleum Engineers.

Cabaleiro, D., Hamze, S., Agresti, F., Estellé, P., Barison, S., Fedele, L., Bobbo, S., 2019. Dynamic viscosity, surface tension and wetting behavior studies of paraffin–In–Water nano–Emulsions. Energies 12, 3334.

Cheraghian, G., Rostami, S., Afrand, M., 2020. Nanotechnology in enhanced oil recovery. Processes 8(9), 1073.

Dantas, T.N.C., Santanna, V.C., Souza, T.T.C., Lucas, C.R.S., Neto, A.D., Aum, P.T.P., 2019. Microemulsions and nanoemulsions applied to well stimulation and enhanced oil recovery (EOR). Brazilian Journal of Petroleum and Gas 12(4), 251–265.

Davies, J., 1987. A physical interpretation of drop sizes in homogenizers and agitated tanks, including the dispersion of viscous oils. Chemical Engineering Science 42, 1671–1676.

El-Diasty, A.I., Ragab, A.M.S., 2013. Applications of Nanotechnology in the Oil & Gas Industry: Latest Trends Worldwide & Future Challenges in Egypt. In North Africa Technical Conference and Exhibition. pp. 1–13.

Esmaeili, A., 2011. Applications of Nanotechnology in Oil and Gas Industry. AIP Conference Proceedings, 133–136.

Forgiarini, A., Esquena J., Gonzalez C., Solans C., 2000. Studies of the Relation Between Phase Behavior and Emulsification Methods with Nanoemulsion Formation, in: Trends in Colloid and Interface Science XIV, pp. 36–39.

Forgiarini, A., Esquena J., Gonzalez C., Solans C., 2001. Formation of nano-emulsions by low-energy emulsification methods at constant temperature. Langmuir 17(7), 2076–2083.

Foroozesh, J., Kumar, S., 2020. Nanoparticles behaviors in porous media: Application to enhanced oil recovery. Journal of Molecular Liquids, 113876.

Gharibshahi, R., Jafari, A., Haghtalab, A., Karambeigi, M.S., 2015. Application of CFD to evaluate the pore morphology effect on nanofluid flooding for enhanced oil recovery. RSC Advances 5(37), 28938–28949.

Griffin, W.C., 1949. Classification of surface-active agents by 'HLB'. Journal of Society of Cosmetic Chemists of Japan 1(5), 311–326.

Hamidi, R., Ghasemi, S., Hosseini, S.R., 2020. Ultrasonic assisted synthesis of Ni3 (VO4) 2-reduced graphene oxide nanocomposite for potential use in electrochemical energy storage. Ultrasonics Sonochemistry 62, 104869.

Hendraningrat, L., Torsæter, O., 2015. Metal oxide-based nanoparticles: Revealing their potential to enhance oil recovery in different wettability systems. Applied Nanoscience 5(2), 181–199.

Hu, Z., Haruna, M., Gao, H., Nourafkan, E., Wen, D., 2017. Rheological properties of partially hydrolyzed polyacrylamide seeded by nanoparticles. Industrial & Engineering Chemistry Research 56(12), 3456–3463.

Huang, X.F., Wang, X.H., Lu, L.J., Liu, J., Peng, K.M., 2016. Technical progress of multiscale study on oil-water emulsion stability. Chemical Industry and Engineering Progress 1, 26–33.

Izquierdo, P., Feng, J., Esquena, J., Tadros, T.F., Dederen, J.C., Garcia, M.J., Azemar, N., Solans, C., 2005. The influence of surfactant mixing ratio on nanoemulsion formation by the PIT method. Journal of Colloid and Interface Science 285(1), 388–394.

Joonaki, E., Ghanaatian, S.J.P.S., 2014. The application of nanofluids for enhanced oil recovery: Effects on interfacial tension and coreflooding process. Petroleum Science and Technology 32(21), 2599–2607.

Kakaç, S., Pramuanjaroenkij, A., 2009. Review of convective heat transfer enhancement with nanofluids. International Journal of Heat and Mass Transfer 52(13–14), 3187–3196.

Kapusta, S., Balzano, L., Te Riele, P.M., 2011, November. Nanotechnology Applications in Oil and Gas Exploration and Production. In International Petroleum Technology Conference. OnePetro.

Karimi, A., Fakhroueian, Z., Bahramian, A., Pour Khiabani, N., Darabad, J.B., Azin, R., Arya, S., 2012. Wettability alteration in carbonates using zirconium oxide nanofluids: EOR implications. Energy & Fuels 26(2), 1028–1036.

Khilar, K.C., Fogler, H.S., 1998. Hydrodynamically Induced Release of Fines in Porous Media, in: Migrations of Fines in Porous Media. Springer, pp. 63–71.

Kim, I., Worthen, A.J., Lotfollahi, M., Johnston, K.P., DiCarlo, D.A., Huh, C., 2017, April. Nanoparticle-Stabilized Emulsions for Improved Mobility Control for Adverse-Mobility Waterflooding. In IOR 2017–19th European Symposium on Improved Oil Recovery (Vol. 2017, No. 1). European Association of Geoscientists & Engineers, pp. 1–13.

Kong, X., Ohadi, M.M., 2010. Applications of Micro and Nano Technologies in the Oil and Gas Industry–An Overview of the Recent Progress. In Society of Petroleum Engineers—14th Abu Dhabi International Petroleum Exhibition and Conference (ADIPEC 2010). pp. 1703–1713.

Kortam, M., Mousa, D., Abdeldayim, I., Santo, G., Lamberti, A., 2017, March. Designing and Implementing Adequate Monitoring Plan for Pilot Project of Polymer Flooding in Belayim Land Field. In Offshore Mediterranean Conference and Exhibition, OMC-2017. OMC.

Kumar, N., Verma, A., Mandal, A., 2021. Formation, characteristics and oil industry applications of nano-emulsions: A review. Journal of Petroleum Science and Engineering, 109042.

Lau, H.C., Yu, M., Nguyen, Q.P., 2017. Nanotechnology for Oilfield Applications: Challenges and Impact. In International Petroleum Exhibition & Conference. pp. 1160–1169.

Levitt, D.B., Pope, G., 2008. Selection and Screening of Polymers for Enhanced-Oil Recovery. In Society of Petroleum Engineers. SPE-113845, pp. 1–18.

Li, S., Hendraningrat, L., Torsæter, O., 2013, March. Improved Oil Recovery By Hydrophilic Silica Nanoparticles Suspension: 2 Phase Flow Experimental Studies. In IPTC 2013: International Petroleum Technology Conference (pp. cp-350). European Association of Geoscientists & Engineers.

Li, Y., Dai, C., Zhou, H., Wang, X., Lv, W., Wu, Y., Zhao, M., 2017. A novel nanofluid based on fluorescent carbon nanoparticles for enhanced oil recovery. Industrial & Engineering Chemistry Research 56(44), 12464–12470.

Liang, F., Al-Muntasheri, G.A., Li, L., 2016. Maximizing Performance of Residue-Free Fracturing Fluids Using Nanomaterials at High Temperatures. In SPE Western Regional Meeting, SPE-180402. SPE.

Mandal, A., Samanta, A., Bera, A., Ojha, K., 2010. Characterization of oil–water emulsion and its use in enhanced oil recovery. Industrial & Engineering Chemistry Research 49(24), 12756–12761.

Matteo, C., Candido, P., Vera, R., Francesca, V., 2012. Current and future nanotech applications in the oil industry. American Journal of Applied Sciences 9(6), p. 784.

Maurya, N.K., Mandal, A., 2016. Studies on behavior of suspension of silica nanoparticle in aqueous poly-acrylamide solution for application in enhanced oil recovery. Petroleum Science and Technology 34(5), 429–436.

Mendes, T.M., Hotza, D., Repette, W.L., 2015. Nanoparticles in cement based materials: A review. Reviews on Advanced Materials Science 40, 89–96.

Mishra, S., Bera, A., Mandal, A., 2014. Effect of polymer adsorption on permeability reduction in enhanced oil recovery. Journal of Petroleum Engineering 2014.

Moslan, M.S., Sulaiman, W.R.W., Ismail, A.R., Jaafar, M.Z., 2017. Applications of aluminium oxide and zirconium oxide nanoparticles in altering dolomite rock wettability using different dispersing medium. Chemical Engineering Transactions 56, 1339–1344.

Liu, J.X., Zhu, H.J., Wang, P., Pan, J.M., 2021. Recent studies of Pickering emulsion system in petroleum treatment: The role of particles. Petroleum Science 18(5), 1551–1563.

Negin, C., Ali, S., Xie, Q., 2016. Application of nanotechnology for enhancing oil recovery—A review. Petroleum 2, 324–333.

Ogolo, N., Olafuyi, O., Onyekonwu, M., 2012. Enhanced Oil Recovery Using Nanoparticles. In Saudi Arabia Section Technical Symposium and Exhibition. pp. 9–17.

Pal, N., Kumar, N., Saw, R.K., Mandal, A., 2019. Gemini surfactant/polymer/silica stabilized oil-in-water nanoemulsions: Design and physicochemical characterization for enhanced oil recovery. Journal of Petroleum Science and Engineering 183, 106464.

Pal, N., Mandal, A., 2020a. Enhanced oil recovery performance of Gemini surfactant-stabilized nanoemulsions functionalized with partially hydrolyzed polymer/silica nanoparticles. Chemical Engineering Science 226, 115887.

Pal, N., Mandal, A., 2020b. Oil recovery mechanisms of Pickering nanoemulsions stabilized by surfactant-polymer-nanoparticle assemblies: A versatile surface energies' approach. Fuel 276, 118138.

Pei, H., Zhang, G., Ge, J., Zhang, J., Zhang, Q., 2015. Investigation of synergy between nanoparticle and surfactant in stabilizing oil-in-water emulsions for improved heavy oil recovery. Colloids and Surfaces A: Physicochemical and Engineering Aspects 484, 478–484.

Ponmani, S., Nagarajan, R., Sangwai, J., 2013. Applications of nanotechnology for upstream oil and gas industry. Journal of Nano Research 24(1), 7–15.

Ponnapati, R., Karazincir, O., Dao, E., Ng, R., Mohanty, K.K., Krishnamoorti, R., 2011. Polymer-functionalized nanoparticles for improving waterflood sweep efficiency: Characterization and transport properties. Industrial & Engineering Chemistry Research 50(23), 3030–13036.

Pu, W.F., Liu, R., Wang, K.Y., Li, K.X., Yan, Z.P., Li, B., Zhao, L., 2015. Water-soluble core–Shell hyperbranched polymers for enhanced oil recovery. Industrial & Engineering Chemistry Research 54(3), 798–807.

Rostami, P., Sharifi, M., Aminshahidy, B., Fahimpour, J., 2019. The effect of nanoparticles on wettability alteration for enhanced oil recovery: Micromodel experimental studies and CFD simulation. Petroleum Science 16(4), 859–873.

Sagala, F., Hethnawi, A., Nassar, N.N., 2020. Hydroxyl-functionalized silicate-based nanofluids for enhanced oil recovery. Fuel 269, 117462.

Sakthipriya, N., Doble, M., Sangwai, J.S., 2015. Enhanced Oil Recovery Techniques for Indian Reservoirs, in: Petroleum Geosciences: Indian Contexts (Springer Geology). Springer International Publishing.

Shalbafan, M., Esmaeilzadeh, F., Safaei, A., 2019. Experimental investigation of wettability alteration and oil recovery enhance in carbonate reservoirs using iron oxide nanoparticles coated with EDTA or SLS. Journal of Petroleum Science and Engineering 180, 559–568.

Sharma, M.M., Zhang, R., Chenevert, M.E., 2012. A New Family of Nanoparticle Based Drilling Fluids. In SPE Annual Technical Conference and Exhibition. pp. 1–13.

Singh, S.K., Ahmed, R.M., Growcock, F., 2010, January. Vital Role of Nanopolymers in Drilling and Stimulations Fluid Applications. In SPE Annual Technical Conference and Exhibition. Society of Petroleum Engineers.

Sun, Q., Li, Z., Li, S., Jiang, L., Wang, J., Wang, P., 2014. Utilization of surfactant-stabilized foam for enhanced oil recovery by adding nanoparticles. Energy & Fuels 28(4), 2384–2394.

Taber, J.J., Martin, F.D., Seright, R.S., 1996. EOR Screening Criteria Revisited. In Proceedings of the SPE/ DOE Tenth Symposium on Improved Oil Recovery (SPE 35385), Tulsa, OK.

Tadros, T.F., Izquierdo, P., Esquena, J., Solans, C., 2004. Formation and stability of nano-emulsions. Advances in Colloid and Interface Science 108, 303–318.

Talebian, S.H., Masoudi, R., Tan, I.M., Zitha, P.L., 2013, July. Foam assisted CO2-EOR; Concepts, Challenges and Applications. In SPE Asia Pacific Enhanced Oil Recovery Conference, SPE-165280. SPE.

Wasan, D., Nikolov, A., Kondiparty, K., 2011. The wetting and spreading of nanofluids on solids: Role of the structural disjoining pressure. Current Opinion in Colloid and Interface Science 16(4), 344–349.

Wooster, T.J., Golding, M., Sanguansri, P., 2008. Impact of oil type on nanoemulsion formation and Ostwald ripening stability. Langmuir 24, 12758–12765.

Ye, Z., Qin, X., Lai, N., Peng, Q., Li, X., Li, C., 2013. Synthesis and performance of an acrylamide copolymer containing nano-SiO_2 as enhanced oil recovery chemical. Journal of Chemistry 2013, 1–33.

Zakaria, M., Husein, M., Harland, G., 2012. Novel Nanoparticle-Based Drilling Fluid with Improved Characteristics. In Proceedings of SPE International Oilfield Nanotechnology Conference. p. 2013.

Zhang, T., Espinosa, D., Yoon, K.Y., Rahmani, A.R., Yu, H., Caldelas, F.M., Ryoo, S., Roberts, M., Prodanovic, M., Johnston, K.P., Milner, T.E., 2011, January. Engineered Nanoparticles as Harsh-Condition Emulsion and Foam Stabilizers and as Novel Sensors. In Offshore Technology Conference.

Zhang, T., Roberts, M.R., Bryant, S.L., Huh, C., 2009, April. Foams and Emulsions Stabilized with Nanoparticles for Potential Conformance Control Applications. In SPE International Conference on Oilfield Chemistry, SPE-121744. SPE.

7 Other EOR Methods

7.1 MICROBIAL ENHANCED OIL RECOVERY

7.1.1 INTRODUCTION

Microbial enhanced oil recovery (MEOR) is an important tertiary oil recovery approach that uses microorganisms and their metabolites, including biosurfactants, biopolymers, biogenic acids, enzymes, solvents, and biogases to mobilize residual oil (Banat, 1995; Lazar et al., 2007; Sen, 2008). In comparison to the different petroleum-based chemical EOR methods, MEOR processes have the advantages of cost independence of crude oil prices, because microorganisms can metabolize metabolites with inexpensive raw materials—for example, molasses, corn syrup, and other agricultural by-products, even agricultural organic waste. In addition, all the additives used in MEOR are biodegradable and, therefore, environmentally friendly in comparison to other chemical EOR processes.

The concept of using microbes for enhancing oil recovery was first proposed by J. W. Beckman in 1926. Later on, the use of microorganisms for oil recovery enhancement was patented by C. E. ZoBell in 1946; since then, the MEOR process has been validated by numerous studies and successful fields tests, the first of them carried out in Arkansas, the United States, in 1954.

MEOR mechanisms are the same mechanisms obtained from other chemical-enhanced oil recovery (CEOR) methods; however, MEOR presents the advantage that microbial metabolites are directly produced in the reservoir rock formation, which makes them more effective. The microbes consume nutrients and reproduce, while beneficial metabolites such as surfactants and polymers are a biological by-product of this process. MEOR processes offer a low-cost approach for improving oil recovery and, hence, is of interest for the oil industry.

7.1.2 MECHANISMS

In 1947, Zobell performed many field tests and found that bacteria help in recovering oil from sedimentary materials (Zobell, 1947). From the observation of the tests, Zobell proposed the following mechanisms of oil recovery by MEOR processes:

(1) Production of gaseous CO_2.
(2) Production of organic acids and detergents.
(3) Dissolution of carbonates in rock.
(4) Physically dislodgement of the oil.

Other mechanisms for enhanced oil recovery in MEOR, as proposed by Lazar et al. (2007), are as follows:

- porosity and permeability modification;
- wettability alteration;
- oil solubilization;
- emulsification;
- interfacial forces alteration;
- lowering oil mobility ratio; and
- microbial metabolic pathways alteration by sodium bicarbonate.

DOI: 10.1201/9781003098850-7

So the use of bacteria to enhance oil recovery (MEOR) involves the formation of different gases (CO_2, CH_4, H_2) in the reservoir and surfactants, polymers, alcohols, and acids or the enzymic breakdown of the hydrocarbon molecules (Zahid et al., 2007). These microbial products have different effects on reservoir containing oil and aid in improving/enhancing oil recovery from it as shown in Table 7.1.

In microbial oil recovery, microbial products undergo a series of very desirable changes in the physicochemical properties of crude oil and significantly improve or almost completely restore the lithology of reservoir rocks, as shown in Figure 7.1 (Niu et al., 2020). Therefore, MEOR can be considered as a complex CEOR that uses microorganisms and composite multiple recovery technologies.

7.1.3 FACTORS TO BE CONSIDERED FOR APPLYING MEOR

The main constraint of MEOR technology is insufficient consideration of the conditions that characterize petroleum reservoirs and the physiology of microorganisms that thrive in these conditions (Sheehy, 1991). The activities of microbes employed in the MEOR process depends on the physical and chemical conditions they encounter in the reservoirs.

7.1.3.1 Selecting the Reservoir

Selecting the reservoir for MEOR processes is a challenging task as the growth of the microbes are sensitive to salinity, temperature, pressure, pH, redox potential, etc., and these reservoir conditions vary a great deal from one reservoir to another. The following common factors should be considered before applying MEOR technology to a certain reservoir.

TABLE 7.1
Microbial Products with Application Routes in MEOR (Niu et al., 2020)

Microbial Products	Application Routes
Biosurfactant	• Reduce surface/interfacial tension • Alter rock wettability • Emulsified crude oil to form oil-water emulsion
Biopolymers	• Improve the viscosity of displacing fluid, and decrease the mobility ratio of water/oil • Improve the sweep areas and efficiency by selective plugging
Biogases (CO_2, CH_4, H_2, N_2)	• Reduce the viscosity of crude oil, and improve its fluidity • Restore reservoir pressure • Partial gas or miscible gas flooding
Biogenic acids (Low molecular weight fatty acids such as formic acid and acetic acid)	• Improve porosity and permeability by dissolving rock in pore throats • Dissolve cementing materials in the formation and, thereby, in flow channels • CO_2 produced during the dissolution of the rock has the same effect as biogases
Solvents (Low molecular weight alcohols and ketones such as ethanol, butanol, acetone)	• Reduce the viscosity of oil by dissolving in crude oil • Alcohols and ketones can participate in the micelle formation process as cosurfactants, which can reduce CMC and interfacial tension to promote the emulsification process • Improve porosity and permeability by dissolving heavy oil in pore throats
Biomass (Microbial cells)	• Alter rock wettability by bacteria cells (as biosurfactants) • Selective plugging to improve sweep efficiency (as biopolymers) • Selective degradation of heavy oil

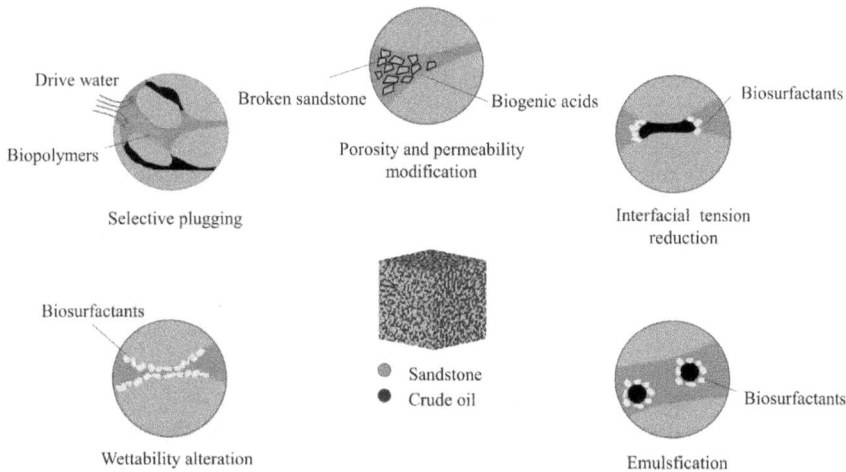

FIGURE 7.1 Schematic illustration of the mechanism in MEOR.

Source: Niu et al., 2020.

7.1.3.2 Structural Analysis

Structural analysis should be done to have an optimized plan before injecting the microbes into the formation. It also helps to identify the drilling uncertainties and operational risks. A detailed information and analysis of initial water saturation, spatial distribution of oil lenses, and special distribution of facies and faults are of utmost importance before adopting MEOR.

7.1.3.3 Geological Complexity

A detailed information of geological complexity is required for proper planning of a MEOR project. Due to several reasons like changes in permeability, porosity, wettability, etc., the microbes might not reach the target zone.

7.1.3.4 Petrophysical Properties

Porosity and permeability are important parameters for selecting the strain of bacteria and its survival. Even though, the pores in rock can be connected in different ways, pore spaces less than 0.5 nm can place severe restrictions on the ability of most bacteria (most bacteria have dimensions of length approximating 0.5–10.0 μm and widths of 0.5–2.0 μm) to be transported through the rock matrix especially for those bacteria whose sizes are comparable to the sizes of the rock pores (Jenneman and Clark, 1992).

7.1.3.5 Temperature

Temperature plays a significant role in bacterial metabolism. With increasing depth, the temperature increases. Therefore, it is certain that bacterial growth and their metabolism will be affected, as increasing temperature can exert negative effects on enzyme function by disruption of important cell activities. So, the reservoir temperature must suit the microorganisms for their survival and growth, and in high-temperature reservoirs, the development of such conditions that can sustain and stimulate the growth of microorganisms is a big challenge.

7.1.3.6 Salinity

Sodium chloride makes up about 90% or more of the total dissolved solids found in reservoir brines, and therefore, tolerance to salt concentrations is one of the most important characteristics needed for microorganisms used in MEOR. The effect to which salinity causes changes in bacterial growth and metabolism depends on the osmotic balance required for such growth, as the solute concentration of the surrounding environment can affect cell growth.

7.1.3.7 Pressure

The effect of pressure on microorganisms depends not only on the magnitude but also on the duration of pressure applied in combination with temperature, pH, oxygen supply, and composition of the culture media (Abe, 2007). The effects of the pressure can be very complex and often difficult to interpret. For example, recent results indicated that lactic acid bacteria *Lactobacillus sanfranciscensis* growth at 50 MPa is 30% less than at atmospheric pressure.

7.1.3.8 pH

pH is one of the major environmental factors that affect microbial growth and is one of the most studied because of its importance in fundamental research. In general, the optimal pH for growth for microorganisms is between 4.0 and 9.0, but at very low pH, the metabolic activities of microorganisms can be affected.

7.1.4 Nutrients

The success of the MEOR process depends on the availability of essential nutrients for bacterial growth and metabolism, and there is a smooth relationship between the growth rate of bacteria and nutrient concentrations (Monod, 1949). Bacterial requirements for growth include sources of energy, mostly organic carbon (i.e., sugars and fatty acids) and mineral ions (e.g., iron and phosphorus). These nutrients are mostly transported in the aqueous phase. Fermentative bacteria use glucose-, sucrose-, or lactose-containing nutrients.

7.1.5 Selection of Bacteria

For MEOR, microbial species selection is crucial. They have to fit the reservoir conditions and produce the required products. Different strains of bacteria have different adaptabilities; therefore, selecting the right bacteria based on reservoir conditions and fluid properties is important. The reservoir ecosystem provides the basis for a positive response from bacteria. Pore colonization by bacteria and the consequent products produced by bacteria need special attention also.

Earlier studies on MEOR showed that both mixed culture and pure strain of bacteria have been used for the process. For example, Hitzman (1983) used pure and mixed cultures of *Bacillus*, *Clostridium*, and *Pseudomonas* in 2%–4% molasses in the United States. Wang et al. (1993), used mixed enriched bacteria cultures of *Bacillus*, *Pseudomonas*, *Eurobacterium*, *Fusobacterium*, and *Bacteriodes* in a 4% residue sugar.

According to the source of the strains, microorganisms can be divided into exogenous microorganisms and indigenous microorganisms. Thus, microbial flooding recovery could be divided into indigenous microbial flooding recovery and exogenous microbial flooding recovery.

7.1.6 Classification of MEOR

The objective of most MEOR projects is to reduce the remaining oil in the reservoir; however, the implementation of the MEOR strategy can differ. There are three methods of MEOR utilization:

- The first injects a solution of microbes and nutrients into the well.
- The second injects only nutrients in the hopes of activating in situ microbes.
- The third involves developing the microbes outside of the reservoir and only injecting their metabolites into the well.

The application can be cyclic (single well simulation), microbial flooding recovery, or selective plugging recovery.

The application of MEOR technology thus can be any of the following (Lazar, 2007):

(a) cyclic (single well simulation);
(b) microbial flooding recovery; or
(c) selective plugging recovery.

(a) Cyclic (single well simulation): The process is similar to cyclic steam injection. In cyclic microbial recovery, microorganisms and nutrients are injected into production wells. The wells are shut in for a period long enough to allow microbial growth and metabolite formation. This can be for a number of days or weeks. The different mechanisms of EOR are already mentioned in Table 7.1. Finally, the oil production phase begins and extends over a period of weeks or months. In cyclic microbial recovery, when production declines, another phase of injection is normally started. In this case, the depth of the area covered by bacteria would be limited by the injection rate and the kinetics of the microbial process (Bryant and Lockhart, 2002).

(b) Microbial flooding recovery: The second type of application is microbial flooding. In microbial flooding, the microbial growth is usually stimulated by adding nutrients to the injection water to encourage the proliferation of microorganisms indigenous to the formation (Jimoh, 2012). If the requisite microbial activity is not present, then microorganisms can be injected into the formation along with the nutrients. In some approaches, injection into the formation is stopped to allow time for in situ growth and metabolism to occur (Youssef et al., 2009). In other approaches, injection of brine is continued after nutrient and/or cell injection. This option would most likely be less expensive, as the growth would be stimulated in larger parts of the reservoir, particularly, where the carbon source (residual oil) is located, which is usually the target of the EOR treatment (Kaster et al., 2012). A schematic of microbial flooding for EOR is shown in Figure 7.2.

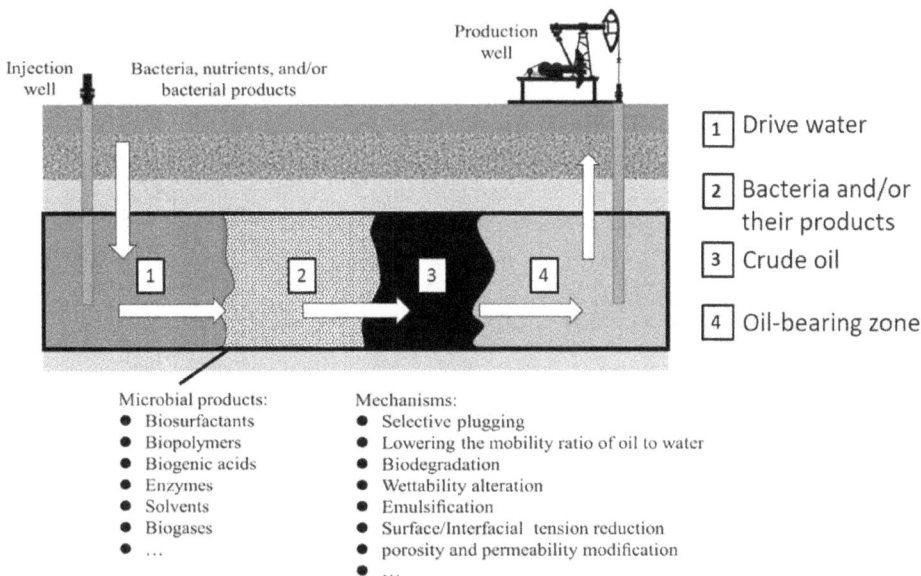

FIGURE 7.2 A sketch of microbial flooding oil recovery.

Source: Niu et al., 2020.

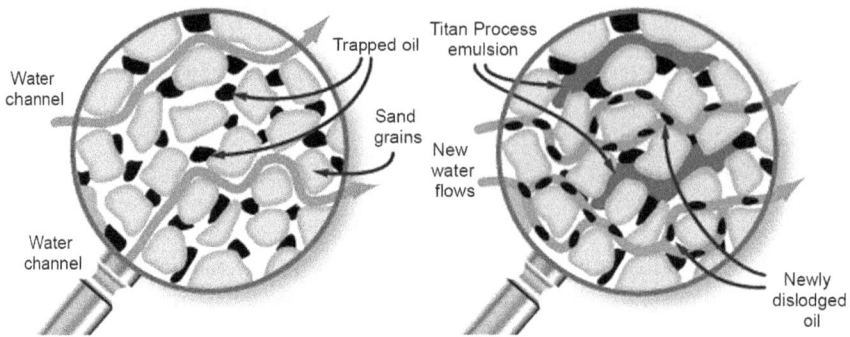

FIGURE 7.3   Schematic of selective plugging in MEOR.

(c) Selective plugging recovery: Microbial selective plugging encompasses a microbial process to divert water into low-permeability regions to block water channels deep in the reservoirs (Jimoh, 2012). With this type of treatment, nutrient preferentially flows into high-permeability regions, which then stimulates biomass and polymer production in these regions; both of which reduce the permeability of the rock (Raiders et al., 1986). In contrast, heavy oil modification is usually by microbial decomposition of long-chain compounds within the formation. A schematic of selective plugging recovery mechanism is shown in Figure 7.3.

7.1.7 Screening Criteria

The screening criteria for a reservoir vary greatly. The activity of microorganisms in the reservoir is affected by the environmental conditions of the reservoir when injected or activated (Safdel et al., 2017). These influencing factors include lithology, reservoir properties, fluid properties, etc., which directly affect the growth, migration, and metabolism of microorganisms (Hong et al., 2019). With developments in technology, changing oil prices, and new strategies being defined every day, it may be difficult to generalize screening criteria. A careful review of the results of a number of MEOR field trials reveals that the projects are successful under the following conditions:

- Minimum reservoir permeability of 75 md to ensure propagation of bacteria and nutrients deep into the reservoir.
- Reservoir temperature < 200°F.
- Brine salinity < 100,000 ppm.
- Reservoir must be under a waterflooding recovery.

7.1.8 MEOR Field Implementation

Prior to field implementation, the MEOR projects should be passed through (a) initial lab tests and then (b) pilot tests. The initial tasks include the following (Zahid et al., 2007):

(i)   Screening of available microbe-nutrient systems that are viable in reservoir conditions in terms of compatibility, competitiveness, and abilities to propagate in porous media.
(ii) Investigation of likely process by-products (polymers, surfactants, and gases) and their effects on oil recovery.
(iii) Screening candidate reservoirs for MEOR application.

In pilot testing, the following parameters require strict analysis (Zahid et al., 2007):

(i)   Oil production
(ii)   Test results on standard cores
(iii)   Huff-n-puff tests
(iv)   Well selection
(v)   Incremental reserves
(vi)   Evidence of new oil in produced fluids
(vii)   Distribution of nutrients throughout the reservoir
(viii) Evidence of microbial proliferation in the reservoir
(ix)   Performance of nutrient-injector wells
(x)   Maintenance and growth media
(xi)   Modeling of microbial enhanced oil recovery
(xii)   Incremental production cost per barrel

The earliest field trials of MEOR were mainly conducted in the United States, former Soviet Union, and Eastern European countries in the 1950s and 1960s. In the past decades, a number of MEOR field trials around the world have achieved varying degrees of success. According to global field test statistics, more than 90% of MEOR field trials produce positive effects (Safdel et al., 2017). In China, MEOR has conducted field tests and applications in several oil fields, such as Daqing, Shengli, Xinjiang, Jilin, Liaohe, Qinghai, and Changqing. China has a total of 678 wells in 12 fields, including the Jilin, Shengli, Zhongyuan, and Daqing oil fields (Niu et al., 2020). Implementation of microbial huff-n-puff in the Shengli and Daqing oil fields produced additional oil recovery of 219,000 tons and 64,000 tons, respectively. In fact, China is one of the leaders in the MEOR field due to the successful application of MEOR in China in recent years. In India, the Oil and Natural Gas Corporation (ONGC) Limited, in collaboration with the Energy and Resources Institute (TERI, New Delhi) and the Institute of Reservoir Studies (IRS), Ahmedabad, conducted some field trials through huff-n-puff. These field trials were based on stringent anaerobic microbial population isolation from the reservoir, and field trials of 12 wells in four fields showed a threefold increase in crude oil production and a significant reduction in water cut (Sen, 2008).

7.2   CARBONATED WATER INJECTION FOR EOR

7.2.1   INTRODUCTION

In the case of CO_2 injection, when CO_2 transfers from brine to oil, then oil starts to swell, and due to this, the swelling viscosity reduces. Due to a density difference between oil and CO_2, the sweep efficiency is very low during the CO_2 injection into the oil reservoir, and gravity segregation takes place. Poor sweep efficiency decreases the contact between the reservoir oil and the injected CO_2 due to a decrease in contact area, and hence, the performance of CO_2 injection decreases. Poor sweep efficiency also has a significant adverse effect on the economical aspect of CO_2-EOR projects because it leads to low oil recovery and premature CO_2 breakthrough, which requires CO_2 separation and reinjection of carbon dioxide. After secondary recovery, the presence of isolated oil ganglia is affected by the water layers present between them, which is called the water shielding effect or the water blocking effect. These water layers negatively impact CO_2 performance. This so-called water shielding or water blocking effect can prevent direct contact between the oil and CO_2, which leads to a reduction in the rate of CO_2 dissolution in the oil.

Therefore, direct injection of CO_2 might not result in improvement of additional oil recovery economically. As an alternative to this, carbonated water injection (CWI) is one of the best options, as carbonated water has a much better sweep efficiency and slower CO_2 breakthrough, because in carbonated water injection, CO_2 is dissolved in water, which leads to an increase in the viscosity of the displacing fluid. And due to an increase in viscosity, the mobility of the displacing fluid decreases, which improves the sweep efficiency and channeling issue also.

Technically, there are two major differences between carbonated water injection and direct CO_2 injection or water alternating gas injection. First, the amount of CO_2 injected is solubility dependent, so the amount of CO_2 is limited. Second, the displacement process is CO_2 mass transfer dependent rather than minimum miscibility pressure.

In the injection of CO_2 in free phase in a reservoir, there is a chance of leakage through the micropores of the caprock, so CO_2 can be better stored as carbonated water injection. CWI can be a very attractive method for CO_2 sequestration also, as we know that CO_2 plays an important role in global warming. So, CO_2 sequestration is becoming a hot topic in terms of environmental issues, which will also be of benefit in the form of a reduction in greenhouse gases. Leakage of CO_2 gas sometimes limits the number of reservoirs available for CO_2 sequestration. On the other hand, CO_2 sequestration can be implemented without volume limitations with a lower risk of gas leakage through the micropores of the caprock. Carbonated water has a higher density and viscosity due to the dissolved CO_2 gas; hence, it will sink into the bottom of the reservoir (Ghosh et al., 2020).

7.2.2 Process Overview of Carbonated Water Injection

In the CWI process, CO_2 remains in the dissolved state before being injected into the reservoir. When carbonated water is injected into the reservoir, it comes into contact with oil. And as we know that the solubility of CO_2 in oil is much higher than in water, so a mass transfer of CO_2 from water to oil will occur. This mass transfer reduces the oil viscosity, lowers the water-oil interfacial tension (IFT), and causes oil swelling, which is responsible for the reconnection of isolated residual oil ganglia, mobilizing the trapped oil toward the production well. Oil mobility thus increases due to a reduction in the viscosity of the oil, which is in favor of enhanced oil sweeping efficiency (Esene et al., 2019).

In a vertical well, the pressure of the carbon dioxide builds up sharply at the start of injection, but it drops afterward when injection of CO_2 starts at constant mass flow rate. In a horizontal well, the pressure of CO_2 increases continuously. On the basis of caprock stability and formation stability, the type of well (vertical or horizontal) can be chosen. But in a horizontal well, CO_2 gives better sweep efficiency as compared to vertical wells. Hence, additional recovery will be more in case of a horizontal well as compared to a vertical well. As expected, more oil sweeping efficiency, stable displacement front, delayed water breakthrough, and less channeling are observed during the displacement process in the CWI process.

From Figure 7.4, it can be observed that in the case of direct CO_2 injection, sweeping is less uniform in comparison to CWI, which leads to a comparatively lower recovery in direct CO_2 injection.

7.2.3 Properties of CO_2 and Brine/Oil Systems

7.2.3.1 Properties of Pure CO_2

CO_2 can be injected as liquid, gas, or supercritical fluid; all of them are strong functions of pressure and temperature, which can be explained by a phase diagram (Fig. 7.5).

7.2.3.2 Properties of CO_2-Brine Systems

CO_2 has much higher solubility in the brine phase as compared to other reservoir gases. As we know that feasibility of the CWI process depends on the solubility of CO_2 in brine. Dissolved CO_2 can react with water and dissociate into HCO_{3-} and CO_3^{2-} ion. Solubility of CO_2 in brine will depend on temperature as well as pressure. Solubility of CO_2 in brine is inversely proportional to salinity.

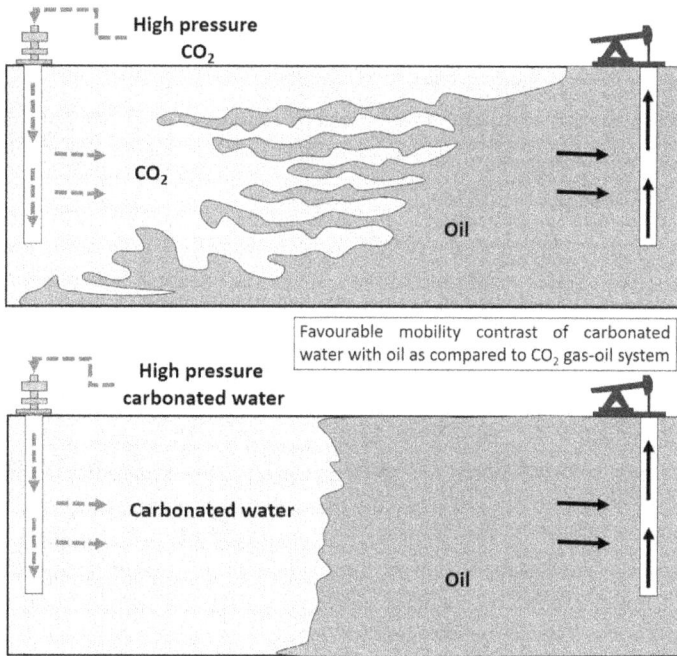

FIGURE 7.4 Comparison of sweeping during direct CO_2 injection and carbonated water injection.

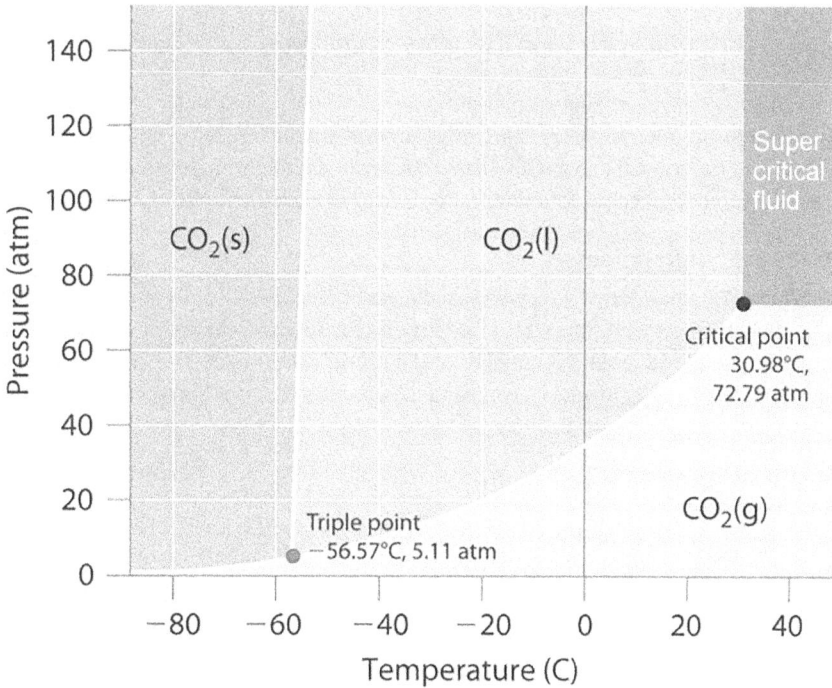

FIGURE 7.5 A typical phase diagram for a substance that exhibits three phases—solid, liquid, and gas—and a supercritical region.

Source: https://chem.libretexts.org.

7.2.3.3 Properties of CO_2-Oil Systems

In carbonated water injection, the production rate of oil depends on the solubility of the CO_2-oil system. If solubility of the CO_2 in oil is high, then swelling in the oil will be much higher, which will lead to more recovery. Solubility of CO_2 in oil depends on temperature and pressure.

7.2.4 Possible Mechanisms and Principle of CWI

During the mutual interactions process, fresh carbonated water comes in contact with live crude oil, the dissolved CO_2 in water will partition between live crude oil and water phases which will result in oil swelling and a change in interphase position. Partition coefficient is defined as the ratio of CO_2 concentration in oil to CO_2 concentration in carbonated water after equilibrium. These are the several mechanisms that can explain the recovery by CWI, as discussed in the following:

7.2.4.1 Oil Swelling and Viscosity Reduction

Dissolution of CO_2 into crude oil causes two visible changes (i) swelling of crude oil (ii) reduction in its viscosity. The swelling factor of oil is inversely proportional to the residual oil left in the formation. The mechanism is explained by CO_2 mass transfer from water phase to oil phase without forming a different CO_2-rich phase. The reason behind the mass transfer can be explained by the solubility effect, as we know that CO_2 is much more soluble in oil as compared to water. This leads to the transfer of CO_2 from water to oil under the same pressure and temperature conditions. When CO_2 is transferred into oil, then the oil swells, its viscosity reduces, which then improves the mobility of the oil. All these subsequent processes together result in greater oil recovery than in conventional methods.

Distribution of CO_2 at equilibrium between the oil and water phases may be expressed in terms of the partition coefficient ($K_{CO_2, ow}$) at the prevailing reservoir pressure and Temperature conditions by Equation 7.1

$$K_{CO_2, ow} = C_{CO_2, o} / C_{CO_2}, w \qquad (7.1)$$

where $C_{CO_2, O}$ and $C_{CO_2, w}$ are the concentration of CO_2 in oil phase and water phase, respectively. The partition coefficient is a function of pressure, temperature and the crude oil composition.

In CWI, the oil starts to swell due to CO_2 mass transfer in the oil through carbonated water. The CO_2 diffusion into oil leads to the reconnection of trapped oil ganglia, which subsequently starts to mobilize.

7.2.4.2 Wettability Alteration

When the surface becomes more water wet, then water thickness on the pore surface increases. When CO_2 is dissolved into oil, then the viscosity of the oil reduces, and it oil swells. Due to the dissolution of CO_2 in oil, the polar component of the oil is destabilized. The destabilized polar components can disperse through the water layers and adsorb onto the rock surface, which then increases the tendency of the pore system to become water wet.

7.2.4.3 Improved Relative Permeability

The effect of injected CO_2 depends on several factors such as the rock chemistry, injected fluid type, injection strategy, and the physical conditions of the reservoir. It is known that when CO_2 is injected into carbonate formation, it leads to the dissolution of the carbonate minerals because of the acidic nature of carbonated water. Dissolution of the rock leads to an initial increase in formation permeability; subsequently, the transportation of these minerals and later precipitation lead to a decrease in permeability and effective porosity (Ghosh et al., 2020).

7.2.4.4 Evolution of Solution Gas

When we inject carbonated water into the reservoir, CO_2 transfers from water to oil. Due to the transfer of CO_2, the solution gas starts to come out from the crude oil, which also favors additional recovery.

7.2.4.5 IFT Reduction

CO_2 solubility is a governing mechanism for controlling the IFT of carbonated brine and oil. At constant pressure and temperature, CO_2 solubility is affected by brine salinity, where the higher the salinity, the lower the solubility (Mohammadian et al., 2015). Due to their low reactivity toward polar water molecules, CO_2 molecules tend to move toward the surface (oil/water interface), which results in IFT reduction. A decrease in IFT increases the oil displacement efficiency, which results in higher recovery.

When CO_2 does not evolve as a separate free gas, the IFT of CW oil contributes to the efficiency of CWI. The IFT measurements of CW crude oil at reservoir conditions reported in the literature show that the IFT of carbonated brine with oil is less than that of brine oil using similar fluids because of CO_2 effects (Manshad et al., 2016). In addition, an increase in temperature and pressure causes a further reduction in the IFT of CW oil (Honarvar et al., 2017).

7.2.5 Factors Affecting CWI Performance

There are several factors that affect CWI performance (Esene, 2019). These are discussed in the following.

7.2.5.1 Effects of Petrophysical Properties on CWI Performance

The interactions between the injected carbonated water and reservoir rock can lead to various chemical reactions, especially in carbonate types of rock. These reactions will influence the fluid flow and oil recovery mechanisms. The performance of carbonated water injection depends on different reservoir conditions, which include reservoir permeability, wettability, formation salinity, crude oil properties, reservoir pressure and temperature, etc. Experimental investigation shows that the overall performance of CWI is not much affected by heterogeneities such as fractures and vugs in a porous medium (Sohrabi et al., 2009). Carbonated water injection is slightly better in the mixed-wet condition as compared to that in the water-wet condition. The presence of continuous oil-wet paths of appreciable lengths in mixed-wet rock causes better oil connectivity, as it allows film flow of the oil along the wetting phase even at a low oil saturation.

7.2.5.2 Effects of Fluid Properties on CWI Performance

CWI performance also depends on the physical properties of the carbonated water and the properties of crude oil. The salinity of carbonated water, CO_2 content, crude oil density, and crude oil viscosity have significant effects on CWI performance. (i) The recovery will be higher in case of low-salinity water injection due to synergistic mechanisms of CO_2 injection and low-salinity waterflooding. (ii) It is observed that the recovery factor of a CWI process with a higher quantity of dissolved CO_2 is more than that with a lower CO_2 concentration. (iii) The higher recovery is generally experienced for the light oil during CWI, which is attributed to a higher oil swelling and more reconnection of the oil ganglia, as compared to the viscous oil.

7.2.5.3 Effects of Operational Parameters on CWI Performance

CWI performance also depends on operational parameters such as pressure, temperature, and injection rate. (i) For higher operating pressures, the solubility of CO_2 in water will be higher, which will lead to extra oil recovery. (ii) If the temperature is low, then the solubility of CO_2 is higher, and the carbonated water holds more CO_2, which can be subsequently stored and produced in the pore spaces of the porous media. (iii) If the injection rate is low, then the contact time between the carbonated water injected and the oil will be longer, which will allow a greater amount of CO_2 transfer across the phases, leading to a higher recovery factor.

7.2.6 New Development: Hybrid CWI

In recent times, several new techniques have been developed to recover even more oil. Smart water flooding, low salinity water injection and nano-fluid injection are some of them. Carbonated water injection, coupled with any of these techniques is considered to be even more efficient EOR option owing to their synergistic impacts. The effect of these smart techniques seems to be even more promising methods to recover additional oil.

Dong et al. (2014) studied the influence of adding active components to carbonated water. The study indicated that active components can reduce the interfacial tension between oil and carbonated water, and more CO_2 would be easier to transfer to oil phase from the carbonated water, which would improve CWI performance. In addition, carbonated water has the ability to replace and disseminate the surfactant adsorbed in formation, so it can improve surfactant flooding performance. Therefore, using active carbonated water (adding surfactant in carbonated water) may have better EOR performance than conventional CWI.

7.3 PLASMA PULSE TECHNOLOGY IN EOR

The productivity and injectivity of a well can be improved by well-practiced stimulation technology. Plasma Pulse Technology (PPT) is a special kind of stimulation technique where oil production can be enhanced without formation damage and without using chemical agents. PPT is based on scientific research in the areas of geology, petrophysics, blasting theory, acoustic, wave theory, and resonance theory (Chellappan et al., 2015). The main mechanism of this technology is to reduce crude oil viscosity. The wave vibration and viscoelastic impact of plasma pulse has a very positive effect on the reduction of crude oil viscosity, especially in heavy oils (Burhanov et al., 2004).

7.3.1 PROCESS

The technology was first introduced to the US industry in 2013. It was invented at St. Petersburg State Mining University in Russia. The technology (Fig. 7.6) works on plasma physics principles where a plasma pulse generator is lowered into the wellbore and controlled by a central unit at the surface. The tool generates high power shock waves, which clean in intervals followed by propagation in the form of elastic waves in the reservoir. The elastic waves initiate resonant frequencies that increase the relative mobility of crude oil to water. The series of impulse waves/vibrations also penetrate deep into the reservoir causing nano-fractures in the matrix, which increase reservoir permeability. PPT is an environmentally friendly technology that allows producers/injectors to obtain sustained higher productivity/injectivity. The tool cleans the perforated intervals and changes the wells' inflow characteristics by fixing near-wellbore damage while increasing the mobility of hydrocarbons within the surrounding reservoir. The cleaning of the near-wellbore region, increased relative oil mobility, and the generation of elastic vibrations and their resonance continues after the well is treated with PPT and can sustain an increased production flow for periods of up to 12 months or more. The resonance vibrations created in the formation make it possible to clean existing filtration channels and to create new ones at a distance over 1,500 meters from the point of initiation of the plasma pulse action. In addition to the large-scale action as discussed earlier, this creation of plasma also allows local problems such as poor well drainage to be resolved as paraffin, asphaltenes, scales, and other materials are cleaned away. This plasma pulse technology can be applied in vertical, deviated, and horizontal wells, with proper conveyance techniques.

The principal operating procedures of PPT are very simple and can be done in a simple way:

- Plasma pulse tool is placed opposite to the perforated interval.
- Initiation of the metallic conductor explosion and formation of plasma accompanied by a compression wave.

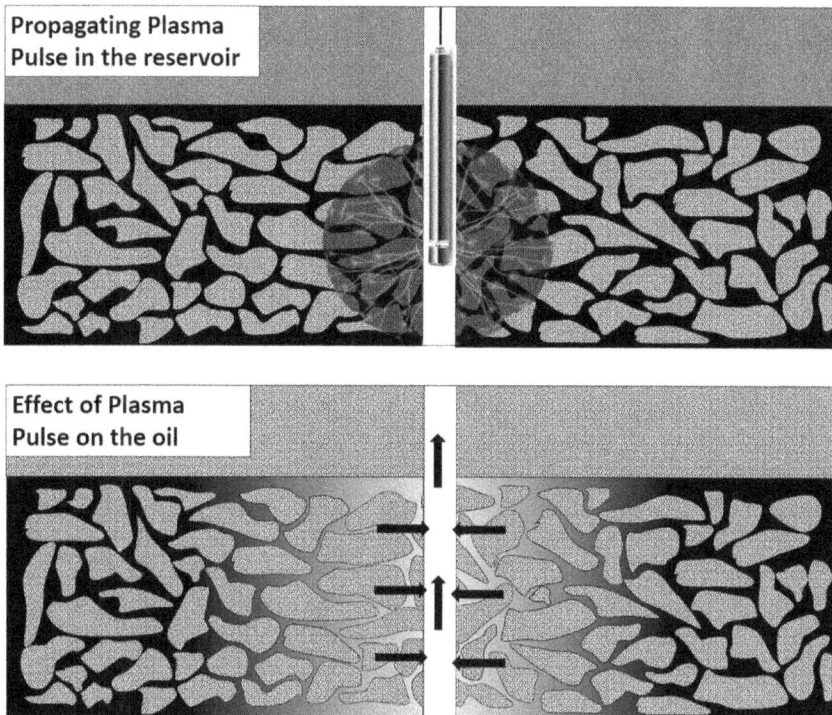

FIGURE 7.6   Schematic mechanism of plasma pulse technology for oil recovery.

- Through the perforated channels, the initiated shock wave penetrates the drainage area and propagates further on into the stratum inducing elastic vibrations.
- Plasma cools down, and the excessive formation pressure forces the sedimentation to flow into the sump of the well, the shock wave altering into flexible volume oscillations.

7.3.2 MECHANISMS

PPT has a good potential to improve oil recovery by energizing the well at the development stage. The mechanisms of PPT depend on the geology of formation and the rock-fluid properties of the reservoirs. Reduction of effective crude oil viscosity is one of the main mechanisms of EOR by PPT. It is highly effective in heavy oil reservoirs. Depending upon the crude oil type, there may be up to 30% reduction in oil viscosity as a result of thixotropic structure destruction. Depending on the composition of crude, their molecular structures can be excited with a particular range of frequency (Patel et al., 2018). Table 7.2 shows the results of the rheological measurement of the high-viscosity oils at the Usinskoye field after PPT (pulse number 10–40); they reflect a reduction of oil viscosity by 30% and manifestation of thixotropic properties up to 40% depending on the viscosity of the treated oil (Pashchenko and Ageev, 2016).

However, there are many other factors that are indirectly involved in EOR, as given here:

- Multiple increase in liquid aggregation.
- Due to flotation, oil drops are forced to the surface.
- Creation of fissures resulting in enhanced permeability.

TABLE 7.2
Rheological Properties of High-Viscosity Oil after PPT (Pashchenko and Ageev, 2016).

Oil Sample #	Effective Viscosity of Oil, MPa*s		Hysteresis Loop Area, J/m²		Thixotropy Energy, J/m³	
	Before Treatment	After Treatment	Before Treatment	After Treatment	Before Treatment	After Treatment
1	682	620	5.72	4.20	1.49×10^4	1.31×10^4
2	342	267	4.32	2.95	1.65×10^3	1.48×10^3
3	212	149	2.11	1.43	1.24×10^3	0.95×10^3
4	235	172	2.53	1.57	1.78×10^3	1.02×10^3
5	14550	9877	87.15	75.72	1.85×10^5	1.33×10^5
6	1792	1367	20.30	15.63	1.17×10^4	0.73×10^3
7	407	329	5.19	3.11	7.8×10^2	4.03×10^2

- Applicable in all types of well profiles with proper conveyance techniques.
- Uses wave velocities that will not damage the well.
- Integration of stimulation and EOR processes in a single step.
- Increase the injection wells' injection capacity.
- Restore the well's yield or production capacity after drop in the hydraulic fracturing impact.
- Redistribute the well's intake/injection capacity profile.

PPT is an environmentally friendly technology that allows producers/injectors to obtain sustained higher productivity/injectivity. The tool cleans the perforated intervals and changes the wells' inflow characteristics by fixing near-wellbore damage while increasing the mobility of hydrocarbons within the surrounding reservoir. PPT successfully tested in Kuwait for the first time in well RA-000A, where the productivity of the tested well increased from 1.5 to 2.25 bbl/day/psi.

REFERENCES

Abe, F., 2007. Exploration of the effects of high hydrostatic pressure on microbial growth, physiology and survival: Perspectives from piezophysiology. Bioscience, Biotechnology and Biochemistry 71(10), 2347–2357.
Banat, I.M., 1995. Biosurfactants production and possible uses in microbial enhanced oil recovery and oil pollution remediation: A review. Bioresource Technology 51, 1–12.
Beckman, J.W., 1926. The action of bacteria on mineral oil. Industrial Engineering Chemical (News edition) 4, 23.
Bryant, S.L., Lockhart, T.P., 2002. Reservoir engineering analysis of microbial enhanced oil recovery. SPE Reservoir Evaluation & Engineering 5, 365–374.
Burhanov, R.N., Hannanov, M.T., 2004. Geology of Natural Bitumen and High-Viscosity, Textbook. AGNI.
Chellappan, S.K., Al Enezi, F., Marafie, H.A., Bibi, A.H., Eremenko, V.B., 2015, October. First Application of Plasma Technology in KOC to Improve Well's Productivity. In SPE Kuwait Oil and Gas Show and Conference. OnePetro.
Dong, Y., Dindoruk, B., Ishizawa, C., Lewis, E., Kubicek, T., 2011, October. An Experimental Investigation of Carbonated Water Flooding. In SPE Annual Technical Conference and Exhibition. OnePetro.
Esene, C., Rezaei, N., Aborig, A., Zendehboudi, S., 2019. Comprehensive review of carbonated water injection for enhanced oil recovery. Fuel 237, 1086–1107.
Esene, C.E., 2019. New insights into transport phenomena involved in carbonated water injection: Effective mathematical modeling strategies (Doctoral Dissertation, Memorial University of Newfoundland).
Ghosh, B., Al-Hamairi, A., Jin, S., 2020. Carbonated water injection: An efficient EOR approach. A review of fundamentals and prospects. Journal of Petroleum Exploration and Production Technology 10(2), 673–685.

Hitzman, D.O., 1983. Petroleum Microbiology and the History of Its Role in EOR, in: Donaldson, E.C., Clarks, J.B., (eds) Proceedings, 1982 International Conference on Microbial Enhancement of Oil Recovery. NTIS, pp. 162–218.

Honarvar, B., Azdarpour, A., Karimi, M., Rahimi, A., Afkhami Karaei, M., Hamidi, H., Ing, J., Mohammadian, E., 2017. Experimental investigation of interfacial tension measurement and oil recovery by carbonated water injection: A case study using core samples from an Iranian carbonate oil reservoir. Energy & Fuels 31(3), 2740–2748.

Hong, E., Jeong, M.S., Lee, K.S., 2019. Optimization of nonisothermal selective plugging with a thermally active biopolymer. Journal of Petroleum Science and Engineering 173, 434–446.

Jenneman, G.E., Clark, J.B., 1992, 22–24 April. The Effect of In-Situ Pore Pressure on MEOR Processes. In SPE 24203 SPE/DOE Enhanced Oil Recovery Symposium, Tulsa, OK, 15 pp.

Jimoh, I.A., 2012. Microbial Enhanced Oil Recovery. Luma Print.

Kaster, K.M., Hiorth, A., Eilertsen, G.K., Boccadoro, K., Lohne, A., Berland, H., Stavland, A., Brakstad, O.G., 2012. Mechanisms involved in microbially enhanced oil recovery. Transport in Porous Media 91(1), 59–79.

Lazar, I., Petrisor, I.G., Yen, T.F., 2007. Microbial enhanced oil recovery. Petroleum Science and Technology 25, 1353–1366.

Manshad, A.K., Olad, M., Taghipour, S.A., Nowrouzi, I., Mohammadi, A.H., 2016. Effects of water soluble ions on interfacial tension (IFT) between oil and brine in smart and carbonated smart water injection process in oil reservoirs. Journal of Molecular Liquids 223, 987–993.

Mohammadian, E., Hamidi, H., Asadullah, M., Azdarpour, A., Motamedi, S., Junin, R., 2015. Measurement of CO_2 solubility in NaCl brine solutions at different temperatures and pressures using the potentiometric titration method. Journal of Chemical & Engineering Data 60(7), 2042–2049.

Monod, J., 1949. The growth of bacterial cultures. Annual Review of Microbiology 3, 371.

Niu, J., Liu, Q., Lv, J., Peng, B., 2020. Review on microbial enhanced oil recovery: Mechanisms, modeling and field trials. Journal of Petroleum Science and Engineering, 107350.

Pashchenko, A.F., Ageev, N.P., 2016, November. Increased Oil Recovery by Application of Plasma Pulse Treatment. In Abu Dhabi International Petroleum Exhibition & Conference. OnePetro.

Patel, K., Shah, M., Sircar, A., 2018. Plasma pulse technology: An uprising EOR technique. Petroleum Research 3(2), 180–188.

Raiders, R.A., McInerney, M.J., Revus, D.E., Torbati, H.M., Knapp, R.M., Jenneman, G.E., 1986. Selectivity and depth of microbial plugging in Berea sandstone cores. Journal of Industrial Microbiology 1, 195–203.

Safdel, M., Anbaz, M.A., Daryasafar, A., Jamialahmadi, M., 2017. Microbial enhanced oil recovery, a critical review on worldwide implemented field trials in different countries. Renewable & Sustainable Energy Reviews 74, 159–172.

Sen, R., 2008. Biotechnology in petroleum recovery: The microbial EOR. Progress in Energy and Combustion Science 34(6), 714–724.

Sheehy, A.J., 1991. Ch. R-1 Microbial Physiology and Enhanced Oil Recovery, in: Developments in Petroleum Science (Vol. 31). Elsevier, pp. 37–44.

Sohrabi, M., Riazi, M., Jamiolahmady, M., Ireland, S., Brown, C., 2009, September. Mechanisms of Oil Recovery by Carbonated Water Injection. In SCA Annual Meeting, pp. 1–12.

Wang, X.Y., Xue, Y.F., Xie, S.H., 1993. Characteristics of Enriched Cultures and Their Application to MEOR Field Tests, in: Development in Petroleum Science (Vol. 39). Elsevier, pp. 335–348.

Youssef, N.H., Elshahed, M.S., McInerney, M.J., 2009. Microbial Processes in Oil Fields: Culprits, Problems, and Opportunities, in: Laskin, A.I., Sariaslani, S., Gadd, G.M., (eds) Advances in Applied Microbiology (Vol. 66). Academic Press, pp. 141–251.

Zahid, S., Khan, H.A., Zahoor, M.K., 2007, March. A Review on Microbial Enhanced Oil Recovery With Special Reference to Marginal/Uneconomical Reserves. In SPE Oklahoma City Oil and Gas Symposium/Production and Operations Symposium, SPE-107052. SPE.

Zobell, C.E., 1946. Action of microörganisms on hydrocarbons. Bacteriological Reviews 10(1–2), 1.

Zobell, C.E., 1947. Bacterial release of oil from sedimentary materials. Oil and Gas Journal 1, 62–65.

8 Low-Salinity Waterflooding

8.1 INTRODUCTION

Waterflooding is frequently used worldwide as a secondary recovery technique. This technology is extensively applied in many oil fields because of its simplicity and low cost. Generally, seawater or produced water of high salinity is injected into the reservoir to displace the oil in place. Over the last two decades, many laboratory experiments have revealed that oil recovery by low-salinity waterflooding (LSWF) is comparatively higher than that by high-salinity waterflooding (HSWF). More than half of the world's proven reserves of hydrocarbon is from carbonate rocks. Oil recovery from these reservoirs is highly challenging because of their complex nature with a low-permeability matrix and high fracture density. The problem becomes more complicated when the wettability of such rock ranges from mixed wet to oil wet. Carbonate rocks are mainly composed of divalent minerals such as calcite ($CaCO_3$), dolomite ($CaMgCO_3$), anhydride ($CaSO_4$), and gypsum ($CaSO_4$, H_2O). A high-salinity sandstone reservoir with substantial clay minerals is also a potential candidature for LSWF. Whether a clean, unaltered sandstone is strongly water wet or not, the associated clay, however, is oil wet. The low-salinity water injection for enhanced recovery of oil is also known as LoSal™ by BP, Smart WaterFlood by Saudi Aramco, Designer Waterflood by Shell, and Advanced Ion Management (AIM[SM]) by ExxonMobil.

Primarily, it is believed that the key mechanism for enhanced oil recovery by LoSal waterflooding is wettability alteration (Alotaibi et al., 2011; Jalili and Tabrizy, 2014). Specific wettability of a reservoir rock is dependent on the mineralogy of rock, fluid properties, interaction between rock and fluids, and history of the fluids exposed to the surface of the rock (Graue et al., 1999). Generally, sandstone reservoirs are mixed wet after oil migration, while most of the carbonate reservoirs are either oil wet or strongly oil wet (Abdallah et al., 1986; Chilingar and Yen, 1983). Thus, the oil recovery from oil-wet sandstone or carbonate reservoirs by LoSal waterflooding is highly promising. Besides wettability alteration, other mechanisms directly or indirectly related to EOR include fine migration and mineral dissolution; pH increase, leading to reduction of interfacial tension (IFT); multi-ion exchange; electric double-layer expansion; formation of thin brine film; salt-in effect; electrokinetic effect of potential determining ions (Ca^{2+}, Mg^{2+}, and SO_4^{2-}) for carbonate reservoirs; etc. (Al-Shalabi and Sepehrnoori, 2016; Sheng, 2014).

8.2 DEFINITION OF LOW-SALINITY WATERFLOODING

Though the practices of LSWF in petroleum reservoirs have been reported long back, recently, the process has been established with profound mechanisms after significant laboratory- and pilot-scale researches. On the basis of experimental observation, it is believed that favorable oil recovery occurs by low-salinity (LoSal) waterflooding because of change of wettability of reservoir rock from oil wet to water wet. Morrow and coworkers (Tang and Morrow, 1999a, 1999b; Zhang et al., 2007) performed a series of laboratory tests to observe the effect of salinity on oil recovery. They mentioned that brine chemistry plays an important role in changing the wettability during the course of waterflooding, leading to enhanced recovery of oil. Lager et al. (2008) stated that cation exchange between the mineral surface and the invading brine is the primary mechanism underlying the improved waterflood recovery observed with LoSal waterflooding. Significantly, the mechanisms of wettability alteration are different for sandstone and carbonate reservoirs. The optimum salinity for LoSal waterflooding in sandstone reservoirs is around 5,000 ppm (Lee et al., 2010), while seawater with salinity in the range of 10,000–20,000 ppm is generally used for LoSal waterflooding in carbonate reservoirs (Mahani et al., 2015). The expansion of the electric double layer (EDL) around

DOI: 10.1201/9781003098850-8

the clay particle of sandstone at low ionic strength and formation of a thin water film around the rock are responsible for the wettability alteration of sandstone reservoirs (Sheng, 2014; Myint and Firoozabadi, 2015). On the other hand, divalent ions, such as Mg^{2+}, Ca^{2+}, and SO_4^{2-} (referred to as potential determining ions, PDIs), play an important role for low-salinity oil recovery from carbonate reservoirs (Fathi et al., 2012).

8.3 CONDITIONS FOR EFFECTIVE RECOVERY BY LOW-SALINITY WATERFLOODING

For the techno-economic feasibility of LSWF, screening criteria similar to other EOR methods must be followed. The parameters that have significant impact on effective recovery, have been pointed out by several researchers (Tang and Morrow, 1999a, 1999b; Lager et al., 2008; Austad et al., 2010; Katende and Sagala, 2019). The details of the conditions that can effectively influence the recovery by LSWF are discussed in the following:

Mineral surface: It is believed that the presence of clay material in the reservoir rock is important for the LSWF effect. There are mainly three different types of clays: (a) discretely dispersed particle clay, (b) pore lining clays, and (c) pore bridging clays. Kaolinite belongs to type (a), whereas illite and chlorite fall into categories (b) and (c). Kaolinite has a very low cationic exchange capacity compared to illite or chlorite. The capacity of a clay to attract and hold cations from a solution is measured as the cation exchange capacity (CEC). Sandstone cores without clay but containing dolomite crystals also show positive results from the injection of low-salinity water (Pu et al., 2010).

Oil: Crude oil contains various organic acids and bases which are polar in nature and are preferentially adsorbed on the rock surfaces resulting oil-wetness of the reservoir rock. As wettability alteration is one of the mechanisms, the LSWF is a preferred EOR technique for these reservoirs. The efficiency of LSWF depends on the initial oil-wetness of the rock.

Formation brine: If the salinity of the formation water is on higher side with rich in divalent cations (Ca^{2+} and Mg^{2+}), LSWF is preferred (Lager et al., 2008).

8.3.1 INJECTION BRINE CONCENTRATION

• The salinity of injection water for a sandstone reservoir is usually between 1,000 and 2,000 ppm, but the effects have been observed up to 5,000 ppm (Austad et al., 2010).
• The low-salinity effect can be observed for carbonate reservoirs with salinity up to 40,000 ppm, where potential determining ions (Ca^{2+}, Mg^2, SO_4^{2-}, etc.) play an important role.

Connate water saturation: The presence of initial or connate water saturation with divalent cations is favorable for the LSWF effect. The typical composition of formation water, seawater, and aquifer water is shown in Table 8.1.

Wettability of reservoir rock: Reservoir rocks with wettability aligned to oil wet or mixed wet hinders recovery with high residual oil saturation. Injection of low-salinity water with less divalent cations changes the wettability to water wet. Thus, the initial oil-wet condition of reservoir rocks favors the better LSWF effect.

Temperature: Reservoir temperature plays an important role in the efficiency of oil recovery by any method, as rock-fluid properties are highly dependent on temperature. The same is true for oil recovery by LSWF also. For carbonate reservoirs, SO_4^{2-} adsorbs more strongly onto the positively charged chalk surface at higher temperatures. At high temperatures, Mg^{2+} is much more reactive and even displaces Ca^{2+} from the chalk surface lattice, leading to different mechanisms of LSWF or smart waterflooding.

TABLE 8.1
Typical Composition of Brines (Nasralla et al., 2011)

Ions	Formation Brine	Seawater	Aquifer Water
Na^+	54,400	16,877	1,504
Ca^{2+}	10,610	664	392
Mg^{2+}	1,610	2,279	66
Sr^{2+}	0	0	5
Cl^-	107,000	31,107	2,577
HCO_3^-	176	193	192
SO_4^{2-}	370	3,560	700
TDS, mg/L	174,156	54,680	5,436

pH of water: The pH of the water is preferentially about 5. Though the aforementioned conditions are desirable for better performance of LSWF, but still, they do not guarantee the effectiveness of the process. The LSWF process is more complex, and there is still no single explanation to fully describe it.

8.4 WETTABILITY CHARACTERISTICS OF SANDSTONE AND CARBONATE RESERVOIR ROCK

The wetting state of a reservoir rock is initially water wet, as the pores of the reservoir rock are initially occupied with water/brine. The crude oil that migrates into the pores of the reservoir from their primary sources interacts with the reservoir rock surface and leads to a change in the wetting state of the reservoir (Rahbar et al., 2012). Thus, the reservoir rocks are initially water wet in nature; however, after the prolonged interaction of crude oil with the reservoir rock, the crude oil gets adsorbed on the rock surface, leading to a change in the wetting state of reservoir rock to oil wet (Saxena et al., 2019). However, the interaction of the crude oil with the surface of the reservoir rock is prevented by the film of water present between the rock/water and oil/water interfaces. The strength of the water film is dependent on rock mineralogy and the composition of crude oil and brine in the reservoir (Sohal et al., 2016). A strong water film maintains the water-wetting state of the reservoir, whereas a weak water film is easily ruptured, leading to contact of the crude oil and rock surface. The rupture of the brine layer occurs due to the presence of different attractive forces such as electrostatic interaction, van der Waals forces, and hydrogen bonding between the crude oil and rock surface (Hirasaki, 1991a, 1991b). The dominance of a specific force depends upon the composition of the crude oil and mineralogy of the reservoir rocks. These forces cause adsorption of specific components of crude oil on the rock surface and the oil-wet state of the reservoir rock. Various interactions between the crude oil and the reservoir rock surface have been shown in Figure 8.1.

The rock mineralogy of a reservoir rock plays an important role in the alteration of rock wettability from water wet to oil wet. Depending on the mineralogical composition, the surfaces of the reservoir rocks are generally positively or negatively charged. The surface of sandstone reservoir rock is negatively charged due to the presence of minerals such as quartz (Santha et al., 2017). The surface of the carbonate reservoir rock is positively charged due to the presence of minerals such as calcite and dolomite (Derkani et al., 2019). These charged surfaces attract the polar components of the crude oil. The negatively charged sandstone rock surface attracts the positively charged basic components of the crude oil. Similarly, attractive forces exist between the positively charged

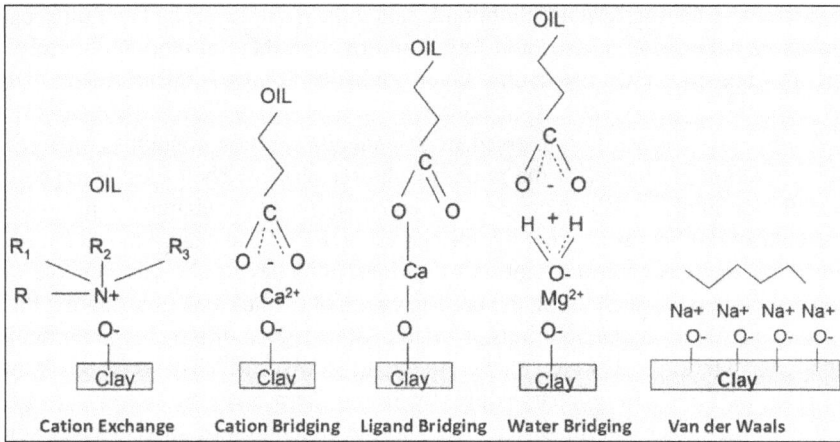

FIGURE 8.1 Adhesion interactions between crude oil and the reservoir rock surface.

Source: Awolayo et al., 2018.

carbonate rock surface and the negatively charged acidic components of the crude oil (Doust et al., 2011; Mohammed and Babadagli, 2016). Thus, the attractive forces between the charged sites of the rock surface and the polar components of the crude oil lead to the adsorption of the crude oil at the rock surface (Kumar and Mandal, 2019). The adhesion of the crude oil components on the rock surface causes the oil-wetting of the reservoir rocks.

The polar components of the crude oil are positively or negatively charged resins and asphaltenes. Thus, crude oil with a higher amount of these polar components leads to a strong oil-wetting nature of the reservoir rock. However, the interaction between the crude oil and the rock surface is further improved due to the presence of multivalent cations in the reservoir brine (Lin et al., 2018). Thus, a high-saline formation water leads to the oil-wetting state of the reservoir rock. The multivalent or divalent ions such as Ca^{2+} and Mg^{2+} strongly adsorb on rock surfaces until the rock is fully saturated. These adsorbed ions form organometallic complexes with the polar compounds of the crude oil and promote the oil-wetness of the reservoir rock (Awolayo et al., 2018).

The high salinity of formation brine also leads to the weakening of the brine film, as higher salinity causes the thinning of the electric double layer (EDL) and lower value of the zeta potential of a charged reservoir rock surface at high salinity (Bonto et al., 2019). Thus, the charged polar components of the crude oil easily adhere to and form attractive forces with the rock surface, leading to promotion of the oil-wetting nature of reservoir rock. Thus, the electrostatic forces between the charged reservoir rock surface and polar crude oil components, along with a weak brine film, lead to the oil-wetting nature of reservoir rocks.

8.5 ELECTRIC DOUBLE LAYER AND ZETA POTENTIAL (ζ)

The mechanisms of oil recovery by LSWF can be best explained by the zeta potential value at the mineral-brine and oil-brine interface. In colloidal chemistry, the zeta potential is usually denoted as ζ-*potential*. From a theoretical viewpoint, the zeta potential is the electric potential in the interfacial double layer at the location of the slipping plane relative to a point in the bulk fluid away from the interface (Fig. 8.2). The electrostatic interaction between rock, brines, and crude oil is best explained by the Derjaguin-Landau-Verwey-Overbeek (DLVO) theory: the electrostatic interactions acting on the oil-brine-rock system, comprising the rock-brine and oil-brine interfaces, lead to the development of the EDL. The double layer around the interface consists of two parts:

an inner region where the ions are tightly bound, called the Stern layer; and an outer region (double layer) where a balance of electrostatic forces and random thermal motion determines the ion distribution. The potential after the double layer gradually decays with increasing distance and approaches zero.

Whenever a mineral surface is immersed in an aqueous solution containing different ions, it leads to the separation of charges at the mineral/solution surface. The excess charge obtained at the mineral/solution interface is balanced by the equal and opposite charges in the solution. This type of arrangement of charges is generally termed as the EDL. The charge of the mineral surface is balanced by the decreasing concentration of co-ions (ions having the same charge as mineral surface) and a simultaneous increase in the concentration of counterions (ions having the opposite charge as mineral surface) in the solutions. The immediate region of the charged mineral surface is surrounded by counterions that are attached to the mineral surface. This region is known as the Stern layer (Fig. 8.2). Generally, charges present in the Stern layer are not able to balance the surface charge of the mineral surface. This leads to the generation of another layer surrounding the Stern layer, called as diffuse layer or Gouy-Chapman layer. The diffuse layer contains the charges, which are a mixture of co-ions and counterions, that are required to maintain the electrical neutrality of the EDL. The main difference between the Stern and diffuse layer is that the ions present in the diffuse layer are not attached to the mineral surface, whereas this is not so in the Stern layer (Fig. 8.2).

Electrokinetic effects arise when one phase is moved past to the second phase, tangentially. When a charged mineral surface is tangentially moved relative to excess charges in the diffuse layer, electrokinetic effects arise. The EDL thickness is a function of charges present at the

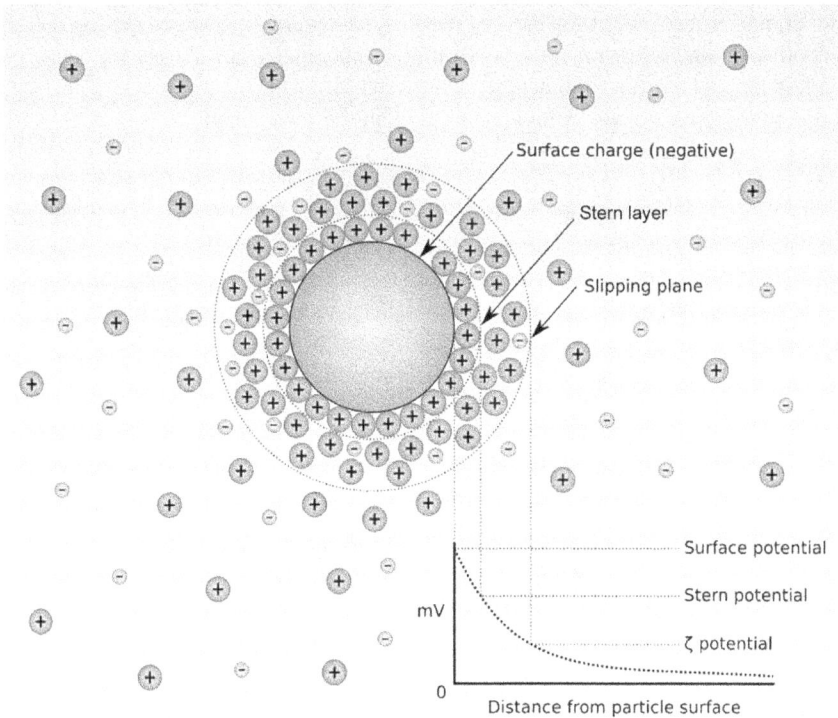

FIGURE 8.2 Schematic of EDL and zeta potential (ζ).

Source: https://en.wikipedia.org/wiki/Zeta_potential.

mineral/brine solution interface and is estimated by the zeta potential. The zeta potential is the electrical potential measured at the shear plane, also known as slipping plane. Generally, there are four different methods to measure the zeta potential: (i) electrophoresis method (EPM); (ii) streaming potential method (SPM); (iii) electro-osmosis method; and (iv) sedimentation potential method. Of these, EPM and SPM are widely used. In the electrophoresis method, a solid particle is suspended in the liquid phase of interest, and suspended particles are made to move in the surrounding liquid phase by the application of an electric field. This is called electrophoresis (Hunter, 1981). Measurement of the velocity of the suspended particles, normalized by the applied electric field, gives the electrophoretic mobility (u_e). This electrophoretic mobility is then used to estimate the zeta potential by the Helmholtz-Smoluchowski equation for electrophoresis (Delgado et al., 2005).

$$u_e = \frac{\epsilon_{rs}\varepsilon_0\varsigma}{\mu} \tag{8.1}$$

where ϵ_{rs} is the relative permittivity of the electrolyte solution, ε_0 is the electric permittivity of vacuum, μ is the dynamic viscosity of solution, and ς is the zeta potential.

The oil reservoirs are classified into two types: i.e., sandstone and carbonate. Both types of reservoirs are very much distinct with respect to their mineralogical compositions and, hence, in their electrokinetics. Sandstone reservoirs are composed of different minerals, mainly quartz, feldspar, and clays. Clay minerals typically consist of swelling clay such as montmorillonite and nonswelling clays such as kaolinite, illite, chlorite (Austad et al., 2010). Clay minerals are negatively charged at reservoir pH range. Sandstone reservoirs are generally found to be negatively charged due to the abundant availability of clay minerals. The isoelectric point (IEP) of sandstone is found to be in the range of a pH of 2–3 (Jaafar et al., 2014). IEP is defined as the pH where no net electrical charge exists on surface. Crude oil is generally negatively charged above a pH value of 3 (Takamura and Chow, 1985). The determination of IEP plays a very important role, as it decides the successfulness of the increased oil recovery, ensuring negative charges at both the interface of mineral/brine and brine/oil surface. The zeta potential is very much dependent on salinity, pH of the suspended medium, and mineralogical composition of the sandstone rocks (Shehata and Nasr-El-Din, 2015). The monovalent cations are better than divalent ions in increasing the absolute value of the zeta potential, and the zeta potential tends to be more negative with a decrease in salinity. Changing the pH of the solution causes a significant change in the charges of different types of sandstones. The water chemistry has an important role to play in zeta potential measurement (Nasralla and Nasr-El-Din, 2014b). Alotaibi et al. (2010) examined the interaction of aquifer and seawater with different clay minerals as well as sandstones as shown in Figure 8.3. Results suggest that the zeta potential is strongly dependent on ionic strength and clay minerals. The impact of brine salinity and composition on zeta potential at the oil/brine interface is shown in Figure 8.4. A higher value of the zeta potential stabilizes the dispersed oil droplets in the aqueous media.

Carbonate rocks are mainly composed of aragonite, glauconite, quartz, ankerite, pyrite, and siderite (Akbar et al., 2000). Carbonate rocks are positively charged at reservoir conditions, and their IEP lies between a pH of 9 and 10 (Jaafar et al., 2014). The electrokinetic effects of carbonate rocks are complex in nature, and in literature, both positive and negative results are reported. It is revealed that below the IEP, calcite particles acquire a positive charge and a negative charge above the isoelectric point (Heberling et al., 2010). The zeta potential of carbonate rock is mainly dependent on salinity, ionic composition, and pH (Alroudhan et al., 2016; Chen et al., 2014). The ionic composition of injected brine salinity has a greater impact on zeta potential, as studies have shown a positive effect on oil recovery by the manipulation of Ca^{2+}, Mg^{2+}, and SO_4^{2-} ions in injection brine (Alroudhan et al., 2016).

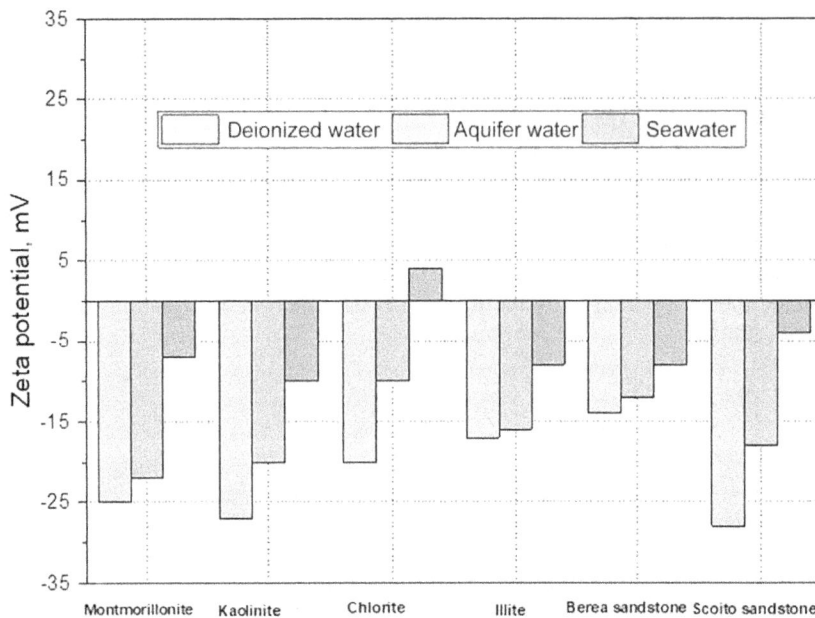

FIGURE 8.3 Typical variation of zeta potential for different clays and sandstones with different brines.

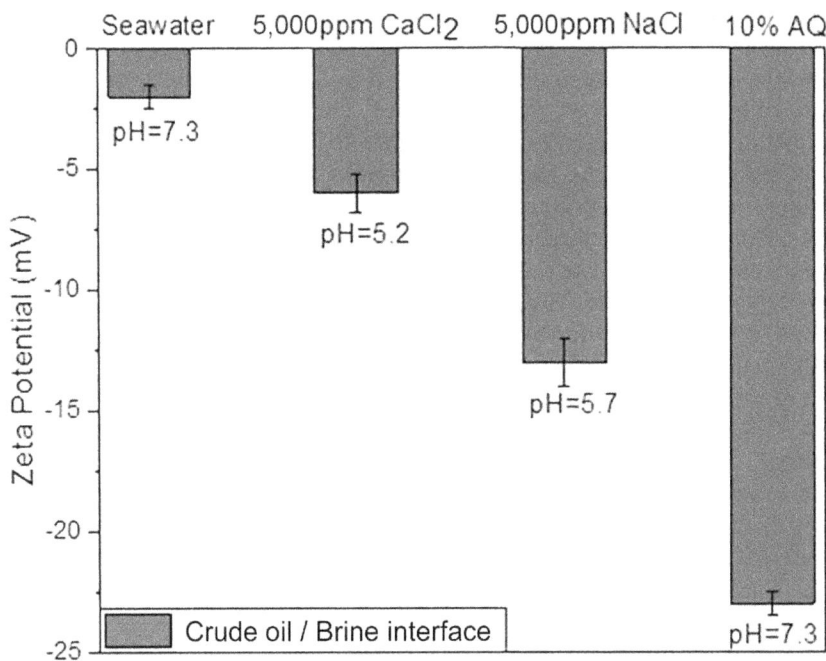

FIGURE 8.4 Typical impact of brine salinity and composition on the zeta potential at the oil/brine interface.

8.6 MECHANISM OF OIL RECOVERY BY LOW-SALINITY WATERFLOODING

Many researchers have attempted to understand the mechanism behind the production of oil through LSWF, and many more are still going on, as a single mechanism is unable to explain the phenomena occurring during LSWF, both in sandstones and carbonate reservoirs. A number of mechanisms have been reported in literature (Tang and Morrow, 1999a, 1999b; Lager et al., 2006; Austad et al., 2010; Sheng, 2014; Katende and Sagala, 2019). Some of the mechanisms are related to others and very specific to reservoir rock mineralogy and crude oil composition. In this section, only the major mechanisms of LSWF are discussed. The mechanisms are as follows:

- i)   Fine migration and mineral dissolution.
- ii)   Wettability alteration.
- iii) Change of pH and IFT.
- iv)   Formation of water film.
- v)   Multi-ion exchange (MIE).
- vi)   Salting-in mechanism.
- vii) Expansion of the EDL.

8.6.1   Fine Migration and Mineral Dissolution

8.6.1.1 Fine Migration

Clay swelling and fine migration are considered a prominent mechanism in sandstones and rock dissolution in carbonate rocks. Clay tends to hydrate and swell when contacted with fresh or low-salinity water that has few divalent ions. Swelling occurs in three steps: crystalline swelling, hydration swelling, and force swelling. Clay is found in nature to be very reactive, having a negatively charged site to be bonded with cations of existing connate brine or newly injected low-salinity water and a positive charge to be bonded with carboxylic groups of crude oil. As low-salinity water is injected, ionic interaction takes place at the surface of the clay, from which the bonded ion gets detached, triggering a release of clay surfaces. The release and, thus, transport of these clay particles is referred to as fine migration. Fine migration occurs if the ionic strength of the injected brine is less than a critical flocculation concentration. A higher ionic strength means a lower zeta potential—i.e., less net electrical charge contained within the region bounded by the slipping plane. The critical flocculation concentration is a strong function of the relative concentration of divalent cations. The presences of divalent cations causes a weaker electrostatic repulsion and a stronger flocculation by lowering the zeta potential (lower the net charge of clay). This leads to a stabilization of the clay material. A less-saline solution destabilizes clay and silt in the formation. The clay and silt, upon dispersion, flow with water. The polar components of crude oil are preferentially adsorbed on the solid particle to form mixed-wet fines (Fig. 8.3a). Figure 8.3b shows the partial stripping of the mixed-wet and water-wet fines from the pore wall during water-flooding, while Figure 8.3c demonstrates the mechanisms of dispersion of crude oil and wettability alteration by fine migration. Water preferentially flows along high-permeability channels or zones. The clay and silt dispersing in water become lodged in smaller pores or pore throats. This leads to a reduction in formation permeability, and the water is forced to take other flow paths. As a result, the sweep efficiency is improved. Poorly cemented clay particles, such as kaolinite and illite, can be detached during aqueous flow during LSWF. According to Tang and Morrow (1999a), a decrease in salinity causes the expansion of the EDL around the mineral/rock surface, which leads to the stripping of clay particles from the rock surface. As most of the sandstone rocks contain clay, this LSWF effect is very common in sandstone reservoirs, but the expansion of the EDL can occur during LSWF in carbonates as well.

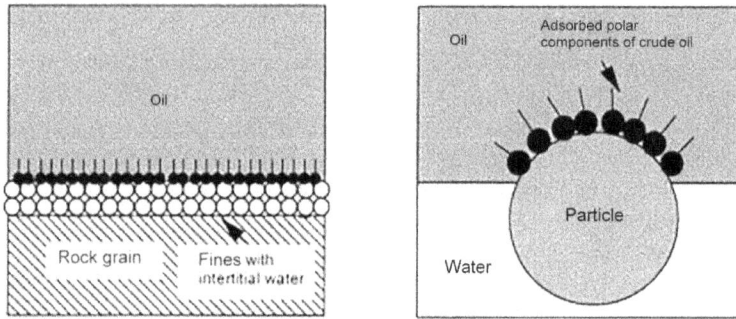

Adsorption onto potentially mobile   Mobilized particle at oil-water
fines at low initial water saturation   interface

(a) Adsorption of Polar Components from Crude Oil to Form
Mixed-Wet Fines

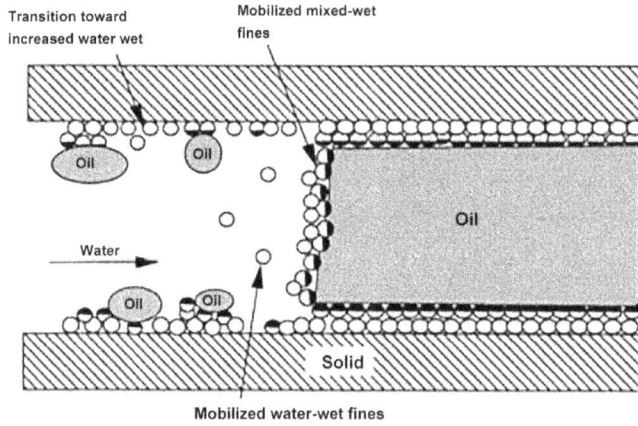

(b) Partial Stripping of Mixed-Wet Fines from Pore Walls
during Waterflooding

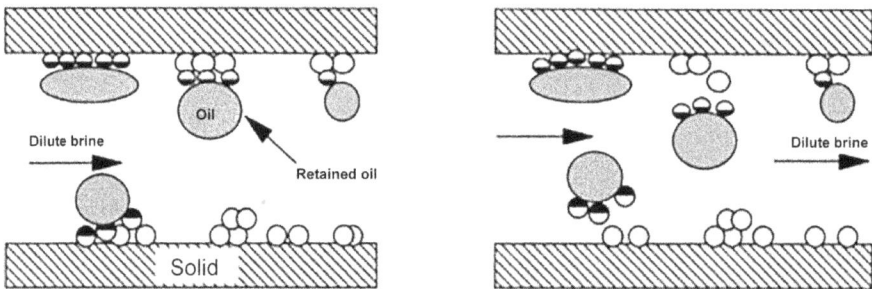

Retained oil before injection of   Partial mobilization of residual oil
dilute brine   through detachment of fines

(c) Mobilization of Trapped Oil

FIGURE 8.5   Effect of clay fine migration on recovery improvement throughout low-salinity waterflooding.
Source: Tang and Morrow, 1999a.

8.6.1.2 Mineral Dissolution

Rock (calcite) dissolution is also one of the mechanisms suggested by various authors and proved to be very much helpful in analyzing the recovery in carbonate rocks by LSWF (Den Ouden et al., 2015). In reservoirs, carbonate rocks remain in equilibrium with the calcium ions present in the formation water. As the low-salinity water with lower calcium ion concentration is injected into the reservoir, the equilibrium gets disturbed; to achieve equilibrium again, calcium ions from the carbonate rock get dissolved in the low-salinity brine accompanied with the polar components of the oil. When this occurs, a new, fresh surface is exposed, which has not been aged with the crude oil. Therefore, this new surface will be less oil wet, which is favorable for oil production. As rock dissolution takes place, the dissolved minerals also get transported and later get precipitated and block the interconnected pores of the rocks, which in turn, forces the low-salinity water to adapt to some new flow channels through unswept areas, thus increasing the sweep efficiency and oil produced. This mechanism is similar to some extent to the mechanism discussed earlier for sandstone reservoirs. Calcite dissolution also causes a change in the porous structure, leading to enhanced connectivity between the pores. As a result, the permeability and sweep efficiency might be increased. Calcite dissolution can induce fine migration as well.

8.6.2 Wettability Alteration

Wettability alteration is one of the most important mechanisms for oil recovery by LSWF (Vledder et al., 2010). The oil-wetness of the reservoir rock in high-salinity conditions have been explained earlier. The interactions between oil, brine, and rock at equilibrium condition for the oil/brine/rock system can be expressed by the augmented Young-Laplace equation:

$$P_c = \pi(h) + 2\sigma_{ow}\cos\theta / r \tag{8.2}$$

where P_c is capillary pressure (Pa) between the wetting (water) and nonwetting phases (oil), $\pi(h)$ is disjoining pressure (Pa), σ_{ow} is interfacial tension (N/m), θ is contact angle (°), and r is mean radius of pores (Ding and Rahman, 2017). The DLVO theory is often used to correlate the electrokinetics of the rock surface to the thermodynamic interaction involved in wettability alteration (Mahani et al., 2015). The force that tends to separate the oil and water phases around the reservoir rock is termed as disjoining pressure. It results from molecular and interionic interactions between the rock, oil, and brine phases (Hirasaki, 1991b). The disjoining pressure ($\Pi(h)$) is the sum of the electrical or electrical double-layer forces, van der Waals forces (vdW), and structural forces.

Electrostatic forces: These forces are the result of the development of the charges between interacting surfaces. The charges can be formed either by dissociation of the surface charges or adsorption of the charges onto an uncharged surface. The EDL force is estimated using zeta potentials and is approximated by (Gregory, 1975):

$$\Pi_{electrical}(h) = n_b k_B T \left(\frac{2\psi_{r1}\psi_{r2}\cosh(\kappa h) - \psi_{r1}^2 - \psi_{r2}^2}{(\sinh(\kappa h))^2} \right) \tag{8.3}$$

where ψ_{r1}, ψ_{r2} are the reduced potential, κ is the reciprocal Debye-Huckel double-layer length, n_b is the ion density in the bulk solution, h is the water film thickness between the calcite/brine and oil/brine interface, and k_B is the Boltzman constant.

Van der Waals forces: These forces play an important role in the stability of a water film around the rock surface. For crude oil/brine/rock systems, the van der Waals forces are considered to be negative.

In the oil/brine/rock system, the *vdW* force for two flat planes can be calculated by Equation 8.4:

$$\Pi_{vdW} = \frac{-A}{12\pi h^3} \tag{8.4}$$

where A is the Hamaker constant (Hamaker, 1937) for the oil/brine/rock system.

Structural or solvation interactions: Structural forces are short-range forces, which include the effect of hydrogen bonding and specific ion/water interactions. These are called solvation or hydration effects or structural forces, because they are the result of the intermolecular structure of the solvent or water. The structural interaction can be calculated from Equation 8.5 (Hirasaki, 1991b):

$$\Pi_{structure} = A_k \exp\left(\frac{h}{h_s}\right) \tag{8.5}$$

where A_k is the coefficient and h_s is the characteristic decay length for the exponential model.

At high salinity, the oil molecules are held on the surface of the negatively charged clay particles mainly by divalent cations. The clay acts as a cation exchanger, with its relatively large surface area. During LSWF, with water of an ionic strength much lower than that in the initial formation brine, the equilibrium associated with the brine-rock interaction is disturbed, and a net desorption of cations, especially Ca^{2+}, occurs. To compensate for the loss of cations, protons, H^+, from the water close to the clay surface adsorb onto the clay, and a substitution of Ca^{2+} by H^+ takes place:

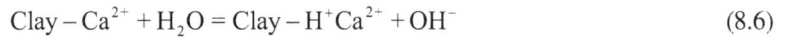

$$Clay - Ca^{2+} + H_2O = Clay - H^+Ca^{2+} + OH^- \tag{8.6}$$

As a result, there is a local increase in pH close to the clay surface, which causes reactions between adsorbed basic and acidic material as in an ordinary acid-base proton transfer reaction as shown here:

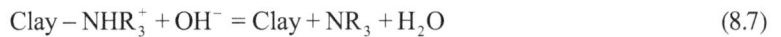

$$Clay - NHR_3^+ + OH^- = Clay + NR_3 + H_2O \tag{8.7}$$

$$Clay - RCOOH + OH^- = Clay + RCOO^- + H_2O \tag{8.8}$$

The schematics of the aforementioned desorption of acidic and basic material from the rock surface are shown in Figure 8.6. Adsorption of basic and acidic material onto clay minerals is very sensitive to the pH of the system.

Several mechanisms such as MIE, polar component desorption, fine migration, EDL expansion, etc. have indirect effect on the wettability alteration of reservoir rocks from oil wet to mixed wet or from mixed wet to water wet, leading to more oil recovery. The stability of water films is very much dependent on the salinity of water injected in the reservoir field; as the salinity of the injected water reduces, stability increases, leading to a more water-wet state.

A schematic of wettability change from the initial state to LSWF through normal waterflooding from a sandstone reservoir is shown in Figure 8.7. After LSWF, the initially adsorbed oil on the clay and rock surface gets detached and moves forward toward the production well under the maintained pressure gradient or viscous force. The change of contact angle, as reported by Mahani et al. (2015), is shown in Figure 8.8. The change of contact angle with time is negligible for HSWF, whereas an enormous change of contact angle with time for LSWF indicates change of wettability from oil wet to water wet. Thus, change of wettability is a very important mechanism of oil recovery by LSWF.

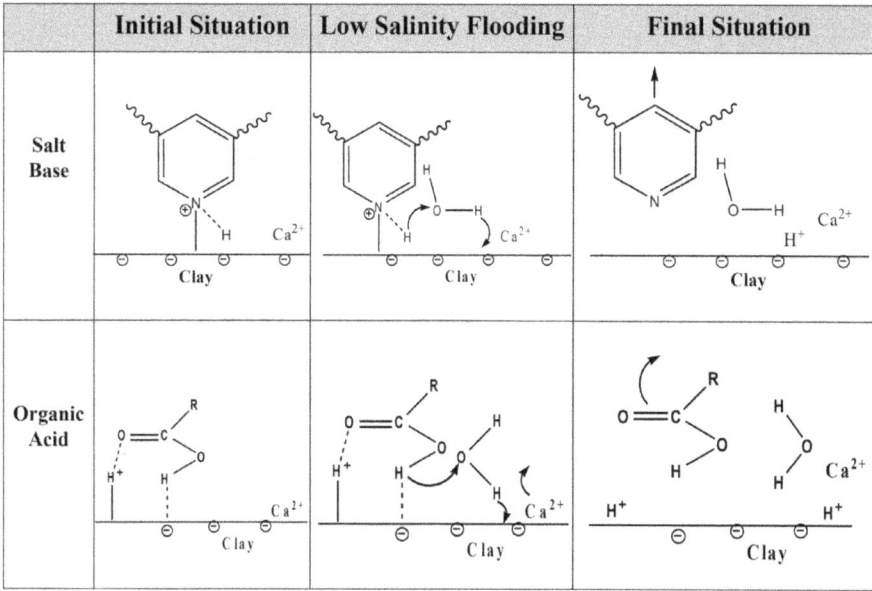

FIGURE 8.6 Ion exchange process during low-salinity flooding: (a) initial state of clay surface, (b) reaction occurring during low-salinity flooding, (c) final state of clay surface.

Source: Jiang et al., 2018.

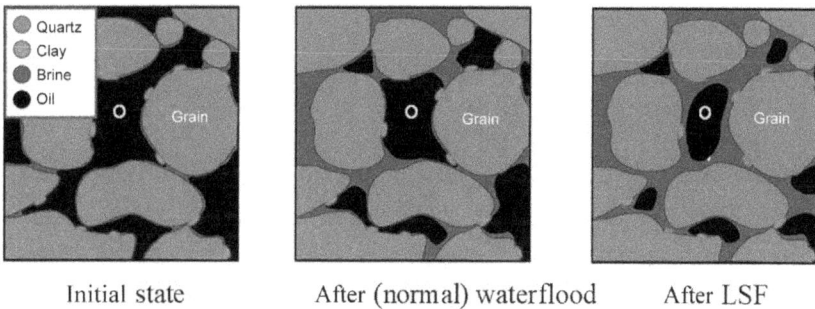

FIGURE 8.7 Schematic of the wettability change in a porous sandstone rock caused by LSWF.

Source: Mahani et al., 2015.

LSWF can be applied in a secondary as well as in a tertiary flood (i.e., after high-salinity flood). In a secondary flood (low-salinity water injection from the beginning), the change of wettability is described phenomenologically using the extended Buckley-Leverett theory (Jerauld et al., 2006). The water saturation profile for an originally oil-wet system as a function of distance from the injector well for high-salinity injection water (no wettability alteration) and for low-salinity water injection (with wettability alteration) is shown in Figure 8.9. Comparing the two cases, the following differences can be noted for the low-salinity injection case: (1) delayed water breakthrough; (2) oil banking ahead of the low-salinity shock front due to the accumulation of de-adsorbed oil, leading to constant water saturation over a finite distance; and (3) reduced remaining oil saturation behind the shock front.

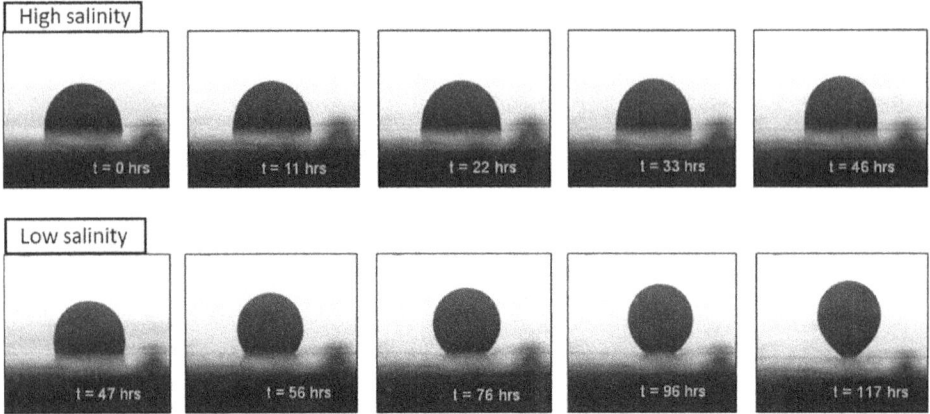

FIGURE 8.8 Change of contact angle with time under high-salinity and low-salinity conditions.

Source: Mahani et al., 2015.

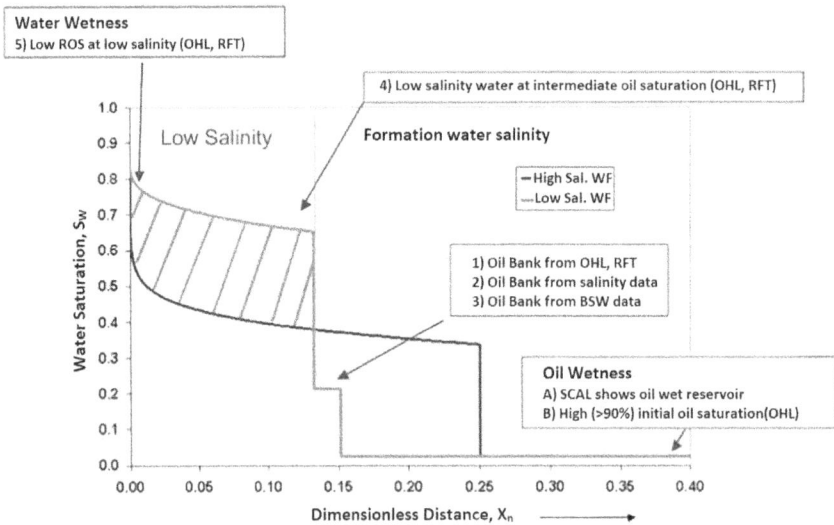

FIGURE 8.9 Water saturation in oil-wet system for high-salinity water and low-salinity water, based on the Buckley theory.

Source: Katende and Sagala, 2019.

8.6.3 CHANGE OF pH AND IFT

Whenever there is an exchange of ions in LSWF, it is always accompanied with a change in pH, depending upon various factors such as rock mineralogy, crude oil composition, etc. Emulsification or snap-off, saponification, and in situ surfactant generation are also related to increased pH and reduced IFT conditions. Several mechanisms such as clay migration, cation exchange, double-layer expansion, MIE are somehow connected to a change in pH. This rise in pH occurs due to two con-comitant reactions: carbonate dissolution and cation exchange.

The dissolution reactions are relatively slow and dependent on the amount of carbonate material present in the rock:

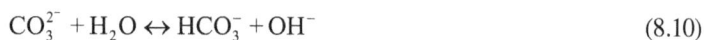

$$CaCO_3 \leftrightarrow Ca^{2+} + CO_3^{2-} \tag{8.9}$$

$$CO_3^{2-} + H_2O \leftrightarrow HCO_3^- + OH^- \tag{8.10}$$

The mineral surface exchanges H$^+$ present in the liquid phase with the cations previously adsorbed (Mohan et al., 1993). If the pH goes above 9 inside a petroleum reservoir, it is equivalent to alkaline flooding. At higher pH values, in situ surfactant may form, which reduces the IFT, and is responsible for wettability alteration and emulsion formation. Nowrouzi et al. (2018) reported almost 78% reduction of IFT between crude oil and water by ten times dilution of seawater from 27.67 mN/m to 6.09 mN/m under 101.4 kPa and 75 °C conditions. The reduction of IFT with dilution of seawater is depicted in Figure 8.10.

However, the mechanism may not be applicable for all reservoirs. It has been proven to be effective where the following conditions prevail (McGuire et al., 2005):

1. The crude oil in the reservoir contains acid components.
2. The reservoir material contains water-sensitive minerals.
3. The reservoir contains an initial water saturation.
4. The injected water contains fewer than about 5,000 ppm TDS.

8.6.4 FORMATION OF WATER FILM

It is believed that all rock surfaces were water wet and water was the wetting phase, but slowly oil displaced the water films present, which is also evident by the presence of irreducible water saturation, and became the wetting phase, leading to the transition of the rock from being water wet to oil wet.

The formation of a water film in the oil/brine/rock system is a very complex phenomenon. Hirasaki investigated the thermodynamics of thin films to determine the interdependence of spreading, contact angle, and capillary pressure using the DLVO theory and the Laplace-Young equation (Hirasaki, 1991b).

Xie et al. (2016) hypothesized that low-salinity water triggers positive disjoining pressure in the crude oil/brine/rock system as a result of double-layer expansion, which leads to the formation of a strong water film as shown in Figure 8.11. The thickness of a water film is basically dependent on the charges of the rock/brine and brine/rock interfaces. If the charges on these two interfaces are of same kind, then repulsion will take place between the two surfaces and there will be a thick film of water, separating the oil and rock surface, leading to the water-wetting nature of the rock. Consequently, if the charges of both the interfaces are of the opposite kind, then attraction will take place, and the thickness of the water film will be very less, whose stability then determines the state of the rock surfaces.

8.6.5 MULTI-ION EXCHANGE

Based on the reported data, Lager et al. (2006) proposed a mechanism of multi-ion exchange (MIE), which is responsible for the increase in oil recovery. At high salinity of the formation water, multivalent ions such as Ca^{2+} and Mg^{2+} are strongly adsorbed on the rock surface until the surface is fully saturated. These multivalent cations on a clay surface generally make bonds with the polar compounds present in the oil phase (resin and asphaltene), forming organometallic complexes. Such complexes have been shown to promote oil-wetness in petroleum reservoirs. At the same time, some

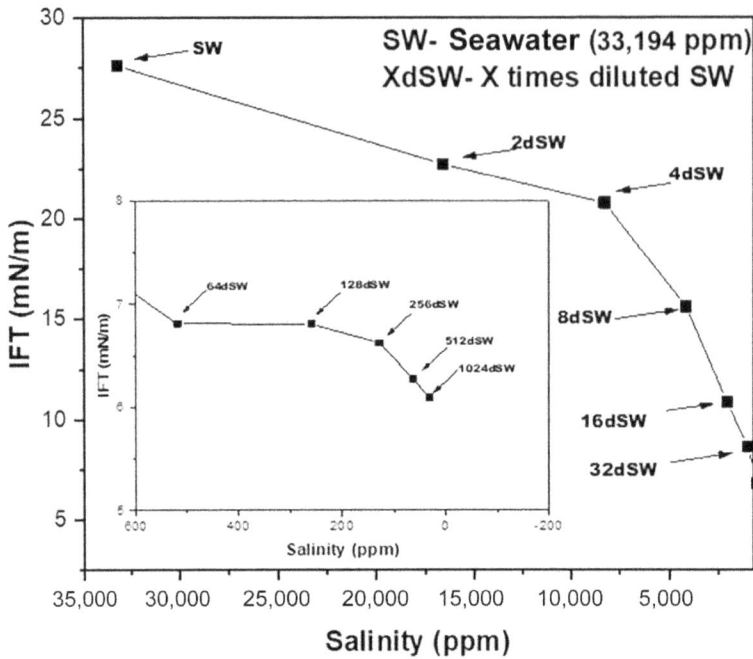

FIGURE 8.10   The reduction of IFT with the dilution of seawater.

Source: Modified from Nowrouzi et al., 2018.

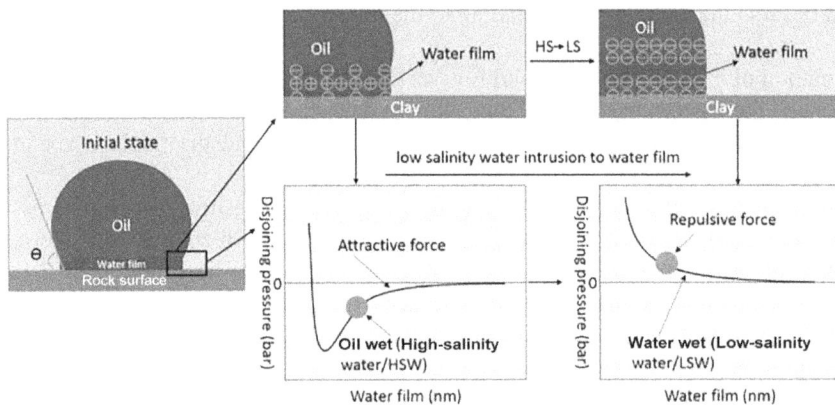

FIGURE 8.11   Schematic of the interaction of crude oil/brine/rock (COBR) system with the presence of either high-salinity water or low-salinity water.

Source: Adapted from: Quan Xie et al., 2016.

organic polar compounds will be adsorbed directly to the mineral surface, displacing the most labile cations present on the clay surface, thus enhancing the oil-wetness of the clay surface.

To establish the mechanism of MIE for LSWF, Lager et al. (2006) performed two sets of flooding experiments, as shown in Figure 8.12. First, they performed an oil recovery experiment by high-salinity brine injection with the core sample, where Ca^{2+} and Mg^{2+} were present at the surface, and

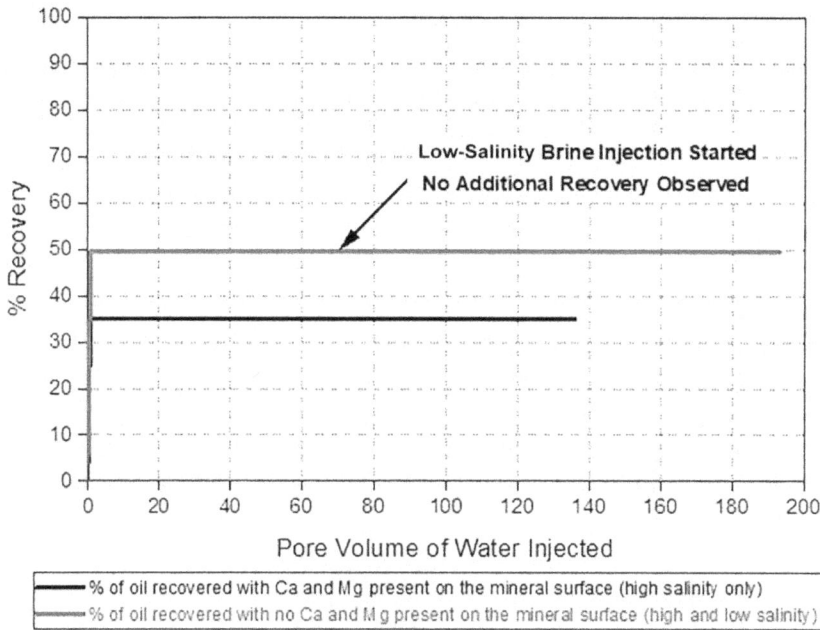

FIGURE 8.12   Typical effect of MIE on the % of oil recovery.

obtained recovery of around 35%. LSWF in tertiary mode resulted in additional 5% oil recovery. In a second set of experiments, they performed flooding with high-salinity brine for the core, where no Ca^{2+} and Mg^{2+} ions were present on the rock surface and found almost 48% recovery. However, low-salinity brine injection on the same core in tertiary mode observed no additional recovery. The oil recovery data of the aforementioned experiments indicates that high-salinity connate brine containing Ca^{2+} and Mg^{2+} ions resulted in poor recovery. Removing the Ca^{2+} and Mg^{2+} ions from the rock surface before waterflooding led to a higher recovery irrespective of salinity. This has confirmed the importance of MIE in the LoSal mechanism. Thus, injection of low-salinity brine causes MIE, which removes the organic polar compounds and organometallic complexes from the surface and replaces them with uncomplex cations. The desorption of polar compounds from the clay surface leads to a more water-wet surface, resulting in an increase in oil recovery.

8.6.6   SALTING-IN MECHANISM

According the mechanism of salting-in, whenever the equilibrium between oil/rock/water is disturbed by injecting low-salinity water, the solubility of polar compounds changes in the water, resulting in the salting-in phenomenon. The solubility of organic material in water can be drastically decreased by adding salt to the solution (i.e., the salting-out effect) and the solubility can be increased by removing salt from the water (i.e, the salting-in effect). Organic materials are solvated in water by the formation of a water structure using hydrogen bonds around the hydrophobic part. Nevertheless, the presence of inorganic material (Ca^{2+}, Mg^{2+}, and Na^+) breaks this water structure and decreases the solubility of these organic molecules (Al-Shalabi and Sepehrnoori, 2016). Thus, at low salinity, the adsorbed oil gets dispersed from the clay's surface into the solution phase. The mechanism is equivalent to wettability alteration, and ultimately, oil recovery is improved. However, this mechanism cannot explain the dependence of mineral composition, pH, salinity shock, etc. in oil recovery by LSWF.

8.6.7 Expansion of the Electric Double Layer

During low-salinity brine injection, with reduced divalent cations, the EDL at both interfaces expand, resulting in an increased electrostatic repulsion between the two interfaces (Fig. 8.13). The EDL is usually compressed at higher salt concentrations, with a lower zeta potential. This perhaps causes a weaker electrostatic repulsion and a stronger flocculation (Möller and Werr, 1972). A lower zeta potential leads to higher stability of the adsorbed oil and, hence, less recovery. On the other hand, the zeta potential becomes higher in magnitude when the salinity of the brine decreases. A higher zeta potential leads to strong repulsion and, hence, better recovery of oil.

8.6.7.1 Cation Exchange Capacity

The cation exchange capacity (CEC) gives an insight into the mechanisms of oil recovery by LSWF, particularly, from sandstone reservoir rock having a significant clay component. The CEC is a measure of the number of mobile counterions, or exchangeable cations, associated with the negative charge deficiency of clay minerals in a reservoir rock. The CEC of rock and formation water composition depend on the clay type, pH, cation valence, etc. The general order of replacement capacity of the cations is as follows:

$$Li^+ < Na^+ < K^+ < Mg^{2+} < Ca^{2+} < H^+$$

During LSWF, depending on the pH and relative concentration of the system, Na^+ and H^+ may replace the bridged divalent cations (Mg^{2+}, Ca^{2+}) from the clay surface. This causes an expansion of the EDL, which reduces the electrostatic attraction forces between the oil and rock and enables the desorption of polar components from the rock surface.

The ζ-potential at the rock/brine and oil/brine interface is significantly affected by the ionic strength of water. The ionic strength (I) is the concentration of ionic charge in the solution and is defined by Equation 8.11:

$$I = \frac{1}{2}\sum_{i=1}^{n} c_i z_i^2 \qquad (8.11)$$

where one half is to include both cations and anions, c_i is the molar concentration of ion i (M, mol/L), z_i is the charge number of that ion, and the total ions in the solution is summed.

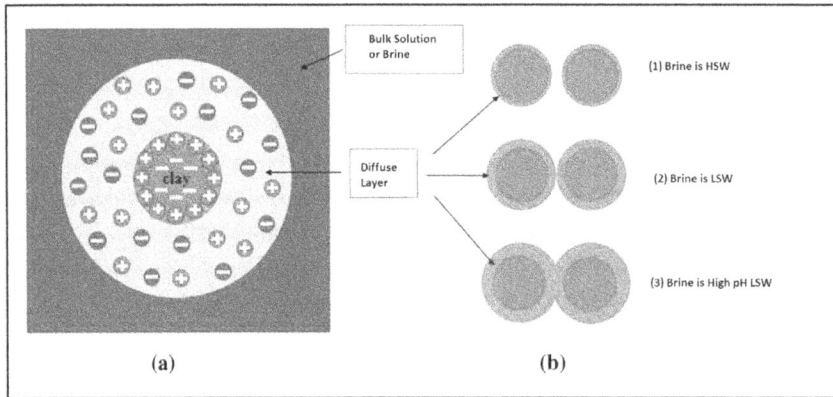

FIGURE 8.13 Impact of salinity and pH on diffuse layer: (a) concept of diffuse layer around clay particle, (b) expansion and approaching of diffuse layer between two clay particles in different brines.

Source: Barnaji et al., 2016.

The surface charge of sandstone in aqueous medium is strongly dependent on salinity and pH. Lowering the brine salinity changes the surface charges of sandstone to strongly negative. Furthermore, the surface charge of solids is affected by the cation type. Ca^{2+} and Mg^{2+} ions result in weak negative charges of Berea sandstone, whereas Na^+ ions make the charges strongly negative (Farooq et al., 2011; Nasralla and Nasr-El-Din, 2011). The effect of pH on the zeta potential of rock/brine system depends on the original salinity and presence of divalent ions. Nasralla and Nasr-El-Din (2014b) observed from their study (Fig. 8.14a) that the lowering of the pH of the brine lowers the magnitude of the negative charge at the rock/brine interface. Crude oils are positively charged at low pH values and become strongly negative at high pH values. The pH at which the charge reaches zero is called the isoelectric point. Charges on rock/brine and oil/brine are the basic entities that govern the stability of aqueous water film on rock surfaces and, hence, the rock wettability. Wettability of the rock surface is mainly dependent on the sign and magnitude of electrical charges at the two interfaces, which governs electrostatic attraction or repulsion, and lead to an oil-wet surface and water-wet surface, respectively; hence, the change of electric charges by dilution results in an alteration of wettability. A study conducted by Nasralla and Nasr-El-Din (2014a) shows that the zeta potential value at the sandstone/brine interface in the presence of 0.5 wt. % NaCl ranges from −21.4 to −30.8 mV, whereas, the zeta potential values for 0.5 wt. % $CaCl_2$ and $MgCl_2$ ranges from −1.5 to −9 mV. An increase in pH increases the magnitude of zeta potential at the oil/brine interface (Fig. 8.14b). Lowering of the brine salinity increases the negative charge at the oil/brine interface, which subsequently, increases the zeta potential value. A higher zeta potential value at the oil/brine interface increases the stability of the dispersed oil droplets detached from the rock surface.

8.6.8 EMULSION FORMATION

Emulsification is an already established mechanism in EOR, as it helps to mobilize the trapped oil with improved mobility ratio and ultralow interfacial tension. As mentioned in Section 8.5, the zeta potential value increases significantly as the salinity of injection water decreases (Fig. 8.14b). Electrokinetic analyses show that the higher zeta potential value with increased repulsive forces stabilizes the emulsions in low-salinity water more than in high-salinity water, which leads to improved oil recovery in LSWF.

8.6.9 OSMOSIS

The movement of a solvent (such as water) through a semipermeable membrane into a solution of higher solute concentration that tends to equalize the concentrations of the solute on both sides of a membrane is known as osmosis. This is also one of the possible mechanisms applicable in a low-salinity environment. According to this mechanism, oil acts as a semipermeable membrane that transports water, but not ions (Sandengen and Arntzen, 2013). Thermodynamically, the driving force is the difference in water activity (i.e., chemical potential of water). Hence, pure water is envisioned to pass through the oil and into the connate water. Connate water, with higher salinity, thereby expands, which expels oil during spontaneous imbibition (Wen and Papadopoulos, 2001; Su et al., 2010). The pore-scale mechanisms of molecular interactions between the high-saline brine, oil phase, and injected low-salinity water are presented in Figure 8.15. The transportation of water occurs due to the difference in chemical potential between the aqueous phases (Fig. 8.15, lower-right inset). The diffusion of water molecules leads to the reestablishment of a new thermodynamic equilibrium condition by equalizing the chemical potential gradient. The transportation of water molecules to high-saline areas causes emulsification and formation of water-in-oil micro dispersions. As the connate water swells and expands the oil phase, mobilizing it from pore to pore in the matrix, it leads to the enhancement of oil recovery.

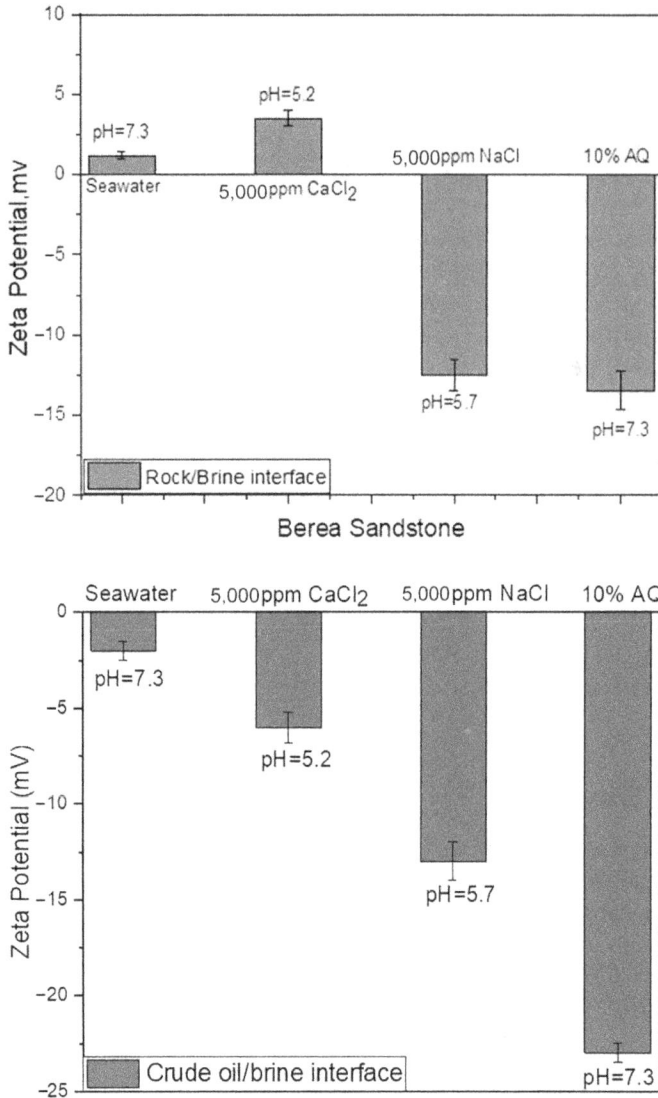

FIGURE 8.14 Typical impact of brine salinity and composition on ζ-potential at (a) Berea-sandstone/brine interfaces and (b) oil/brine interfaces.

Example 8.1: Calculate the ionic strength of a solution of 0.1 M NaCl, 0.02 M Na_2SO_4, and 0.01 M $CaSO_4$.

Solution: $I = 0.5[\{$(Concentration of NaCl in M) × (Number of Na^+) × (Charge of Na)$^2\}$ + $\{$(Concentration of NaCl in M) × (Number of Cl^-) × (Charge of Cl)$^2\}$+$[\{$(Concentration of Na_2SO_4 in M) × (Number of Na^+) × (Charge of Na)$^2\}$+$[\{$(Concentration of Na_2SO_4 in M) × (Number of SO_4^{2-}) × (Charge of SO_4)$^2\}$+$[\{$(Concentration of $CaSO_4$ in M) × (Number of Ca^{2+}) × (Charge of Ca)$^2\}$+$\}$+$\{$(Concentration of $CaSO_4$ in M) × (Number of SO_4^{2-}) × (Charge of SO_4)$^2\}]$

$= 0.5[\{(0.1\ M) \times (1) \times (+1)^2\} + \{(0.1\ M) \times (1) \times (-1)^2\}+[\{(0.02\ M) \times (2) \times (+1)^2\}+[\{(0.02\ M) \times (1) \times (-2)^2\}+[\{(0.01\ M) \times (1) \times (+2)^2\} +\}+\{(0.01\ M) \times (1) \times (+2)^2\}]$

$= 0.5\ [0.1 + 0.1 + 0.04 + 0.08 + 0.04 + 0.04]\ M$

$= 0.20\ M$

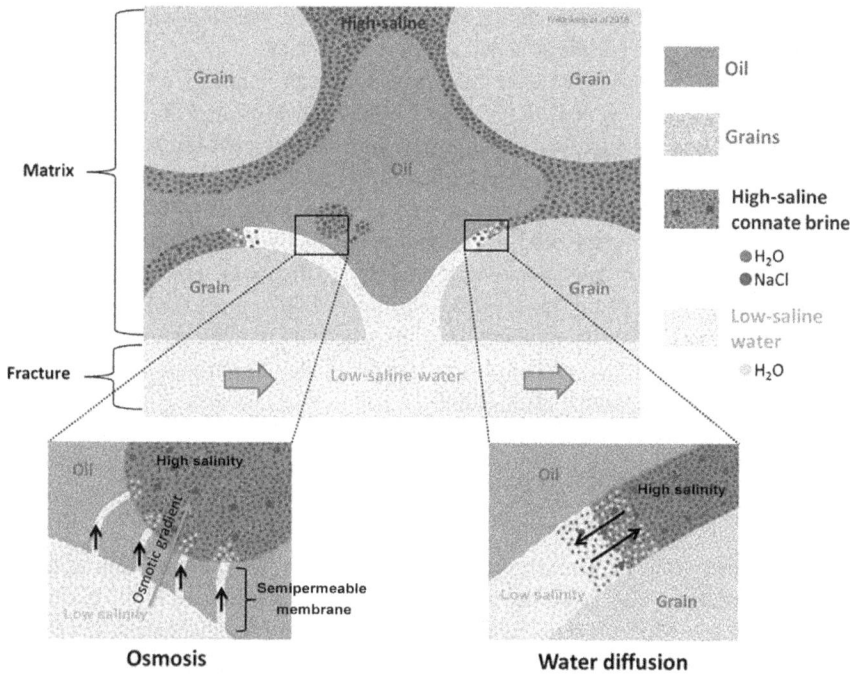

FIGURE 8.15 Occurrence of osmosis with the oil-phase acting as a semipermeable membrane separating the two aqueous phases.

Source: Fredriksen et al., 2018.

8.7 SMART WATERFLOODING FOR CARBONATE RESERVOIRS

Of all the proven reserves found in the world, carbonate reservoirs contain almost more than half of it. Carbonate reservoirs are mainly composed of calcite and dolomite with impurities such as quartz, anhydrite, aragonite, and clay minerals. After primary recovery, most of the oil remains trapped in the reservoirs due to the complex wettability of porous media and rock-fluid interactions, which motivate the development of techniques that are both cost effective and environment friendly for recovering as much oil as possible.

Carbonate reservoirs are mostly composed of calcium carbonate (limestone or chalk), either without or with the presence of magnesium (e.g., dolomite). Due to the lack of clay minerals in carbonate reservoirs, the mechanisms of LSWF in carbonate reservoirs are, to some extent, different than for sandstone reservoirs. The different mechanisms of LSWF in sandstone reservoirs are also applicable for carbonate reservoirs, but the exact chemical composition of the brine plays a very important role for carbonate reservoirs. In sandstone, the low-salinity effect is usually observed below a threshold value of 5,000–7,000 ppm. On the other hand, the low-salinity effect for a carbonate reservoir can be observed with brine salinity between 30,000 and 45,000 ppm. It has been reported by many researchers that certain ions (Ca^{2+}, Mg^{2+}, and SO_4^{2-}) in brine may promote wettability alteration of the carbonate rock surface (Rezaei Gomari and Hamouda, 2006; Mahani et al., 2015). These ions are called potential determining ions (PDIs). The injection water with low salinity and a manipulated concentration of PDIs is called smart water.

The main mechanisms of oil recovery by smart waterflooding is wettability alteration and rock dissolution. The main reason for altering the wetting surface of the rock surface is multicomponent ionic exchange. In carbonate rocks, PDIs such as Ca^{2+}, Mg^{2+}, and SO_4^{2-} are the driving ions in changing the wettability. The divalent anion SO_4^{2-}, present in formation water, competes with the carboxylic acid of crude oil, which is attached to the rock surface.

Temperature plays an important role in the oil recovery mechanisms for carbonate reservoirs. Mutual interactions between Ca^{2+} and SO_4^{2-} and also between Mg^{2+} and SO_4^{2-} at the chalk surface cause the displacement of the adsorbed organic materials. At a higher temperature, SO_4^{2-} ions are strongly adsorbed on the positively charged carbonate rock surface. This causes a net decrease in positive charge on the rock surface, and thus, more Ca^{2+} is attached to the surface, probably in the Stern layer. The sulfate ions are solvated in the bulk water by hydrogen bonds. At higher temperatures, the reactivity of the SO_4^{2-} ions increases, partly due to the breakage of these hydrogen bonds. The increased reactivity of SO_4^{2-} toward the chalk surface at high temperature is partly due to breakage of these hydrogen bonds. At low temperatures, the reactivity of Mg^{2+} decreases, as it is hydrated in water, but its reactivity increases in the presence of SO_4^{2-} due to ion-pair formation according to the following equation (RezaeiDoust et al., 2009):

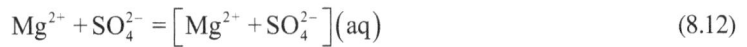

$$Mg^{2+} + SO_4^{2-} = \left[Mg^{2+} + SO_4^{2-} \right](aq) \tag{8.12}$$

As the temperature increases, the equilibrium moves to the right. In equimolar solutions of Ca^{2+} and Mg^{2+} at room temperature, Ca^{2+} adsorbs more strongly on the carbonate rock surface than Mg^{2+}, but at high temperatures, Mg^{2+} become much more reactive and even displaces Ca^{2+} from the chalk surface lattice. The substitution reaction on the chalk surface can be depicted by Equation 8.13:

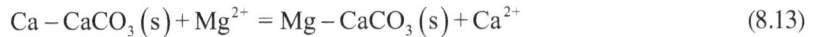

$$Ca - CaCO_3(s) + Mg^{2+} = Mg - CaCO_3(s) + Ca^{2+} \tag{8.13}$$

8.7.1 Ca^{2+} AND SO_4^{2-} AS THE WETTABILITY MODIFIERS

The mechanisms of wettability alteration by Ca^{2+} and SO_4^{2-} present in smart water is schematically presented in Figure 8.16. Here, SO_4^{2-} as a PDI plays an important role. SO_4^{2-} absorbs on the positively charged chalk surface and decreases the net positive charge of the chalk. This causes attraction of Ca^{2+} toward the chalk surface due to a lower electrostatic repulsion. Ca^{2+} can react with carboxylic material and displace it from the surface according to the following reaction:

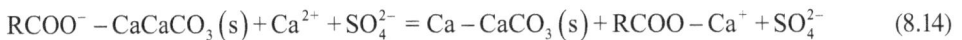

$$RCOO^- - CaCaCO_3(s) + Ca^{2+} + SO_4^{2-} = Ca - CaCO_3(s) + RCOO - Ca^+ + SO_4^{2-} \tag{8.14}$$

In this reaction, SO_4^{2-} only acts as an important catalyst in promoting an increase in the concentration of Ca^{2+} close to the surface.

8.7.2 Mg^{2+} AND SO_4^{2-} AS THE WETTABILITY MODIFIERS

At higher temperatures, Mg^{2+} is more reactive than Ca^{2+} and, hence, can displace the Ca^{2+} ion, which is connected to the carboxylic group at the chalk surface (Fig. 8.16). In this case also, this reaction is catalyzed by SO_4^{2-}. This substitution can be illustrated by the following reaction:

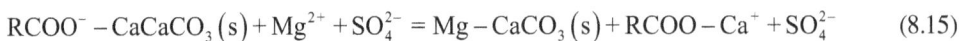

$$RCOO^- - CaCaCO_3(s) + Mg^{2+} + SO_4^{2-} = Mg - CaCO_3(s) + RCOO - Ca^+ + SO_4^{2-} \tag{8.15}$$

Ca^{2+} nor Mg^{2+} cannot improve the oil recovery without the presence of SO_4^{2-} in the injection water, and SO_4^{2-} is not active without the presence of either Ca^{2+} or Mg^{2+} (Zhang et al., 2007).

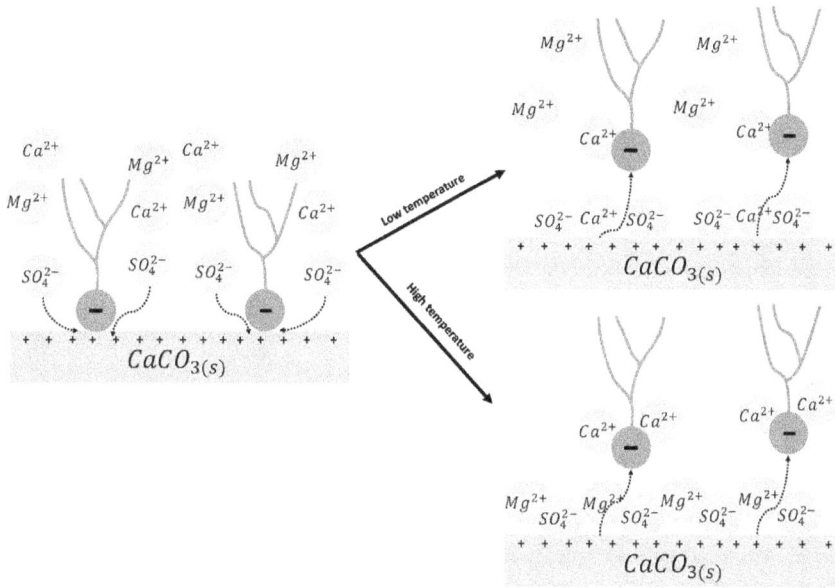

FIGURE 8.16 An illustration of the proposed mechanism of wettability alteration by "MIE" in carbonate reservoirs showing the polar oil component displacement from the carbonate rock surface through PDIs competition: (Left) Original state, (Right-upper) Low-temperature state, and (Left-lower) High-temperature state above 100°C.

Source: Awolayo et al., 2018.

Therefore, the conditions of improved oil recovery with the mechanism of wettability alteration by smart water in carbonate reservoirs are as follows:

- The injected water must contain SO_4^{2-} in addition to Ca^{2+} or Mg^{2+} or both.
- High temperature, usually > 90°C.

8.7.3 Effect of Non-potential Determining Ions

Fathi et al. (2012) reported that not only the concentration of PDIs (Ca^{2+}, Mg^{2+}, and SO_4^{2+}) but also amounts of other non-active salts also have indirect effects on oil recovery from carbonates. If the ionic double layer near a positively charged calcite surface contains a low amount of a non-active salt, such as NaCl, access of sulfate to the calcite surface increases. In this way, sulfate, which is the key ion to change the wettability of a rock surface, can act more efficiently in the absence of non-active ions.

8.8 SIMILARITIES AND DIFFERENCES IN MECHANISM FOR SANDSTONE AND CARBONATE

LSWF is applicable for both sandstone and carbonate rocks, as discussed earlier, but the mechanism that governs the recovery of oil is very much different in both types of rocks due to the difference in mineralogy, surface charges, brine composition, etc. Clay, which is thought to be an important factor that governs oil recovery, is a desirable requirement for sandstone rock, but this is not true for carbonate rocks, as there are evidences of oil recovery without clay content in both rocks (Shehata and Nasr-El-Din, 2014). The most striking difference between the wetting properties in carbonate and sandstone is the adsorption strength of the organic materials attached to the rock. The carboxylic material adsorbs strongly onto the calcite surface, and it is very difficult to remove it by traditional solvent cleaning. It can be removed by increasing the surface reactivity of the potential determining ions (Ca^{2+}, Mg^{2+} and SO_4^{2}) at high temperature by MIE mechanisms. Whenever low-salinity water is injected into the reservoir, the equilibrium between rock and formation brine gets disturbed, and to get back this equilibrium, ion interaction take place. The phenomenon of increased oil recovery and wettability alteration is very much dependent on this reaction; if this ion interaction reaction is slow, there will be no improvement in wettability. To speed up this reaction, sufficient activation energy has to be provided, in which temperature plays the role of a catalyst. This activation energy is very much dependent on the bonding between rock and oil components, which is more in the case of carbonate rocks than sandstones. It is also stated that the dilution of seawater is a more common practice in sandstone; however, tuning of injected water salinity is more common for carbonate.

8.9   OIL RECOVERY BY LSWF

Many studies on low-salinity brine injection confirm that this method can improve oil recovery by 2%–42% depending on the brine composition, crude oil composition, and rock type. However, there are some laboratory and field studies that do not show any increase in oil recovery by low-salinity brine injection. A typical oil recovery by high-salinity and low-salinity waterflooding in a secondary mode is shown in Figure 8.17. Some researchers have studied the effect of oil recovery by low-salinity brine injection in secondary or tertiary recovery modes (Ashraf et al., 2010). A laboratory study was conducted by Yousef et al. (2010) on the applicability of low-salinity water injection (smart waterflood) through carbonate rocks for improving oil recovery using seawater and different dilutions of seawater. They observed increased oil recovery, with a stepwise dilution of seawater, up to an 18% incremental oil recovery, due to tertiary water injection, as reported in Figure 8.18.

8.9.1   SECONDARY VERSUS TERTIARY MODE OF INJECTION

Conventionally, formation water (FW) is used as injection water for waterflooding, since a huge amount of water is produced during oil production; the reinjection of this produced water is cost

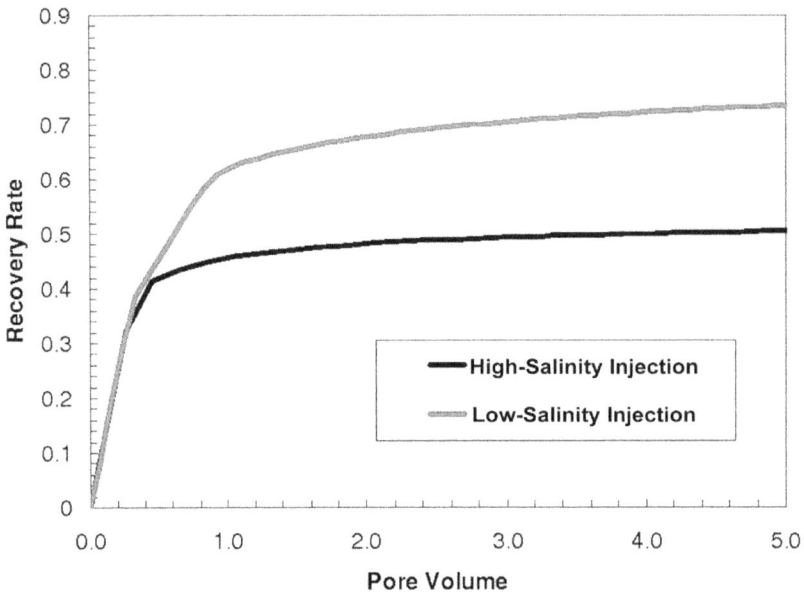

FIGURE 8.17   Typical oil recovery by high-salinity and low-salinity waterflooding (secondary).

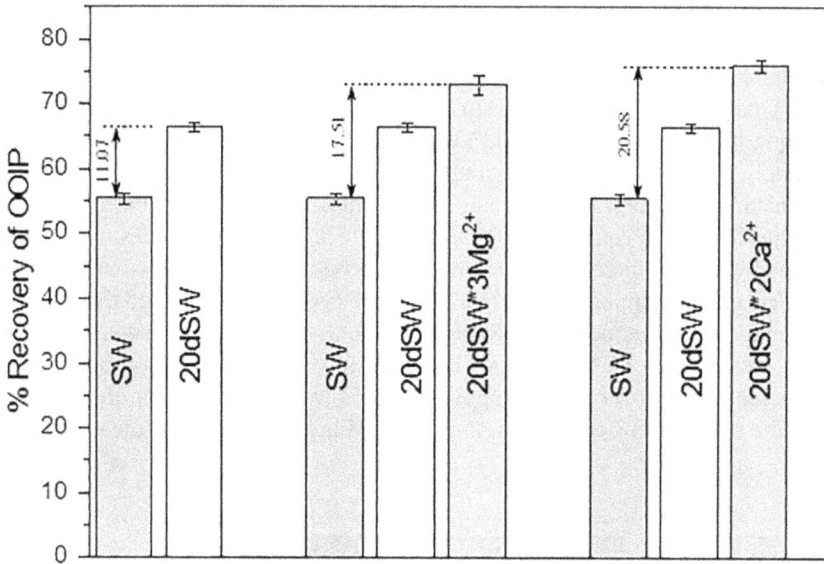

FIGURE 8.18 Typical comparison of oil recovery obtained both by dilution and smart waterflooding of SW, 20dSW, 20dS*3Mg^{2+}, and 20dSW*2Ca^{2+} brines.

effective and compatible with the formation. Recent studies in the field of LSWF have shown that the injection of optimum low-salinity water can produce an increased amount of oil than the injection of FW at secondary mode. Generally, the process of waterflooding is based on two different approaches:

(i) Injection of FW at secondary mode and then successive injection of diluted/ion-tuned brine in tertiary mode.

(ii) Injection of low-salinity water/ion-tuned seawater (ITSW) brine in secondary mode and then successive injection of diluted/ion-tuned seawater until all the movable oil has been recovered.

It is generally found that the secondary mode of injection always produces more oil than the tertiary mode of water injection. This might be due to the reason that there is a continuous film of oil layer present in the secondary mode of injection, while in the tertiary mode the oil ganglia are in discontinuous form, which have to form an oil bank for better recovery. The tertiary water injected has to produce these discontinuous oil ganglia, which is often problematic. Another possible reason for the increased oil recovery in the secondary mode of LSWF is that, low-salinity water gets enough time to interact and mobilize the oil present in the reservoir due to the various LSWF mechanisms discussed than in the tertiary mode of LSWF.

Sometimes, it is also observed that the tertiary mode of injecting the reservoir doesn't produce any increased oil recovery; this may be due to fact that the tertiary water injected was not able to invade the unswept zone that was left behind during the secondary waterflood due to various reasons such as fingering, or the tertiary water injected was not able to move the oil to the wellbore and failed to make the oil bank.

A schematic comparison of the oil sweeping of FW and LSWF in a secondary mode is shown in the following figures. In Figure 8.19a, the transition of the rock phase from the water-wet to the oil-wet region with irreducible water saturation has been shown. In Figure 8.19b, the application of FW at secondary mode has been depicted, and in Figure 8.19c, the application of LSWF/ITSW has been shown. It can be seen that after the FW flooding in secondary mode, the amount of discontinuous oil (which further has to be targeted in tertiary recovery and has to be made in the form of oil banks to be produced) that is present in the reservoir is more as compared to the amount of discontinuous oil present during the injection of LSWF/ITSW in secondary mode as shown in Figures 8.19b and 8.19c, respectively. The less amount of discontinuous oil present in Figure 8.19c may be due to the various mechanism of LSWF/ITSW that took effect right from the start, as discussed earlier.

Various studies have been conducted to assess the effect of different modes of injection of low-salinity/ion-tuned water for increased oil recovery purposes. Zhang and Morrow (2006) conducted a series of experiments pertaining to LSWF in sandstone cores. They observed that LSWF is highly dependent on crude oil/brine/rock (COBR) interaction, as in their experiment, they observed no increased oil recovery upon the injection of low-salinity water in secondary mode, whereas they observed an increase in recovery on the application of low-salinity brine in the tertiary mode of injection with the same COBR conditions, thus establishing the important role of complex COBR

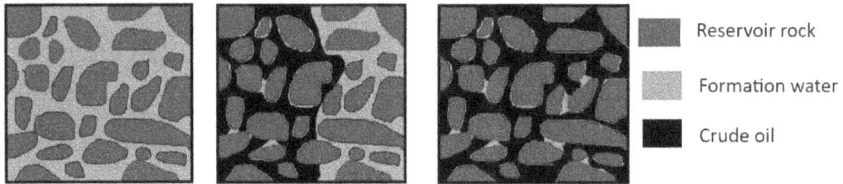

FIGURE 8.19   (a) Transition of rock from water-wet to oil-wet region.

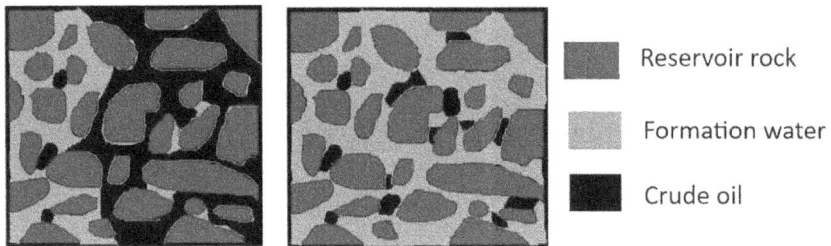

FIGURE 8.19   (b) Application of secondary waterflood with FW and status after FW flooding.

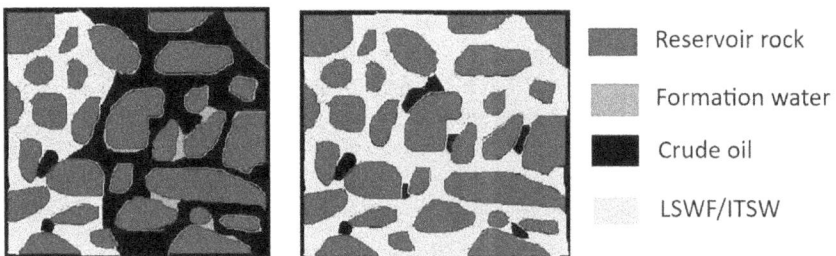

FIGURE 8.19   (c) Application of secondary waterflood with LSWF/ITSW and status after LSWF/ITSW flooding.

TABLE 8.2

Comparison of Oil Recovery by Low-Salinity Waterflooding in Secondary and Tertiary Modes

References	Cores	Recovery Mode	Brine	Additional Oil Recovery (%)
Bernard (1967)	Synthetic and Berea sandstone	Tertiary	10,000–1,000 NaCl	2.63–6.98
Zhang et al. (2007)	Sandstone	Secondary, tertiary	1,479 SW, 1,500 NaCl	29.2, 7–14
Gamage and Thyne (2011)	Berea and field Sandstone	Secondary, tertiary	1% FW	10–22, 2–6
Piñerez Torrijos et al. (2016)	Sandstone	Secondary	1% FW	24
Mohanty and Chandrasekhar (2013)	Limestone	Secondary, tertiary	FW (179,730 ppm), 20dSW (872 ppm)	40, 32
Al-Attar et al. (2013)	Carbonate	Secondary	1,000–5,000 SW	21.5
Awolayo et al. (2014)	Carbonate	Tertiary	$43,000 SW \text{ to } 0.5–8 SO_4^{2-}$	10
Saw and Mandal (2020)	Carbonate	Secondary	$20dSW*2Ca^{2+}$ (1,639 ppm)	20.58

interaction at different stages of flooding. They also stated that reservoir sandstone shows a better response to LSWF than the outcrop sandstone. Gamage and Thyne (2011) conducted recovery test by LSWF in secondary and tertiary mode by various sets of experiments and observed that there is an increase in oil recovery in both Berea and Minnelusa cores. The secondary mode of injection produced more oil than the tertiary mode of injection—thus, showing the efficiency of LSWF in oil recovery irrespective of the mode of injection. In another study on sandstone core, Piñerez Torrijos et al. (2016) also found that secondary mode of injection of LSWF was far superior than the tertiary mode. They also stated that the tertiary mode of injection of low-salinity polymer solution after the injection of LSWF at secondary mode further enhanced the ultimate recovery to 86% of OOIP. Gandomkar and Rahimpour (2015) also conducted several core floods in limestone cores with three different crude oils and observed no incremental oil recovery by LSWF in the tertiary mode of injection; however, they also stated that further investigation should be done in this regard. They also confirmed that seawater would act as a better EOR fluid than the low-salinity FW injected in secondary mode in limestone reservoirs. A comparison of various studies has been shown in Table 8.2. The aforementioned literature indicates that to harness the potential of LSWF, injection mode should be decided carefully, keeping in mind the complex interaction of COBR, as it depends upon each specific reservoir case.

8.10 HYBRID LOW-SALINITY WATERFLOODING

To increase the efficiency of LSWF, the process may be combined with other established mechanisms of recovery such as injection of surfactants, polymers, foams, nanoparticles, or even injection steam. The synergistic effects of LoSal and other methods improve oil recovery significantly. Different such hybrid methods are discussed in the following sections.

8.10.1 POLYMER-AUGMENTED LOW-SALINITY WATERFLOODING

LSWF can be an unstable process due to the low mobility ratio (Tripathi and Mohanty, 2007). Polymer injection improves the sweep efficiency with a reduction of the mobility ratio. The adsorption of polymer is also decreased, as the salinity of water decreases (Sorbie, 2013). Adding polymers to low-salinity water can improve sweep efficiency and mobilize part of the detached oil by a low-salinity wettability alteration mechanism. Thus, the effectiveness of oil recovery from homogeneous and heterogeneous reservoirs may be improved significantly by hybrid polymer and low-salinity waterflooding (Mohammadi, 2012). Borazjani et al. (2016) presented an analytical solution for polymer slug injection with varying water salinity. They indicated that salinity reduction in the polymer slug slows the waterfront, resulting in higher oil cuts being observed after water breakthrough for low-salinity polymer flooding than for high-salinity polymer flooding. It has been observed that the low salinity and polymer injection at low concentrations show superior results in secondary injection mode compared to the tertiary injection mode (Skauge and Shaker Shiran, 2013).

8.10.2 LOW-SALINITY SURFACTANT

The injection of surfactants enhances the mobilization of trapped oil by lowering the capillary forces due to a reduction in the interfacial tension (IFT) between the oleic and aqueous phases. It also alters the wettability of oil-wet rock to water-wet rock by rock-surfactant interactions and subsequent stabilization of the detached oil by micellization. However, the mechanisms of oil recovery by surfactants are highly dependent on the salinity of the system. High salinity causes precipitation of the surfactant from the bulk solution if salinity is higher than the salt tolerance limit of the surfactant. Further, an increase in salinity results in the formation of water-in-oil microemulsions, at which the surfactant tends to stay in the oil phase (Hosseinzade Khanamiri et al., 2016). On the other hand, the effectiveness of the surfactant is maximum at optimum salinity, which is in the low-salinity range. Combination of low-salinity water injection with surfactant injection seems to have environmental and economic advantages. Another advantage of surfactant injection at low ionic strength is the reduction in surfactant retention by adsorption. This may create an opportunity to inject surfactants at lower concentrations. Alagic and Skauge (2010) performed a series of core flooding experiments and reported high tertiary oil recovery by surfactant injection after establishing a low-salinity environment by injection of low-salinity water. They also observed lower tertiary oil recovery by surfactant without a low-salinity preflush.

8.10.3 OTHER HYBRID METHODS

Other LSWF-based hybrid methods are low salinity water–assisted foam flooding (LSWAF), low salinity alternating steam flooding (LSASF). LSWAF consists of injecting low-salinity water (LSW) followed by an alternate injection of a surfactant aqueous solution and CO_2 gas (Shabib-AsL et al., 2019). The mechanism of LSWF is coupled with the beneficial effects of mobility control by foam and IFT reduction by surfactant. In LSASF, steam can reduce oil viscosity. LSW can alter the wettability of the reservoir toward more water wet. This combined technology can resolve steam problems and alter the steam by using LSW, which has the ability to increase oil recovery. Al-Saedi et al. (2018) reported promising results of more than 15% recovery of original oil in place after secondary and tertiary treatments by LSASF.

8.11 CASE STUDIES

8.11.1 CASE STUDY 1

Norne is an oil field located around 80 kilometers north of the Heidrun oil field in the Norwegian Sea. The field consists of two separate oil compartments, the Norne Main Structure and the Northeast Segment. The reservoir, rock, and fluid properties of the Norne Field are shown in Table 8.3. The Norne Main Structure (Norne C, D, and E segments) was discovered in December 1991 and includes 97% of the oil in place. The Norne Field's E segment produced about 40% of the oil in

TABLE 8.3

Reservoir, Rock, and Fluid Properties of the Norne Field (Islam et al., 2018)

Property	Value	Property	Value
Initial pressure	273 bar @ 2639 m TVD	Oil density	859.5 kg/m³
Temperature	98.3°C	Gas density	0.854 kg/m³
Permeability	20–2,500 mD	Water density	1,033 kg/m³
Porosity	0.25–0.30	Gravity	32.7 API
Formation type	Sandstone	Oil formation volume factor	1.32
Depth	2,500–2,700 m	Gas formation volume factor	0.0047
Net thickness	110 m	Water formation volume factor	1.038
Bubble-point pressure	251 bar (C, D, and E segments), and 216 bar (G segment)	Pore compressibility	4.84×10^{-5} 1/bar @ 277 bar
Rock wettability	Mixed wet	Water compressibility	4.67×10^{-5} 1/bar @ 277 bar

place over a period of seven years from 1997 to 2004 through a combination of primary recovery and seawater flooding as a secondary recovery technique (Islam et al., 2018). Then LSW simulation studies using original wells indicated that water injection with optimal salt concentration of 1,000 ppm TDS yields a significant improvement in recovery. In the comparison of cumulative oil for the six cases (Case 1: Base case with seawater flooding using the existing wells; Case 2: LSWF using the original wells; Case 3: LSWF using the original wells in addition to a new producing well; Case 4: LSWF using the original wells in addition to a recompleted producing well; Case 5: LSWF using the original wells in addition to a new injection well; Case 6: LSWF using the original wells in addition to a recompleted injection well), as reported by Islam et al. (2018), it is observed that the cumulative oil recovery is substantially higher for the five different LSWF cases than for the base case of seawater flooding.

8.11.2 CASE STUDY 2

A systematic assessment of the potential of LSWF for the Dong-He-Tang reservoir in the Tarim oil field, China, is reported by Liu et al. (2016). This reservoir has a high reservoir temperature of 140°C, high FW salinity of 234,000 ppm total dissolved solids, and an in situ oil viscosity of 2.2 cp. The porosity and permeability of the reservoir are in a range of 11%–16% and 1–400 mD, respectively. First, they investigated the wettability changes by LSW with the manipulation of divalent cations and found the reservoir's wettability changed from oil wet to water wet. Spontaneous imbibition result showed that LSW gave 47% of ultimate oil recovery, whereas formation brine only exhibited 12% ultimate oil recovery. LSW increased oil recovery in both secondary and tertiary modes at temperatures up to 140°C. Also, LSW shifted the relative permeability curve and imbibition capillary pressure curves toward lower residual oil saturation than formation brine.

The parameters derived from laboratory experiments were used as input for reservoir simulation models to investigate the potential of LSWF in the reservoir using two layered box models. Findings showed that LSW accelerated oil production by increasing the oil's relative permeability, thus resulting in a higher recovery factor with only a fraction of pore volume of LSW injection.

8.11.3 CASE STUDY 3

The Omar oil field in Syria is a potential candidature for LSWF. Its initial oil volumes are roughly evenly split between the cretaceous sheetlike shallow marine Lower Rutbah and the triassic coastal fluvial plane Mulussa formation. (Vledder et al., 2010). The Lower Rutbah consists of well-developed shallow marine and tidal channel sands with a high net to gross ratio (~73%). The Mulussa formations consist of fluvial channel sands and flood plain shales with a total thickness of up to 350 m. The reservoir contains light oil with viscosity of around 0.3 cP. Formation water in Omar has a salinity of around 90,000 mg/L, with a high content of bivalent cations of around 5,000 mg/L. Scanning electron microscope pictures reveal kaolinite lining up the rock surface. Petrography experiments (XRD) reveal that 0.5%–4% of rock composition is clay of which 95%–100% is kaolinite. SCAL measurements on the core collected from the field showed oil-wetness. Low-salinity effect observed the change of wettability from oil wet to water wet, with a change of the wettability index (W) from 1 to 0.2–0.4. The low-salinity flooding led to an incremental recovery of 10%–15% of the stock tank oil initially in place.

8.12   LSW SCREENING, DESIGN, AND OPTIMIZATION

There are so many advantages of oil recovery by LSWF over conventional HSWF and other EOR methods. The major advantages of LSWF include: (1) considerable recovery benefit; (2) lower costs and relatively simpler; (3) easier to be implemented in both onshore and offshore reservoirs; (4) possible utilization of on-site facilities without requiring a large quantity of chemicals or gases for EOR projects; (5) more environmentally friendly.

LSWF is a strongly geological-dependent process, but most of the study has been reported only on a laboratory scale in the literature so far. For implementing LSWF in a larger and more successful manner, a comprehensive and specific development plan must be followed. The development strategy for LSW implementation in the field scale consists of four main steps: screening, design, optimization, and economics evaluation.

8.12.1   LSW Screening

In the initial stage of the LSW development project, prescreening is done. The reservoir lithology, rock, fluid properties, facilities, and operating conditions play a vital role. It is also essential to find which EOR/IOR method suits the best for the reservoir candidate. Based on a few pilot tests and lab analyses, there are few prescreening criteria to be kept in mind before applying a LSW method. Based on the core flooding studies performed in the laboratory and pilot tests of LSWF, the pre-screening criteria are reported in Table 8.4 (Dang et al., 2015).

8.12.2   LSW Design

Once a reservoir is classified as a promising candidate for LSWF implementation based on the aforementioned screening criteria, a detailed LSWF design plan is required next to apply this emerging technology on a field scale. There is a systematic procedure for designing and testing LSW implementation, namely:

- Reservoir characterization: geochemistry, minerology, petrophysics.
- Fluid design: injected composition ($Na+$, Ca^{2+}, Mg^{2+}).
- Core flooding: relative permeability analysis, recovery factor, effluent ions, pH.
- Modeling: core flooding, history-match, field-scale simulation, sensitivity analysis.
- Pilot test: logistics, facilities, single-well chemical tracer tests, full-field scale.

8.12.3   LSW Optimization

LSWF is an emerging EOR technique in which people try to modify the injection brine composition for achieving a higher recovery factor. There are a number of parameters, and the optimization of

TABLE 8.4
Prescreening Conditions for LSW Implementation (Dang et al., 2015)

Property	Preferred Condition
Reservoir	• Sandstones • Carbonates (possibility)
Crude oil	• Must contain polar components (not effective with synthetic oil) • Viscosity is not too high for waterflooding
Clay minerals	• Reservoir must contain sufficient amount of clay • Medium sand with high CEC clay, porosity, permeability is preferred
Reservoir minerals	• Calcite • Dolomite
Formation water	• Presence of divalent ions such as Ca^{++} and Mg^{++}
Initial wettability	• Oil-wet or mixed-wet reservoir • Small or ineffective in strong water-wet reservoir
Reservoir temperature	• Not limited
Reservoir depth	• Not limited
Reservoir energy	• Sufficiently high pressure for achieving miscibility conditions
Injection fluid	• Lower salinity concentration than formation water • Must contain divalent ions • Injected compositions must promote the adsorption of divalent ions • Sufficient CO_2 or chemical sources for hybrid LSWF implementation

these parameters ultimately improves the recovery of a LSWF. The details of these parameters are discussed in the following:

- The first attempts to optimize the LSWF process is to find an optimal injection recipe for the injection fluid, which is very much reservoir specific. It can be obtained by laboratory studies, which include both dilution of high-salinity injection water and ion-tuning of the optimized diluted water.
- In addition to the injected brine composition, LSWF can be effectively optimized by well placement under geological uncertainties. This optimization definitely helps to maximize the oil recovery by selecting the best location for LSW injection that promotes ion exchange and wettability alteration and enhances the water/oil displacement efficiency.
- Another important factor in LSWF optimization is to determine the optimal time for LSWF application. LSWF is effective in both secondary and tertiary recovery processes, though the laboratory and simulation results show that the secondary LSWF is more effective than the conventional HSWF and the tertiary LSWF in terms of timing and oil recovery. Thus, proper planning provides not only a better oil recovery but also reduces the cost and complexity of the process.
- Additionally, one can utilize the synergy between LSWF and conventional EOR methods such as polymer flooding, surfactant flooding, and miscible gas flooding to (1) promote the synergy between different EOR techniques; (2) overcome the current technical challenges inside these methods; and (3) achieve the highest oil recovery factor and maximize project profits (Dang et al., 2015).

8.12.4 LSW Implementation and Economic Evaluation

In terms of economic evaluation, a LSWF project can utilize the current facilities of conventional waterflooding; thus, the major difference comes from formation or seawater desalination costs. There are obvious expenditures for LSW desalination facilities that depend on several important factors such as salinity of source water, salinity of injected water, field location, project scale, energy costs, and oil price. By combining the expected amount and timing of the incremental oil recovery, one can estimate the economic efficiency of the LSW project for a reservoir candidate.

REFERENCES

Abdallah, W., Buckley, J.S., Carnegie, A., Edwards, J., Herold, B., Fordham, E., Graue, A., Habashy, T., Seleznev, N., Signer, C., Hussain, H., 1986. Fundamentals of wettability. Technology 38(1125–1144), 268.

Akbar, M., Vissapragada, B., Alghamdi, A.H., Allen, D., Herron, M., Carnegie, A., Dutta, D., Olesen, J., Chourasiya, R., Logan, D., Stief, D., Netherwood, R., Russell, S.D., Saxena, K., 2000. A snapshot of carbonate reservoir evaluation. Oilfield Review 12, 20–41.

Alagic, E., Skauge, A., 2010. Combined low salinity brine injection and surfactant flooding in mixed–Wet sandstone cores. Energy & Fuels 24, 3551–3559.

Al-Attar, H.H., Mahmoud, M.Y., Zekri, A.Y., Almehaideb, R., Ghannam, M., 2013. Low-salinity flooding in a selected carbonate reservoir: Experimental approach. Journal of Petroleum Exploration and Production Technology 3, 139–149.

Alotaibi, M.B., Azmy, R., Nasr-El-Din, H.A., 2010, 1 January. A Comprehensive EOR Study Using Low Salinity Water in Sandstone Reservoirs. In SPE Improved Oil Recovery Symposium. Society of Petroleum Engineers.

Alotaibi, M.B., Nasralla, R.A., Nasr-El-Din, H.A., 2011. Wettability studies using low-salinity water in sandstone reservoirs. SPE Reservoir Evaluation & Engineering 14(6), 713–725.

Alroudhan, A., Vinogradov, J., Jackson, M.D., 2016. Zeta potential of intact natural limestone: Impact of potential-determining ions Ca, Mg and SO_4. Colloids Surfaces A: Physicochemical and Engineering Aspects 493, 83–98.

Al-Saedi, H.N., Flori, R.E., Alkhamis, M., Brady, P.V., 2018, August. Coupling Low Salinity Water Flooding and Steam Flooding for Sandstone Reservoirs; Low Salinity-Alternating-Steam Flooding (LSASF). In SPE Kingdom of Saudi Arabia Annual Technical Symposium and Exhibition. Society of Petroleum Engineers.

Al-Shalabi, E.W., Sepehrnoori, K., 2016. A comprehensive review of low salinity/engineered water injections and their applications in sandstone and carbonate rocks. Journal of Petroleum Science and Engineering 139, 137–161.

Ashraf, A., Hadia, N.J., Torsæter, O., Tweheyo, M.T., 2010, January. Laboratory Investigation of Low Salinity Waterflooding as Secondary Recovery Process: Effect of Wettability. In SPE Oil and Gas India Conference and Exhibition. OnePetro.

Austad, T., RezaeiDoust, A., Puntervold, T., 2010. Chemical Mechanism of Low Salinity Water Flooding in Sandstone Reservoirs. In SPE Improved Oil Recovery Conference (SPE 129767), pp. 19–22.

Awolayo, A., Sarma, H., AlSumaiti, A.M., 2014. A Laboratory Study of Ionic Effect of Smart Water for Enhancing Oil Recovery in Carbonate Reservoirs. In SPE EOR Conference at Oil and Gas West Asia 2014 Driving Integrated and Innovative EOR, pp. 46–69.

Awolayo, A., Sarma, H., Nghiem, L., 2018. Brine-dependent recovery processes in carbonate and sandstone petroleum reservoirs: Review of laboratory-field studies, interfacial mechanisms and modeling attempts. Energies 11, 3020.

Barnaji, M.J., Pourafshary, P., Rasaie, M.R., 2016. Visual investigation of the effects of clay minerals on enhancement of oil recovery by low salinity water flooding. Fuel 184, 826–835.

Bernard, G.G., 1967. Effect of Floodwater Salinity on Recovery of Oil from Cores Containing Clays. In Annual SPE of AIME California Regular Meeting, pp. 1–8.

Bonto, M., Eftekhari, A.A., Nick, H.M., 2019. An overview of the oil-brine interfacial behavior and a new surface complexation model. Science Reports 9, 6072.

Borazjani, S., Bedrikovetsky, P., Farajzadeh, R., 2016. Analytical solutions of oil displacement by a polymer slug with varying salinity. Journal of Petroleum Science and Engineering 140, 28–40.

Chen, L., Zhang, G., Wang, L., Wu, W., Ge, J., 2014. Zeta potential of limestone in a large range of salinity. Colloids and Surfaces A: Physicochemical and Engineering Aspects 450, 1–8.

Chilingar, G.V., Yen, T.F., 1983. Some notes on wettability and relative permeabilities of carbonate reservoir rocks, II. Energy Sources 7(1), 67–75.

Dang, C.T., Nguyen, N.T., Chen, Z., 2015, April. Practical Concerns and Principle Guidelines for Screening, Implementation, Design, and Optimization of Low Salinity Waterflooding. In SPE Western Regional Meeting. Society of Petroleum Engineers.

Delgado, A.V., González-Caballero, F., Hunter, R.J., Koopal, L.K., Lyklema, J., 2005. Measurement and interpretation of electrokinetic phenomena (IUPAC Technical Report). Pure and Applied Chemistry 77, 1753–1805.

Den Ouden, L., Nasralla, R.A., Guo, H., Bruining, H., Van Kruijsdijk, C.P., 2015, April. Calcite Dissolution Behaviour During Low Salinity Water Flooding in Carbonate Rock. In IOR 2015–18th European Symposium on Improved Oil Recovery, cp-445. European Association of Geoscientists & Engineers.

Derkani, M.H., Fletcher, A.J., Fedorov, M., Abdallah, W., Sauerer, B., Anderson, J., Zhang, Z.J., 2019. Mechanisms of surface charge modification of carbonates in aqueous electrolyte solutions. Colloids and Interfaces 3, 62.

Ding, H., Rahman, S., 2017. Experimental and theoretical study of wettability alteration during low salinity water flooding–An state of the art review. Colloids and Surfaces A: Physicochemical and Engineering Aspects 520, 622–639.

Doust, A.R., Puntervold, T., Austad, T., 2011. Chemical verification of the EOR mechanism by using low saline/smart water in sandstone. Energy & Fuels 25, 2151–2162.

Farooq, U., Tweheyo, M.T., Sjöblom, J., Øye, G., 2011. Surface characterization of model, outcrop, and reservoir samples in low salinity aqueous solutions. Journal of Dispersion Science and Technology 32(4), 519–531.

Fathi, S.J., Austad, T., Strand, S., 2012, 16–18 April. Water-Based Enhanced Oil Recovery (EOR) by "Smart Water" in Carbonate Reservoirs. In Proceedings of the SPE EOR Conference at Oil and Gas West Asia (Paper SPE 154570), Muscat.

Fredriksen, S.B., Rognmo, A.U., Fernø, M.A., 2018. Pore-scale mechanisms during low salinity waterflooding: Oil mobilization by diffusion and osmosis. Journal of Petroleum Science and Engineering 163, 650–660.

Gamage, P., Thyne, G.D., 2011. Comparison of Oil Recovery by Low Salinity Waterflooding in Secondary and Tertiary Recovery Modes. In SPE Annual Technical Conference and Exhibition.

Gandomkar, A., Rahimpour, M.R., 2015. Investigation of low-salinity waterflooding in secondary and tertiary enhanced oil recovery in limestone reservoirs. Energy & Fuels 29, 7781–7792.

Graue, A., Viksund, B.G., Eilertsen, T., Moe, R., 1999. Systematic wettability alteration by aging sandstone and carbonate rock in crude oil. Journal of Petroleum Science and Engineering 24(2–4), 85–97.

Gregory, J., 1975. Interaction of unequal double layers at constant charge. Journal of Colloid and Interface Science 51(1), 44–51.

Hamaker, H., 1937. The London—Van der Waals attraction between spherical particles. Physica 4, 1058–1072.

Heberling, F., Trainor, T.P., Lutzenkirchen, J., Eng, P., Denecke, M.A., Bosbach, D., 2010. Structure and reactivity of the calcite– Water interface. Journal of Colloid and Interface Science 354, 843–857.

Hirasaki, G.J., 1991a. Interfacial Phenomena in Petroleum Recovery (Chapter 3). Marcel Dekker.

Hirasaki, G.J., 1991b. Wettability: Fundamentals and surface forces. SPE Formation Evaluation 6, 217–226.

Hosseinzade Khanamiri, H., Baltzersen Enge, I., Nourani, M., Stensen, J.Å., Torsæter, O., Hadia, N., 2016. EOR by low salinity water and surfactant at low concentration: Impact of injection and in situ brine composition. Energy & Fuels 30(4), 2705–2713.

Hunter, R.J., 1981. Zeta Potential in Colloid Science: Principles and Applications. Academic Press.

Islam, M.S., Kleppe, J., Rahman, M.M., Abbasi, F., 2018. An Evaluation of IOR Potential for the Norne Field's E-Segment Using Low Salinity Water-Flooding: A Case Study. In SPE Kingdom of Saudi Arabia Annual Technical Symposium and Exhibition. Society of Petroleum Engineers.

Jaafar, M.Z., Nasir, A.M., Hamid, M.F., 2014. Measurement of isoelectric point of sandstone and carbonate rock for monitoring water encroachment. Journal of Applied Sciences 14(23), 3349–3353.

Jalili, Z., Tabrizy, V.A., 2014. Mechanistic study of the wettability modification in carbonate and sandstone reservoirs during water/low salinity water flooding. Energy & Environment Research 4(3), 78.

Jerauld, G.R., Lin, C.Y., Webb, K.J., Seccombe, J.C., 2006, September. Modeling Low-Salinity Waterflooding. In SPE 102239, Paper Presented at the 2006 SPE Annual Technical Conference and Exhibition, San Antonio, TX, pp. 24–27.

Jiang, S., Liang, P., Han, Y., 2018. Effect of clay mineral composition on low-salinity water flooding. Energies 11(12), 3317.

Katende, A., Sagala, F., 2019. A critical review of low salinity water flooding: Mechanism, laboratory and field application. Journal of Molecular Liquids 278, 627–649.

Kumar, A., Mandal, A., 2019. Critical investigation of zwitterionic surfactant for enhanced oil recovery from both sandstone and carbonate reservoirs: Adsorption, wettability alteration and imbibition studies. Chemical Engineering Science 209, 115222.

Lager, A., Webb, K.J., Black, C.J.J., Singleton, M., Sorbie, K.S., 2006. September. Low Salinity Oil Recovery – An Experimental Investigation. In Presented at the International Symposium of the Society of Core Analysts, Trondheim.

Lager, A., Webb, K.J., Black, C.J.J., Singleton, M., Sorbie, K.S., 2008. Low salinity oil recovery—An experimental investigation. Petrophysics 49(1), 28–35.

Lee, S.Y., Webb, K.J., Collins, I.R., Lager, A., Clarke, S.M., O'Sullivan, M., Routh, A.F., Wang, X., 2010, April. Low Salinity Oil Recovery–Increasing Understanding of the Underlying Mechanisms. In SPE Improved Oil Recovery Symposium. OnePetro.

Lin, M., Hua, Z., Li, M., 2018. Surface wettability control of reservoir rocks by brine. Petroleum Exploration and Development 45, 145–153.

Liu, Y., Jiang, T., Zhou, D., Zhao, J., Xie, Q., Saeedi, A., 2016, October. Evaluation of the Potential of Low Salinity Water Flooding in the High Temperature and High Salinity Dong-He-Tang Reservoir in the Tarim Oilfield, China: Experimental and Reservoir Simulation Results. In SPE Asia Pacific Oil & Gas Conference and Exhibition. Society of Petroleum Engineers.

Mahani, H., Keya, A.L., Berg, S., Bartels, W.B., Nasralla, R., Rossen, W.R., 2015. Insights into the mechanism of wettability alteration by low-salinity flooding (LSF) in carbonates. Energy & Fuels 29(3), 1352–1367.

McGuire, P.L., Chatham, J.R., Paskvan, F.K., Sommer, D.M., Carini, F.H., 2005, 1 January. Low Salinity Oil Recovery: An Exciting New EOR Opportunity for Alaska's North Slope. In SPE Western Regional Meeting. Society of Petroleum Engineers.

Mohammadi, H., Jerauld, G.R., 2012, 14–18 April. Mechanistic Modeling of Benefit of Combining Polymer with Low Salinity Water for Enhanced Oil Recovery. In Paper SPE 153161, Presented at the Eighteenth SPE Improved Oil Recovery Symposium, Tulsa, OK.

Mohammed, M.A., Babadagli, T., 2016. Experimental investigation of wettability alteration in oil-wet reservoirs containing heavy oil. SPE Reservoir Evaluation & Engineering 19, 633–644.

Mohan, K.K., Fogler, H.S., Vaidya, R.N., Reed, M.G., 1993. Water Sensitivity of Sandstones Containing Swelling and Non-Swelling Clays, in: Colloids in the Aquatic Environment. Elsevier, pp. 237–254.

Mohanty, K.K., Chandrasekhar, S., 2013. Wettability Alteration with Brine Composition in High Temperature Carbonate Reservoirs. In SPE Annual Technical Conference and Exhibition.

Möller, P., Werr, G., 1972. Influence of anions on Ca^{2+}-Mg^{2+} surface exchange process on calcite in artificial sea water. Radiochimica Acta 18(3), 144–147.

Myint, P.C., Firoozabadi, A., 2015. Thin liquid films in improved oil recovery from low-salinity brine. Current Opinion in Colloid & Interface Science 20(2), 105–114.

Nasralla, R.A., Alotaibi, M.B., Nasr-El-Din, H.A., 2011, 1 January. Efficiency of Oil Recovery by Low Salinity Water Flooding in Sandstone Reservoirs. In SPE Western North American Region Meeting. Society of Petroleum Engineers.

Nasralla, R.A., Nasr-El-Din, H.A., 2011, January. Impact of Electrical Surface Charges and Cation Exchange on Oil Recovery by Low Salinity Water. In SPE Asia Pacific Oil and Gas Conference and Exhibition. Society of Petroleum Engineers.

Nasralla, R.A., Nasr-El-Din, H.A., 2014a. Double-layer expansion: Is it a primary mechanism of improved oil recovery by low-salinity waterflooding? SPE Reservoir Evaluation & Engineering 17, 49–59.

Nasralla, R.A., Nasr-El-Din, H.A., 2014b. Impact of cation type and concentration in injected brine on oil recovery in sandstone reservoirs. Journal of Petroleum Science and Engineering 122, 384–395.

Nowrouzi, I., Manshad, A.K., Mohammadi, A.H., 2018. Effects of dissolved binary ionic compounds and different densities of brine on interfacial tension (IFT), wettability alteration, and contact angle in smart water and carbonated smart water injection processes in carbonate oil reservoirs. Journal of Molecular Liquids 254, 83–92.

Piñerez Torrijos, I.D., Puntervold, T., Strand, S., Austad, T., Abdullah, H.I., Olsen, K., 2016. Experimental study of the response time of the low-salinity enhanced oil recovery effect during secondary and tertiary low-salinity waterflooding. Energy & Fuels 30, 4733–4739.

Pu, H., Xie, X., Yin, P., Morrow, N.R., 2010, 1 January. Low-Salinity Waterflooding and Mineral Dissolution. In SPE Annual Technical Conference and Exhibition. Society of Petroleum Engineers.

Rahbar, M., Roosta, A., Ayatollahi, S., Ghatee, M.H., 2012. Prediction of three-dimensional (3-D) adhesion maps, using the stability of the thin wetting film during the wettability alteration process. Energy & Fuels 26, 2182–2190.

Rezaei Gomari, K.A., Hamouda, A.A., 2006. Effect of fatty acids, water composition and pH on the wettability alteration of calcite surface. Journal of Petroleum Science and Engineering 50, 140–150.

RezaeiDoust, A., Puntervold, T., Strand, S., Austad, T., 2009. Smart water as wettability modifier in carbonate and sandstone: A discussion of similarities/differences in the chemical mechanisms. Energy & Fuels 23(9), 4479–4485.

Sandengen, K., Arntzen, O.J., 2013, 16–18 April. Osmosis During Low-Salinity Water Flooding. In Presented at the 17th European Symposium on Improved Oil Recovery, St. Petersburg.

Santha, N., Cubillas, P., Saw, A., Brooksbank, H., Greenwell, H.C., 2017. Chemical force microscopy study on the interactions of COOH functional groups with kaolinite surfaces: Implications for enhanced oil recovery. Minerals 7, 250.

Saw, R.K., Mandal, A., 2020. A mechanistic investigation of low salinity water flooding coupled with ion tuning for enhanced oil recovery. RSC Advances 10, 42570–42583.

Saxena, N., Kumar, A., Mandal, A., 2019. Adsorption analysis of natural anionic surfactant for enhanced oil recovery: The role of mineralogy, salinity, alkalinity and nanoparticles. Journal of Petroleum Science and Engineering 173, 1264–1283.

Shabib-AsL, A., Abdalla Ayoub, M., Abdalla Elraies, K., 2019, March. Combined Low Salinity Water Injection and Foam Flooding in Sandstone Reservoir Rock: A New Hybrid EOR. In SPE Middle East Oil and Gas Show and Conference. Society of Petroleum Engineers.

Shehata, A.M., Nasr-El-Din, H.A., 2014. Role of Sandstone Mineral Compositions and Rock Quality on the Performance of Low-Salinity Waterflooding. In International Petroleum Technology Conference, pp. 10–12.

Shehata, A.M., Nasr-El-Din, H.A., 2015. Zeta Potential Measurements: Impact of Salinity on Sandstone Minerals. In SPE International Symposium on Oilfield Chemistry, pp. 13–15.

Sheng, J.J., 2014. Critical review of low-salinity waterflooding. Journal of Petroleum Science and Engineering 120, 216–224.

Skauge, A., Shaker Shiran, B., 2013, 16–18 April. Low Salinity Polymer Flooding. In Paper A14 Presented at 17th European Symposium on Improved Oil Recovery, St. Petersburg.

Sohal, M.A., Thyne, G., Søgaard, E.G., 2016. Review of recovery mechanisms of ionically modified waterflood in carbonate reservoirs. Energy & Fuels 30, 1904–1914.

Sorbie, K.S., 2013. Polymer-Improved Oil Recovery. Springer Science & Business Media.

Su, J.T., Duncan, P.B., Momaya, A., Jutila, A., Needham, D., 2010. The effect of hydrogen bonding on the diffusion of water in n-alkanes and n-alcohols measured with a novel single microdroplet method. The Journal of Chemical Physics 132(4).

Takamura, K., Chow, R.S., 1985. The electric properties of the bitumen/water interface Part II. Application of the ionizable surface-group model. Colloids and Surfaces 15, 35–48.

Tang, G., Morrow, N.R., 1999a, 1–4 August. Oil Recovery By Waterflooding and Imbibition—Invading Brine Cation Valency and Salinity. In Paper SCA9911 Presented at the International Symposium of the Society of Core Analysts, Golden, CO.

Tang, G-Q., Morrow, N.R., 1999b. Influence of brine composition and fines migration on crude oil/brine/rock interactions and oil recovery. Journal of Petroleum Science and Engineering 24, 99–111.

Tripathi, I., Mohanty, K.K., 2007, 11–14 November. Flow Instability Associated with Wettability Alteration. In Paper SPE 110202 Presented at the SPE Annual Technical Conference and Exhibition, Anaheim, CA.

Vledder, P., Gonzalez, I.E., Carrera Fonseca, J.C., Wells, T., Ligthelm, D.J., 2010, January. Low Salinity Water Flooding: Proof of Wettability Alteration on a Field Wide Scale. In SPE Improved Oil Recovery Symposium. Society of Petroleum Engineers.

Wen, L., Papadopoulos, K.D., 2001. Effects of Osmotic Pressure on Water Transport in $W_1/O/W_2$ Emulsions. Journal of Colloid and Interface Science 235(2), 398–404.

Xie, Q., Saeedi, A., Pooryousefy, E., Liu, Y., 2016. Extended DLVO-based estimates of surface force in low salinity water flooding. Journal of Molecular Liquids 221, 658–665.

Yousef, A.A., Al-Saleh, S., Al-Kaabi, A., Al-Jawfi, M., 2010, October. Laboratory Investigation of Novel Oil Recovery Method for Carbonate Reservoirs. In Canadian Unconventional Resources and International Petroleum Conference. OnePetro.

Zhang, Y., Morrow, N.R., 2006. Comparison of Secondary and Tertiary Recovery with Change in Injection Brine Composition for Crude-Oil/Sandstone Combinations. In SPE/DOE Symposium on Improved Oil Recovery.

Zhang, Y., Xie, X., Morrow, N.R., 2007, November. Waterflood Performance by Injection of Brine with Different Salinity for Reservoir Cores. In SPE Annual Technical Conference and Exhibition, SPE-109849. SPE.

9 Techno-economic Feasibility Analysis

9.1 INTRODUCTION

Crude oil and natural gas are found in large subterranean deposits of sedimentary basins. Hydrocarbon extraction is usually possible from the connected pore spaces within the rock strata by creating pressure gradients and altering in situ fluid properties. It is important to recognize the technical aspects and limitations of enhanced oil recovery (EOR) processes so that they can be precisely employed in times of crisis. This has been achieved with the improvement in seismic analyses, the range of chemicals and pre-operation tools. Hence, new capabilities in EOR methods could target trapped and bypassed accumulations, and drain the crude oil phases (Ambastha, 2008; Green et al., 2011). An EOR project involves long-term planning and capital/operating investments on the part of the petroleum company. Therefore, a systematic, methodical strategy must be developed to expect long payback periods. Most of the world's EOR activities are associated with production benefits to maximize returns for the government or national companies. Incentives from oil production may improve a nation's economy by reducing the tax burden, improving the stock prices, and bringing about capital growth.

The statistics of worldwide oil production indicate that the average overall recovery from light and medium gravity oils is around 30%–45% of the original oil in place (OOIP) by conventional (primary/secondary) methods, while from heavy oil deposits on the average, only 10%–15% OOIP is recoverable. Hence, a substantial amount of oil is unrecoverable by conventional methods. However, these remaining reserves can be recovered by different efficient EOR methods to increase the recovery percentage. EOR operations are still being improved to suit different oil field specifications.

Enhanced oil recovery has been vulnerable due to an unfavorable economy and also technically, due to a lack of knowledge of the reservoir characteristics. As the different EOR methods are generally associated with the injection of huge amounts of chemicals, solvents, or energy (heat), economic considerations prohibit their massive use. Most EOR projects also require a substantial infill drilling program. The main bottleneck of the EOR process is the additional cost of the different EOR projects. Thus, the EOR process will be economically feasible only when the crude oil price remains on the higher side.

9.2 EOR ECONOMIC MODEL

An economic model should evaluate various production strategy schemes. Economic models are designed to simulate the development and operation of actual EOR projects. The characteristics of the reservoir and the cost of producing EOR oil in that reservoir are entered into the model, which then generates and estimate (Zekri and Jerbi, 2002):

- the amount of crude oil that can be recovered from the project;
- a price sufficient to reimburse the full cost of the project and provide an adequate return on investment (ROI);
- the timing at which reserves in the reservoir will be produced.

These estimates are then accumulated for overall estimates of daily production, cumulative production, and ultimate recovery.

DOI: 10.1201/9781003098850-9

9.2.1 GENERAL STRUCTURE OF THE ECONOMIC MODEL

The estimation of oil recovery by any EOR process is a function of many parameters—oil saturation, pore volume, and previous primary and secondary recoveries—but the actual recovery by different technologies differs significantly from the estimated value.

This estimate indicates the total incremental EOR production from the start of the project and annual incremental production. The oil recoveries can be predicted by using a compositional reservoir simulation model. The price estimate is based on the projection of cash flows and a set rate of return. Cash inflows are made by the production of oil. Cash outflow comprises the following investment and operating costs: field development expenditures, equipment expenditures, operating and maintenance costs, injection material costs, and other costs (Zekri and Jerbi, 2002).

To calculate the required price for the oil, the production estimate is harmonized with the investment and operating costs and various rates of return. Conversely, the models compute the rate of return yielded at a series of fixed prices. The quantities of oil are combined by price to construct the price-supply curves. Individual price-supply curves for each technique and an overall price-supply curve for EOR recovery are generated. Based on selected prices and development assumption, these price-supply curves are converted to the timing at which reserves become proved and are produced. These curves are then extrapolated based on the remaining oil in place, and then the aggregated quantities of oil are aggregated by price to construct the price-supply curves (Zekri and Jerbi, 2002).

The cost components of a typical miscible EOR project are given in the following:

i. Field development expenditure: drilling and completion; workover and conversion.
ii. Equipment expenditure: well, lease, and field production equipment; injection equipment; separation and compression equipment.
iii. Operating and maintenance costs: normal operating and maintenance costs; incremental injection operation and maintenance costs.
iv. Injection material costs: purchase of injectants; recycle of injection fluids.
v. Other costs: field study, engineering, and supervision.

9.2.2 FEASIBILITY STUDY

The purpose of EOR feasibility study is to fully understand the fields in the first place, whether the fields are economically feasible to be developed by the implementation of EOR methods. The feasibility study consists of screening the EOR method, laboratory study, process facility study, and then a full-scale economic evaluation. If categorized as feasible, then the EOR method needs to be tested by field trial.

9.2.3 ECONOMIC ANALYSIS

The project profitability of a particular design can be evaluated by the production data obtained from the simulation results, which are imported into an economic model. Project profitability measures are used as the decision-making variables in an iterative approach to optimize the design. The essential variables for an economic model are (1) time (yrs), (2) pore volumes injected, (3) cumulative oil recovery, and (4) total fluid production (bbls). A discounted cash flow analysis is used to economically evaluate each design (Zekri and Jerbi, 2002).

The profitability is a measure of the success of an EOR project. It may happen that the project is technically sound to recover additional oil but not profitable. EOR processes are long-term projects, with profit depending strongly on oil prices. The long-term prediction of the profit of an EOR project may be difficult if the oil price changes significantly during the project life. However, in the short term, the project cost and the profit can be predictable.

The measures used in the economic model for profitability are the net present value (NPV) and internal rate of return (IRR). These profitability measures are calculated using the constant-dollar net cash flows until an economic limit is reached, which is defined as the production rate when EOR production costs exceed revenue from production. EOR production costs include (1) startup costs, (2) operating costs, and (3) temporarily fixed costs.

9.3 ECONOMICS OF CHEMICAL-ENHANCED OIL RECOVERY

Chemical-enhanced oil recovery technologies are well-established methods that improve overall oil recovery significantly. High oil prices and falling reserves replacement have stimulated interest in these technologies, particularly, for application in mature waterfloods. The decision to proceed with an oil field project, depending upon the perspective of investors, may rest upon the projected return on investment, rate of return, $/bbl of added reserves, cumulative cash flow, or combinations of these (Wyatt et al., 2008).

The key features of chemical flooding for EOR from an economic point of view are (1) high initial investment, including cost of drilling of new wells, (2) the cost of building the EOR storage facilities, and (3) high operating costs, as huge amounts of chemicals (polymer, surfactant, alkali) are required over a long period. There are several major risks surrounding investment in chemical EOR projects. They include (1) degradation of the chemicals under the harsh condition of the reservoir and (2) loss of chemicals because of adsorption on the rock surface, particularly for surfactants.

The basis of costs for a typical economic run is reported in Tables 9.1 and 9.2, which can also be used to estimate the cost for a particular project. For instance, from Table 9.1, the gross chemical cost for applying the MP (micellar polymer) process in an example of 10,000 Mbbl reservoir is:

$$10^7 \times \$3.75 = \$37,500,000$$

For recovery of 0.1 PV of oil, the chemical cost per barrel of incremental oil is:

$$\$3.75/0.1 = \$37.5/bbl$$

In this example, the 10,000 Mbbl field has 6,000 Mbbl OOIP. For recovery of 0.1 PV, i.e., 1000Mbbl, the recovery factor will be 16.7% $\{(1000/6000) \times 100\%\}$.

TABLE 9.1
Injected Solution Volume and Cost (Wyatt et al., 2008)

Process	Low-Tension "Slug" Injected Volume	Softening $/bbl	Injected Cost $/bbl	Polymer Flush Injected Volume	Injected Cost $/bbl	Injected Cost $/bbl * PV
MP	0.07		49.00	0.63	0.50	3.75
SP	0.30		8.20	0.40	0.50	2.66
	Na$_2$CO$_3$	**Fresh water**				
AP	0.30	0.03	0.92	0.40	0.50	0.49
ASP	0.30	0.03	2.21	0.40	0.50	0.87
	NaOH	**Brine**				
AP	0.30	1.00	3.11	0.40	0.50	1.44
ASP	0.30	1.00	4.41	0.40	0.50	1.82

TABLE 9.2

Model Economics of Chemical Flooding Process at $50 Oil (Wyatt et al., 2008)

Process	Incremental Oil %OOIP	Incremental Cost M$	Present Worth @ 15% M$	Added Oil Cost $/net BBL	Payout Years	ROR, %
MP	25.0	45,677	−9,806	38.07	7.3	6.90
SP	25.0	30,777	4,948	28.20	5.8	21.80
	Na₂CO₃	Fresh water				
AP	25.0	10,250	21,462	8.54	3.4	74.50
ASP	25.0	14,331	18,102	11.94	3.9	56.00
	NaOH	Brine				
AP	25.0	17,235	15,680	14.37	4.4	46.20
ASP	25.0	21,321	12,317	17.77	4.8	36.10

The estimated economics of costs and assumptions for the various processes are depicted in Table 9.2. In the first comparison, one can hold oil recovery constant to compare economic parameters over a range of oil prices. The typical results are briefed in Table 9.2 for $50/bbl oil price. Thus, it may be concluded that the chemical flooding processes can be economic if applied to the proper reservoir setting, and the economic performance is subjugated by oil price, cost per barrel of injected solution, and duration of chemical injection.

9.4 ECONOMICS OF IN SITU COMBUSTION PROJECT

Recent technical advances in the understanding of in situ combustion have made it worthwhile to consider this process as an important method of recovering heavy oil. However, it must first prove itself to be economically feasible, particularly, in contrast with steam-based recovery processes.

The potential economic success of an in situ combustion project is very sensitive to the combustion characteristics (reflected in the air requirement) and the amount of oil available for recovery (reflected in the product of the initial oil saturation and porosity).

9.5 ECONOMICS OF LOW-SALINITY WATERFLOODING

From an economic point of view, a low-salinity project involves a considerable initial investment followed by a modest annual production spread over a very long time period. The result is that the project payback period is also very long. In absolute terms, the extra operating costs are substantial, particularly, in terms of requirements for manpower and offshore beds. The key risks of the projects relate to (1) the effectiveness of the waterflood technology in enhancing oil production, (2) the commissioning of the low-salinity kit, (3) the additional complexity of managing the reservoir, and (4) the extra problems regarding well integrity. A further feature relates to the extra weight on the platform from the low-salinity kit, which reduces the flexibility of other activities on the platform (Kemp and Stephen, 2014).

9.6 ECONOMICS OF CO₂ MISCIBLE FLOODING

The oil recovery by injection of CO_2 into oil reservoirs is a widely used EOR technique and has been used in the oil industry for more than 40 years. In general, CO_2 is not first-contact miscible with reservoir fluids at most reservoir pressures. When the reservoir pressure is more than the MMP (minimum miscibility pressure), the miscibility may be achieved by multiple contact. Generally, for a given injection-gas composition, there is an MMP above which dynamic miscibility can be accomplished. The ability to achieve dynamic miscibility at achievable pressures in a wide range of reservoirs is a major advantage of the CO_2 miscible process.

The advantages of a CO_2 flood include the following (Ibukunoluwa and Onyekonwu, 2010):

- CO_2 is miscible with crude oil at relatively low pressures.
- Displacement efficiency of miscible CO_2 injection is high.
- Recovery of oil is aided by using a solution gas drive.
- Useful over a wider range of crude oils than hydrocarbon injection methods.
- Miscibility can be regenerated if lost.

The disadvantages of a CO_2 flood may include the following:

- CO_2 is expensive to transport and not always available.
- Poor sweep and gravity segregation can happen under certain conditions.
- The risk of corrosion is increased.
- Special handling and recycling of produced gas is necessary.

Oil reservoirs with higher capital cost requirements and less favorable ratios of CO_2 injected to incremental oil produced will not achieve an economically justifiable return on investment without advanced, high-efficiency CO_2 EOR technology and/or fiscal/tax incentives for storing CO_2. The NPV increases with increasing injection rate, but there is a limit where increasing the injection rate will reduce the NPV. Both capital and operating costs for an EOR project can vary over a range, and the value of CO_2 behaves as a commodity, priced at pressure, pipeline quality, and accessibility, so it is important for an operator to understand how these factors might change. The operating expenditure cost (OPEX) of CO_2 injection is associated with operation and maintenance of the installation, while the capital expenditure cost (CAPEX) of CO_2 injection should be considered as CO_2 extraction, processing, compression and transportation supply/injection/recycling scheme, corrosion-resistant field production infrastructure, and drilling or reworking wells to serve as both injectors and producers. There are several economic indicators to compare investment possibilities, which have value in oil field development through CO_2 injection. These economic indicators can be included such as unit development cost (UDC), unit technical cost (UTC), gas utilization factor (GUF), cash flow, payback period, and NPV.

UDC and UTC: The UDC in $/bbl is the total cost divided by the total incremental cumulative oil produced, while the UTC ($/bbl) represents the discounted total cost divided by discounted total incremental oil produced. The total costs include the OPEX (operational expenditures) and CAPEX (capital expenditures) (Badru et al., 2003).

GUF: In most cases, gas injection leads to increased production cost (although this is not a rule), especially when gas has to be purchased and transported to the injection site. The GUF is an important parameter to understand the project economics of the WAG (water alternating gas) displacement process. The higher the number, the less efficient is the flood. GUF is represented as the injected gas required at standard conditions to produces one barrel of incremental oil from the WAG process and can be defined as follows:

$$GUF = \frac{\left[\text{Solvent}\,(\text{gas})\,injected - Solvent\,(gas)\,produced\right], Mscf}{\text{Oil produced, stb}} \qquad (9.1)$$

Cash flow: It is the movement of cash into or out of a business, project, or financial product. It is usually measured during a specified, finite period of time. Measurement of cash flow can be used to determine a project's rate of return or value. The time of cash flows into and out of projects are used as inputs in financial models such as IRR and NPV.

Payback period: It refers to the period of time required for the return on an investment to "repay" the sum of the original investment. The time value of money is not taken into account. Payback period intuitively measures how long something takes to "pay for itself." All else being equal, shorter payback periods are preferable compared to longer payback periods.

NPV: The NPV is an important indicator of profitability for projected investment in the WAG displacement process. The reservoir heterogeneity plays a vital role in the project economy (Bhatia et al., 2014). Ettehadtavakkol et al. (2014) shows that increase in WAG ratio provides better NPV but after a further increase in WAG ratio shows a decline in NPV, while NPV will be highest at optimum WAG ratio in both carbonate and sandstone reservoirs. However, permeability heterogeneities have a significant impact on WAG ratio to achieve optimum value. The NPV is computed from the summing of all the present values from outgoing and incoming cash flows over a period (Ganesh and Satter, 1994) as presented here:

$$NPV = \sum_{t=1}^{t} \frac{(C_{in} - C_{out})_t}{(1+r)^t} - C_o \tag{9.2}$$

where C_{in}, C_{out}, C_o, r, and t are the net cash inflow (revenue) during period t, net cash investment during period t, investment costs, discount rate, and number of years, respectively.

In WAG displacement, the availability of gas at field level greatly affects the economics. However, the reservoir heterogeneity also plays a role in economics due to the effect on oil recovery, solvent utilization factor, WAG response time, GOR, and water production. A homogeneous reservoir shows higher oil recovery compared to a heterogeneous reservoir at 1 PVI, which is expected due to less gas gravity override. However, the heterogeneous reservoir has a higher amount gas injected and higher solvent recovery, which is also expected due to high matrix permeability, and higher gas gravity override through upper-high-permeability streaks layers. The homogeneous reservoir requires double the number of wells due to tight permeability and, hence, a higher capital cost to be invested compared to the heterogeneous reservoir. Simulation results show that the heterogeneous reservoir shows faster response with higher initial oil acceleration production but with lower initial solvent utilization factor compared to the homogeneous reservoir in miscible WAG displacement. Table 9.3 shows the economic indicators for heterogeneous reservoirs.

TABLE 9.3

Economic Indicators for CO_2 EOR Flooding in a Typical Heterogeneous Reservoir

Economic indicators	CO_2 EOR economics after 30 years of oil production assuming $3/ MSCF CO_2 gas price and different oil price		
	$20/bbl	**$30/bbl**	**$50/bbl**
Cumulative oil production (Bbbls)	1	1	1
CAPEX ($MM)	1,060	1,060	1,060
UDC ($/bbl)	9.5	9.5	9.5
UTC ($/bbl)	19	19	19
Value investment ratio (VIR)	0.011	0.528	1.558
Rate of return—RROR (%)	8%	13%	21%
Payback period (years)	21	16	13
NPV ($MM)	9.1	559.4	1,658

Conceptually, the economics of a CO_2 injection project are computed with a revenue, project cost, operating income, tax, and profit balance sheet on an annual basis as follows:

Project Revenue (Year)

Net oil sold (less royalty)
Net gas sold (less royalty) – *Total Revenue (R)*

Project Cost

Fixed operating cost
Variable operating cost
CO_2 purchase cost
Produced water treatment and injection cost – *Total operating cost (O)*

Capital Costs

Well workover
New wells (as required)
CO_2 injection distribution system
CO_2 delivery pipeline
Water injection plant (as required)
CO_2 recycling plant – Capital cost (C)

Taxes

Profit tax
State severance tax
Other state and local taxes
Federal and state income taxes – Tax cost (T)

That is, $P = R - O - C - T$, with the sum of the after tax cash flow, P, discounted to present value defined as the after tax DCF (discounted tax flow), sometimes called present value profit. The value of DCF for a project and its confidence interval are usually the primary criteria upon which a project is judged. Economic criteria such as P/I ratio, $bbl profit, $bbl cost, and ROR provide additional sensitivity and decision-risk information. However, the primary indicator of project incremental economics is usually DCF.

REFERENCES

Ambastha, A., 2008. Heavy-Oil Recovery. Society of Petroleum Engineers.
Badru, O., Kabir, C.S., 2003, January. Well Placement Optimization in Field Development. In SPE Annual Technical Conference and Exhibition. Society of Petroleum Engineers.
Bhatia, J., Srivastava, J.P., Sharma, A. Jitendra S., Sangwai, J.S., 2014. Production performance of water alternate gas injection techniques for enhanced oil recovery: Effect of WAG ratio, number of WAG cycles and the type of injection gas. International Journal of Oil Gas and Coal Technology 7(2), 132–151.
Ettehadtavakkol, A., Lake, L.W., Bryant, S.L., 2014. CO_2-EOR and storage design optimization. International Journal of Greenhouse Gas Control 25, 79–92.
Ganesh CT, Satter A., 1994. Integrated Petroleum Reservoir Management: A Team Approach. PennWell Books Publishing Company.
Green, D.W., Hirasaki, G., Pope, G.A., Willhite, G.P., 2011. Surfactant Flooding. Society of Petroleum Engineers.
Ibukunoluwa, A.O., Onyekonwu, M.O., 2010, July. Prospects and Economics of Miscible-CO_2 Enhanced Oil Recovery in Some Nigerian Reservoirs. In Nigeria Annual International Conference and Exhibition. OnePetro.

Kemp, A.G., Stephen, L., 2014. The Economics of Enhanced Oil Recovery (EOR) in the UKCS and the Tax Review. Aberdeen Centre for Research in Energy Economics and Finance, University of Aberdeen.

Wyatt, K., Pitts, M.J., Surkalo, H., 2008, 20 April. Economics of Field Proven Chemical Flooding Technologies. In SPE Symposium on Improved Oil Recovery. OnePetro.

Zekri, A., Jerbi, K.K., 2002. Economic evaluation of enhanced oil recovery. Oil & Gas Science and Technology 57(3), 259–267.

10 Profile Modification and Water Shutoff

10.1 INTRODUCTION

The process of controlling undesirable water production from a well by conducting treatments to prevent coning or cresting at the injection well is called profile modification. On the other hand, the water shutoff treatment can increase the thickness of the oil production interval and decrease the thickness of the high water cut interval, which in turn eliminates unwanted water production. The water shutoff and profile modification both plug the channels of high porosity and high permeability to improve the macroscopic sweep efficiency using the synthetic composite gel system, which had a low initial viscosity and high strength after gelation. The profile modification treatment is conducted at the injection end, while the water shutoff treatment is conducted at the production end. Therefore, the combined operation of profile modification and water shutoff could plug the high permeability layer and treat the reservoir heterogeneity. For this, a variety of treatment options are available, the majority of which are designed to lower the permeability of the water-bearing zones to increase preferential flow through the oil-bearing formation. A common treatment involves injecting polymers or chemicals into the formation matrix to form a rigid gel.

Excess water production can be dealt with using a variety of established technologies. Reservoir management, well design, wellbore intervention, and other methods can be used to control excess water (Vasquez et al., 2009; Vossoughi, 2000). Wellbore water control technologies include bridge plug and cement squeeze. These techniques are useful when there is an excess of water in the channel behind the casing or due to casing leakage. In some cases, these techniques can be extended to isolate watered-out zones too.

The two broad categories of treatment are as follows.

10.1.1 CHEMICAL METHODS

These methods work well inside the matrix, where there is a higher surface area to volume ratio. The treatment solution, which resembles water, is prepared on the surface and then injected into the matrix. It increases viscosity while decreasing matrix permeability in a specified time period. This method allows for a significant depth of penetration and placement.

10.1.2 MECHANICAL METHODS

Downhole separation and coning avoidance (reverse coning) are two of these technologies. A disposal unit is installed near the wellbore in the downhole separation process, which separates oil and water on the spot. While oil is produced, the water is injected back into the lower water-bearing zone. Although this unit is pricey, it saves money on lifting and water disposal.

10.2 SOURCES OF UNWANTED WATER PRODUCTION

1. In some cases, the water injection well happens to be connected with the production well through an open fracture or features known also as "thief zones."
2. If open fractures are connected to the aquifer, they can also result in an excessive amount of water.

 DOI: 10.1201/9781003098850-10

3. Coning, which arises due to the drawdown of the pressure.
4. Casing leaks or poor cementing behind the casings can lead to channels connecting unwanted water production formations/sources to the wellbore.

10.3   IDENTIFYING THE PROBLEM

Gathering all available reservoir and production data is usually the first step in reducing excessive water production. The water entry points are then located using logging tools. Finally, a proper profile modification method is used based on the results. The most crucial aspect of any procedure is a precise diagnosis of the problem. It's critical to understand the water entry point, reservoir rock heterogeneity, dominant production mechanisms, and wellbore schematics. In fact, all available information about the well, such as drilling operations reports, logs, and production history, is considered valuable. Because each well has its own workflow based on its properties, history, and reservoir heterogeneity, this is the case. Accurate investigation leads to water shutoff success, increased oil production, and cost-savings in water handling.

Water production zones are typically identified using production logging tools in production wells, which is an important step in planning for an optimized water shutoff operation. Water flow logs are used to identify thief zones in water injection wells. Horizontal wells, on the other hand, pose difficulties in both identifying the problem and in intervening. This is due to the wellbore's involvement, as well as flow regimes and their effects on obtaining the required data.

Running cement bond logs or ultrasonic pulse-echo logs behind the casings is critical for ensuring the integrity of the cement job behind the casing. These logs evaluate the cement job's bonding properties behind the casing and point out any bad cement areas. Production, temperature, and noise logs can all be used to pinpoint the source of leaks in casings.

10.4   TREATMENTS

10.4.1   Chemical Methods

Water shutoff operations can be performed far from the wellbore, in the reservoir, or near the wellbore, using a variety of chemical treatments. These chemical solutions improve reservoir conformance while also blocking unwanted water production zones. The idea is to reduce the permeability of the paths of least resistance in front of the water to prevent water from reaching the wellbore through them.

To put it another way, the goal is to block open features and high-permeability channels, forcing water to take the more difficult path of sweeping oil from the matrix rock, which yields higher overall economic returns than producing oil from fractures.

Some common chemical treatments are as follows:

1. Gel
   One of the most well-known chemical solutions for water shutoff operations is gel injection (Bai et al., 2015; Abdulbaki et al., 2014). It's used to lower the water to oil ratio and improve pattern conformance. This is accomplished by the gel's ability to reduce permeability and block open features, fractures, and water zones with high permeability. Water, small amounts of polymers, and cross-linking chemical agents make up the injected gel, which can completely seal off layers.
   Injecting gel into the reservoir is based on four factors. The first is the viscosity of the gel at injection time, which aids in directing the gel to the least resistant and largest paths. Second is the nature of the gel phase, which is usually chosen to be the aqueous phase, as the water is the desired phase to be shut off. The density of the gel comes in third. To avoid losing the gel treatment's effectiveness, it must be carefully designed and based on the density of the formation water. A schematic of in-depth diversion in the heterogeneous

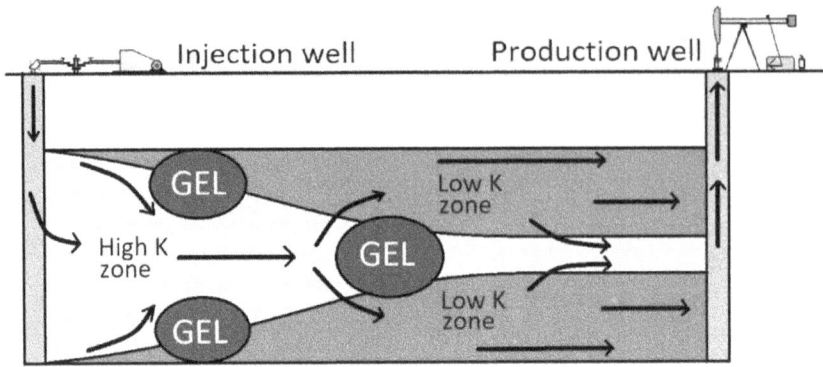

FIGURE 10.1   In-depth diversion in the heterogeneous reservoir.

reservoir gel injection is shown in Figure 10.1. The high-permeable zones are blocked to allow the flow through low-permeable zones (Lenchenkov, 2017).

2. Polymer Flooding

The use of polymer flooding to increase the viscosity of the water is another common technique for profile modification operations. This technique is used to make the drive fluid (water) more viscous, which aids in mobilizing and displacing the oil in the reservoir matrix rock. To achieve better sweeping efficiency in the reservoir, this technique is usually used in the reservoir far from the production wells through water injection wells. This eventually leads to a reduction in the amount of water produced.

3. Foam Injection

In some cases, foams may perform better than polymer solutions in terms of sweep. These conditions necessitate the following: (1) foam forms in high-permeability paths but not in low-permeability paths, (2) there is no cross flow between high- and low-permeability strata, and (3) the foam resistance factor in high-permeability strata is high enough to overcome the permeability contrast and the unfavorable mobility ratio between the gas bank and the oil/water bank in the less permeable strata. Under most circumstances, foams will not be superior to polymers unless gravity effects provide a fortuitous benefit.

4. Microorganisms, Emulsions, Particulates, Precipitates, and Nanoparticles

Other materials, such as microorganisms, emulsions, particulates, precipitates, and nanoparticles, have been proposed for use in conformance improvement, particularly for in-depth profile modification. These materials' potential and claims must be viewed in the same light as conformance-improvement gels. The materials listed in the section's title are still very much in the research and development stage. None of them should be regarded as tried and true.

10.4.2   MECHANICAL METHODS

Within the wellbore, there are available technologies that can successfully shut off unwanted water production. Mechanically controlling water production is well known for its quick results and low costs. It's usually a rigless job, so it's less expensive.

Some common mechanical solutions are as follows:

1. Plugs and Packers

Packers and plugs are one of the most well-known mechanical solutions for water shut-off and isolation operations inside the wellbore. They have been used successfully in removing production from undesirable water zones. They're frequently used to improve the well's performance and stop excessive water production. Because it can be installed without pulling the production tubing or using a drilling rig, this hardware is known for being cost effective and dependable in achieving isolation. Coiled tubing can be used to run them through the wellbore to install them.

In comparison to chemical injection solutions, the results can be obtained relatively quickly, in the range of a few hours to days. Simply put, packers and plugs are small-diameter elements, typically rubber, that can expand downhole into larger diameters, forming a seal and isolating the well from unwanted features or zones.

2. Tubing Patches

This method is primarily used to address well integrity issues, specifically casing leaks. Casing leaks are common in older wells and wells completed in formations containing corrosive gases such as H_2S. If the unwanted water was discovered to be coming from a leak in the casing, squeezing cement or resin patches is a viable solution. Only after pinpointing the exact location of the leak can this method be used.

REFERENCES

Abdulbaki, M., Huh, C., Sepehrnoori, K., Delshad, M., Varavei, A., 2014. A critical review on use of polymer microgels for conformance control purposes. Journal of Petroleum Science and Engineering 122, 741–753.

Bai, B., Zhou, J., Yin, M., 2015. A comprehensive review of polyacrylamide polymer gels for conformance control. Petroleum Exploration and Development 42(4), 525–532.

Lenchenkov, N., 2017. Conformance control in heterogeneous oil reservoirs with polymer gels and nanospheres (Doctoral dissertation, Delft University of Technology).

Vasquez, J.E., Eoff, L.S., Dalrymple, E.D., 2009, May. Laboratory Evaluation of a Relative Permeability Modifier for Profile Modification in Injection Wells. In Latin American and Caribbean Petroleum Engineering Conference. OnePetro.

Vossoughi, S., 2000. Profile modification using in situ gelation technology—A review. Journal of Petroleum Science and Engineering 26(1–4), 199–209.

Index

For Product Safety Concerns and Information please contact our EU
representative GPSR@taylorandfrancis.com
Taylor & Francis Verlag GmbH, Kaufingerstraße 24, 80331 München, Germany

www.ingramcontent.com/pod-product-compliance
Lightning Source LLC
Chambersburg PA
CBHW080914220326
41598CB00034B/5567